URBAN FORESTRY
Planning and Managing Urban Greenspaces

SECOND EDITION

ROBERT W. MILLER

University of Wisconsin, Stevens Point

PRENTICE HALL, Upper Saddle River, New Jersey 07458

Library of Congress Cataloging-in-Publication Data

Miller, Robert W., 1940–
Urban forestry : planning and managing urban greenspaces / Robert W. Miller.—2nd ed.
p. cm.
Includes bibliographical references and index.
ISBN 0–13–458522–4
1. Urban forestry. I. Title.
SB436.M55 1996
635.9'77'091732—dc20 96–725
CIP

Acquisitions Editor: Charles Stewart
Production Editor: Walsh Associates
Director of Manufacturing & Production: Bruce Johnson
Manufacturing Manager: Ilene Sanford
Production Manager: Mary Carnis
Formatting/Page Makeup: Pine Tree Composition, Inc.
Cover Director: Jayne Conte
Cover Designer: Bruce Kenselaar

Simon & Schuster/A Viacom Company
Upper Saddle River, New Jersey 07458

Printed in the United States of America
10 9 8 7 6 5 4 3 2 1

ISBN: 0-13-458522-4

Prentice-Hall International (UK) Limited, *London*
Prentice-Hall of Australia Pty. Limited, *Sydney*
Prentice-Hall Canada Inc., *Toronto*
Prentice-Hall Hispanoamericana, S.A., *Mexico*
Prentice-Hall of India Private Limited, *New Delhi*
Prentice-Hall of Japan, Inc., *Tokyo*
Simon & Schuster Asia Pte. Ltd., *Singapore*
Editora Prentice-Hall do Brasil, Ltda., *Rio de Janeiro*

Contents

Preface

Trees and related vegetation have long been planted in cities for a variety of reasons. From trees with special religious significance in ancient temples to trees captured in planters adjacent to our newest office buildings, we have sought to accompany our urban lives with some representation of nature. During the past few decades individuals and society have placed a much greater emphasis on urban vegetation in an attempt to improve the quality of life in our communities. This emphasis has created a demand for professionals to manage urban trees and forests.

Trees and other vegetation can be found throughout our communities from small farm villages to the largest metropolitan areas, in developing countries and developed countries, and on lands occupied by both the poor and the wealthy. Trees line city streets, fill residential yards and vacant land, surround factories, and are the most dominant feature of most parks and other greenspaces. Trees exist in cities through careful design and cultivation, by poor or no design, and by accident and neglect. In many cities trees on private property outnumber public trees ten to one. We must increasingly come to regard and understand cities as ecosystems with the same processes going on as can be found in the countyside, and to understand energy and material links within the city and between the city and nature. The intent of this book is to address how to plan for, establish, and manage urban and community trees, forests, and other elements of nature in the urban ecosystem.

Part 1 discusses why we have trees in cities and how we use them; Part 2 deals with appraisal and inventory of urban vegetation; Part 3, the final and most extensive section, addresses planning and management. Planning and management find strong application to public vegetation, especially street trees, park vegetation, and forested greenbelts. However, the same principles of planning and management apply to the management of privately owned vegetation as addressed in the last chapter, dealing with commercial and utility arboriculture.

This book is meant to serve as a textbook for students of urban forestry and arboriculture who are specifically interested in the management of tree populations and other greenspaces. It is also the intent of this text to serve as a reference book for city foresters, greenbelt managers, and commercial and utility arborists.

I am indebted to many for their assistance in the preparation of the first edition of this book. I am equally indebted to many for their assistance in this edition. Dick Rideout, Wisconsin Urban Forestry Coordinator, provided a thorough review of the first edition and many helpful suggestions for this edition. Lengthy discussions with Ken Ottman, Milwaukee Bureau of Forestry, and Bob Skiera, Consulting Urban Forester, helped me to focus on many changes in this edition. Thanks to Larry King for the Highland Park Ordinance, Larry Hall for the Hendrickson customer newsletter, Bruce Allison for the Madison Tree Walks, Dealga O'Callaghan for the Amenity Valuation of Tree and Woodlands, and Gary Moll and Jill Mahon for American Forests' City Green ecosystem assessment material.

Much appreciation goes to the International Society of Arboriculture, American Forests, the City of Milwaukee, the University of Florida, and the Arboricultural Association for materials used in this book. Thanks to the many city foresters, scientists, and arborists for their positive comments on the first edition and their encouragement in the preparation of this edition. As with the first edition, my students have continued to provide me with ideas that have stimulated much of what is new here. I am indebted to Prentice-Hall for encouraging me to write a second edition, especially editors Frank Burrows, Catherine Rossbach, and Charles Stewart.

Finally, I wish to express to my wife, Marlene, and son, Casey, my sincere appreciation for their support and patience during the long process of preparing both editions. To them, and to all of my students, past, present, and future, I dedicate this book.

PART 1
Introduction to Urban Forestry

1

Evolution of Cities

INTRODUCTION

The term *urban forestry* at first seems a contradiction in terminology. Historically, forestry as a profession has a long tradition of activity in the more remote regions of the globe. However, as foresters were busy developing silvicultural systems and forest management plans for rural woodlands, people were shifting from rural agrarian societies to urban industrial social systems. Indeed, it was, and is, the urbanization of the world that created a high demand for forest products, leading to the overexploitation of forests and the transition to managed forests.

Three events occurred that led to the concept of urban forestry, all of them directly related to the urbanization process. First, as more and more people concentrated in cities, urban centers expanded and interfaced with rural woodlands. Second, social values shifted to reflect urban living, and these values exert a strong influence on the management of rural land. Third, the urbanization process has had, and continues to have, a negative impact on vegetation within cities, at the urban/rural interface, and rural forests.

To understand fully the impact of urban development on vegetation and other natural systems, it is important to understand the history of urbanization, how it influenced development patterns in the urban landscape, what interaction urban residents have with rural landscapes, and what may be projected for the future of urban living. In this chapter we address the first of the three events described above, with the latter two being discussed in subsequent chapters.

A BRIEF URBAN HISTORY

Agricultural Revolution

Anthropologists estimate that *Homo sapiens sapiens* has existed in its present form for between 40,000 and 200,000 years. As a species we have spent most of our existence living as nomadic hunters and gatherers. This being the case, we did not develop permanent settlements until the development of agriculture some 15,000 years ago. For most of our history our population was dispersed across the landscape in low numbers, with the extended family and small bands or tribes serving as the social unit. We survived by our intelligence, and the ecosystem controlled our numbers. We had not yet begun to shape our environment to any great degree (Mumford, 1961).

Early attempts at agriculture took place in the Middle East and North Africa 10,000 to 15,000 years ago. Along the banks of the Tigris, Euphrates, Indus, and Nile rivers human beings started cultivating crops in rich alluvial soils. The soils were easily irrigated by shallow ditches from the rivers and were enriched by annual flooding (Mumford, 1961).

The immediate impact of early agriculture was the end to nomadic living, since food could be produced in sufficient quantity to allow permanent settlement. Small villages were built and social interaction became more complex. The political structure of these early settlements was based on rule by the strongest. Rivalry grew between villages and food reserves in villages were attractive to residents of other settlements and to nomadic groups. Walls were constructed for protection and the continuous threat of warfare created a need for protective military, thus consolidating power of the rulers. With the passage of time villages expanded to small cities, and through warfare, city-states consolidated to form empires (Gallion and Eisner, 1975).

The effects of agriculture were manyfold. Each farm family produced a little more food than it needed. This surplus allowed a small percentage of the population to survive by means other than farming. Craftsmanship and metallurgy improved, and cultural evolution accelerated. In time, writing and the recording of events developed, trade between cities was initiated (in both goods and ideas), and technological innovations such as the wheel and the sail improved transportation (Gallion and Eisner, 1975).

Ancient Cities

The Egyptian civilization reached its apex around 3000 B.C. with the construction of the pyramids and other monuments. Egyptian settlements certainly did not reflect the grandeur of the monuments; rather, they indicate that early Egyptian citizens and slaves lived in sparse quarters. Walls surrounded the towns, and homes consisted of several small rooms with an inner court. These homes were arranged in rectangular blocks along narrow streets that served as both access and sewers (Figure 1–1).

A series of early civilizations arose along the Tigris and Euphrates rivers. As in Egypt, palaces and temples were constructed for the rulers while most of the population lived in crowded quarters of two- and three-story buildings. The walled city of Babylon was surrounded by a moat, and contained the first recorded use of urban vegetation, the Hanging Gardens (Figure 1–2).

The Aegean civilization flourished on islands in the Aegean Sea from 2500 to 1400 B.C. The cities were constructed without walls, because difficulty of access to the islands provided adequate protection. The palaces of the rulers were open to the public and served as public gathering places and markets. Aegean cities were attractive and were constructed in harmony with the rugged terrain of the islands.

The cities of ancient Greece during the Hellenic Age reflected the development of a democratic way of life. This age lasted from 1200 to 338 B.C., and reached its apex around 400 B.C. The city of Athens represented the highest development of the

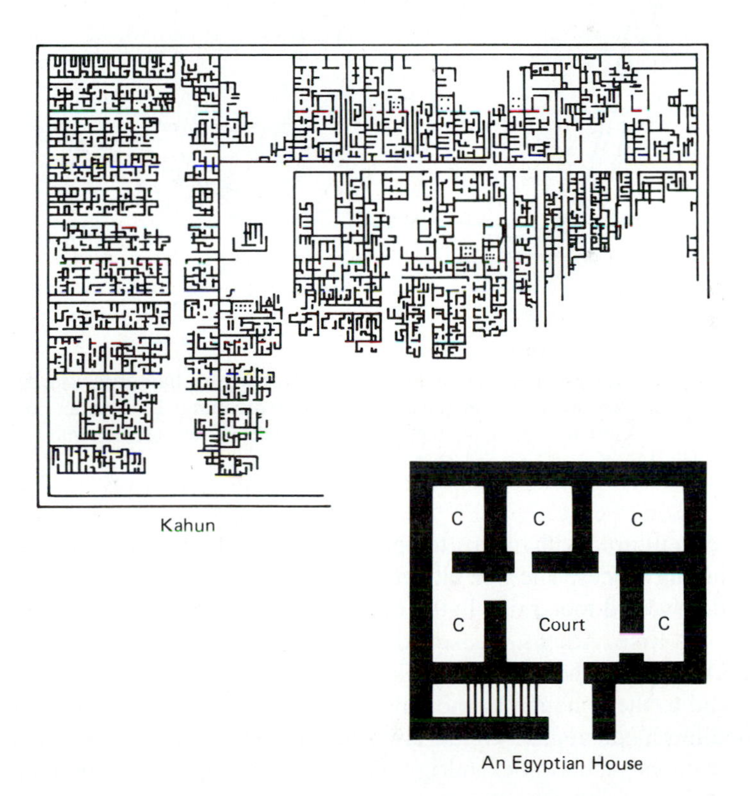

Figure 1–1 The city of Kahun, Egypt, was built for the slaves and artisans who were used to build the Illahun pyramid around 3000 B.C. Houses were an assembly of cells surrounding an open court, with sleeping and living quarters on the roof. (From Gallion and Eisner, 1975.)

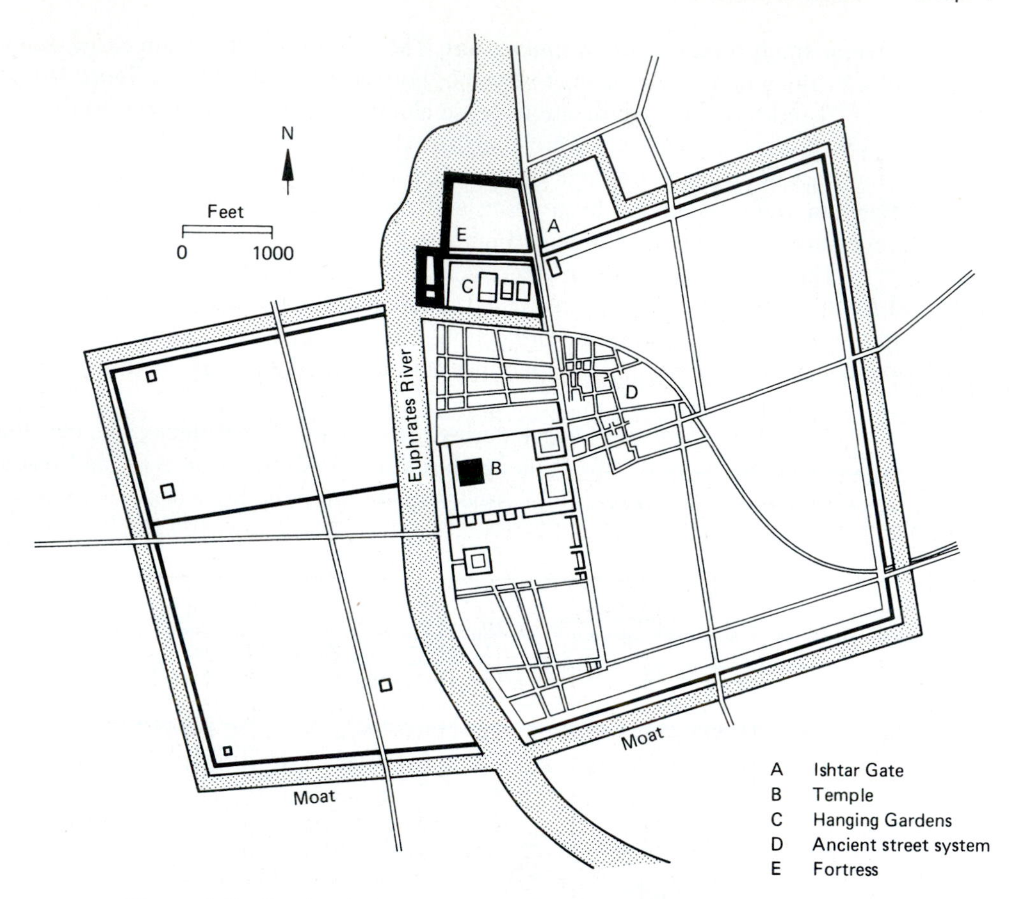

Figure 1–2 By the sixth century B.C. Babylon had grown to be a large city consisting of a rectangular street system, dwellings for a variety of classes, and monuments, including the Hanging Gardens. (From Gallion and Eisner, 1975.)

Greek culture, with many temples, public buildings, public markets, and political gathering places. The free citizens accepted responsibility for their government, lived modestly, and took pride in their public temples and buildings (Figure 1–3).

In time, Athens, weakened by the Peloponnesian Wars and political corruption, succumbed to the armies of Alexander the Great. However, the culture of Greece spread to the conquerors and gave rise to the Hellenistic civilization. Throughout the Mediterranean region Greek architecture and ideas influenced the development of large cities such as Alexandria, Syracuse, and Pergamon. These cities had many public buildings and monuments, and elaborate homes for the rulers and wealthy classes. However, democracy was abandoned, and, once again, rule was by the strongest. The Hellenistic civilization persisted from 338 to 146 B.C.

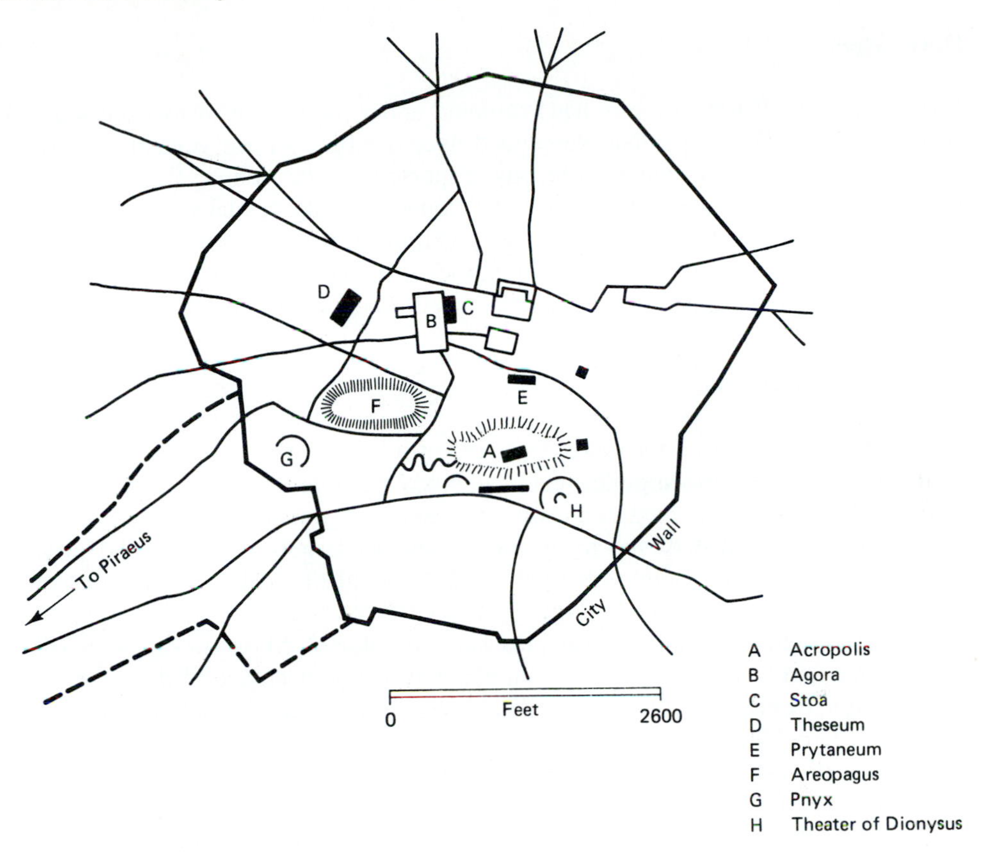

Figure 1–3 Ancient Athens around 400 B.C. (From Gallion and Eisner, 1975.)

While new cities were being built during the Hellenistic era, Rome began to emerge as a civilization between 300 and 200 B.C. The Roman Empire eventually conquered most of Europe and the Mediterranean region, with the Romans, too, adopting much of the Greek culture and architecture. Roman cities grew large and rich, with many advances in engineering. Highways, aqueducts, and sewers serviced the largest cities, and extravagant monuments and public buildings were constructed. However, Roman cities became increasingly congested, and the wealth of the empire was squandered on the building of monuments to its leaders, rather than on urban improvements. The wealthy classes moved out of the cities to extravagant villas, while most of the population lived in urban squalor (Gallion and Eisner, 1975).

Civilizations have grown and declined in other parts of the ancient world. The Aztecs, Incas, and Mayans developed agriculture and large urban centers in the Americas. Likewise, civilizations grew and developed empires and urban centers in many parts of Asia.

Dark Ages

By A.D. 500 the Roman Empire had crumbled under the weight of its excesses and corruption. The Dark Ages had descended upon Europe, with a general decline in Western civilization. Urban populations dispersed to the countryside, and cities shrank in size and influence. City-states developed, with the feudal system ruling the day. Constant warfare between city-states precipitated the need for protective walls around cities. As time passed, urban life in the cities became more attractive, and by A.D. 1100 trade between cities reestablished itself (Gallion and Eisner, 1975).

The Medieval Town

As Europe emerged from the Dark Ages, urban life improved considerably. Craftsmen and merchants formed guilds and challenged the nobility for economic and political power. The typical medieval town was walled and contained a castle, a market, and a church. The rest of the town was composed largely of shops and private dwellings. The medieval city was small, and the residents had ready access to the countryside.

Life in the medieval town was pleasant and colorful. Marketplaces were filled with people and festivals were common. The physical appearance of the towns was one of subtle beauty. The location of the cities was influenced initially by the ability to defend them. Hilltops and other easily defended sites selected for strategic reasons provided aesthetically pleasing urban settings and buildings were constructed in harmony with the form of the landscape. Streets were narrow, homes connected, and small yards behind the homes provided space for gardens and livestock (Figure 1–4).

The Middle Ages

The rise of mercantilism brought an end to the medieval town. The number of towns in Europe increased dramatically during the Middle Ages, but most towns remained small due to confining walls. However, by the fourteenth century the population of Florence reached 30,000, while Paris grew to 240,000.

The power of the feudal lords and guilds declined, while the power of the merchants and landowners increased. Cities became congested, as garden spaces gave way to buildings. Second stories of homes and other buildings were extended over streets as more people crammed behind city walls. The quality of life declined due to overcrowding and poor sanitation. Plagues swept across Europe, extracting their heaviest toll in the crowded cities.

The introduction of gunpowder by the fourteenth century created the demand for even larger armies and stronger fortifications around cities. New, stronger, and more elaborate fortifications were constructed, which effectively cut off access to the countryside for most urban residents. Areas adjacent to city walls were cleared for protection and became a "no-man's-land" (Gallion and Eisner, 1975).

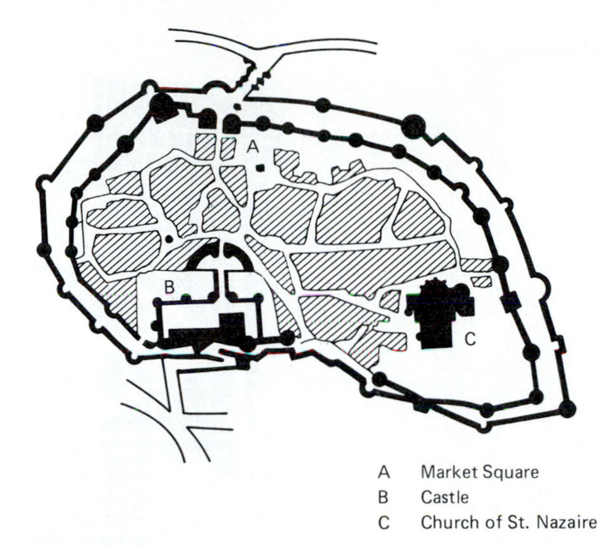

Figure 1–4 The medieval city of Carcassonne around the thirteenth century. Carcassone, like other medieval cities, was walled for protection. (From Gallion and Eisner, 1975.)

The Renaissance

Consolidation of power under monarchs and development of long-range artillery signaled the end of the walled city. The arts and sciences expanded, and a desire to reintroduce monumental architecture into European cities prevailed. Without walls, cities expanded rapidly, and gardens, parks, and open space were the rule during the Baroque period. New homes were built for the wealthy and broad avenues provided access to new churches and public buildings (Figure 1–5). In spite of these additions to cities, most residents were poor and lived in deplorable conditions (Gallion and Eisner, 1975).

Early American Villages

Colonial expansion to the New World yielded small settlements along the east coast of North America. Early towns in New England consisted of the town commons, town hall, commercial district, and the homes of colonists who farmed lands adjacent to villages. These settlements were, for the most part, built without fortifications and served primarily as commercial centers (Gallion and Eisner, 1975).

The Industrial Revolution

Prior to the Industrial Revolution most of Europe's population consisted of tenant farmers who tilled the soil for wealthy landowners. To be sure, some urban centers had become large, but agriculture did not produce enough food to support a large segment of the population in cities. The primary sources of energy to grow and process food consisted of human and animal muscle, and to a lesser degree, water and wind power.

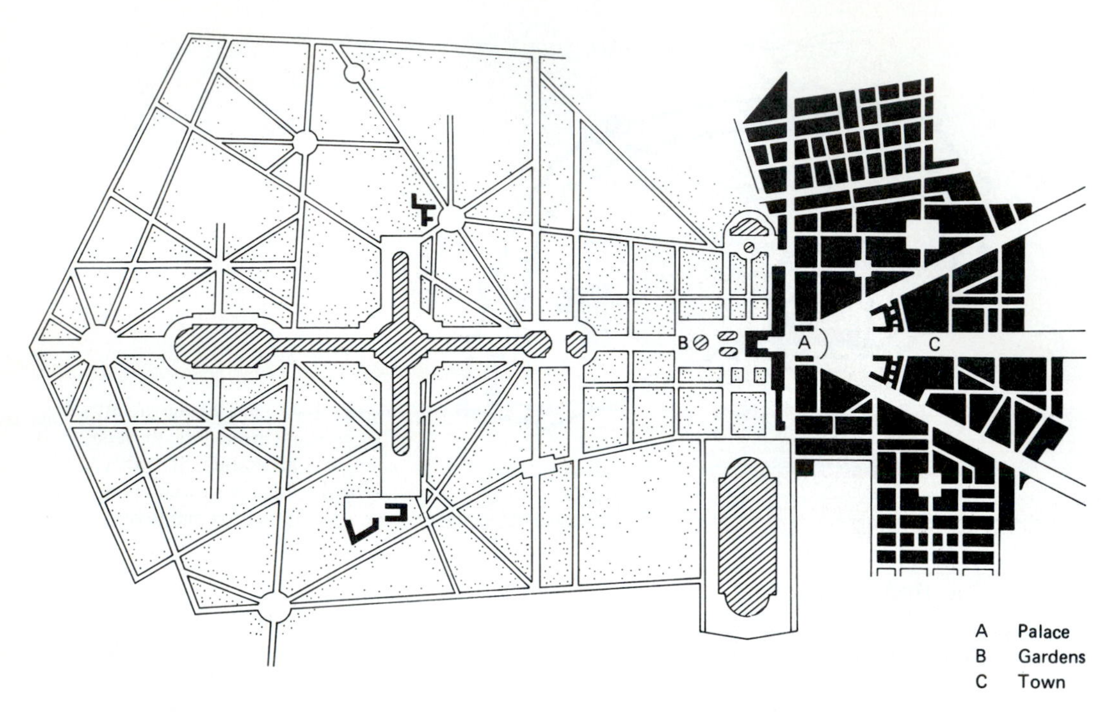

Figure 1–5 The baroque city of Versailles provided the model for European cities by the eighteenth century. (From Gallion and Eisner, 1975.)

Following the Renaissance, a great awakening in the sciences took place throughout much of Europe. The pace of scientific discovery accelerated, and new knowledge fueled rapid technological development. By the mid-eighteenth century, improved metallurgy and the steam engine ushered in the Industrial Revolution.

Factories produced large quantities of cheap goods, steam locomotives moved raw materials and finished goods, agriculture mechanized, and advances in public health and sanitation allowed cities to support larger populations. The mechanization of agriculture reduced the demand for labor in the farm economy, while new factories in the cities demanded labor to run the machines of industry. The countryside was depopulated, while urban centers grew in population at rates faster than the ability of the urban infrastructure to provide adequate services. Cities became overcrowded, dismal, and polluted by the wastes of industry. For the first time in human history, the bulk of the human population lived separated from nature (Figure 1–6).

It took about 100 years for the Industrial Revolution to reach the United States. In 1850, most American citizens were farmers, with less than 20 percent of the population living in cities. By 1920, 50 percent of the population was urban, composed of former farm families and newly arrived immigrants. Today, less than 3 percent of the population is engaged in food production and over 80 percent of American citizens can be classified as functionally urban.

Figure 1–6 Nineteenth-century industrial city.

Spatial Development of American Cities

Transportation is the key to both the location of cities and the spatial patterns of land use within cities. At the time of settlement in the United States the primary mode of transportation was water. Towns were located in natural ports and along navigable waterways. Land transportation was limited to walking and riding on horses; thus towns and cities were compact, with residents living in close proximity to stores and jobs. Cities were small with narrow streets, and residents had easy access to the surrounding countryside (Figure 1–7a).

The coming of the Industrial Revolution in the mid-nineteenth century brought the same problems of overcrowding and pollution to American cities as was found in European cities. Railroads and trolleys linked cities and provided the means for the newly emerging industrial elite classes to escape the industrial squalor and pollution for more amenable surroundings in new suburbs. Urban centers were dominated by industry and associated housing for immigrant workers. Suburban communities sprung up at stations along railroad lines, and populated strands developed along trolley corridors (Figure 1–7b).

This first wave of suburban sprawl took place in the 1920s, reflecting general negative social attitudes toward cities. The single-family dwelling and the desire for open space became the social norm for the middle class. People lived within walking distance of train stations and trolley lines, thus limiting the size of suburban communities.

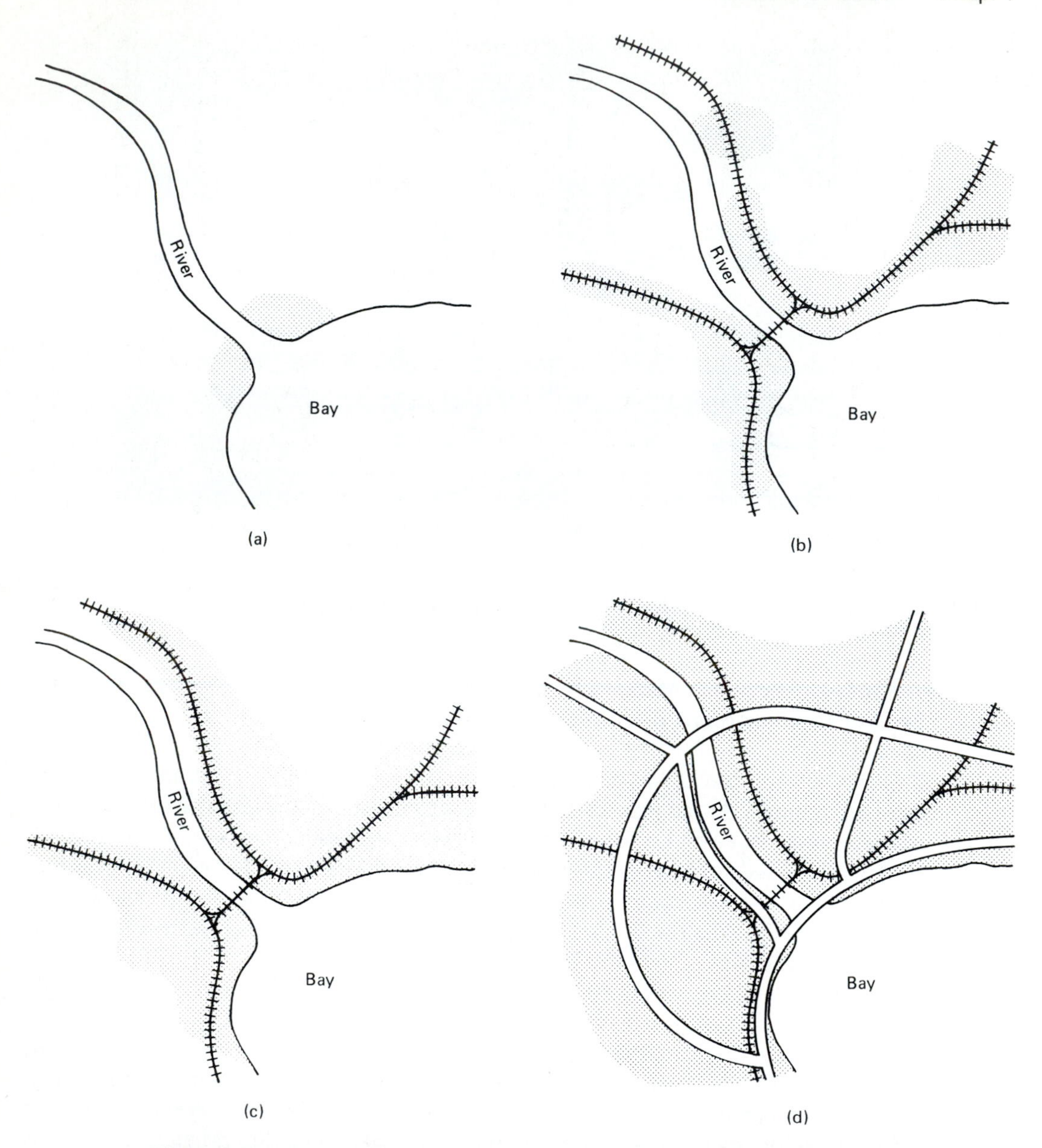

Figure 1–7 The evolution of American cities was strongly influenced by transportation modes. (a) Colonial cities were compact to accommodate foot and horse travel, and were located on waterways and bays. (b) The Industrial Revolution spread cities to nodes and strands along railroads and trolley lanes. (c) Automobiles allowed more mobility, thus open space between nodes and strands began to fill in. (d) The freeway, built to relieve congestion, actually hastened the movement of the population from high-density urban dwellings to low-density suburban housing.

The spatial pattern of the American city in the 1920s was one of a densely populated urban node surrounded by smaller, less densely populated suburban nodes. These nodes were connected by populated strands that followed trolley lines. Between the nodes and strands lay a countryside of woodlands and farms, all within access of the residents of the new suburbs, and to a lesser degree, urban residents (Figure 1–7c).

By the end of World War II, the automobile started to change the face of American cities. Millions of returning servicemen sought to marry, find jobs, and settle down in a suburban home. Post–World War II subdivisions were no longer tied to the old fixed mass transportation routes, thanks to the automobile. De facto greenspace between old transportation corridors was within easy access, by automobile, of urban centers. New subdivisions spread rapidly to these areas as developers sought land for housing. Typically, these developments followed a hopscotch pattern across the landscape, as demand for land bid prices up and builders sought cheaper parcels farther and farther from urban centers. In time, the remaining open space filled in with more housing, shopping centers, and new industry (Figure 1–7d).

New suburban residents abandoned mass transit for the automobile because of the attractiveness of personal transportation and because the new low-density developments were not well served by mass transit. By the mid-1950s, traffic congestion reached the crisis stage in most American cities.

Freeway Cities

In 1956, Congress authorized the construction of a vast system of interstate highways in the United States. The initial plan was to construct 36,000 miles of freeways across the country for the purpose of national defense. These highways were originally intended to approach but not penetrate cities. On the other hand, urban governments, struggling with traffic congestion, were ill equipped to deal with the problem financially. Members of Congress from urban districts and urbanized states pressed for federal assistance, and the Congress authorized use of federal funds to extend the interstate systems to urban centers.

Unfortunately, most urban interstates were obsolete by the time of ribbon-cutting ceremonies. Rather than allow better access to central cities, the new expressways encouraged more people to flee to the suburbs. Low-income minority groups moved to decaying inner cities from the south and from other parts of the globe, seeking improved economic opportunities. Racial prejudice accelerated the white flight to the suburbs and declining property values bankrupted urban governments. The suburbs flourished while the cities decayed (Figure 1–7d).

As the American population became more affluent, house lots in the suburbs grew larger. The shift from high-density housing in the cities to low-density housing in the suburbs resulted in unprecedented urban sprawl. The Springfield, Massachusetts, Standard Metropolitan Statistical Area (SMSA) grew in population from 1950 to 1965 by 18 percent. During the same period the amount of land used for urban purposes increased by 136 percent (Lindsay, 1972). The seven-county southeastern region of Wisconsin (including Milwaukee SMSA) increased in population by 42 percent from 1950 to 1970, while land dedicated to urban use rose by 216 percent (League of Women Voters of Wisconsin, 1975).

In a similar study, LaGro (1994) reported that a geographic information analysis of land use in a nonmetropolitan county in southeastern New York State determined urban land use increased from 6.7 percent in 1968 to 17.8 percent in 1985. This growth in urban land use was eight times faster than the population increase. "The number of urban patches, or contiguous urban areas, decreased from 936 to 829, although the average size of the urban patches increased from 9.4 to 25.4 ha."

Open space was lost to the suburbs, and suburban dwellers had to be content with what they had in their yards. Minimal land-use planning was done, and what was done was usually too little and too late. Suburban residents at least had the opportunity to escape to the countryside, but residents of the central cities were effectively cut off from nature. Remaining greenspaces near or in the cities were gone or abandoned to undesirable elements of society, and mass transportation to suburban parks was virtually nonexistent.

High-Technology Information Age

By the middle to late 1960s and early 1970s American cities had reached their lowest ebb. The affluent had abandoned the central cities for the suburbs, leaving urban governments to survive on declining property tax income. Riots, triggered by racial inequality, had decimated sections of the largest urban centers. Central cities could be characterized by declining populations, poor transportation, pollution, increasing noise levels, blighted neighborhoods, crime, and bankrupt municipal governments.

The federal government had initiated a number of programs following World War II aimed at improving housing and commercial centers of cities, but there are those who will argue that these programs made problems worse rather than improving the situation. All too frequently, good housing was destroyed along with bad, and low-income housing projects increased the population density of the poor, thus intensifying social problems.

For a time predictions were made by urban planners that the sprawl of the suburbs would not stop until cities in proximity to one another merged to become huge urban masses. The urban corridor from Boston to Washington, D.C., was dubbed Bos-Wash; Chicago to Pittsburgh, Chi-Pits; and San Diego to San Francisco, San-San. However, the urbanization of these regions was misleading when viewed from the highway corridors connecting them. For example, southern New England, which contains the northeastern end of Bos-Wash, was 9 percent urban and 65 percent forested in 1970 (Smith, 1970).

The predicted merging of cities within urban regions has not yet materialized, and probably will not. The reasons are rooted in socioeconomic changes that started in the late 1950s and early 1960s. The United States reached its apex as an industrial nation during the first half of the twentieth century. However, in 1956 the number of workers in white-collar jobs outnumbered blue-collar workers for the first time. Since 1956, there has been a slow but accelerating shift in employment toward white-collar jobs, with most of the work force today being in nonheavy industrial employment (Naisbitt, 1984). This shift has been primarily the result of the introduction of computers and accompanying high technology into the economic system. Industries that rely on new information and high technology no longer need to be tied to old metropolitan centers, and frequently locate in small cities. For the first time in the history of the United States, the U.S. Census Bureau now reports that population is growing faster in nonurban areas than in urban areas.

Increasing congestion in the suburbs has caused many residents to reevaluate the commuting costs associated with living in the suburbs and working in the city. Old urban neighborhoods with sound housing stock are being renovated by people seeking a return to the city. Corporate headquarters that moved to the suburbs have found that locations in close proximity to other corporate headquarters are more desirable than originally perceived. Public and private capital is once again being invested in urban centers, and this trend should continue for the foreseeable future.

Future of the American City

What, then, does the future hold for American cities? First and foremost, the city of the future will probably appear much as the city of today. The migration of the middle class back to old and new urban dwellings should continue as people seek to be closer to jobs and urban amenities. This will improve the financial situation of urban government as the tax base increases in value. Old industries will continue to shift to parts of the globe that are in early stages of industrialization, thus changing the character of urban employment. Mass transportation will improve as the new, higher-income urban residents demand better service. Suburban growth will decline as national population growth stabilizes and more people choose to live in central cities. Costs associated with low-density suburban living will continue to increase, making the suburbs a less attractive option.

The revitalization of urban America will bring with it a demand for more urban amenities. Among these demands will be the desire for more open space, parks, and more attractive urban landscapes. Urban vegetation will be prized by urban residents, and management of this vegetation will receive a higher priority by urban government.

The low population density of the suburbs brought with it ample room for private landscapes of the residents. As this vegetation ages, there will be an increased demand for services aimed at the maintenance of mature landscapes. Public spaces will receive more attention by public agencies as remaining private open spaces are developed for other uses. Preservation of open space will receive higher priority as parcels are purchased outright or are protected by easements and/or zoning.

European Cities

European cities responded to the pressures of industrialization and sprawl differently than did American cities. Following early excesses of industrialization, cities in Great Britain made strong efforts at controlling land use and size by developing greenbelts around cities, regulating land use, and developing satellite communities beyond urban centers. Many of these new communities were initially designed to be self-sufficient, with residential, commercial, and industrial land available for development. As industrialization spread across Europe similar patterns of urban development occurred. European cities today tend to use land efficiently, make good use of mass transit, and exert strong control over land use and urban development. This has limited the size of urban centers and has provided urban residents with access to greenbelt forests and other undeveloped lands.

Cities in Developing Countries

Population growth rates are very high in most developing countries, especially rural areas. Surplus population in rural areas cannot find the means of supporting themselves and consequently migrate to urban centers. Many cities throughout the developing world now support enormous populations, with new residents building squatters' communities on the urban periphery. These new communities often lack adequate water, sanitation, and other essential urban services, and generally have little land available for parks and other greenspace. Mexico City evolved from a major city of the Aztec civilization (population of 300,000) through the colonial period as a capital, to its present-day status as capital of Mexico with a population over 17 million (Deloya, 1993). As a colonial capital Mexico City developed with parks, gardens, and tree-lined boulevards, a tradition continued after independence. However, rapid population growth during the latter part of the twentieth century has reduced the amount of greenspace to 2.2 percent of the total metropolitan area. Planned development of land for more affluent residents has included greenspace, but the poor suburbs contain virtually no parks or street trees.

LITERATURE CITED

DELOYA, M. C. 1993. "Urban Forestry in Mexico City." *Unasylva* 44: 28–32.

GALLION, A. B., and S. EISNER. 1975. *The Urban Pattern: City Planning and Design.* New York: Van Nostrand Reinhold Company.

LAGRO J. A., JR., 1994. "Population Growth Beyond the Urban Fringe: Implications for Rural Land Use Policy." *Landscape and Urban Planning* 28:143–158.

LEAGUE OF WOMEN VOTERS OF WISCONSIN. 1975. *Wisconsin's Land Facts and Issues.* Madison: League of Women Voters of Wisconsin.

LINDSAY, B. E. 1972. "The Influence of Selected Variables on Land Values in the Rural-Urban Interface." M.S. thesis, University of Massachusetts at Amherst.

MUMFORD, L. 1961. *The City in History.* New York: Harcourt Brace & World.

NAISBITT, B. E. 1984. *Megatrends.* New York: Warner Books.

SMITH, D. M. 1970. "Adapting Forestry to Megapolitan Southern New England." *J. Forestry* 67:372–377, January.

2

Social Needs and Values of Urban Society

Human attitudes and values are not and have never been static. A given set of cultural values may persist for a period of time, but history has shown that change is the rule rather than the exception, particularly in modern times. Cultural values direct our interaction with one another, other cultures, and our physical environment. Cultural values toward the physical environment include the perception of humankind's relationship with the natural world. In this chapter we deal with the human relationship to nature, its historic evolution, and changes taking place at the present time. We also discuss how these changes affect social needs and values in relation to urban trees, forests, and open space.

EVOLVING RELATIONSHIP TO NATURE

Recent human history may be characterized as a gradual shift from living in nature to living in cities. As this shift took place human beings redefined their relationship with the natural world repeatedly, to accommodate new discoveries and new modes of economic interaction. Hunting and gathering societies lived with an understanding of their total dependence on nature and a realization of consequences associated with violation of natural laws. Deities were found in nature and religions based on what were perceived as natural laws. An individual was born into nature, derived sustenance from nature, and was claimed by nature in the end (Toeffler, 1980). Ancient peoples considered nature and culture to be intimately connected. "Daily activities such as prayer, music, dance, play and work were originally woven together in sacred ceremonies which spoke to our relationship with the natural world" (Berry, 1993). The Kayapo people of the Amazon rainforest in Brazil dance to preserve the integrity of the natural world, believing that if they do not, nature will be destroyed (Berry, 1993).

The Agricultural Revolution changed how we human beings sustained ourselves, and changed forever our relationship with nature. No longer was survival dependent on what could be found to eat in the ecosystem. We could grow food, cease nomadic ways, and settle in permanent dwellings. But to manipulate nature demanded a redefinition of the relationship with nature. No longer could deities be in the rocks, trees, and animals, for they had to be moved or destroyed to grow crops. The gods were shifted elsewhere, and this provided freedom to alter nature as needed. However, we were still very much aware that we were under the control of nature. Too little rain, too much rain, insects, and a host of other problems served as constant reminders of our dependence. Agrarian societies lived by natural rhythms, with a time to do all things as dictated by the natural world (Toeffler, 1980).

Industrialization brought a new economy, and yet another definition of our relationship with nature. Toeffler (1980) describes three beliefs that were necessary for humankind to accept the excesses of the Industrial Revolution.

1. War with nature
2. Social Darwinism
3. Progress principle

To exploit nature on a grand scale, it is necessary to assume no dependence on nature and, in fact, to declare a war of survival for the species. This line of thinking provided the freedom to cut and burn forests, rape hillsides for minerals, and dump human and industrial wastes into natural systems.

Social Darwinism is defined as the survival of the fittest individual and/or society, and the subjugation and/or elimination of all inferior individuals or cultures. Human history prior to the Industrial Revolution is certainly filled with examples of the elimination of one culture by another, but Social Darwinism provided "scientific" justification for global exploitation of resources by industrial nations and the elimination of all who stood in the way of progress.

The belief that all progress is good and that bigger is better is the basis for the progress principle. If a small city provides certain amenities, a larger city will provide more. A dam on a river provides water storage and flood control; therefore, a series of dams will improve the river further. Expansion of the human population provides economic growth; thus, continued expansion of the human population will improve the world economy even more.

As we near the end of the twentieth century, we carry with us most of the belief system developed at the start of the Industrial Revolution over 200 years ago. As the

Industrial Revolution gives way to the Information/High-Technology Era, so then will our values change to reflect a different socioeconomic system and a different perception of our relationship to nature.

TODAY'S HUMAN ENVIRONMENT

Many species actively alter their physical environments to provide optimums in temperature, humidity, light, air movement, spatial structure, and social order. This is done through the construction of nests, burrows, colonies, hives, and so on. In the human species, alteration of the physical environment has reached its highest development in our homes, places of work, and transportation systems. People live in climate-controlled housing, work in air-conditioned offices and factories, entertain themselves in indoor arenas, and travel about encapsulated in personal microclimates (Poole, 1971).

Socialization of children in an urban society consists, among other things, of weaning them away from the natural world to prepare them for lives in this new environment. Young children have an intense curiosity about the world around them. Casual observation finds them on top, under, around, tasting, touching, smelling, and watching the things of nature (Figure 2–1). Too often parents and other adults warn children of real and imagined hazards in nature. They are chastised for getting dirty, told not to climb, warned not to touch insects, toads, and plants, and in general made cautious (if not terrified) of the natural world.

Although the process of weaning children from the natural world may help to prepare them for urban living, it may also create negative values toward nature that they will carry throughout their lives. The transfer of negative values toward nature to children is also a function of the belief system adopted at the start of the Industrial Age: that of being at war with nature for survival.

Sinton (1971) describes current social attitudes held by an urban society toward nature as a continuum of values. These attitudes may be divided into four categories: love and dependency on nature, renewal in nature, tamed nature, and nature haters. The love and dependency attitude is rare in an urbanized world. People with this attitude must live immersed in nature, draw subsistence from it, and would suffer emotional if not physical death without nature. This attitude is best reflected in the values of those few remaining hunting and gathering societies left in the world, and a few individuals who leave the city to live out their lives in the world's few remaining wilderness areas.

Many city dwellers reflect the second attitude, that of renewal through contact with nature. Most of their lives are spent in urban areas at urban tasks, but they renew themselves spiritually through periodic contact with nature. This renewal consists of activities such as backpacking, cross-country skiing, canoeing, and other pursuits that allow them to immerse themselves briefly in the natural world (Figure 2–2).

The third attitude, controlled nature, is the feeling that occasional contact with nature is acceptable, but not on nature's terms. People with this value surround themselves with pieces of tamed nature in their yards, camp in campers parked in high-

Figure 2–1 Children have an affinity with the natural world and immerse themselves in it at every opportunity.

Figure 2–2 Some urban residents find renewal through intimate contact with nature. (R. L. Geesey.)

Figure 2–3 For many urban residents a well-serviced campground provides sufficient contact with nature.

density campgrounds, build elaborate vacation homes, enjoy motorized outdoor recreation, but feel extremely uncomfortable off a well-marked trail (Figure 2–3).

Individuals with the fourth value, nature haters, couldn't care less about nature and are quire comfortable living in apartments and recreating in the city. They find nothing offensive about large areas of asphalt, but may enjoy an occasional stroll in a safe, well-landscaped city park (Figure 2–4).

Figure 2–4 While most urban residents show a distinct preference for trees and other natural features in the landscape, some are satisfied with environments totally devoid of nature.

Differences exist between urban and rural dwellers as to their perceptions of the natural environment. Brush and More (1976) reported results of a study in which rural, suburban, and inner-city children were asked, "What does a tree mean to you?" Rural and suburban children tended to view trees as places to play, while inner-city children thought of trees as sources of wood for construction materials. The authors conclude that the lack of trees in the inner city prevents children from seeing them as sources of recreation and aesthetics.

A wide range of values concerning trees and forests exists throughout the population. These values range from a great interest in nature to scarcely any interest at all, and may be linked to overall social values in reference to the process of shifting from a rural society to an urban society. However, as urban society continues to evolve, social values can be expected to shift to accommodate new knowledge and new ways of viewing our relationship to nature.

CHANGING VALUES

There exists in most people a need for nature in some form. As a species that has spent most of its existence hunting, gathering, and/or growing food, it is doubtful that we have been able psychologically to remove ourselves from nature in the 200 years since the onset of the Industrial Revolution. One of the reasons for the exodus to the suburbs is the desire to be close to nature. At a time of the greatest urbanization in the world there is renewed interest in nature. The proliferation of books about nature, the interest in house plants, the marketing of sophisticated wilderness camping equipment, and the environmental movement have all come about following the apex of industrialization. According to the National Gardening Association, 78 percent of households in America have gardens, spending $35 billion per year (Gibbs, 1988).

Yi Fu Tuan, in his book *Topophilia* (1974), describes the shift in values toward wilderness by human society through the process of urbanization. Tuan points out that the first cities were centered around temples and other religious monuments, and urban design followed a culture's perception of the universe. Cities were attempts to duplicate the perfection of the gods in their heavens, while the objects of the earth were associated with the underworld and imperfections of nature. The city represented the gods, order, perfection, safety, and nature tamed. Wilderness, on the other hand, represented the profane, disorder, fear, and the untamed. Today, many would argue the same attributes of cities versus the wilderness, with only the headings reversed.

PSYCHOLOGICAL VALUES

Additional support for the premise that it may not be wise to divorce ourselves from nature comes from a study by Talbott et al. (1978). The researchers found that social interaction, food consumption, and time spent in the dining room all increased in a psychiatric hospital following the introduction of flowering plants. Users of the

Morton Arboretum near Chicago reported that the most common feelings they recalled following a visit were peacefulness, quiet, serenity, or tranquility (Schroeder, 1986). In another study, Ulrich (1978) showed a series of slides depicting various scenes of urban and rural landscapes and recorded the alpha brain waves of the viewers. (Alpha waves are an indication of a person's wakeful relaxation.) Ulrich found that the greater percentage of green and water in the slide, the greater the state of relaxation observed in the subjects.

There is some evidence of the restorative benefits of nature on humans. Hartig et al. (1987) studied subjects who sought to relieve stress by walking for 40 minutes in a natural area with trees, walking for 40 minutes in an urban setting, listening to music, or reading to relax. Participants who had walked in the natural area reported more positively toned feelings than those who had engaged in the other two activities.

A group of hospital patients recovering from surgery whose windows looked into a small stand of trees was compared to a similar group of patients whose windows looked at a brick wall (Ulrich, 1984). "Individuals with tree views had significantly shorter postoperative hospital stays, had far fewer negative evaluation comments in nurses notes, and tended to have lower scores for minor post surgical complications such as persistent headache or nausea" than patients with the wall view.

E. O. Wilson (1992) uses the term *biophilia* to describe our subconscious connection to the rest of life, and postulates that it has its basis in human evolution. Biophilia refers to both negative and positive responses to elements of the natural world that had survival value to our evolutionary predecessors. For example, there seems to be universal human fear of spiders, snakes, wolves, heights, and closed spaces, to name a few common phobias. Because these species and circumstances are often dangerous to humans there is survival value in their avoidance. Likewise, given the opportunity, most humans prefer to live on a prominence near water overlooking savannalike parkland. In our evolutionary past these kinds of locations had survival value for obtaining food and water, and to watch for game and enemies. Suburban lawns and shade trees are a kind of savanna, and we certainly pay premium prices for waterfront property (Wilson, 1992).

Although it is generally recognized that there is a need for some contact with nature for most people, there is also a shift in values taking place as the result of the Information/High-Technology Revolution. This shift in values is taking many forms, including a change in the perception of humankind's relationship to nature. As we become more aware of our dependence on the biosphere, in spite of our attempts to convince ourselves otherwise, we are becoming more cautious of the potential impact of our technologies on living systems. We are beginning to see ourselves as part of a system rather than as independent entities. Our cities are ecosystems, and they interface and are supported by the biosphere. The structure of science is changing from a linear approach to problem solving to an ecologic or holistic approach that considers all social and scientific ramifications of new discoveries and subsequent technologies (Ferguson, 1980).

SOCIAL NEEDS, VALUES, AND URBAN FORESTRY

As social values shift to a more holistic perception of our relationship to nature, so will our expectations shift to a greater desire to complement our urban lifestyle with elements of nature. The concept of urban forestry is one of establishing and maintaining individual trees and tree communities or ecosystems in and around urban areas. To establish and maintain woody vegetation in cities, it is necessary to understand what trees and associated vegetation mean to urban residents. Lipkis and Lipkis observe in their book *The Simple Act of Planting a Tree* (1990), "When we plant and care for trees, alone or together, we begin to build an internal place of peace, beauty, safety, joy, simplicity, caring, and satisfaction."

Appleyard (1978) groups the many meanings of trees under three categories as they apply to our perception of them in the environment: sensory, instrumental, and symbolic. As sensory features in the landscape, trees are appreciated as natural forms, ornaments, incense, sound and visual screens, and as contrast to the harsh lines of the urban landscape. The instrumental functions of urban trees are to provide shade, shelter, environmental protection, recreation, fruits, and to a limited degree, wood. Trees serve as symbols of nature in the urban scene and symbols of one's self or group when identified with a particular area or neighborhood. Trees can be alien symbols to some when planted by outsiders seeking to bestow their values on others, or as social tokens when planted in low-income neighborhoods under the guise of improving the lot of the poor (Figure 2–5).

A study by Getz, Karow, and Kielbaso (1982) of tree preferences by inner-city residents in Detroit found that two-thirds of the residents felt that more money should be spent on trees in parks and on residential streets. Furthermore, residents placed a higher priority on street trees than on trees in parking lots and as industrial screens. When asked what attributes they preferred more of in city parks, respondents gave the highest priority to trees and the shade they provide.

Reporting on the results of a study concerning the sociology of urban tree planting, Ames (1980) found that the benefits go beyond establishing more neighborhood trees. Residents who participated in a program of street tree planting in a low-income neighborhood in Oakland, California, felt an enhanced sense of community, a better understanding of their neighbors, and greater control over their environment.

Purchasers of wooded lots for home construction in a Wisconsin city indicated that the presence of trees and other vegetation, the countrylike atmosphere, and the resale value of the parcel were the most important attributes considered in selecting the lot. These homeowners valued their vegetation primarily for aesthetics, although they did recognize other attributes, such as contribution to property values, shade, and attraction of desirable wildlife (Vander Weit and Miller, 1986).

Driver and Rosenthal (1978), in a survey of empirical studies describing perceived psychological benefits of urban forests and related greenspaces by urban residents, summarized the following as important functions:

Figure 2–5 Inner-city trees may serve as alien symbols of the affluent when planted without neighborhood involvement.

1. Developing, applying, and testing skills and abilities for a better sense of self-worth
2. Exercising to stay physically fit
3. Resting both physically and mentally
4. Association with close friends and other users to develop new friendships and a better sense of social place
5. Gaining social recognition to enhance self-esteem
6. Enhancing a feeling of family kinship or solidarity
7. Teaching and leading others, especially to help direct the growth, learning, and development of one's children
8. Reflecting on personal and social values
9. Feeling free, independent, and more in control than is possible in a more structured home and work environment
10. Growing spiritually

Urban residents desire urban trees, parks, and forests for a variety of reasons. We are a species that evolved from the natural world and find it difficult to live without some contact with nature. As we become more urban in our ways of living, the desire for contact with nature will increase rather than decrease. This desire will manifest itself in an increased demand for urban tree and forest amenities as we seek to enhance the quality of life in our cities.

LITERATURE CITED

AMES, R. G. 1980. "The Sociology of Urban Tree Planting." *J. Arbor.* 6(5):120–123.

APPLEYARD, D. 1978. "Urban Trees; Urban Forests: What Do They Mean?" *Proc. Natl. Urban For. Conf.,* ESF Pub. 80–003. Syracuse: SUNY, pp. 138–155.

BERRY, R. 1993. "Nature and Culture: Building Community from the Ground Up." *Proc. 6th. Natl. Urban For. Conf.,* American Forests, Washington, D.C., pp. 226–228.

BRUSH, R. O., and T. A. MORE. 1976. "Some Psychological and Social Aspects of Trees in the City." *Better Trees for Metropolitan Landscapes,* USDA-For. Serv., Gen. Tech. Rep. NE-22, pp. 25–29.

DRIVER, B. L., and D. ROSENTHAL. 1978. "Social Benefits of Urban Forests and Related Green Spaces in Cities." *Proc. Natl. Urban For. Conf.,* ESF Pub. 80–003. Syracuse: SUNY, pp. 98–103.

FERGESON, M. 1980. *The Aquarian Conspiracy.* Los Angeles: Jeremy P. Tarcher.

GETZ, D. A., A. KAROW, and J. J. KIELBASO. 1982. "Innercity Preferences for Trees and Urban Forestry Programs." *J. Arbor.* 8(10):258–263.

GIBBS, N. R. 1988. "Paradise Found." *Time,* June 20, pp. 62–71.

HARTIG, T., M. MANG, and G. W. EVANS. 1987. *Perspectives on Wilderness: Testing the Theory of Restorative Environments.* Paper presented at the 4th World Wilderness Conference, Estes Park, Colorado.

LIPKIS, A., and K. LIPKIS. 1990. *The Simple Act of Planting a Tree.* Los Angeles: Jeremy P. Tarcher.

POOLE, W. R. 1971. "Social Comforts of Trees to an Urbanizing Environment." *Trees and Forests in an Urbanizing Environment,* Plann. Res. Dev. Ser. No. 17. Amherst: Univ. Mass., pp. 77–78.

SCHROEDER, H. W. 1986. "Psychological Value of Urban Trees: Measurement, Meaning and Imagination." *Proc. 3rd Natl. Urban For. Conf.,* American Forests, Washington, D.C., pp. 55–60.

SINTON, J. 1971. "The Social Value of Trees and Forests for Recreation and Enjoyment of Wildlife." *Trees and Forests in an Urbanizing Environment,* Plann. Res. Dev. Ser. No. 17. Amherst: Univ. Mass., pp. 71–76.

TALBOTT, J. A., et al. 1978. "Flowering Plants as Therapeutic/Environmental Agents in a Psychiatric Hospital." *HortScience* 11(4):365–366.

TOEFFLER, A. 1980. *The Third Wave.* New York: Bantam Books.

ULRICH, R. S. 1978. *Psycho-physiological Effects of Nature vs. Urban Scenes.* Stockholm, Sweden: Report prepared for Natl. Counc. Build. Res.

ULRICH, R. S. 1984. "View Through a Window May Influence Recovery from Surgery." *Science* 224:420–421.

VANDER WEIT, W., and R. W. MILLER. 1986. "The Wooded Lot: Homeowner and Builder Knowledge and Perception." *J. Arbor.* 12(5):129–134.

WILSON, E. O. 1992. *The Diversity of Life.* Cambridge, Mass.: Harvard University Press, pp. 349–350.

YI FU TUAN. 1974. *Topophilia.* Englewood Cliffs, N.J.: Prentice-Hall.

3

The Urban Forest

NATURE OF THE URBAN FOREST

The urban forest may be defined as the sum of all woody and associated vegetation in and around dense human settlements, ranging from small communities in rural settings to metropolitan regions. Clegg (1982) describes this forest in the United States as covering an area of 69 million acres, and including "not only tree lined streets, but parking lots and school yards, downtown parks and riverbanks, cemeteries, and freeway interchanges. It includes everything from the dwarf dogwood tree we tenderly nurture to the huge oaks and pines of the natural forest at the end of the furthest suburban street." Kielbaso et al. (1988), in a national survey of municipal street programs, reported that there are at least 60 million city-owned street trees in the United States, with an additional 60 million vacant spaces where trees can be planted.

The city of Chicago has an estimated 4.1 million trees, of which 10 percent are street trees that hold 24 percent of the leaf area surface in the city. The remaining trees are located in parks and on private land. There are an estimated 50.8 million trees in Chicago and its suburbs, an area consisting of Cook and DuPage counties (McPherson, Nowak, and Rowntree, 1994).

More specifically, this forest is the sum of street trees, residential trees, park trees, and greenbelt vegetation. It includes trees on unused public and private land, trees in transportation and utility corridors, and forests on watershed lands. Some of these trees and forests were willfully planted and are carefully managed by their owners, while others are accidents of land-use decisions, economics, topography, and neglect.

A great diversity of vegetation may be found in cities, ranging from non-forested wetlands and grass- or brush-covered sites to natural forests, mature parks, selected cultivars on streets, and landscaped private property. The urban forest is part and parcel of the human environment, and, according to DeGraaf (1974), provides habitat for a great diversity of wildlife in the city. The idea of the city as a habitat for people, consisting of structures, vegetation, and other animal species, leads to the concept of the ecosystem. In recent years a new field of study has emerged, urban ecology, that is based on this concept. Urban ecology is defined as a blend of elements from biological ecology (natural science) and human ecology (social science). Smith (1971) describes urban ecology as including the "natural processes of weather, vegetation, and animal life, their interaction with man made environments, the behavioral, physiological, and developmental processes of man that seem directly related to aspects of the natural environment, and the value systems that affect the inclusion of natural elements in the creation of urban environments." In their definition of urban ecosystems Moll and Petit (1994) state, "Cities are ecosystems of many species of plants and animals; urban forests and urban ecosystems are the areas in and around those cities." They describe the nonbiological component as consisting of streets and buildings as well as rocks and topography, and suggest that full understanding of urban ecosystems must include descriptions of both natural- and human-directed material cycling and energy flows.

The concept of an urban forest is best understood when viewing cities from the air. In the eastern United States, cities are surrounded by forests and almost appear to be threatened with disappearance beneath the forest canopy (Figure 3–1). New

Figure 3–1 Forests surround cities in regions with adequate rainfall but marginal agricultural opportunities.

subdivisions spread into woodlands, and efforts are made to preserve residual trees. Neighborhoods built on open land quickly fill with saplings and in time geometric lines fade beneath the natural contours of tree crowns. In agricultural and arid regions the presence of trees on the horizon frequently signals the presence of human settlements (Figure 3–2). With the exception of commercial and industrial districts, cities often appear to be located in forests, as residential and park trees, and trees on undeveloped land, form a nearly continuous canopy of green.

The density of the urban forest varies with patterns of land use in urban areas (Figure 3–3). Land use in and around cities may be divided into four distinct zones from the city center outward; urban, suburban, exurban, and rural. The urban zone consists of commercial districts, old industrial sites, and high- and medium-density residential neighborhoods. Generally, this zone has the fewest number of trees and the lowest percent of land dedicated to park and open space. In many instances the urban zone will contain the oldest and most decadent vegetation.

Land in the suburban zone is used primarily for low-density housing, with other land dedicated to shopping centers, new industrial parks, and transportation corridors.

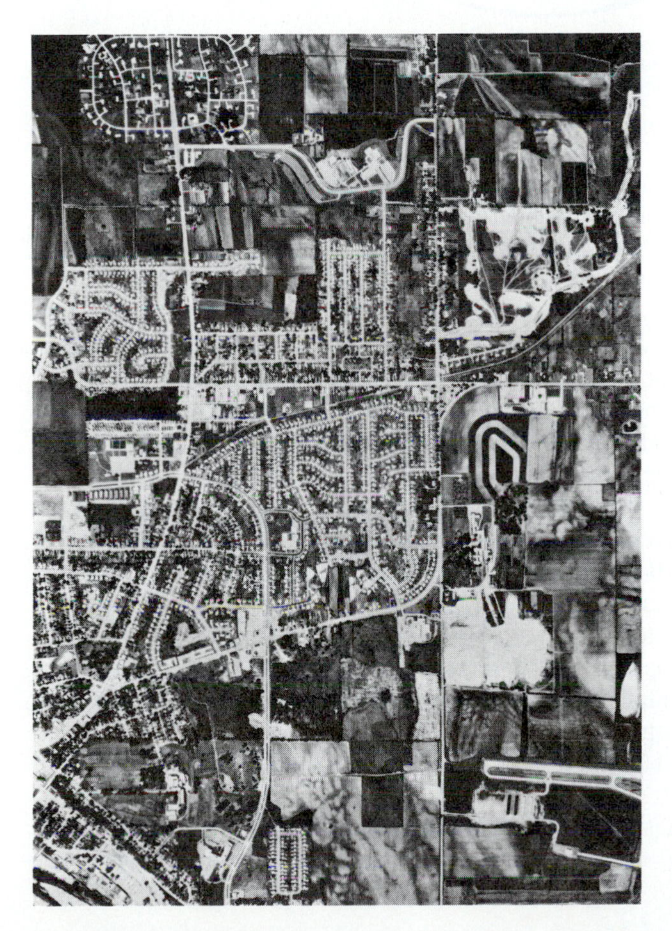

Figure 3–2 Agricultural or arid regions are largely devoid of trees and forests except in communities.

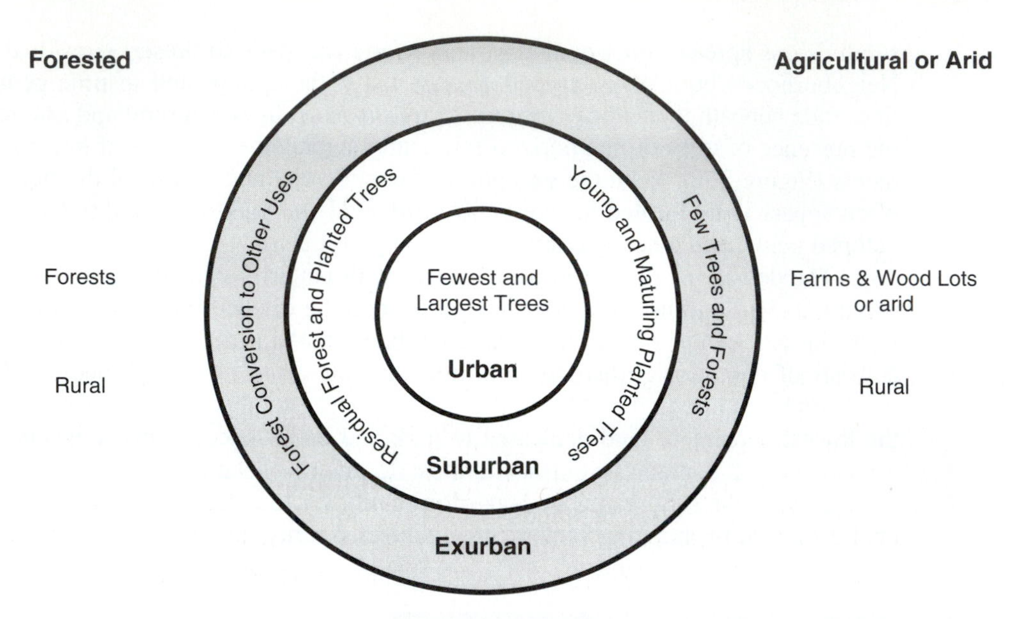

Figure 3–3 Land use and the urban forest in a forested region and agricultural/arid region. (From Miller, 1994.)

Usually, more trees, parks, and other green spaces are found in suburban zones, and this vegetation consists of more recently established plantings, along with residual trees and forests, especially in humid nonagricultural regions.

The exurban zone is located at the interface of the suburbs and the hinterlands. Often, this zone contains hobby farms, residual agricultural and forest land, land held in speculation, and assorted commercial land uses generally not permitted in the suburbs. In forested regions, most of this land will be wooded, while in arid or agricultural regions, this land will contain few if any trees. As urban growth continues this land will shift to suburban land uses. In forested regions the number of trees will decline, while in nonforested regions the number of trees will increase. For example, the site of Oakland, California, had only 2 percent tree cover before the city was established, but today trees cover 19 percent of the landscape (Nowak, 1993). Rural zones near cities shift to exurban land use in time, and vegetation changes follow the same patterns in the exurban zone.

Nowak (1994) reports that urban tree cover in the United States ranges from 55 percent in Baton Rouge, Louisiana, to 1 percent in Lancaster, California. He further reports that the highest percentage of tree cover is usually found on vacant, park, and residential areas. Street trees generally account for one in ten trees in cities, but in Chicago street trees account for one in four trees in one- to three-family residential neighborhoods.

Rowntree (1984) suggests that three factors determine overall urban forest structure: urban morphology, natural factors, and management. Urban morphology provides the space for vegetation and is determined by how we build our communities. Natural factors control what will grow there and include factors such as climate, soils, and topography. Management is what species and ecosystems humans elect to include and manage for in the urban landscape.

MANAGING THE URBAN FOREST

If the urban forest is the sum of all woody and associated vegetation, then urban forest management is the establishment and care of this resource. The USDA Forest Service (1990) defines the management of the urban forest as "the planning for and management of a community's forest resources to enhance the quality of life. The process integrates the economic, environmental, political and social values of the community to develop a comprehensive management plan for the urban forest." Stewart (1974) describes the management of urban forest resources in the Chicago area as falling under the authority of a number of agencies. The street tree population is maintained by the Chicago Bureau of Forestry, while over 7000 acres in city parks are managed by the Chicago Park District. At the county level, the Cook County Forest Preserve District manages over 66,000 acres of land in and around the city of Chicago, primarily for recreation. Stewart also includes the management of state and national forests and parks in Illinois and neighboring states as part of urban forest management, due to intense recreation pressure placed on these lands by urban residents.

The management of public trees and public land in and around urban areas is one part of urban forestry, but publicly owned trees and forests typically represents only 10 percent of the urban forest in the United States (Moll and Kollin, 1993). The remaining 90 percent of urban trees are found on private property such as residential yards, wood lots, industrial sites, and corporate parks, and are managed by their owners or agents of owners, or are not managed at all. This is the realm of the commercial arborist, landscape contractor, consulting forester, and other professionals who provide services for both private citizens and public agencies. In total, the structure, composition, and maintenance of the urban forest is influenced far more by individuals and private contractors than by public agencies.

Arboriculture and Urban Forestry

Just as the urban forest is a complex mosaic of vegetation and urban development, so the management of the urban forest is a mosaic composed of a number of professions interacting with the landscape and one another. The two professions most frequently mentioned when dealing with the urban forest are arboriculture and urban forestry. Other professions, such as landscape architecture, horticulture, forestry, recreation, turf management, and wildlife management, most certainly affect the urban forest, and are discussed later in the chapter.

Harris (1992) defines arboriculture as being "primarily concerned with the planting and care of trees and more peripherally concerned with shrubs, woody vines, and ground-cover plants." Arboriculture is an area of specialization within the overall field of horticulture, which is defined as the cultivation of plants for food and aesthetic purposes. Harris (1992) defines urban forestry as a specialized branch of forestry that involves the "management of trees in urban areas on larger than an individual basis."

Just as arboriculture is a specialized discipline within horticulture, so urban forestry is a specialized discipline within the profession of forestry. Forestry in North America evolved during this century through three distinct phases: classical forestry, economic forestry, and ecosystem management. Classical forestry was brought to North America from Europe around the turn of the century by American foresters trained in the methods of European silviculture. This approach to forest management persisted primarily on public land during the time that public forests were timber reserves, and the role of the forester was that of custodian of these reserves. Silviculture was carried out in public forests on an experimental basis, while private forests were being cleared of old-growth timber. World War II saw a great demand for wood in the war effort, and by the end of the war private stocks of old-growth timber were largely depleted. Public forests were increasingly expected to take up the slack in timber supply during the postwar years, and this ushered in the era of economic forestry.

Field and academic foresters increasingly applied economic analyses to classical forestry techniques, and frequently opted for economy over other resource considerations. Cutting on public lands accelerated, and good silviculture was the technique that ensured high yields and satisfactory regeneration. However, the public had different ideas, and during the late 1960s the profession of forestry came under public criticism for timber management policies, especially on public lands.

Some of the criticism was not warranted, but some was indeed valid. The profession of forestry responded to public pressure, and thus ushered in the current phase in forestry, ecosystem management. Ecosystem management is concerned with economics, but it is also concerned with other aspects of silviculture such as biodiversity, aesthetics, wildlife habitat, watershed protection, recreation, and soil protection. The overall approach to management starts with broadly defined landscapes, and considers the spatial and temporal arrangement of different habitat types in that landscape. The end result is sustainable forests and associated ecosystems that consist of both managed lands and reserves, and function with no net loss in productivity and diversity. Ecosystem management is practiced on public lands across agency boundaries, and influences the management of private lands.

As foresters communicated with proponents of the environment movement, it became obvious that not only could the profession of forestry adapt to public demands in rural areas, but foresters could contribute to solving environmental problems in urban areas. Thus a specialized branch of forestry evolved, environmental forestry, and subsequently, urban forestry. Foresters recognized two things: that they had to deal with urban residents, and that the political power base had shifted to cities. Silviculture had to include public acceptance if it was to remain the prerogative of the forest manager.

Environmental and Urban Forestry

Although foresters and arborists had been managing urban trees and forests since the early 1900s, it wasn't until the 1970s that urban forestry became a recognized discipline within the forestry profession. In June 1967, the Citizens Committee on Recreation and Natural Beauty recommended to the president that an urban and community forestry program be created in the U.S. Forest Service to provide technical assistance, training, and research (Archibald, 1973). The Pinchot Institute for Environmental Forestry Studies of the U.S. Forest Service was created in 1970, and gave the following definition of environmental forestry (Pinchot Institute, 1973):

> Environmental forestry involves those aspects of resource management dealing with man's need for, and association with, the tangible and intangible values of forest vegetation in and around metropolitan areas. Such forested vegetation involves a wide range of forested conditions—ranging from city park environments to greenbelts and woodlands in the rural areas that intersperse the huge, sprawling, urban complexes throughout Megalopolis.

Since its formation, the Pinchot Institute (now the Consortium for Environmental Forestry Research) has provided leadership in urban and environmental forestry research.

In 1968 the Bureau of Outdoor Recreation presented an urban forestry proposal that would provide educational and training assistance to communities. However, it was not until 1971 that Congressman Sikes of Florida introduced the Urban Forestry Act to Congress (Archibald, 1973). This bill, passed May 5, 1972, amended the Cooperative Forestry Assistance Act of 1950 to read:

> SECTION 1: The Secretary of Agriculture is hereby authorized to cooperate with State foresters or appropriate officials of the several States, Territories, and possessions to provide technical services to private landowners, forest operators, wood processors, and public agencies, with respect to the multiple-use management and environmental protection and improvement of forest lands, the harvesting, marketing, and processing of forest products, and the protection, improvement, and establishment of trees and shrubs in urban areas, communities, and open spaces. All such technical services shall be provided in each State, Territory, or possession in accordance with a plan agreed upon in advance between the Secretary and the State forester or appropriate official of the State, Territory, or possession. The provisions of this Act and the plan agreed upon for each State, Territory, or possession shall be carried out in such manner as to encourage the utilization of private agencies and individuals furnishing services of the type described in this action.

Following passage of the Sikes bill, many states amended their own cooperative forestry laws with provisions for urban forestry programs. In 1971 the Florida legislature amended the County Forestry Law, adding "to improve the beauty of urban and suburban areas by helping to create in them an attractive and healthy environment through the proper use of trees and related plant associations." The amendment allowed county and city governments to enter into contractural agreements with the Florida Division of Forestry for urban forestry assistance (Harrell, 1978).

The Cooperative Forestry Assistance Act of 1978 expanded the commitment to urban forestry by the federal government by authorizing the Secretary of Agriculture to provide financial and technical assistance to state foresters. An allocation of $3.5 million was made to provide urban and community forestry assistance in 1978. However, the commitment to urban forestry on the part of the federal government did not change for more than a decade, and in fact declined to a low of $1.5 million in community forestry assistance support by 1984 (Casey and Miller, 1988). A much greater commitment to urban forestry by the federal government was part of the 1990 Farm Bill, which amended the Cooperative Forestry Assistance Act to:

- Expand the authority of the Forest Service to work with States to administer grants and technical assistance
- Raise funding from $2.7 million in 1990 to $25 million in 1993.
- Create a fifteen-member Urban and Community Forestry Advisory Council appointed by the Secretary of Agriculture.

To qualify for grants and technical assistance states are required to establish an urban forestry coordinator in state government, and to appoint a state urban forestry advisory council.

The Society of American Foresters created an Urban Forestry Working Group in 1972, and this group offered the following definition of urban forestry:

> Urban Forestry is a specialized branch of forestry that has as its objective the cultivation and management of trees for their present and potential contribution to the physiological, sociological and economic well-being of urban society. Inherent in this function is a comprehensive program designed to educate the urban populace on the role of trees and related plants in the urban environment. In its broadest sense, urban forestry embraces a multi-managerial system that includes municipal watersheds, wildlife habitats, outdoor recreation opportunities, landscape design, recycling of municipal wastes, tree can in general, and the future production of wood fiber as raw material.

Stewart (1974) gives a briefer definition of urban forestry, stating that "Urban forestry is the application of basic forest management principles in areas subject to concentrations of population." Carlozzi (1971) simply states that "all forestry is urban forestry" in an urban society.

Arboriculture and urban forestry as defined so far appear to split rather neatly at the individual tree level. Arboriculture deals with the tree as the basic management unit, while urban forestry treats the stand as the basic management unit. Andresen (1978) supports this concept, stating: "In simplistic terms . . . arboriculture deals with individual trees and urban forestry addresses tree aggregates." However, the Cooperative Forestry Act of 1978 defines urban forestry as follows:

> Urban Forestry means the planning, establishment, protection and management of trees and associated plants, individually, in small groups, or under forest conditions within cities, their suburbs, and towns.

This definition includes the individual tree, and it is at this point that urban forestry and arboriculture become one. Nobles (1980) elaborates on this theme by stating: "Urban forestry can and should complement arboriculture." He further remarks: "Many of us feel that urban forestry is more than arboriculture. We do not feel the two terms have the same meaning. Arboriculture is an important function of urban forestry, probably the most important one."

Forest Continuum

Conceptually, one approach to understanding how various resource professionals interact with the forest is to consider the forest as existing on a continuum, ranging from rural landscapes to the urban centers (Figure 3–4). Arboriculture exerts its major influence on the individual tree in the urban and suburban landscape, and plays a lesser role in the exurban zone. Recreation deals with all aspects of the forest continuum, with municipal recreation in the urban zone and wild land recreation in the rural zone. Traditionally, outdoor recreation was the realm of the person who left the city seeking rural solitude. However, at the urban end of the continuum, Dwyer (1982) considers the urban forest extremely important to urban residents who have limited access to rural areas. The landscape architect is most important in the urban landscape both as a primary designer and as a secondary influence on urban planners

Forest Continuum

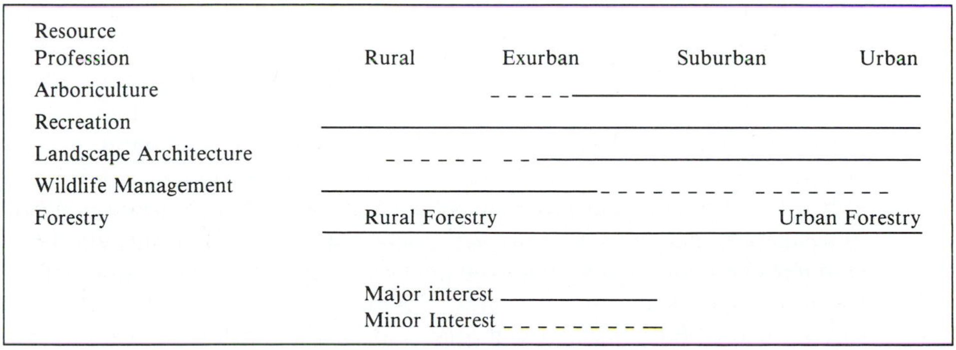

Resource Profession	Rural	Exurban	Suburban	Urban
Arboriculture		- - - - - ———	————————	————————
Recreation	————————	————————	————————	————————
Landscape Architecture	- - - - - -	- - ———	————————	————————
Wildlife Management	————————	————————	- - - - - - - -	- - - - - - - -
Forestry	Rural Forestry			Urban Forestry

Major interest ————————
Minor Interest - - - - - - - - - -

Figure 3–4 Relationship of various resource professions to the forest continuum.

and architects. At the rural end of the continuum the landscape architect is involved in recreation site design and in timber management of public land but does not exert a major impact on the overall forested rural landscape. Wildlife management has historically been concerned with management of rural wildlife populations. During the past two decades wildlife agencies and managers have become increasingly involved with urban wildlife populations and have been exerting effort in nonconsumptive wildlife management in rural areas, primarily for an urban constitutency.

As discussed previously, the forestry profession has become more and more involved with urban vegetation. At the urban end of the continuum, urban foresters manage tree populations in parks, greenbelts, and on city streets. Arboriculture is the management of the tree on an individual basis, and urban forestry is the management of tree populations. At its most intensive level, urban forestry must deal with the individual tree as it exists in a population of trees. On an individual tree level, arboriculture and urban forestry are synonymous. They separate only when dealing with populations of trees, such as managing street tree populations and greenbelt forests.

Rural and Urban Forestry

At the rural end of the continuum is rural forestry, with no division between rural forestry and urban forestry. There cannot be a division between urban forestry and rural forestry in an urban society because urban values permeate our collective value system. According to Rhodes (1971), "The future course of resource and environmental management will be determined largely by city dwellers through their legislators." Caldwell (1971) supports this, and adds: "The first concern of ecology oriented citizens with the forests is their place in the biosphere, their place in the economy is an important but secondary consideration." Rural forests are managed for an urban population, and rural foresters must be keenly aware of urban values and needs that go beyond wood fiber. Urban forestry and urban foresters provide the link to an urban population, and communication must go both ways. Maintaining open lines of communication is the only means of preventing the objectives of forest managers from being at odds with the objectives of society.

Urban forestry does much to enhance the quality of life in our urban regions. We are a highly technological society and will continue to live in urban environments, be they large cities with populations numbering in the millions, or small cities with populations numbering in the thousands. We are attempting to improve our communities on all fronts, and aesthetics play an important role in this improvement. Urban forestry and other professions that deal with urban vegetation play a key role in aesthetics, and a key role in other quality-of-life issues. Caldwell (1971) considered trees and other vegetation to be important elements in the restoration and preservation of open space in our cities. DeGraaf (1974) observed that "When urban forestry and wildlife management become part of a holistic effort to give people jobs, self respect, safety, and health, urban residents will probably have greater respect for tees and wildlife." Many urban residents no longer understand their connection to the land and its resources, and have only vague notions regarding the dynamic nature of ecosystems. Resource professionals must provide urban residents with a basic understanding of ecology and ecosystem management if urban residents are to understand the role resource management plays in their lives and well being. Finally, Dana (1971) urges us to "make sure that improvements in the physical environment and in the social environment, go forward together, with trees and forests playing a significant part in that advance."

Municipal Forestry

Municipal forestry may be defined as the establishment, protection, and maintenance of trees and associated vegetation on public land in communities. It includes private vegetation in matters of public safety and in control of insect and disease outbreaks that endanger vegetation throughout the community. A city forestry department may exist as an autonomous program, or as part of a larger city agency such as a park and recreation department.

Miller and Bate (1978) found that the Wisconsin cities most likely to have an urban forestry program have populations in excess of 10,000, a higher-than-average per capita income, a large number of community-owned trees, and a problem with Dutch elm disease. Although many communities have well-established municipal tree care programs, many do not. A 1986 survey of municipal forestry programs in the United States by Kielbaso et al. (1988) found that only 39 percent of respondents exercised any form of systematic tree care of public trees, a decline from 50 percent of respondents in a similar survey in 1980. Similarly, only 38 percent of respondents knew the number of trees under their jurisdiction any degree of confidence, and just 17 percent had an urban forestry management plan. The average per capita expenditure for forestry in 1986 was $2.60, or $10.62 per tree. A similar survey of urban forestry programs in the State of Pennsylvania by Reeder and Gerhold (1993) found that only 28 percent of responding communities had a street tree inventory, 48 percent plant less than fifty trees per year, and 40 percent provided systematic inspections for removal needs. These communities spent 61 percent of their budgets on street trees and 26 percent on park trees.

Greenbelt Forestry

Greenbelt forestry is defined as the management of forest and other vegetative communities in and near cities for the primary purpose of providing open space, recreational opportunities, and other environmental amenities. These forests may be used for wood fiber production, but in most cases this is of secondary importance to the functions listed above. Greenbelt lands may be publicly owned or may be subject to public control through easements and/or zoning. These lands are usually managed at the ecosystem level, with the typical management unit defined as a community consisting of similar vegetation on a relatively uniform site.

As mentioned previously, the Cook County Forest Preserve District manages over 66,000 acres of greenbelt forest in the Chicago metropolitan area (Figure 3–5).

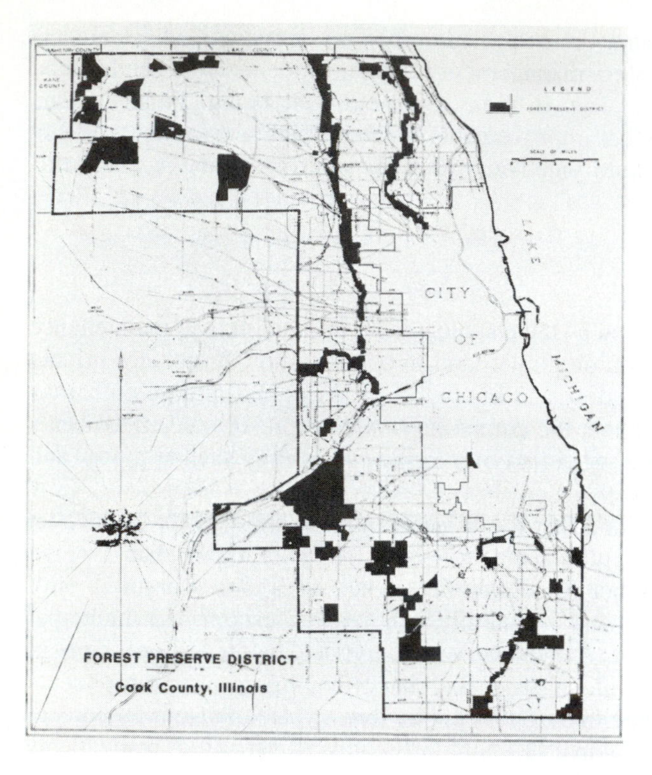

Figure 3–5 The Cook County Forest Preserve District, Cook County, Illinois, provides forest belts throughout the Chicago metropolitan area.

This land is held "for the purpose of protecting and preserving the flora, fauna and scenic beauties within such District and to restore, restock, protect and preserve the natural forests and such lands together with their flora and fauna as nearly as may be in their natural condition for the purpose of the education, pleasure and recreation of the public" (Buck, 1983).

Open space is also preserved in planned unit developments. This land is held in public trust and is managed by associations of homeowners or their agents. Other open space is held by easements on private property, and these easements may mandate maintenance of specific vegetation types by the landowners. Zoning is sometimes used to protect agriculture and forest land, as well as rare and endangered species and ecosystems.

Green Industry

The green industry consists of individuals and organizations that provide vegetation establishment, maintenance, and removal on a commercial basis. This includes, but is not limited to, commercial arborists, consulting urban foresters, landscape contractors, landscape architects, commercial nurseries, and tree, shrub, and lawn care firms.

Although the following discussion treats each of these as separate entities, it is important to remember that services offered by these industries may overlap considerably.

Commercial arboriculture is the management of individual trees and other woody plants on a commercial basis. This includes servicing both private and public owners of vegetation, and involves planting, protection, maintenance, and removal of trees and other woody plants. Commercial arborists are involved primarily with residential, commercial, industrial, and institutional tree care. However, they frequently contract with public agencies to supply vegetation management services. Smaller communities are especially prone to contract services of commercial arborists, primarily for pruning and removal of public trees.

Consulting urban foresters offer services that involve the development of urban forestry management plans. Typically, this is in smaller communities that desire to provide good public care but are not large enough to support a city forestry program on a full-time basis. These services may include street tree inventories, master street tree plans, park management plans, writing tree ordinances, and developing standards for contractural services.

Landscape architects and contractors provide for the design and installation of landforms, vegetation, and structures in the urban environment. Commercial nurseries cultivate plant materials for landscape contractors, homeowners, and public agencies. Tree, shrub, and lawn care firms provide maintenance services such as fertilization and pest management for residential, commercial, and industrial customers.

Utility Forestry

Thousands of miles of utility rights-of-way exist across the countryside and penetrate to the very core of our cities. These corridors contain high-voltage transmission lines designed to carry electricity cross-country, and connect to distribution lines designed to carry electrical service to customers. Maintenance of these rights-of-way is an enormous task. For example, Ontario Hydro in Canada manages a distribution system of 101,000 km (60,000 miles) and serves 760,000 customers (Griffiths, 1984).

Vegetation in cross-country corridors must be managed to provide access to, and clearance for, utility lines. This involves maintenance of vegetation such as grass and/or shrub communities that do not interfere with transmission lines, and removal of vegetation such as trees that have the potential to come in contact with energized lines. These lines also provide green corridors that connect peripheral green areas to open space in the city.

Distribution lines in communities generally run along streets and alleys, and pose the potential for conflict between wires and both public and private trees. The interest of public safety and service reliability necessitates the pruning or removal of trees that interfere with distribution lines.

Other Public Agencies in Urban Forestry

Beyond municipal and federal forestry programs a number of additional public agencies are involved in urban forestry, including county governments, state governments, and state universities. At the county level, efforts in urban vegetation management exist primarily on park and greenbelt land, as discussed previously. However, in some instances counties go beyond the management of public land. For example, Sarasota County, Florida, operates a county urban forestry program that is charged with the protection of residual vegetation during land development, overall countywide beautification efforts, and management of watershed lands.

Most state governments are actively involved in community forestry programs. This involvement includes serving as coordinator for federal urban forestry assistance monies, sponsorship of educational programs, and state assistance to communities. In the late 1970s and early 1980s the state of Minnesota provided grants to municipalities with a severe Dutch elm disease problem. These monies were used to assist in removal efforts, disease control, and replanting.

Universities are actively involved in urban forestry research. Plant pathologists and entomologists devote considerable energy to urban tree pest problems. Geneticists provide improved varieties of plants for urban planting, while other university scientists study techniques to provide more effective technical assistance to communities, and serve as a conduit for information from universities.

Urban Forestry Worldwide

Urban forestry is an international effort. As discussed in the next chapter, the planting of trees in communities is as old as the first human development of permanent communities. Throughout the globe residents of cities in developed and developing countries are expanding their efforts to include trees as part of their communities. Beijing, China, had a forest cover of 3.2 percent in 1949, and has expanded its greenspace cover to 26.9 percent (Profous, 1992). A recent inventory of Dublin, Ireland, using remote sensing and planting records estimated that there are over two million trees now in the city, one million of which have been planted in the past twenty years (Boylan, 1992). An additional one million trees are to be planted in the near future.

Mexico City now has the highest relative population density in the world, and as a result experiences serious environmental degradation (Caballero, 1986). Among other measures, both the national and city governments have increased efforts in urban forestry to improve environmental conditions in the city, including maintaining old trees in the central city, planting street and park trees in new urban communities, and trying to establish green areas and parks in the poor communities on the edge of the city.

National governments are increasingly involved in promoting urban forestry. For more than twenty years the government of China has been involved in a national effort to expand the area of forest cover. This includes the planting of trees in cities, greenbelts along waterways, and around communities and farms, and connecting forests to cities with roadside plantings (Yang, 1995). In Sweden the University of Agricultural Science provides intensive courses in tree care and maintenance to municipal park department employees (Vollbrecht, 1988). The Ministry of the Environment in the Canadian Province of Quebec offers technical assistance to communities on tree management (Desbiens, 1988).

Education in Urban Forestry

The education of professionals involved with urban forest management varies from the school of experience to advanced university degrees. At the formal level professional education is typically in arboriculture, urban forestry, landscape horticulture, or forestry. Graduates hold technical degrees from one-year programs, associate degrees from two-year programs, and bachelor's degrees from four-year programs. A survey by McPherson (1984) found that prospective employers felt that arborists should have at least two years of academic training, and urban foresters, at least four years. In both cases, respondents indicated that a minimum of six months of field experience is highly desirable prior to employment. He further found that employers expected arborists to be well trained in basic tree care skills, and urban foresters to have competencies in tree selection, plant materials, public relations, and budgeting, as well as traditional tree care techniques.

Informal education may be gathered on the job and from professional organizations. These organizations include the International Society of Arboriculture, the Society of Municipal Arborists, the National Arborist Association (commercial arborists), and the Utility Arborists Association. Organizations that also have an interest in urban vegetation management include the American Association of Botanic Gardens and Arboreta, the American Forestry Association, the American Society for Horticultural Science, the American Entomological Society, the American Society of Landscape Architects, the Society of American Foresters, the American Associations of Nurserymen, and the Phytopathological Society (Harris, 1983).

LITERATURE CITED

ANDRESEN, J. W. 1978. "Urban Foresters and Planners as Managers." *Proc. Natl. Urban For. Conf.,* Cincinnati, Ohio. Washington, D.C.: Am. For. Assoc., pp. 152–155.

ARCHIBALD, P. L. 1973. "Urban and Community Forestry—Past and Present." *Proc. Urban For. Conf.* Syracuse, N.Y.: SUNY, pp. 4–9.

BOYLAN, C. 1992. "Trees of Dublin." *Arbori. J.* 16:327–341.

BUCK, R. 1983. "The Forest Preserve District of Cook County, Illinois, an Inlying Forest with an Outlying Purpose." *Proc. Semin. Manage. Outlying For. Metropolitan Populations,* Milwaukee, Wis.: Man and the Biosphere, pp. 7–9.

CABALLERO, M. 1986. "Urban Forestry Activities in Mexico." *J. Arbor.* 12(10):251–256.

CALDWELL, L. K. 1971. "Social and Political Trends Affecting Environmental Forestry Programs." *Trees and Forests in an Urbanizing Environment,* Plann. Res. Dev. Ser. No. 17. Amherst: Univ. Mass., pp. 151–155.

CARLOZZI, C. A. 1971. "Forestry, Ecology, and Urbanization." *Trees and Forests in a Urbanizing Environment,* Plann. Res. Dev. Ser. No. 17. Amherst: Univ. Mass., pp. 97–100.

CASEY, C. J., and R. W. MILLER. 1988. "State Government in Community Forestry: A Survey." *J. Arbor.* 14(6):141–144.

CLEGG, D. 1982. "Urban and Community Forestry—The Delivery." *Proc. Sec. Natl. Urban For. Conf.,* Cincinnati, Ohio. Washington, D.C.: Am. For. Assoc., pp. 13–17.

DANA, S. T. 1971. "The Challenge." *Trees and Forests in an Urbanizing Environment,* Plann. Res. Dev. Ser. No. 17. Amherst: Univ. Mass., pp. 165–168.

DEGRAAF, R. M. 1974. "Wildlife Considerations in Metropolitan Environments." *Forestry Issues in Urban America.* Proc. Natl. Conv. Soc. Am. For., pp. 97–102.

DESBIENS, E. 1988. "Urban Forestry in Quebec." *J. Arbor.* 14(1):24–26.

DWYER, J. W. 1982. "Challenges in Managing Urban Forest Recreation Resources." *Proc. Sec. Natl. Urban For. Conf.,* Cincinnati, Ohio. Washington, D.C.: Am. For. Assoc., pp. 152–156.

GRIFFITHS, S. T. 1984. "A Managed System for Distribution Forestry." *J. Arbor.* 10(6):184–187.

HARRELL, J. B. 1978. "Florida's Urban Forestry Program." *J. Arbor.* 4(9):202–207.

HARRIS, R. W. 1992. *Arboriculture: Integrated Management of Landscape Trees, Shrubs, and Vines.* 2nd Edition. Englewood Cliffs, N.J.: Prentice-Hall, pp. 2–3.

KIELBASO, J., B. BEAUCHAMP, L. LARISON, and C. RANDALL. 1988. "Trends in Urban Forestry Management." *Baseline Data Report* 20(1), Int. City Mgmt. Assoc., Washington, D.C.

KIELBASO, J. J., G. HASTON, and D. PAWL. 1982. "Municipal Tree Management in the U.S.—1980." *J. Arbor.* 8(10):253–257.

KIELBASO, J. J., and K. A. OTTMAN. 1978. "Public Tree Care Programs." *Proc. Natl. Urban For. Conf.* Syracuse, NY.: SUNY, pp. 437–447.

MCPHERSON, E. G. 1984. "Employer Perspectives on Arboricultural Education." *J. Arbor.* 10(5):137–141.

MCPHERSON, E. G., D. J. NOWAK, and R. A. ROWNTREE, eds. 1994. *Chicago's Urban Forest Ecosystem: Results of the Chicago Urban Forest Climate Project.* Gen. Tech. Rep. NE-186. Radnor, Pa.: USDA Forest Service, Northeast Forest Experiment Station.

MILLER, R. W. 1994. *Introduction to Urban and Community Forestry: An Independent Study Course by Correspondence.* Gainesville, Dept. of Ind. Study, Univ. Fla.

MILLER, R. W., and T. R. BATE. 1978. "National Implications of an Urban Forestry Survey in Wisconsin." *J. Arbor.* 4(6):125–127.

MOLL, G., and C. KOLLIN. 1993. "A New Way to See Our City Forests." *American Forests* 99(9–10):29–31.

MOLL G., and J. PETIT. 1994. "The Urban Ecosystem: Putting Nature Back in the Picture." *Urban Forests* 14(5):8–15.

NOBLES, B. 1980. "Urban Forestry/Arboriculture Program." *J. Arbor.* 6(2):53–56.

NOWAK, D. J. 1993. "Historical Vegetation Change in Oakland and Its Implications for Urban Forest Management. *J Arbor.* 19(5):313–319.

NOWAK, D. J. 1994. "Understanding the Structure of the Urban Forest." *J. Forestry* 92(10):42–46.

PINCHOT INSTITUTE. 1973. *The Pinchot System for Environmental Forestry Studies,* USDA For. Serv. Gen. Tech. Rep. NE-2.

PROFOUS, G. V. 1992. "Trees and Urban Forestry in Beijing, China." *J. Arbor.* 18(3):145–153.

REEDER, E. C., and H. D. GERHOLD. 1993. "Municipal Tree Programs in Pennsylvania." *J. Arbor.* 19(1):12–19.

RHODES, A. D. 1971. "Research Needs in Urban-Related Environmental Forestry." *Trees and Forests in an Urbanizing Environment,* Plann. Res. Dev. Ser. No. 17. Amherst: Univ. Mass., pp. 157–163.

ROWNTREE, R. A. 1984. "Ecology of the Urban Forest-Introduction to Part I." *Urban Ecology* 8:1–11.

SMITH, F. E. 1971. "Trees and Urban Ecology." *Proc. Symp. Role Trees South's Urban Environ.* Athens: Univ. Georgia, pp. 1–5.

STEWART, C. A. 1974. "Management and Utilization of Urban Forests." *Forestry Issues in Urban America.* Proc. Natl. Conv. Soc. Am. For., pp. 85–91.

USDA FOREST SERVICE. 1990. *Urban Forestry Five-Year Plan: 1900 through 1994.* Northeastern Area, State and Private Forestry.

VOLLBRECHT, K. 1988. "Tree Care in Scandinavia." *J. Arbor.* 14(1):3–6.

YANG, S. 1995. Institute for Applied Ecology, Shenyang, China. Personal communication.

4

Uses of Urban Vegetation

Urban trees, forests, and associated vegetation have numerous uses and functions in the urban environment. These uses range from the obvious to the obscure, and present opportunities for improvement of urban living conditions. Certainly, the architectural and aesthetic use of urban vegetation is obvious even to the casual observer. At some point during the design of urban landscapes and structures the landscape architect is brought into the process and directed to include vegetation and, sometimes, landform in the final design, the end result being the introduction of a substantial quantity of vegetation into our cities. Many outdoor recreation experiences in cities take place in urban parks and woodlands, and urban wildlife in these areas become a part of this experience.

Beyond these uses, urban vegetation performs a number of engineering functions. Trees and forests modify urban microclimates, which in turn affect human comfort and interior energy budgets. Vegetation has been found to reduce certain air pollutants, but will sometimes sustain damage by these pollutants. Unwanted sound (noise) is reduced by dense screens of trees and shrubs. Smooth, light-colored surfaces along with artificial lighting can be bothersome to people, and urban vegetation assists in solving this problem. Construction and overuse exposes urban soils to the ravages of wind and water erosion. The appropriate vegetation type, along with control of pedestrian traffic patterns, can substantially reduce erosion problems. Finally, urban greenbelts and forests have been found to cleanse partially treated wastewater from municipalities and industries.

THE HISTORY OF TREES IN THE CITY

Trees have probably been a part of cities since their first development. Since agriculture led to the first permanent settlements, it stands to reason that domesticated plants were a part of the community, including trees cultivated for food. The early Egyptians described trees being transplanted with balls of soil more than 4000 years ago (Chadwick, 1971). In thirteenth-century China Kublia Khan required tree planting along all public roads in and around Beijing for shade and to mark them in the snow (Profous, 1992). However, it is not likely trees were abundant in ancient cities except in the gardens of rulers and on the grounds of temples.

During the Middle Ages cities in Europe contained some trees and other plants in the private gardens of the ruling classes, but these were primarily for utilitarian purposes such as fruit rather than for ornament. The Renaissance in Italy during the sixteenth century saw the first development of villas on the periphery of cities. These villas had walled gardens and tree-lined paths, called allées, for walking and relaxation. The villa soon spread to France and Spain. In some towns allées of trees were planted as public promenades along the tops of earthwork fortifications, and by the beginning of the seventeenth century the upper classes began to develop allées of trees in and near cities for recreation activities such as archery and bowling. Allées of trees for both pedestrian and vehicular traffic were planted along and on the fortifications of French cities. The first planting in Paris was on the bulwark called the Grand Boulevart, which became known as the Grands Boulevards (Lawrence, 1993).

In the Netherlands allées of trees were planted along canals and their adjacent streets. The plan for the expansion of Amsterdam called for one tree for each standard building width. In London trees and lawns were planted in enclosed squares surrounded by new residences for the exclusive use by the occupants. However, by the end of the seventeenth century trees and landscaped spaces were still uncommon in European cities, being primarily available to the upper classes (Lawrence, 1993).

The eighteenth century saw the rise of merchant and professional classes throughout Europe, and their tastes in housing and landscapes emulated that of the aristocrats. They too built houses with gardens, and with the upper classes created more public places in the city. More trees were planted along boulevards, and public gardens became popular. As old fortifications around cities were dismantled they were often turned into parks, or walks and boulevards for carriages. Large, landscaped parks were built on the outskirts of London for the upper classes, but throughout Europe the lower classes did not have access to parks and gardens, often being excluded by entrance fees or for improper dress (Lawrence, 1993).

Baroque gardens in France were first developed in hunting preserves, with wide pathways radiating from clearings in the forest for shooting game during the hunt (Zube, 1971). These radiating pathways lined with trees came to influence the design of eighteenth-century villages and towns, and ultimately produced the radial street pattern of Washington, D.C. (Figure 4–1).

By the early nineteenth century there was concern in the British House of Commons over the lack of public parks for the "humbler" classes. Members felt that access to public parks would have a civilizing effect on the lower classes, so more parks were constructed and the lower classes were allowed access to them. In a similar fashion more public parks were built in cities throughout much of Europe. Entrepreneurs also recognized that trees and open space added to the market value of

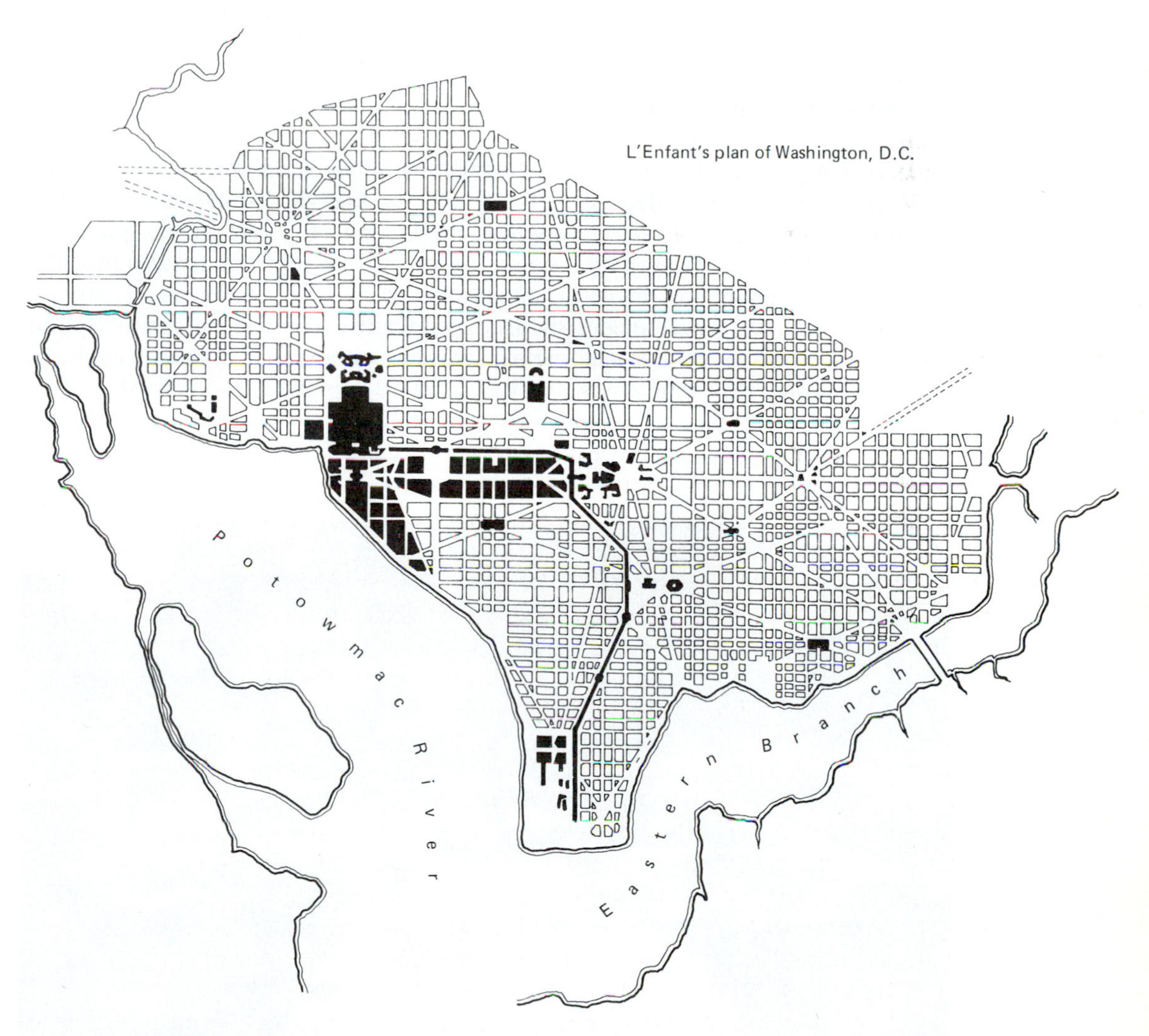

Figure 4–1 The plan for Washington, D.C., was strongly influenced by the radial street pattern developed in Europe. (From Gallion and Eisner, 1975.)

property, and increasingly added these amenities in the construction of residential neighborhoods.

Napoleon III transformed Paris in the 1850s and 1860s by imposing a radial street pattern of tree-lined boulevards on the existing city, and by building public parks, squares, monuments and gardens. The new wide boulevards were designed to allow better access to city neighborhoods for times of both celebration and riot. Cities throughout Europe adopted the renovation designs of Paris, adding more boulevards, tree-lined streets, and parks. The influence of European urban architecture complete with trees and public parks spread throughout much of the world through settlement and colonization by Europeans, and through widespread adoption of European-style architecture (Lawrence, 1993; Zube, 1971).

The Romantic Landscape

Industrialism in Great Britain produced deplorable urban conditions, as discussed in Chapter 1. The resulting decline in the quality of life yielded efforts to improve urban conditions, one of which was the introduction of public green spaces. As stated previously, trees were introduced into London in a series of squares placed in residential neighborhoods. Industrialization also spawned the Romantic Landscape movement with the development of the first suburbs to escape urban conditions (Zube, 1971). While the Baroque garden was the epitome of the formal landscape (Figure 4–2), the Romantic landscape was informal in nature, and consisted of the

Figure 4–2 Baroque garden.

"natural" arrangement of plants and structures in the landscape. In its most idealistic sense, the romantic landscape embodied a mix of what was considered the best from the city and nature.

American Cities

Early colonial villages in New England were built around a village green. The function of the green had nothing to do with aesthetics, but rather served as a place to muster militia and keep livestock during times of attack. It was not until late in the eighteenth century that trees and lawns were intentionally established in village greens (Figure 4–3). Philadelphia, on the other hand, was designed in 1682 by William Penn and contained five open spaces of 5 to 10 acres each, filled with trees. Trees on streets and in yards were uncommon, and it was not until after 1784 that insurance companies began to insure homes with trees near them (Zube, 1971).

Following the Revolution, Americans sought to create a new identity for themselves. Thomas Jefferson believed in a country governed by "sturdy yeoman farmers" and regarded the city dweller with a certain amount of suspicion. This attitude was embraced by the populace, and influenced early attempts to incorporate nature in the urban design as well as identifying nature as a source of moral virtue (Schmitt, 1973).

Examples of early legislation concerning urban vegetation were found in Michigan and Mississippi. A Territory of Michigan law in 1807 specified that trees be planted on boulevards in the city of Detroit, and that squares be established and planted with trees. The commission charged with selecting a capital for the state of

Figure 4–3 Village green in a New England town.

Figure 4–4 Central Park, New York, was on the urban fringe when originally established.

Mississippi in 1821 recommended that the new capital have every other block filled with native vegetation or be planted with groves of trees. The commission felt that this would provide a more healthy environment and provide easier fire control in a city constructed of wood (Zube, 1971).

The introduction of industrialism into the United States in the mid-nineteenth century was followed by the importation of the Romantic landscape movement from England. This manifested itself in city beautification efforts, including the introduction of trees on streets, and construction of city parks and civic centers. The city park movement, led in America by Frederick Law Olmstead, had as its goal the introduction of naturally landscaped parks into rapidly growing industrial cities. Olmstead, designer of New York's Central Park (Figure 4–4), stated, "The park should, as far as possible, complement the town . . . what we want is a simple, broad, open space with sufficient play of surface and a sufficient number of trees about it to supply a variety of light and shade" (Gardescu, 1976).

Suburbs for the emerging middle and upper classes sprang up on the outskirts of cities following the spirit of the Romantic landscape. Streets were laid out in curvilinear patterns, and homes built on rolling wooded parcels. Communities such as Llewellyn Park in New Jersey, Roland Park in Baltimore, Ridley Park in Pennsylvania, and Lake Forest and Riverside (Figure 4–5) in Illinois were designed in the latter half of the nineteenth century and served as models for twentieth-century suburbs. However, as mass housing became the rule from the 1930s on, smaller lots were subdivided and, to accommodate construction, existing trees were removed and not replaced (Zube, 1971). This trend did not begin to reverse itself until the late 1960s, when home buyers started placing higher premiums on wooded parcels.

Figure 4–5 Riverside, Illinois, a Chicago suburb, is a totally planned community designed by Frederick Law Olmstead.

Tree planting became of national interest by the end of the nineteenth century. Arbor Day was first observed in Nebraska in 1872, following its inception by J. Sterling Morton of the Nebraska Board of Agriculture. The first Arbor Day in Nebraska witnessed the planting of over a million trees. The observance of Arbor Day spread to the rest of the country, and each April tree planting celebrations are held, with millions of trees being planted in cities, suburbs, farms, and forests.

During the early twentieth century most large cities and many medium-sized communities initiated city forestry programs to plant and care for street and park trees. Milwaukee's forestry program, started in 1926, had as its goal the planting and maintenance of trees on city streets. Smaller cities, towns, and villages engaged in tree planting projects, but many did not establish community forestry programs until Dutch elm (*Ceratocystis ulmi*) disease ravaged their tree populations.

ARCHITECTURAL AND AESTHETIC USES

A poll by Harris in 1970 reported that 95 percent of Americans listed "green grass and trees around me" as an important part of their physical environment. Hooper (1970), reporting on the survey results, felt that "Americans have an almost Jeffersonian picture of their aspirations; green grass and trees, friendly neighbors, churches, schools and good stores nearby." The interest in attractive landscapes seems to be a major concern of the population, as the green industry has experienced unprecedented growth since the Harris poll and was recognized as one of the top growth industries by the *Wall Street Journal* in 1981. This interest has resulted in a demand for services to provide attractive landscapes at both the private and public levels.

The introduction of vegetation into the urban scene occurs with a great deal of careful planning to establish attractive landscapes. It also occurs as the result of accidents of nature, haphazard planning, or no planning whatsoever. The desirability of vegetation does not necessarily mean that all vegetation in cities is desirable or aesthetically pleasing. Urban landscapes are frequently poorly designed, with the feeling that some vegetation is better than none. Proper planning of the landscape is the realm of the landscape architect, a person all too often ignored when making decisions that alter landscapes. It is important for the vegetation manager to understand some basic concepts of landscape use and design in order to communicate with landscape architects and planners.

The architectural and aesthetic use of vegetation consists of two different approaches in the use of plant materials. Robinette (1972) defined architectural use stating: "Plants, singly or in groups, form walls, canopies, or floors of varying heights and densities; these are architectural characteristics." He further defined the architectural use of plant materials as important in creating comfortable landscapes for human beings by providing privacy and by screening undesirable scenes. Plants have different forms, colors, textures, sizes, densities, and growth rates. As elements of design, these attributes are mixed with structures and landforms to reinforce design and complete landscapes. Plants can direct vision, break up large spaces, and define important components in the environment. They can be used to frame scenes and to provide foregrounds and backgrounds for landscape features (Robinette, 1972).

From an architectural point of view, plants are desired as occupying space and defining space. Aesthetic uses are similar, except that each plant is an object possessing its own aesthetic quality. Robinette (1972) feels that aesthetics have been a prime determinant in the use of vegetation, and further stated: "A plant, whether specimen, topiary, bonsaied, or wind carved is effective as a piece of sculpture in creating interest." Vegetation forms aesthetic backdrops for desirable landscape features, and can be used to create unity in urban landscapes consisting of otherwise inharmonious structures and land uses. Plants also attract birds and other wildlife, and these add aesthetic interest to the landscape. Vegetation creates different moods depending on season, time of day, and weather conditions. Plants have their own sounds, odors, and feel, and these elements put the urban resident in touch with a part of themselves too easily lost in the din of urban living (Robinette, 1972).

Appleyard (1978) describes a number of visual and sensory functions of urban trees. Trees as natural forms and masses introduce the natural into the urban environment in the form of individual trees or as masses of vegetation. Trees as shaped ornaments are carved and pruned into shapes not found in nature, but serve as evidence of our domination over nature. Trees as incense and music produce their own odors, in contrast to the smells of the city, and make natural sounds as the wind shakes their leaves and birds sing in their crowns. Trees to mask ugliness serve as screens and visual filters for unsightly and sometimes necessary urban activities. Trees as a refreshing contrast in color and shape to buildings and harsh textures are easy on the eyes and provide a contrast to indoor living.

Finally, trees as a threat to urbanism is a point of view held by those who admire urban forms and architecture and are disturbed when trees blur these forms and cover signs, artificial lighting, and other details in the urban landscape. Owens (1971) suggests that selecting and planting trees be based on the following fundamental principles:

1. **1.** Remember that every tree, except one in a natural woodland, is an accent point. This will serve as a precaution not to use too many trees.
2. **2.** Select the trees that are just the right size to be in scale with the scene in which the trees are to be placed.
3. **3.** Do not let any tree encroach upon another's sphere of influence.

As elements of design and aesthetics in the landscape, trees and other vegetation should be established following a specific plan that considers the principles of design and the intended use of the area. Forested landscapes in the urban region create their own aesthetics and the manager can manipulate these by disturbing existing vegetation or by letting plant succession take its course. When encountering a more open landscape, the choices are: to leave it as it is, to introduce new plant materials as specimens, or to subdivide the area into smaller units with vegetation as walls, ceilings, and floors. Plantings on boulevards and streets should follow principles of design constrained by the economics of management. All urban vegetation planning should be guided by aesthetic considerations, designated uses, and management constraints.

Recreation and Wildlife

The architectural and aesthetic use of urban vegetation in the urban environment considers plants as elements that form walls, ceilings, and floors, and that possess certain aesthetic features. Spaces are designed by the landscape architect for enjoyment by direct observation or through participation in activities, with the landscape defining the space and providing the setting for recreational pursuits (Figure 4–6).

The fact that urban residents seek outdoor recreational experiences beyond city limits has been well documented. However, many urban residents make infrequent use of forests beyond the city when compared to time spent in urban woodlands. Dwyer (1982) felt that urban forests provide many urban residents with most of their forest-related experiences, especially the elderly, the handicapped, the very young, and the poor. He defined urban forest recreation resource management as involving "the making and carrying out of decisions concerning the location, characteristics, and use of urban forest resources."

Increasing costs associated with transportation to rural forests beyond the city will make recreation in urban forests more attractive in the future. This, as part of overall efforts to improve the quality of life in cities, will provide more and higher-quality greenspaces for the enjoyment by urban residents. Dwyer (1982) describes many changes taking place in waterfronts, new transportation developments, abandonment of railroads and other rights-of-way, creation of new residential developments, renovation of urban parks, and shifts in the composition of urban neighborhoods. These developments will provide spaces for more recreational activities at a time when many people are changing their preferences in recreation. Wellness and the interest in physical fitness has generated a large population of urban forest users

Figure 4–6 A well-designed urban park provides spaces for a variety of activities.

who demand walking, skiing, jogging, and bicycling trails, along with fields for sports activities (Figure 4–7). Of course, not all users desire these sorts of activities, and the urban forest manager must provide for a diverse population with diverse interests. Inner-city residents may have different expectations than those of suburban residents regarding greenbelts. Residents in Chicago were found to prefer intensively

Figure 4–7 Active recreation demands large areas in urban parks.

developed and managed parks for social interaction (Figure 4–8), while their suburban counterparts preferred natural undeveloped sites to get away from people (Dwyer, 1982).

Trees and associated vegetation provide the setting for the activities described above, but they also provide recreational enjoyment in their own right. Trees as colors, shapes, sounds, textures, odors, tastes, and touch certainly provide the urban dweller with pleasures not found in other environments. Trees also provide homes for a multitude of wildlife that could not survive the alien environment of the city. While trees provide sensory stimuli as described above, they lack the animation of things that crawl, walk, or fly. Urban wildlife provides this animation, and cities abound with mammals, birds, insects, reptiles, and amphibians.

The most common wild mammal in North American cities is probably the gray squirrel. The list of other common mammals includes deer, rabbits, skunks, mice, Norway rats, chipmunks, moles, shrews, bats, and the coyote (Franklin, 1982). Williamson (1973) found abundant bird life in Washington, D.C., but there was a difference in species composition between affluent suburbs and the inner city. Suburban neighborhoods were populated with native species such as the cardinal, robin, and mockingbird, while the inner-city neighborhood contained exotic species such as pigeons, starlings, and house sparrows. This difference was related to the abundant vegetation in the suburban neighborhood and the lack of vegetation in the inner-city neighborhood. Sayler and Cooper (1975) report that more than 1800 Canadian geese share the environment with 2 million people in the Twin Cities, Minnesota, taking advantage of the numerous lakes, most of which are surrounded by parks. Urban residents greatly appreciate urban bird life; a study in 1974 found that approximately 20

Figure 4–8 Passive social interaction is encouraged by city park design.

percent of all U.S. households purchase wild bird seed, at a cost of $170 million per year (DeGraaf and Payne, 1975).

Wildlife in cities is a reflection of available habitat and an indication of the condition of the urban ecosystem. Ecologists use diversity as a measure of system stability, with lower diversity indicating less stability. A diverse city in terms of abundance, type, and distribution of vegetation will provide habitat for a diverse wildlife population and a more stable environment for our species. Wildlife will use the habitat we provide in the city, and we use the wildlife to enhance the quality of our lives.

CLIMATOLOGICAL USES

The climate of a given location may be described from three aspects: macroclimate, mesoclimate, and microclimate (Hiesler and Herrington, 1976). Macroclimate is the general climate for a given region covering hundreds of square miles and consists of such parameters as precipitation, the amount, distribution, and mean; temperatures, the maximum, minimum, and means; and winds, the mean direction and velocities. Mesoclimate is on a smaller scale covering tens of square miles, and describes variations from the macroclimate due to effects caused by topographic features, bodies of water, and other influences. Microclimate is smaller yet, consisting of areas of hundreds of square feet and variations in elevation in the tens of feet (Hiesler and Herrington, 1976). These local modifications are due primarily to deviations in the surface, such as soil type, plant cover, and structures. Cities do not exert an influence on the macroclimates, and the same is true for trees. Trees and other vegetation affect urban meso- and microclimates on three levels: human comfort, building energy budgets, and urban mesoclimates.

Human Comfort

Four elements affect human comfort: solar radiation, air movement, air temperature, and humidity and precipitation. Individual trees, groups of trees, and trees in combination with other vegetation, landforms, and structures will have varying degrees of impact on these elements (Robinette, 1972).

Solar radiation. Solar radiation affects human comfort from both a positive and a negative standpoint. Infrared radiation from the sun heats the human body directly when striking the skin or clothing, and indirectly when reflected or reradiated from other objects. Vegetation should be designed to intercept solar radiation when it causes excessive heating, and not to interfere when the opposite effect is desired. Plants as interceptors of solar energy do two things: They block radiation from striking surfaces, and they convert some solar energy into chemical bonds through photosynthesis (Robinette, 1972).

Plants may be used to intercept solar energy directly by providing shade in areas where it is desired, and indirectly by covering surfaces that reflect or reradiate solar energy (Figure 4–9). In temperate climates deciduous trees provide shade during months when solar radiation is a problem, and allow some energy through when it is desirable. On an individual-tree basis the impact on human comfort of shade is due to blocking solar energy, not to lower air temperatures in the shade. A person feels cooler in the shade even though the air temperature may be the same as it is a few feet away in the sun (Robinette, 1972). Hiesler and Herrington (1976) report that controlling infrared radiation is the most important function that trees perform in affecting human comfort, and that columnar tree varieties with dense foliage are the most effective at intercepting these wavelengths.

Air movement. As with solar radiation, wind may exert a positive or a negative impact on human comfort. Moving air will cool the skin's surface in the summer by evaporating perspiration, or chill the skin in winter by rapidly replacing air warmed by the body with cold air. Plants may be used to manipulate air movement by obstruction, guidance, deflection, and filtration (Robinette, 1972).

Obstruction is used to block undesirable winds by arranging dense plantings at a 90° angle to the prevailing wind (Figure 4–10). These barriers may consist of a single row of trees or multiple rows of trees and shrubs. The effect on airspeed is determined by the density of planting, the distance of the planting from the site being protected, and the height of the vegetation. Combinations of trees and shrubs provide the most effective barriers, with shade-tolerant species having the greatest density. Vegetation may also be combined with landforms and structures to serve as wind breaks. Deciduous and evergreen species may be used to provide barriers in the summer, while evergreens are obviously best in winter (Robinette, 1972).

Desirable wind can be guided by vegetation to provide maximum cooling in summer months. In this case walls of vegetation are used to direct air to sites where maximum cooling is desired. Wind may be accelerated by the use of vegetation to

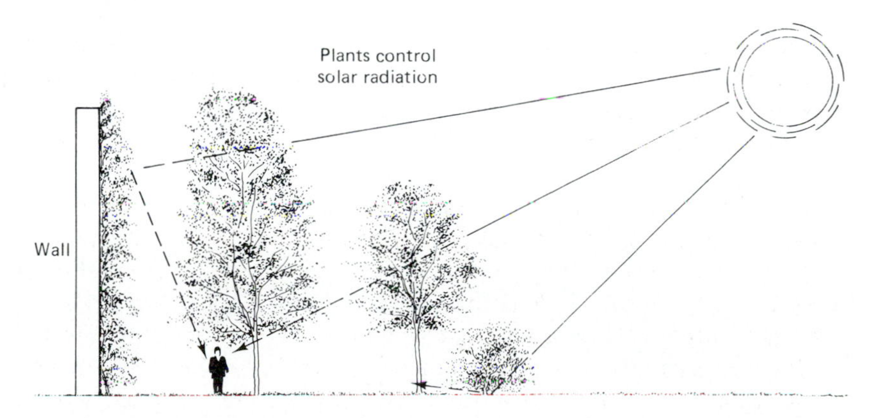

Figure 4–9 Plants can be used to intercept, filter, or block unwanted solar radiation.

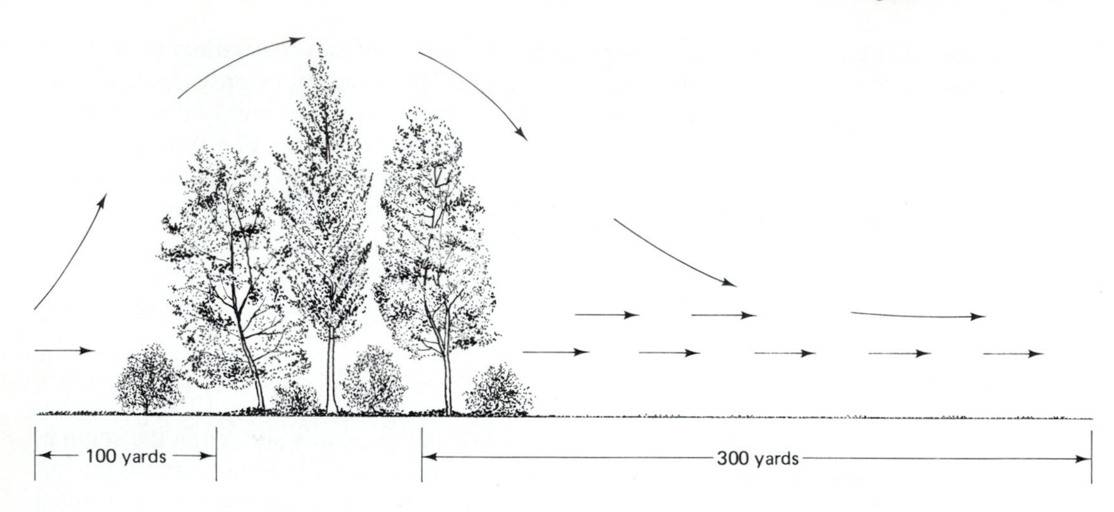

Figure 4–10 Plants act as barriers to unwanted wind. In this case a 30-foot-high (9 m) shelter-belt affects air speed 100 yards (90 m) upwind and 300 yards (270 m) downwind.

constrict airflow, thus creating a vortex effect (Figure 4–11). The use of vegetation to direct air movement includes landforms and structures in the overall landscape design (Robinette, 1972).

Vegetative walls are used to deflect wind away from specific areas (Figure 4–12). This is similar to using vegetation as a barrier, except that the air is guided away from sites where it is undesirable rather than blocking it entirely. The purpose of this approach is one of design, such as when a barrier is not feasible due to exist-

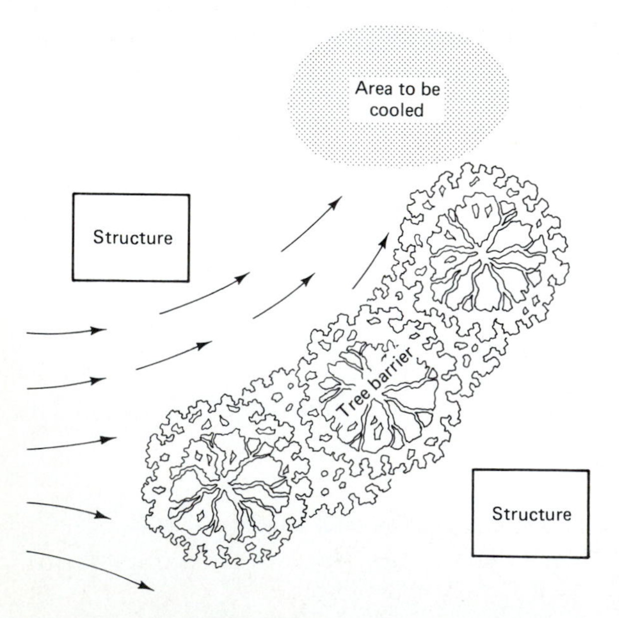

Figure 4–11 Plants may be used to guide and accelerate air movement.

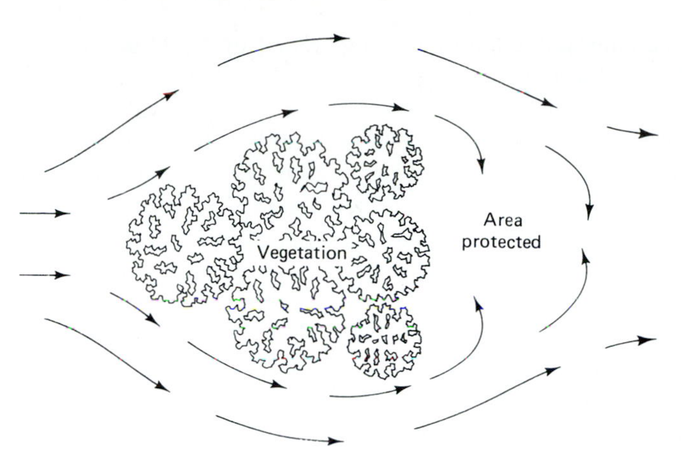

Figure 4–12 Wind deflected by vegetation.

ing landscape features, or when desiring to guide air away from a site where it is undesirable to where it is wanted (Robinette, 1972).

Finally, wind breaks may be designed to slow the velocity of wind by filtration when some, but not too much, is desirable (Figure 4–13). In this case loose arrangements of shade-intolerant species will be most effective. The foliage of intolerant species is not dense and will slow, but not block, air movement (Robinette, 1972).

Air temperature. There has been much speculation concerning the impact of urban trees on air temperature. It is well documented that forest stands moderate air temperature extremes compared with adjacent open areas, and that urban woodlands and parks, and suburban communities, are cooler than central cities. Hiesler and Herrington (1976) report that a series of studies in Syracuse, New York, found the presence of trees on city streets had no measurable effect on air temperature when comparing city blocks. However, Souch and Souch (1993) found statistically significant cooling beneath urban tree canopies of 0.7 to 1.3°C in early afternoon, while Parker

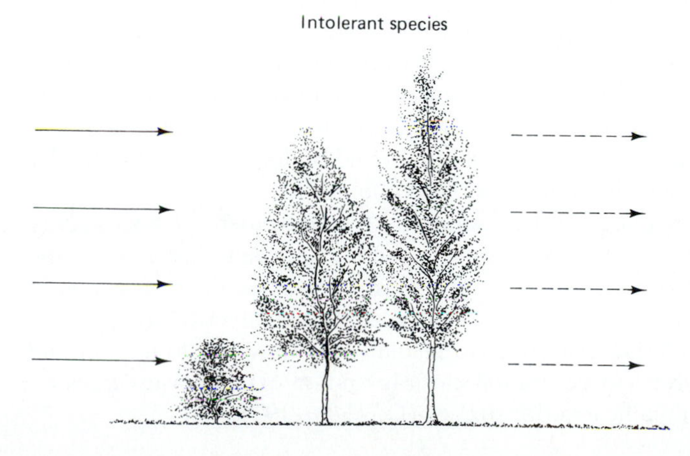

Figure 4–13 Vegetation screen as a wind filter.

(1989) reports an average summer reduction in air temperature of 3.6°C beneath the canopy of a large tree in Miami, Florida.

Humidity and precipitation. Summer relative humidity is higher within forest stands compared to open areas. Hiesler and Herrington (1976) report that the same series of studies in Syracuse, New York, described above also considered relative humidity. Trees were found to have no measurable impact on relative humidity when streets with abundant trees were compared to streets with few trees.

The impact of trees on precipitation as it affects human comfort is one of interception or providing temporary shelter during downpours. Trees do have a more important role in terms of precipitation when available moisture reaching the soil is considered. The foliage of a coniferous stand will intercept about 40 percent of the annual precipitation and allow it to reevaporate back into the atmosphere. Hardwood stands intercept about 20 percent of the annual precipitation before it reaches the ground. This means that it may be more desirable to cultivate hardwood forests on watershed lands when desiring to increase water yield. Coniferous forests may be more desirable in areas of excessive precipitation (Robinette, 1972).

Building Energy Budgets

Trees and other vegetation exert significant impacts on building energy budgets. Heating and cooling costs can be reduced by appropriate use of vegetation or increased by careless use.

Heating. Homes and other buildings lose heat through three mechanisms: air infiltration, heat conduction, and radiation transmission. A well-designed landscape can reduce heat loss by all three mechanisms, while enhancing the attractiveness of the property.

Air infiltration is the function of two forces: wind and pressure differences due to warm interior temperatures and cold exterior temperatures. Wind creates high pressure on the windward side of the building and low pressure on the lee side. Outside air is forced through cracks around doors, windows, and in walls on the upwind side, and interior air is drawn out in similar fashion on the downwind side of the building. Warm air in the home exerts an upward force in the home, pushing air out through upper walls and ceilings, and is replaced by cold-air infiltration through lower walls and floors. On cold, windy days infiltration can be responsible for as much as 50 percent of heat loss from a building. Proper arrangement of vegetation around buildings can substantially reduce infiltration by reducing wind velocities (DeWalle, 1978).

Heat conduction takes place through solids by the movement of heat energy from a warm surface to a cold surface. Heat conduction proceeds faster when the temperature difference between surfaces is greater. In homes, conduction results in heat loss across walls, ceilings, floors, doors, and windows. A layer of still air provides the best building insulating material, and insulation is designed to trap air and prevent it from moving. Layers of vegetation close to and on exterior walls traps air and has an insulating effect on the exterior surface (DeWalle, 1978).

Radiation transmission takes place through windows. Solar radiation entering windows will warm interior surfaces, and blocking this radiation with vegetation will increase heating costs. Radiation through windows from the interior results in heat loss, and exterior vegetation will not affect this loss (DeWalle, 1978).

In summary, vegetation can significantly affect building heating budgets. Windbreaks have been found to reduce home heating costs by 4 to 22 percent, depending on site windiness and how airtight the structure is. On the other hand, vegetation that shades a home in winter can increase heating costs (DeWalle, 1978).

Cooling. Trees and other vegetation affect interior summer temperatures by making non-air-conditioned buildings more comfortable and by reducing energy demands of air-conditioned structures. Solar energy heats buildings by two mechanisms: radiation through windows that heats interior surfaces, or conduction through walls. Trees will intercept up to 90 percent of summer solar energy and provide substantial reduction in interior temperatures (Youngberg, 1983). Interior temperatures in non-air-conditioned homes in California were found to be as much as 20°F cooler in summer months (Deering, 1954). DeWalle (1978) reported that mobile home cooling costs can be reduced up to 60 percent by summer shading, while Parker (1983) found that trees reduced home cooling costs in Florida by up to 50 percent.

Overall home energy conservation can be realized through proper landscaping. McPherson and Rowntree (1993) used computer simulations to estimate energy savings gained by a single 25-foot (7.7 m) tree properly located on a residential property. They found reductions in heating and cooling costs of 8 to 12 percent per household. Hiesler (1989) likewise found home energy savings from solar irradiance and wind speed reductions by shade trees in the northeastern United States (Table 4–1).

TABLE 4–1. MEASURED WIND REDUCTIONS, ESTIMATED REDUCTIONS OF IRRADIANCE BY TREE SHADE, AND ESTIMATED AVERAGE ENERGY SAVINGS IN AIR-CONDITIONED HOMES BY TREES IN RESIDENTIAL NEIGHBORHOODS IN UNITED STATES.[a] (From Heisler, 1989).

		Measured Wind Reductions, % of Open[b]		Estimated Average Irradiance Reductions		Annual Energy Savings		
Tree Density	Tree Cover	Heat	AC	Heat	AC	Madison	Salt Lake	Tucson
none	0	22	22	10	8			
low	24	44	54	15	22	2	4	6
medium	67	60	67	25	40	8	12	20
high	77	66	75	30	50	9	17	23

[a]All values in percent.

[b]Heat indicates reductions in winter, the heating season, and AC indicates reductions in summer, the air-conditioning season.

In hot sunny climates summer shade is the most effective for energy savings, while in cold climates reducing wind speed provides the greatest savings. Heisler (1989) suggests that scattered trees throughout a residential neighborhood have the effect of one large windbreak in reducing wind speeds. He also measured the solar irradiance interception of medium-sized sugar maples (*Acer saccharum*) and other deciduous trees, and found 80 percent interception in summer and 40 percent interception in winter.

For maximum energy savings in cold climates of the northern hemisphere, Sand (1993) recommends deciduous trees on the west and east side of homes, no trees on the south side, windbreaks on the north and west, plantings to shade air conditioners, and an ovє rall increase in tree canopy over communities. Figure 4–14 illustrates an idealized landscape for home energy conservation in northern climates.

Solar energy and trees. Both passive and active solar energy systems are becoming more popular for residential, commercial, and industrial structures, and this trend will continue as traditional energy sources become more costly. Many states and countries have passed, or are considering, solar access laws to protect the property owners' right to solar energy. Solar access laws are generally concerned with two problems: shading of property by structures, and shading by trees. Even deciduous trees in winter can intercept 25 to 60 percent of solar radiation (Youngberg, 1983). Through computer modeling, Thayer and Maeda (1985) determined that rows of deciduous trees to the south of houses with solar collectors constituted an energy penalty in all but the hottest climates, and recommend that communities with

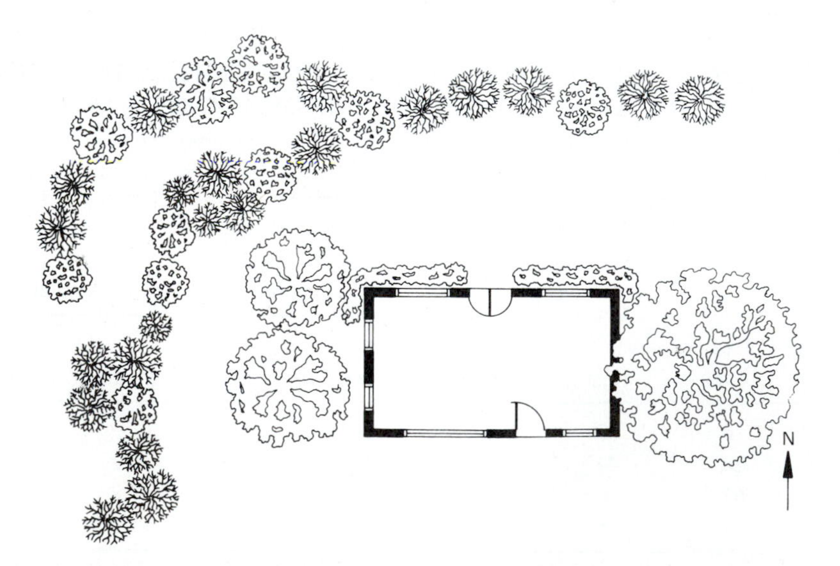

Figure 4–14 Optimum landscape concept for a temperature climate with winter wind predominantly from the west and northwest. (From Hiesler, 1986).

widespread or increasing use of solar collectors reexamine their tree planting policies. They also proposed a mature tree canopy plan for a solar access neighborhood, as depicted in Figure 4–15.

Urban Mesoclimate: The Urban Heat Island

Cities are often referred to as urban heat islands, with the central city having the highest temperatures (Figure 4–16). This is primarily due to the low amount of vegetation in central cities when compared to the suburbs and beyond. Buildings, asphalt, and concrete absorb solar radiation, and emit long-wave radiation that heats the atmosphere. Cities also use large amounts of energy and emit this energy as waste heat, further warming urban heat islands. Figure 4–17 shows the increasing temperature difference between rural and urban areas in California (Akbari et al., 1990).

Trees and other vegetation use some solar energy to make chemical bonds in photosynthesis, and large amounts of solar energy evaporating water to cool leaf surfaces. Trees and other vegetation can reduce urban energy consumption two ways; by intercepting and using solar energy, and by reducing building energy demand through shading and reducing wind speed. Computer simulations predict that increasing the tree cover by 25 percent in the cities of Sacramento, California, and Phoenix,

Figure 4–15 Mature tree canopy in a planned solar access neighborhood. (From Thayer and Maeda, 1985.)

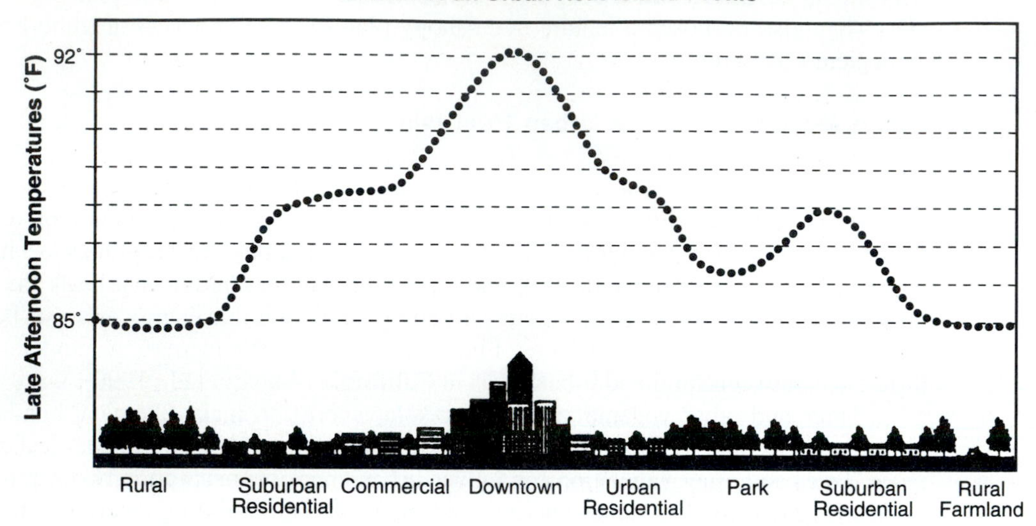

Figure 4–16 Sketch of a typical urban heat-island profile: This graph of the heat island profile in a hypothetical metropolitan area shows temperature changes (given in degrees Fahrenheit) correlated to the density of development and trees. (From Akbari, Davis, Dorsano, Huang, and Winnett, 1992.)

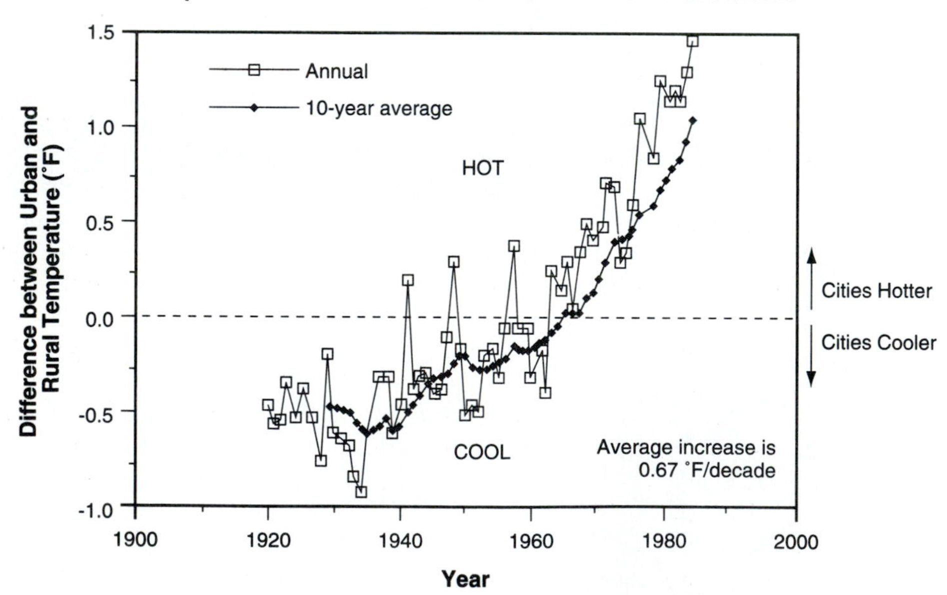

Figure 4–17 Urban areas are getting warmer: Since 1940, the temperature difference between urban and rural stations has shown an increase of 0.67°F per decade. (From Akbari, Davis, Dorsano, Huang, and Winnett, 1992.)

Arizona, will decrease the 2 P.M. air temperature in July by 6 to 10°F (Akbari et al., 1992).

Many atmospheric scientists have also expressed concern over the buildup of carbon dioxide in the atmosphere and the potential for global climate change. Much of this carbon dioxide comes from burning fossil fuels for heating and cooling buildings. Planting more trees and other vegetation in our cities can reduce carbon emissions by reducing the amount of energy needed for heating and cooling, and by sequestering carbon dioxide as biomass. Studies of combining tree planting with making the surfaces of buildings and pavements light colored estimate an annual savings of 25 percent (50 billion kilowatt hours) in electricity consumption for air conditioning in the United States (National Academy of Science, 1991). Nowak (1994b) estimates that carbon emissions avoided because of trees in the Chicago area is 11,400 metric tons per year, and that the trees themselves have sequestered about 5.6 million metric tons of carbon.

Forest belts surrounding Stuttgart, Germany, have significantly cooler summer temperatures than those of the urban heat island. However, thermals over the central city were found to draw cooler air from peripheral forest belts and cool downtown areas. Forest managers of Stuttgart's peripheral forests have substantiated this through the use of infrared aerial photography, and use this information to discourage encroachment on these forests by urban sprawl (Miller, 1983).

ENGINEERING USES

Air Pollution Reduction

The study and control of air pollution is a highly complex science involving meterology, climatology, and organic and inorganic chemistry. The atmosphere has received contaminants from natural sources such as vulcanism and wildfires since the earth formed some 4.6 billion years ago. Natural processes in the atmosphere and biosphere reduce and remove contaminants through dilution, precipitation, filtration, and chemical reactions. It is when human activities exceed the ability of natural processes to remove or reduce contaminants that air pollution becomes a concern. Solving air pollution problems involves political, economic, ecologic, sociologic, and scientific considerations.

Since processes within the atmosphere and biosphere remove pollutants, these processes can assist in solving the overall air pollution problem. It must be cautioned, however, that everything has a cost. To rely on natural processes excessively will result in damage to the biosphere, and ultimately impose unwanted and sometimes unexpected costs on present and future generations. An example of such a problem is acid rain. The atmosphere cleanses itself of acids through precipitation and the biosphere neutralizes these acids chemically. The ultimate costs include ecosystem damage or collapse, and structural damage to our artifacts.

Forest ecosystems will assist in cleansing pollutants from the atmosphere when coupled with other control measures. To understand how vegetation can be used to reduce air pollution it is necessary to discuss the major components of air pollution, how individual plants react with these components, how forest ecosystems react with urban air masses, and what tree species are best adapted to polluted air masses.

The chemistry of air pollution is highly complex and involves hundreds of compounds as they are emitted from their sources and additional compounds that form through reactions in the atmosphere. For simplicity the following outline describes the main components of air pollution (Smith and Dochinger, 1976):

I. Particulates
  A. Solids (organic, inorganic)
  B. Liquids (aerosols)
II. Gasses
  A. Primary (released directly into the atmosphere)
    1. Inorganic
      a. Oxides
        Nitrogen
        Sulfur
        Carbon
      b. Halogens
        Fluorine
        Chlorine
      c. Other
        Hydrogen sulfide
        Ammonia
    2. Organic
      a. Hydrocarbons
      b. Aldehydes
      c. Mercaptans
  B. Secondary (synthesized in the atmosphere)
    1. Ozone
    2. Peroxyacetylnitrate (PAN)

The concentration of these pollutants in the atmosphere depends on the quantity produced, the local climate, and day-to-day variations in weather. In general, most urban air masses are chronically polluted. Air pollution problems are compounded further by air inversions (warm air over cold air) that trap pollutants over cities for extended periods, and polluted urban air masses that move beyond the city to contaminate rural areas.

Urban vegetation interacts with air pollutants and has been found to reduce significantly a number of the contaminants listed above. Following is a summary of research that has investigated the effect of woody plants on urban air pollution.

Particulates. Particulates, both solids and aerosols, are removed by plants through three mechanisms: sedimentation under the influence of gravity, impaction under the influence of eddy currents, and deposition under the influence of precipitation. Wind velocity is reduced by plants, allowing heavier particles to settle out under the influence of gravity. Particulates impact on plant surfaces as air currents divide to pass around plant structures. The soil is the ultimate sink for particulates, as precipitation washes them from plant surfaces (Smith, 1978).

Nitrogen oxides. Nitrogen oxides (NO, NO_2) have been found to be removed by foliar uptake, with some utilization of nitrogen by plants (Smith and Dochinger, 1976).

Sulfure dioxide. While some SO_2 is reduced and utilized, Lampadius (1963) reported that the amount removed by woody plants was negligible when compared to atmospheric concentrations, and that many species encounter foliar damage through exposure.

Carbon monoxide. Some reduction of CO by woody vegetation has been reported by investigators (Smith, 1978).

Halogens. Both chlorine and fluorine are reduced by woody vegetation, with chlorine being removed at a much higher rate. Many species of plants are susceptible to damage by the halogens (Smith, 1978).

Hydrogen sulfide. No reported impact on concentrations of H_2S by woody plants.

Ammonia. Plants will absorb and utilize some ammonia for its nitrogen content (Smith and Dochinger, 1976; Smith, 1978).

Hydrocarbons, aldehydes, and mercaptans. No reported impact on concentrations of these gasses by woody vegetation.

Ozone. Ozone was found to be removed from the atmosphere at a rapid rate. Smith and Dochinger (1976) stated, "in theory a forest could remove about one-eighth of the O_3 content of the atmosphere in about one hour." However, ozone, being reactive, has the capacity to damage foliage on many species.

PAN. PAN forms in the atmosphere as the result of the reaction of hydrocarbons and oxides of nitrogen in the presence of sunlight. Low levels of PAN will be reduced by woody plants, but PAN is a reactive agent and will damage foliage in high concentrations (Smith and Dochinger, 1976).

In the summer of 1991 the urban forest of Cook and DuPage counties (Chicago region) removed an average of 1.2 metric tons (t)/day of carbon monoxide, 3.7 t/day of sulfur dioxide, 4.2 t/day of nitrogen dioxide, 10.8 t/day of ozone, and 8.9 t/day of particulate matter smaller than 10 μm (Nowak, 1994a).

DeSanto et al. (1976), McCurdy (1978), and Smith (1978) all point out that soil beneath the forest is important in reducing atmospheric contaminants. The soil itself neutralizes some pollutants through chemical reactions and direct deposition. The soil is also the ultimate sink for most pollutants filtered by vegetation above it.

Smith (1978), in a review of available literature concerning the effectiveness of plants in reducing air pollution, offered the following generalizations concerning the ideal "tree and forest" for filtration of atmospheric contaminants:

1. For particulate removal, species with high ratios of leaf circumference to area and surface to volume, and with leaf surface roughness, should be favored.

2. Coniferous species and deciduous species with abundant branch and twig structure provide particulate removal during winter months.
3. For gaseous pollutant removal, tree species with high tolerance for urban environments should be favored. These species have the highest probability of maximum metabolic rates and, therefore, stomatal opening. It is particularly critical to favor species with a high resistance to drought.
4. Only relatively large natural areas have significant potential to improve air quality in urban areas. The minimum width of a greenbelt may approximate 150 m but varies greatly about this depending on local conditions.
5. Urban forest density and structure influence pollutant removal capability. A balance must be struck between a stratified forest and a forest impervious to air mass movement. A multilayered forest—soil, herb, shrub, and tree layers—is a more effective pollutant sink than an unstratified forest. If the edge strata are overlapping and dense, however, and the stand of trees may force the air mass up and over and be a relatively ineffective sink. Careful silvicultural practice will be necessary to maintain appropriate structure and density.
6. Mixed plantings of coniferous and deciduous species provide maximum insurance against sudden loss of function due to adverse environmental, entomological, or pathological stress imposed on one species.

DeSanto et al. (1976) made additional observations and recommendations in studying the effect of greenbelts in reducing air pollution. They found that herbaceous species absorbed more gaseous pollutants than did woody species, and that the upper portions of a forest canopy removed more pollutants than did the understory. DeSanto further recommended that a belt of vegetation of medium density is best for gaseous pollutants, while dense vegetation is best for removing particulates.

Selecting the proper tree species for pollution resistance involves among other considerations the resistance of the species to air pollution. Davis and Gerhold (1976) compiled a list of common ornamental tree species in terms of their relative susceptibility to damage by the two most serious urban air pollutants, sulfur dioxide and ozone (Tables 4–2 and 4–3). They recommend that plant breeders consider pollution resistance when developing new cultivars for urban planting.

Cultivars within a species exhibit differing levels of resistance to air pollutants. Smith and Brennan (1984) studied the resistance of five cultivars of honeylocust (*Gleditsia triacanthos inermis*) to ozone damage. They report that the "Imperial" cultivar is highly resistant, followed by "Sunburst," "Skyline," and "Shademaster" cultivars.

It is important to reemphasize that the ultimate sink for pollutants is the soil beneath the trees and water percolating through the forest and soil. Some pollutants are rendered nontoxic through interaction with the forest ecosystem, and others are not. It is quite possible that many pollutants introduced into the ecosystem will persist for extended periods. The urban forest will help solve air pollution problems, but only as part of a greater effort to control contaminants at their source.

TABLE 4–2 RELATIVE SUSCEPTIBILITY OF TREES TO SULFUR DIOXIDE
(From Davis and Gerhold, 1976.)

Sensitive	Intermediate	Tolerant
Acer negundo var. *interius*	*Abies balsamea*	*Abies amabilis*
Amelanchier alnifolia	*Abies grandis*	*Abies concolor*
Betula alleghaniensis	*Acer glabrum*	*Acer platanoides*
Betula papyrifera	*Acer negundo*	*Acer saccharinum*
Betula pendula	*Acer rubrum*	*Acer saccharum*
Betula populifolia	*Alnus tenuifolia*	*Crataegus douglasii*
Fraxinus pennsylvanica	*Betula occidentalis*	*Ginkgo biloba*
Larix occidentalis	*Picea engelmannii*	*Juniperus occidentalis*
Pinus banksiana	*Picea glauca*	*Juniperus osteosperma*
Pinus resinosa	*Pinus contorta*	*Juniperus scopulorum*
Pinus strobus	*Pinus monticola*	*Picea pungens*
Populus grandidentata	*Pinus nigra*	*Pinus edulis*
Populus nigra 'Italica'	*Pinus ponderosa*	*Pinus flexilis*
Populus tremuloides	*Populus angustifolia*	*Platanus* X *acerifolia*
Rhus typhina	*Populus balsamifera*	*Populus* X *canadensis*
Salix nigra	*Populus deltoides*	*Quercus gambelii*
Sorbus sitchensis	*Populus trichocarpa*	*Quercus palustris*
Ulmus parvifolia	*Prunus armeniaca*	*Quercus rubra*
	Prunus virginiana	*Rhus glabra*
	Pseudotsuga menziesii	*Thuja occidentalis*
	Quercus alba	*Thuja plicata*
	Sorbus aucuparia	*Tilia cordata*
	Syringa vulgaris	
	Tilia americana	
	Tsuga heterophylla	
	Ulmus americana	

Sound Control

Unwanted sound, or noise, has been a part of urban living since the earliest cites. Nero passed a law in ancient Rome that prohibited the movement of horse-drawn carts during the night due to the excessive noise of hooves and wooden wheels on cobbled streets. The problem is much greater in the modern city, as ground, air, and water traffic service the city twenty-four hours a day, and other noise, from construction, repair, and main-

TABLE 4–3 RELATIVE SUSCEPTIBILITY OF TREES TO OZONE (From Davis and Gerhold, 1976.)

Sensitive	Intermediate	Resistant
Ailanthus altissima	*Acer negundo*	*Abies balsamea*
Amelanchier alnifolia	*Cercis canadensis*	*Abies concolor*
		Acer grandidentatun
Fraxinus americana	*Larix leptolepis*	*Acer platanoides*
Fraxinus pennsylvanica	*Libocedrus decurrens*	*Acer rubrum*
		Acer saccharum
Gleditsia triacanthos	*Liquidambar styraciflua*	
Juglans regia	*Pinus attenuata*	*Betula pendula*
		Cornus florida
Larix decidua	*Pinus contorta*	*Fagus sylvatica*
Liriodendron tulipifera	*Pinus echinata*	*Ilex opaca*
		Juglans nigra
Pinus banksiana	*Pinus elliottii*	*Juniperus occidentalis*
Pinus coulteri	*Pinus lambertiana*	
Pinus jeffreyi	*Pinus rigida*	*Nyssa sylvatica*
Pinus nigra	*Pinus strobus*	*Persea americana*
Pinus ponderosa	*Pinus sylvestris*	*Picea abies*
Pinus radiata	*Pinus torreyana*	*Picea glauca*
Pinus taeda		*Picea pungens*
Pinus virginiana	*Quercus coccinea*	
	Quercus palustris	*Pinus resinosa*
Platanus occidentalis	*Quercus velutina*	*Pinus sabiniana*
Populus maximowiczii X trichocarpa		*Pseudotsuga menziesii*
	Syringa vulgaris	*Pyrus communis*
Populus tremuloides		*Quercus imbricaria*
	Ulmus parvifolia	*Quercus macrocarpa*
Quercus alba		*Quercus robur*
Quercus gambelii		*Quercus rubra*
Sorbus aucuparia		*Robinia pseudoacacia*
Syringa X chinensis		*Sequoia sempervirens*
		Sequoiadendron giganteum
		Thuja occidentalis
		Tilia americana
		Tilia cordata
		Tsuga canadensis

tenance of the city, adds to the din. Researchers have suggested that intense, persistent noise causes psychological disturbance and is a threat to community life.

Sound usually originates with the vibration or disturbance of a solid, liquid, or gas, and passes through the atmosphere as a series of pressure waves. Sound radiates from its source spherically unless it encounters an object that absorbs it or changes its direction. Sound intensity is measured in decibels (dB), which starts at the threshold of human hearing (0 dB) and increases on a logarithmic scale. For example, 10 dB is 10 times louder than 0 db, 20 dB is 100 times louder than 0 dB, 30 dB is 1000 times louder than 0 dB, and so on (Table 4–4). The frequency of a sound wave is a measurement of the distance between peaks in the wave, and determines its pitch or tone.

TABLE 4–4 SOUND PRESSURE LEVELS OF TYPICAL NOISE SOURCES (From Reethof and McDaniel, 1978.)

Source	Sound-pressure-level range (*dBA*)	Distance (*ft*)
Whisper	25–40	5
Residential area at night	35–45	
Large transformer	48–55	200
Automobile		
< 35 mph	62–72	25
> 35 mph	70–78	25
Accelerating	82–88	
Truck with muffler		
< 35 mph	74–78	50
> 35 mph	80–85	50
Accelerating	82–90	50
Up 4% grade	85–88	50
Truck without muffler up 4% grade	93–100	50
Motorcycles		
< 35 mph	75–83	20
Accelerating	90–97	25
Earthmoving scraper	80–92	50
Concrete mixer	75–88	50
Pile driver	95–105	50
Air compressor	75–85	50
Rock drill	81–98	50
Lawn mower	68	50
Freight train 3000-hp diesel, 30–40 mph	85–95	100
High-speed turbo train	100	100
Can-manufacturing plant	52–65	
Oil refinery	51–68	

High frequencies (short wavelengths) yield high pitches, and low frequencies (long wavelengths), low pitches.

Sound attenuation is the reduction of sound intensity. Normal attenuation takes place as the energy of sound dissipates over distance until not enough energy is left to vibrate air molecules. This type of attenuation occurs at a rate equal to the square of the distance because the surface of a sphere increases by a factor of 4 for each unit increase in radius. Excess attenuation occurs when the reduction in loudness takes place in excess of normal attenuation, and involves five mechanisms: absorption, deflection, reflection, refraction, and masking. Absorption occurs when the energy of sound is transferred to some other object. Deflection describes when the direction of sound is altered somewhat, and reflection occurs when sound is bounced back to its source. Refraction is the bending of sound waves as they pass around an object, resulting in dissipation. Masking is not true attenuation, but rather, the covering up of unwanted sound with sounds more pleasing to the human ear (Robinette, 1972).

Plants and other objects in the landscape can serve as excess attenuators. Leaves, twigs, and branches on trees and shrubs will absorb sound energy, as will grass and herbaceous growth. Plants have been found to absorb high frequencies at a greater rate than low frequencies. Human hearing is more sensitive to high frequencies, so plants selectively filter out the most bothersome frequencies. Barriers of trees will deflect sound away from areas where it is not desired, and when at right angles to the source will bounce sound back to its source. Refraction occurs when sound passes through vegetative barriers and bends around plant structures, resulting in dissipation of sound energy (Robinette, 1972).

Cook (1978) reported that belts of trees 30 m wide and 15 m tall reduce highway noise by as much as 10 dB, although reductions of 6 to 8 dB were more common. Since decibels are on a logarithmic scale, this amounts to a reduction of nearly 50 percent in sound energy. Belts of 30 m width are unlikely in cities, where land costs are prohibitively high. However, Reethof and McDaniel (1978) report that narrow dense belts of trees provide effective noise barriers by reducing sound 3 to 5 dBA (dBA is decibels corrected for human hearing, i.e., human beings perceive increases in sound at a lower rate than that at which these increases actually occur) (Figure 4–18). Trees and other vegetation in conjunction with landforms and solid barriers can make significant reductions in unwanted highway noise. Trees in combination with landforms resulted in a reduction of 6 to 15 dB, while trees in combination with solid barriers (Figure 4–19) reduced noise levels by 5 to 8 dB (Cook, 1978).

Masking of sound by plants involves the substitution of desirable sounds to cover unwanted noise. Music piped into banks and busy offices provides a mask for noise generated by office machines and human conversation. Vegetation generates its own sounds, as wind rustles leaves and birds attracted by food and cover sing. Human beings have the capacity to filter unwanted noise in preference to more desirable sounds and will selectively listen to the sounds of nature over the sounds of the city (Robinette, 1972).

Human perception of sound is also important. People have reported significant

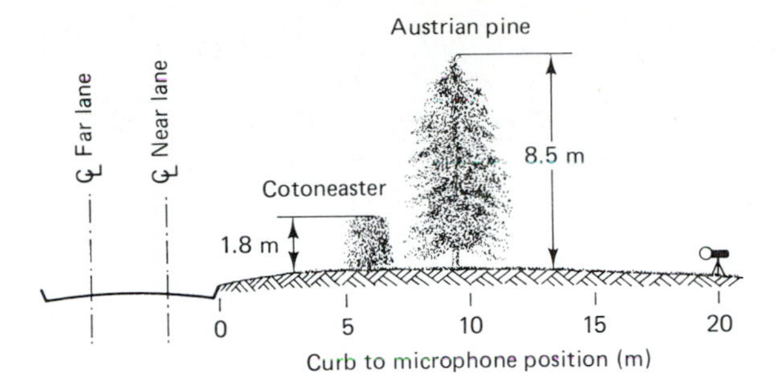

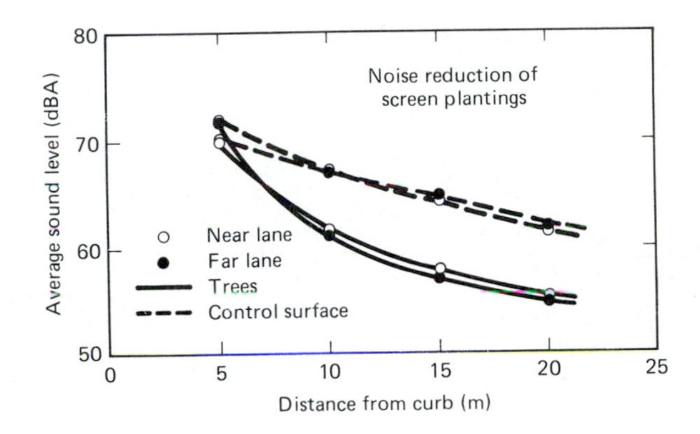

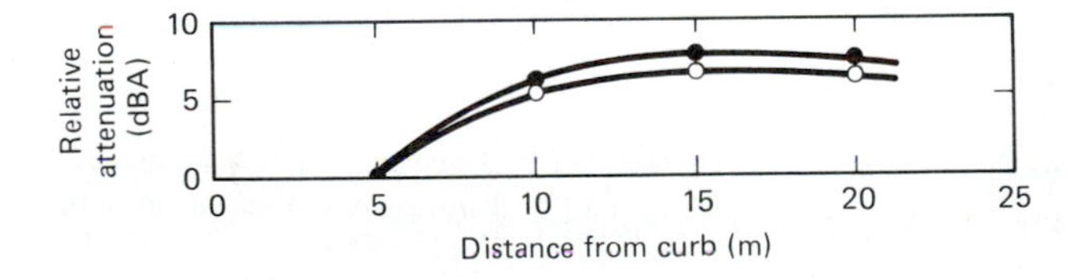

Figure 4–18 Excess sound attenuation on a residential street using trees and shrubs. (From Cook, 1978.)

reductions in noise behind sparse hedges when instruments measure scarcely any reduction at all. Sometimes merely screening the source of unwanted noise from view alters a person's perception of the quietness of a site. People also have different tolerances of noise levels based on their expectations. They were found to be more tolerant of high noise levels when they expected it, and less tolerant when they did not expect it (Anderson et al., 1984).

Glare and Reflection Reduction

Buildings and roadways characteristically have smooth, light-colored surfaces. This, coupled with natural and artificial lighting, results in serious glare and reflection problems in urban areas. Glare consists of a primary light source and atmospheric

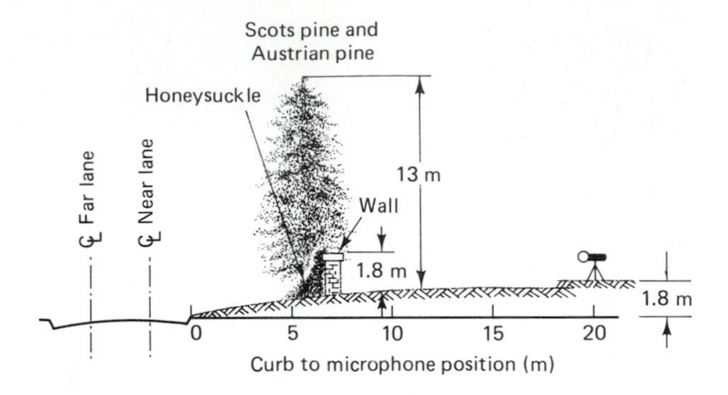

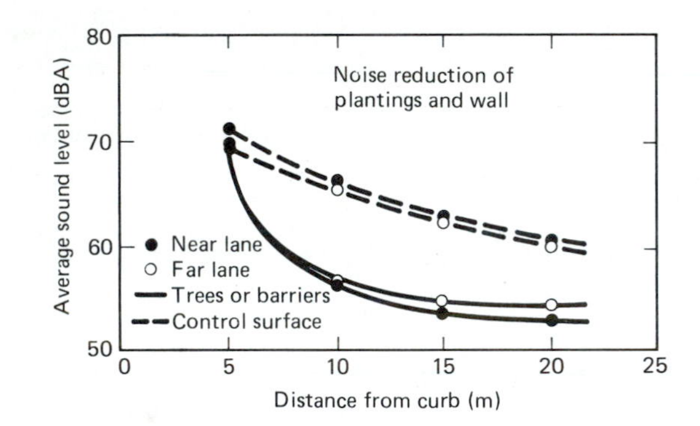

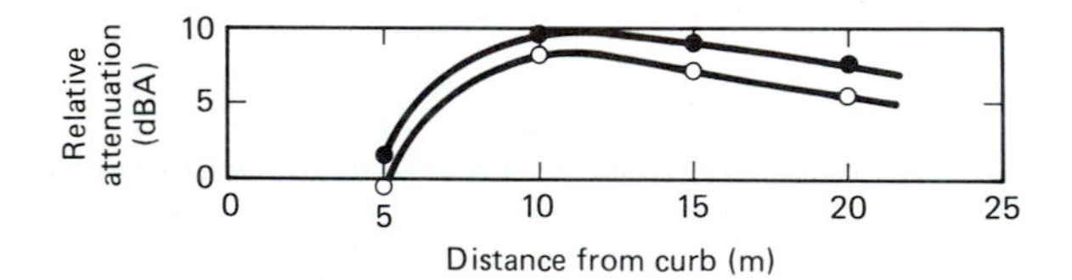

Figure 4–19 Excess sound attenuation using trees and shrubs in combination with a wall. (From Cook, 1978.)

hindrances that intensify it. For example, brilliant sunlight on a clear day is a source of glare, but sunlight filtered through a thin layer of clouds may be an even more intense source of light. Reflection is the bouncing of light rays from a primary source to where it is unwanted (Robinette, 1972).

Control of glare and reflection may be accomplished through the use of vegetation. Once the source of unwanted light has been identified, vegetation may be used to filter or block the source, or to cover a surface that is reflecting the light. It is important when designing for glare and reflection control to consider how sources of unwanted light and vegetation may change with time of day and season, and how vegetation changes in size, shape, and density over time. Vegetation may be used by itself, or in combination with barriers and landforms, to control unwanted light and glare (Robinette, 1972).

Erosion Control

Soil erosion in urban areas is frequently a serious problem. Urban soils are exposed to erosive forces in the atmosphere due to construction, neglect, or overuse of surfaces in public recreation and gathering places. In addition, surfaces covered with buildings, streets, and parking lots are impervious to water infiltration, and concentrate surface flows across exposed soil. Soil erosion results in serious water pollution problems and siltation of vital urban waterways. Controlling soil erosion involves legislative action to protect soils and waterways during construction, and proper management on the part of those responsible for vegetated surfaces and public use areas.

Managers should select the appropriate vegetative or nonvegetative cover for surfaces subject to intensive use. Frequent inspection of these surfaces should be made to determine if the cover is withstanding use, or if it should be renovated, rested, or changed to a more resistant surface. Infiltration may be enhanced by adding organic material to soil surfaces, perforation of lawns, gravel drainage systems for surface runoff, terracing, or use of porous paving blocks on heavy-use areas and parking lots.

Pedestrian traffic control may be necessary to protect vegetated surfaces from overuse and subsequent erosion. Plants may be used to block a line of vision that leads pedestrians from one point to another with disregard for existing walkways. Vegetation may also be planted as barriers to foot traffic and to guide people through the landscape (Robinette, 1972).

Urban Hydrology

Stormwater runoff from urban areas is greater than from rural areas covered with forests and/or farms. Streets, roofs, parking lots, compacted soil, and other impervious surfaces all contribute to runoff by preventing or reducing water infiltration. Central cities have less vegetation and more impervious surface than the suburbs, making the runoff problem more serious. Storm runoff contributes to local and downstream flooding, and has been identified as major source of water pollution. Many communities have combined storm and sanitary sewer systems leading to inadequate waste treatment during periods of high runoff, or expensive expansion of treatment facilities to meet water-quality standards.

Urban trees and other vegetation can contribute to reducing runoff two ways: by maintaining natural or park vegetation to allow infiltration, and by tree crowns intercepting and evaporating water before it reaches the ground. Wetlands are especially effective in reducing runoff and flooding, and are being restored in many communities for that purpose. Sanders (1986) and Lormand (1988) studied the interception and evaporation of rainwater by urban trees and report that urban tree canopies reduce stormwater runoff by 4 to 6 percent, and that increasing canopy cover can further reduce runoff.

Urban Wastewater

Cleansing of urban wastewater has been mandated by a number of federal laws designed to protect waterways from pollution. Costs associated with control of water pollution are high and are getting higher. At the same time the demand for clean water is increasing, taxing existing sources of water and creating demands for new sources. Urban forests can assist in alleviating these problems by providing cleansing of partially treated municipal wastewater and by recharging urban water supply aquifers.

In 1962 the community of State College, Pennsylvania, faced two problems confronting many municipalities. Their community water supply was inadequate to meet demands, and their sewage treatment facility was polluting a local river. Rather than purchase additional watershed lands and upgrade their sewage facility to tertiary treatment, the community, in cooperation with an interdisciplinary team of scientists from Pennsylvania State University, elected to use their secondarily treated sewage effluent to irrigate municipal watershed lands.

Sopper and Kerr (1978) reported that sixteen years of monitoring determined that the application of partially treated sewage to the watershed did not contaminate groundwater and did return 90 percent of the water to the aquifer. The wastewater was sprayed on three forest ecosystems at rates of 2.5 to 5 cm per week, and the groundwater was monitored for nitrate-nitrogen. As long as application rates on the red pine (*Pinus resinosa*) and white spruce (*Picea glauca*) stands were kept below 5 cm per week, nitrates in the groundwater remained below 10 mg/liter (standard for drinking water). The mixed hardwood forest had to be kept below an application rate of 2.5 cm per week in order not to exceed 10 mg per liter of nitrates in the groundwater. Within all three ecosystems tree growth increased, wildlife became more abundant, and levels of nitrogen increased in the soil. The ecosystems became stressed with a resulting reduction in tree seedlings and increase in herbaceous growth. Heavy metals were monitored, but significant contamination was not detected in the soil, plants, or wildlife (Sopper and Kerr, 1978).

In an earlier report, Sopper (1972) made a series of recommendations when selecting forest sites for disposal of partially treated municipal wastewater. Following is a summary of his recommendations.

1. The soil must have a sufficient infiltration and percolation capacity to accommodate the application of wastewater.
2. The soil must have sufficient chemical-absorptive and water-retention capacity and depth to groundwater to retain dissolved minerals temporally for use by vegetative cover.
3. The site must have low relief, vegetative cover, and surface organic matter to prevent and minimize surface runoff, particularly during winter.
4. A groundwater aquifer with sufficient capacity and a fairly deep water table must be present to accommodate subsequent changes in groundwater storage.

Urban woodlands do have the potential to serve as a living filter for partially treated wastewater, and a number of municipalities are now irrigating nearby forest and crop lands with partially treated effluent. Sludge from municipal treatment plant is being used as a soil conditioner on farmland near cities and for reclamation of mine spoils. Municipal sludge from the city of Chicago is sprayed as a slurry on crops grown as livestock feed, and tilled into coal mine spoils as a conditioner prior to planting.

LITERATURE CITED

AKBARI, H., S. D. DAVIS, J. HUANG, and S. WINNETT, eds. 1992. *Cooling Our Communities.* LBL Report 31587, Berkeley, Calif.

AKBARI, H., A. H. ROSENFELD, and H. G. TAHA. 1990. "Summer Heat Islands, Urban Trees, and White Surfaces." *1990 ASHRAE Transactions (Atlanta, GA, January 1990).* LBL Report 28308, Berkeley, Calif.

ANDERSON, L. M., B. E. MULLIGAN, and L. S. GOODMAN. 1984. "Effects of Vegetation on Human Response to Sound." *J. Arbor.* 10(2):45–49.

APPLEYARD, D. 1978. "Urban Trees, Urban Forests: What Do They Mean?" *Proc. Natl. Urban For. Conf.,* ESF Pub. 80-003. Syracuse: SUNY, pp. 138–155.

CHADWICK, L. C. 1971. "3000 Years of Arboriculture: Past, Present and the Future." *Arborists News* 36(6):73a–78a.

COOK, D. I. 1978. "Trees, Solid Barriers, and Combinations: Alternatives for Noise Control." *Proc. Natl. Urban For. Conf.,* ESF Pub. 80-003. Syracuse: SUNY, pp. 330–334.

DAVIS, D. D., and H. D. GERHOLD. 1976. "Selection of Trees for Tolerance of Air Pollutants." *Better Trees for Metropolitan Landscapes,* USDA-For. Serv., Gen. Tech. Rep. NE-22, pp. 61–66.

DEERING, R. B. 1954. "Effect of Living Shade on House Temperatures." *J. For.* 54:399–400.

DEGRAAF, R. M., and B. R. PAYNE. 1975. "Economic Values of Nongame Birds and Some Urban Wildlife Research Needs." *Trans. 40th N. Am. Wildl. Nat. Res. Conf.,* pp. 281–287.

DESANTO, R. S. et al. 1976. *Open Space as an Air Resource Management Measure,* Vol. 2, *Design Criteria,* EPA-450/3-76-028b. U.S. Environ. Prot Agency.

DEWALLE, D. R. 1978. "Manipulating Urban Vegetation for Residential Energy Conservation." *Proc. Natl. Urban For. Conf.,* ESF Pub. 80-003. Syracuse: SUNY. pp. 267–283.

DWYER, J. F. 1982. "Challenges in Managing Urban Forest Recreation Resources." *Proc. Sec. Natl. Urban For. Conf.* Washington, D.C.: Am. For. Assoc., pp. 152–156.

FRANKLIN, T. M. 1982. "Managing the Urban Forest for Wildlife." *Proc. Sec. Natl. Urban For. Conf.* Washington, D.C.: Am. For. Assoc., pp. 145–151.

GALLION, A. B., and S. EISNER. 1975. *The Urban Pattern: City Planning and Design.* New York: Van Nostrand Reinhold Company.

GARDESCU, P. 1976. "A Landscape Architect's View of Better Trees for Urban Spaces." *Better Trees for Metropolitan Landscapes,* USDA-For. Serv., Gen. Tech. Rep. NE-22, pp. 135–142.

HIESLER, G. M. 1986. "Energy Savings with Trees." *J. Arbor.* 12(5):113–125.

HIESLER, G. M. 1989. "Tree Plantings that Save Energy." *Proc. Fourth Urban Forestry Conf.* American Forests, Washington, D.C., pp. 58–62.

HIESLER, G. M., and L. P. HERRINGTON. 1976. "Selection of Trees for Modifying Metropolitan Climates." *Better Trees for Metropolitan Landscapes,* USDA-For. Serv., Gen. Tech. Rep. NE-22, pp. 31–37.

HOOPER, B. 1970. "The Real Challenge Has Just Begun." *Life* 68(1):102–106.

LAMPADIUS, F. 1963. "The Air Hygienic Significance of Forest and Its Modification by Smoke Damage." *Angew. Meterol.* 4:248–249.

LAWRENCE, H. W. 1993. "The Neoclassical Origins of Modern Urban Forests." *Forest and Conservation History* 37:26–36.

LORMAND, J. R. 1988. "The Effects of Urban Vegetation on Stormwater Runoff in Arid Environments." Univ. Ariz., Tucson, M.S. thesis.

MCCURDY, T. 1978. "Open Space as an Air Resource Management Strategy." *Proc. Natl. Urban For. Conf.,* ESF Pub. 80-003. Syracuse: SUNY, pp. 306–320.

MCPHERSON, E. G., and R. A. ROWNTREE. 1993. "Energy Conservation Potential of Urban Tree Planting." *J. Arbor.* 19(6):321–331.

MILLER, R. W. 1983. "Multiple Use Urban Forest Management in the Federal Republic of Germany." *Management of Outlying Forests for Metropolitan Populations.* Man and the Biosphere Seminar, Milwaukee, Wis., pp. 21–24.

NATIONAL ACADEMY OF SCIENCE. 1991. *Policy Implications of Greenhouse Warming.* Report of the Mitigation Panel. Washington, D.C.: National Academy Press.

NOWAK, D. J. 1994a. "Air Pollution Removal by Chicago's Urban Forest." In E. G. McPherson, D. J. Nowak, and R. A. Rowntree, eds., *Chicago's Urban Forest Ecosystem: Results of the Chicago Urban Forest Climate Project.* Gen. Tech. Rep. NE-186. Radnor, Pa.: USDA Forest Service, NEFES.

NOWAK, D. J. 1994b. "Atmospheric Carbon Dioxide Reduction by Chicago's Urban Forest." In E. G. McPherson, D. J. Nowak, and R. A. Rowntree, eds., *Chicago's Urban Forest Ecosystem: Results of the Chicago Urban Forest Climate Project.* Gen. Tech. Rep. NE-186. Radnor, Pa.: USDA Forest Service, NEFES.

OWENS, H. 1971. "Selecting Trees for Environmental Capability and Aesthetic Appeal." *Symp. Role Trees South's Urban Environ.* Athens: Univ. Georgia, pp. 63–65.

PARKER, J. 1983. "Do Energy Conserving Landscapes Work?" *Landscape Archit.* July:89–90.

PARKER, J. R. 1989. "The Impact of Vegetation on Air Conditioning Consumption." *Controlling Summer Heat Islands: Proc. of the Workshop on Saving Energy and Reducing Atmospheric Pollution by Controlling Summer Heat Islands.* Berkeley: Univ. Calif., pp. 45–52.

PROFOUS, G. V. 1992. "Trees and Urban Forestry in Beijing, China." *J. Arbor.* 18(3):145–153.

REETHOF, G., and G. M. HIESLER 1976. "Trees and Forests for Noise Abatement and Visual Screening." *Better Trees for Metropolitan Landscapes,* USDA-FS Gen. Tech. Rep. NE-22, pp. 39–47.

REETHOF, G., and O. H. MCDANIEL. 1978. "Acoustics and the Urban Forest." *Proc. Natl. Urban For. Conf.,* ESF Pub. 80-003. Syracuse: SUNY, pp. 321–329.

ROBINETTE, G. O. 1972. *Plants/People/and Environmental Quality.* USDI-Natl. Park Serv.

SAND, P. 1993. "Design and Species Selection to Reduce Urban Heat Island and Conserve Energy." *Proc. Sixth National Urban Forestry Conf.* American Forests, Washington, D.C., pp. 148–151.

SANDERS, R. A. 1986. "Urban Vegetation Impacts on the Hydrology of Dayton, Ohio." *Urban Ecology* 9:361–376.

SAYLER, R. D., and J. A. COOPER. 1975. "Status and Productivity of Canada Geese Breeding in the Twin Cities of Minnesota." *Thirty-sixth Midwest Fish Wildl. Conf.,* Indianapolis.

SCHMITT, P. J. 1973. "Back to Nature." In A. B. Callow, Jr., ed., *American Urban History.* New York: Oxford University Press, pp. 454–468.

SMITH, W. H. 1978. "Urban Vegetation and Air Quality." *Proc Natl. Urban For. Conf.,* ESF Pub. 80-003. Syracuse: SUNY, pp. 284–305.

SMITH, G. C., and F. G. BRENNAN. 1984. "Response of Honeylocust Cultivars to Air Pollution Stress in an Urban Environment." *J. Arbor.* 10(11):289–293.

SMITH, W. H., and S. DOCHINGER. 1976. "Capability of Metropolitan Trees to Reduce Atmospheric Contaminants." *Better Trees for Metropolitan Landscapes,* USDA-For. Serv., Gen. Tech. Rep. NE-22, pp. 49–59.

SOPPER, W. E. 1972. "Effects of Trees and Forests in Neutralizing Waste." *Trees and Forests in an Urbanizing Environment,* Plann. Res. Dev. Ser. No. 17. Amherst: Univ. Mass., pp. 43–57.

SOPPER, W. E., and S. N. KERR. 1978. "Potential Use of Forest Land for Recycling Municipal Waste Water and Sludge." *Proc. Natl. Urban For. Conf.,* ESF Pub. 80-003. Syracuse: SUNY, pp. 392–409.

SOUCH, C. A., and C. SOUCH. 1993. "The Effect of Trees on Summertime Below Canopy Urban Climates: A Case Study Bloomington, Indiana." *J. Arbor.* 19(5):303–312.

THAYER, R. L., and B. T. MAEDA. 1985. "Measuring Tree Impact on Solar Performance: A Five Year Computer Modeling Study." *J. Arbor.* 11(1):1–12.

WILLIAMSON, R. D. 1973. "Bird- and People-Neighborhoods." *Nat. Hist.* 82(9):55–57.

YOUNGBERG, R. J. 1983. "Shading Effects of Deciduous Trees." *J. Arbor.* 9(11):295–297.

ZUBE, E. H. 1971. "Trees and Woodlands in the Design of the Urban Environment." *Trees and Forest in an Urbanizing Environment.* Plann. Res. Dev. Ser. No. 17. Amherst: Univ. Mass., pp. 145–150.

PART 2
Appraisal and Inventory of Urban Vegetation

5

Values of Urban Vegetation

Urban residents place high value on city trees and other urban vegetation. Following the devastation of Charleston, South Carolina, by Hurricane Hugo in 1989, residents were asked what was the most significant feature of their community damaged by the storm. Over 30 percent identified damage to the urban forest as the most significant loss (Hull, 1992). In earlier chapters we discussed social needs and values as they pertain to urban trees, and important amenity, climatological, and engineering uses of vegetation. All of these uses give value to urban vegetation, and this value can be described in both economic and legal terms. Economic and legal values are closely related, and in many instances economic methodologies are used to ascertain legal values.

ECONOMIC VALUES

Economic values of urban vegetation may be determined by a number of approaches, all of which pertain to both municipal and commercial arboriculture. Trees may be appraised individually, or collectively in woodlands and forests, and may be appraised based on their residue and, sometimes, their wood value.

Individual Trees

City-owned trees have been assigned values as city assets, had their values determined by maintenance costs, or have been given values based on wood products or residues. Both public and private trees can be appraised using formulas, preservation costs, replacement costs, or their contribution to property value. Each of these methods is discussed in terms of technique and relative merit as determinants of tree value.

City assets. City governments keep records of all city-owned assets based on their appraised valuation. City-owned trees are sometimes valued in this manner, based on the cost of investing in city trees. However, city trees are unique compared to other city assets, as other assets usually depreciate in value over time, whereas trees appreciate in value over time. According to Dressel (1963): "Street trees represent one of the half dozen most valuable assets the city has in improvements and is not far below in dollar value the investment in schools, streets, sewers, and water supply." Kielbaso (1974) reports asset values ranging from $71 to $500 per city-owned street tree, based on estimates by city foresters. Ranges in value such as this certainly do not allow much confidence in using asset values for city tree valuation. The problem is that it is not known how individual cities arrived at their appraised tree values.

Maintenance costs. A second approach to tree valuation is that of considering the investment in the tree population. This approach is used in some cases when establishing trees as city assets. The initial investment in a tree is establishment cost, including growing or purchasing planting stock and planting it, and may include fertilization and watering during the first growing season. Periodic maintenance costs follow, including pruning and perhaps fertilization and insect or disease control. Finally, the tree must be removed when it becomes a hazard or when it dies. If a 2-in.-diameter tree was planted at a cost of $100, pruned seven times over a 40-year life span, and finally removed, the investment in that tree would be $324 (Table 5–1). If an interest rate of 6 percent were applied to each operation, the tree would then have a value of $2751.

Two problems are evident with the maintenance cost approach to tree appraisal. First, tree species that demand extensive maintenance will be appraised at higher values, and this is in opposition to the desire of city foresters to plant low-maintenance species. The second problem has to do with the nature of compound interest. As trees live for extended periods of time, their values can inflate to a point of being unreasonably high.

Wood products or residue. Wood as a forest product has value to society, so when no longer needed for other uses, city trees should have some value as forest products. Foster (1965) estimated that street trees in Boston grow 100,000 board fee

TABLE 5–1 SUM OF TREE MANAGEMENT COSTS AND TREE MANAGEMENT COSTS PLUS 6 PERCENT COMPOUND INTEREST OVER 40 YEARS

	Tree Management Costs			Tree Management Costs Plus 6% Interest to 40 Years		
Year	Planting	Pruning	Removal	Planting	Pruning	Removal
0	$100			$2173		
5		$ 6			$89	
10		8			80	
15		12			82	
20		15			70	
25		18			57	
30		21			45	
35		24			35	
40			$120			$ 120
Total			$324			$2751

of wood per year, and suggested that this could be harvested on an annual sustained-yield basis. Kielbaso (1974) assumed a value of $40 per thousand board feet, or an annual return of $4000 per year and a 6 percent discount rate on all future returns. The problem with this approach is that it is extremely unlikely that the public would permit a healthy street tree to be cut for its wood content, and by the time a street tree has declined to the point where it must be removed, much deterioration of the wood will probably have taken place. The exception to this is in the case of a catastrophic event such as Dutch elm disease *(Ceratocystis ulmi).* When comparing the value of street trees as wood fiber to their amenity value, virtually all approaches to apprais-ing urban trees will yield higher values than stumpage, with the possible exception of the occasional veneer-quality log.

As far as street trees as wood products are concerned, it is a matter of tree residues having some market value rather than producing forest products. In dis-cussing the disposal of elm waste wood in Minneapolis/St. Paul, Minnesota, DeVoto (1979) stated: "It must be completely understood that our tree debris is debris, not product. It is something terribly expensive to dispose of; our concern should simply be to find the most economical method of disposal." The city of Dayton, Ohio, at-tempts to utilize fully their urban tree residues. Sawlog and veneer-quality trunks are stockpiled and sold to the highest bidder. Lower-quality wood is sold as firewood, while woodchips and composted leaves are given away. LaRue (1982) reports that the Dayton approach is cheaper than landfilling, and that income from log and fire-wood sales is used to purchase forestry equipment. A guide (Cesa et al., 1994) to marketing sawlogs from street tree removals recommends:

1. Identify one or more sawmills in your area that may be interested in purchasing your material.
2. Learn what their sawlog requirements are and decide whether your street tree logs fit these requirements.
3. Locate and remove metal and other foreign material in the logs.
4. Store sawlogs until a salable quantity is accumulated.
5. Be flexible and persistent to try this concept.

Landscape residue has been identified as a major contributor to the solid waste stream, leading many state and national governments to prohibit its landfilling as part of recycling legislation. For example, landscape waste from the city of Urbana, Illinois, was found to contribute 35 percent to the total amount of waste generated by the community (Darling, 1989). The city now composts succulent material and sells it as a garden mulch, and sells wood chips for landscape applications. Costs exceed revenues, but the material is no longer permitted in landfills. Loggens (1978) suggests that since wood is a clean-burning fuel when properly combusted, and has a low ash and sulfur content, that it be burned for its energy content. Tree residue generated by a private tree care firm in Houston, Texas, was chipped and sold as fuel rather than landfilled. Analysis of costs and financial returns found that energy conversion was an economically more attractive alternative than landfilling (Murphy et al., 1980).

Tree residue chips can also be used as surface mulch in landscaping. Walker (1979) found that compared to pine straw, wood chips were cheaper, longer lasting, less flammable, and retained soil moisture better. Wood-chip-mulched areas were considered as attractive as areas covered with other mulch covers. Fertilization was not necessary as long as only surface applications were used.

Council of Tree and Landscape Appraisers' valuation. The most widely used method of tree and landscape plant appraisal in North America is the system described in the *Guide for Plant Appraisal* (CTLA, 1992), published by the International Society of Arboriculture (ISA). This system had its beginning in 1947 when the National Shade Tree Conference (later becoming the ISA) established a joint committee with the National Arborist Association to develop a method for determining the value of shade trees. A basic method was accepted in 1951 and a booklet published by the Conference in 1957 titled *Shade Tree Evaluation.* Revisions were published by the Conference in 1965 and 1970.

The ISA, American Society of Consulting Arborists, National Arborist Association, American Society of Nurserymen, and later, Associated Landscape Contractors of America formed the Council of Tree and Landscape Appraisers (CTLA). The CTLA has since produced five more editions and these were published by the ISA in 1975, 1979, 1983, 1988, and 1992. There have been modifications of the system over the years, but the essential method has remained the same since first accepted by the National Shade Tree Conference in 1951 (Chadwick, 1980).

The CTLA (1992) valuation method relies on a formula to compute tree and other landscape plant values. A basic value is established as a maximum for a plant, and multiplied by percent ratings for species, specimen condition, and plant location using the formula

$$\text{plant value} = \text{basic value} \times (\text{species classification }\%) \times (\text{condition }\%) \times (\text{location }\%)$$

The basic value is determined by two methods, replacement cost and formula. Replacement cost is used for shrubs, vines, hedge plants, and transplantable-sized trees. The replacement cost is the installed cost of the species (or similar species) of plant lost, adjusted by the condition and location of the plant being replaced. In the case of shrubs, vines, and hedge plants too large to find replacements for, the installed cost is adjusted for condition and location and then compounded by the current interest rate for the number of years it will take the replacement to reach the size of the lost plant.

The appraised value of a tree too large to be replaced with a transplant is calculated using a formula that assigns a base value to a tree based on the cross sectional area of the trunk at 6 in. (15 cm) above ground for trees up to 4 in. (10 cm) in diameter, 12 in. (30 cm) above ground for trees 4 to 12 in. (10 to 30 cm) in diameter, and 4.5 ft. (1.4 m) above ground for trees larger than 12 in. in diameter. The trunk area is calculated using the formula

$$\text{area (sq. in.)} = 0.7854 \times \text{diameter squared}$$

The formula will overvalue trees greater than 30 in. (75 cm) in diameter so adjusted trunk area is used for larger trees (Figure 5–1). The first step is to determine the replacement cost of the largest commonly available transplantable specimen as described above. The value for that portion of the tree larger than the transplant is based on the wholesale price of the transplant without the installation cost. This cost is divided by the cross-sectional area in square inches (cm) of the transplant to determine the per square inch (cm) value of the tree. This value is then multiplied by the remaining cross-sectional area (total area – transplant area) and adjusted for species, condition, and location. The replacement value is then added to the formula value to determine the appraised value of the tree.

The formula method will not work with palms because they are monocots and do not increase in diameter with growth. However, palms increase in height with age, so the CTLA recommends using clear or brown trunk height (clear of fronds) to determine the base value.

Species factor or class is based on a percentage rating from 5 to 100 percent, usually calculated at 5 to 10 percent intervals. Establishment of species class is made by the appraiser, and the appraiser may rely on recommendations by local chapters of the ISA (Table 5–2).

Condition factor or class is also expressed as a percentage rating from 0 to

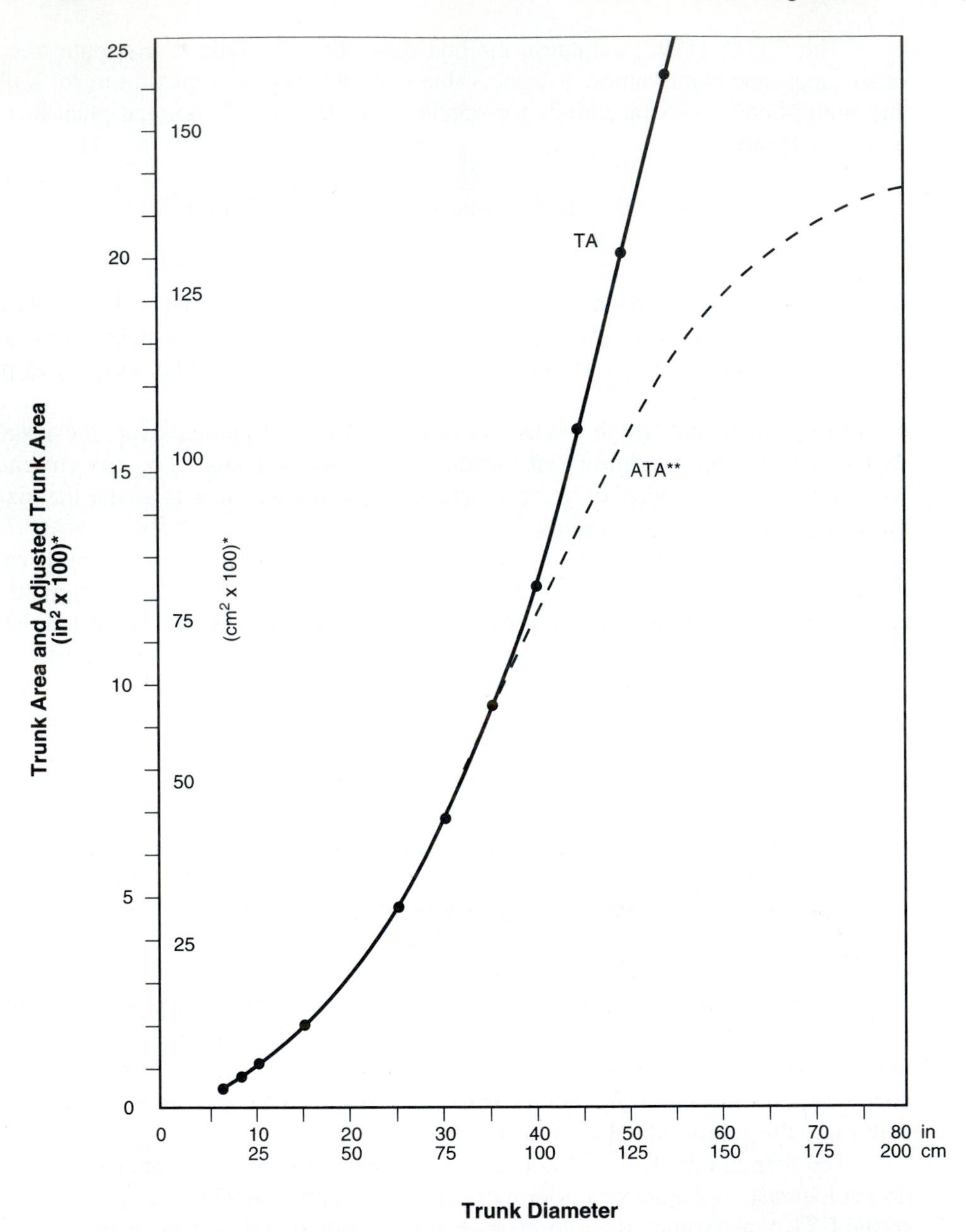

Figure 5–1 Curve **TA** depicts the increase in **Trunk Area** with increasing diameter. Curve **ATA** represents the rate of **Adjusted Trunk Area** increase at the trunk diameters above 30 inches (75 cm).

*Measurements taken at 4.5 feet (1.4 m) above the ground.

****Adjusted Trunk Area (ATA)** values considered to be reasonable for trees with trunk diameters greater than 30 inches (75 cm) were estimated to obtain curve **ATA.** Quadratic equations were calculated to represent the curve of ATA values. (From CTLA, 1992.)

TABLE 5–2 FACTORS TO CONSIDER IN RATING PLANT SPECIES AND CULTIVARS (SEE STATE OR REGIONAL RATING LISTS IF AVAILABLE.) (From CTLA, 1992.)

Climate Adaptability	*Resistance or Tolerance*	*Aesthetic Values*
1. Cold hardiness	1. Diseases	1. Branches and Tree Form
2. Frost tolerance	2. Insects	Growth habit
3. Drought tolerance	3. Air Pollution	Bark color and texture
4. Storms: resistance to ice, snow, wind		2. Foliage
	Maintenance Requirements	Density
	1. Training and pruning	Color
Growth Characteristics	2. Cleanliness: flowers, fruit, leaves, twigs, duration of leaf fall	Texture
1. Tolerance of difficult sites		Duration
2. Vigor		3. Flowers
3. Structural strength	3. Root problems	Prominence
4. Life expectancy	4. Pests	Fragrance/odor
	5. Structural problems (cabling and bracing)	Duration
Soil Adaptability		Color
1. Structure and texture		Size
2. Drainage	*Allergenic Properties*	4. Fruit
3. Moisture excess or deficiency	1. Pollen	Prominence
4. Acidity and alkalinity	2. Dermal toxins	Duration
5. Nutritional deficiencies or excesses		Use
		Color
		Size
		Fragrance/odor

100 percent, usually in increments of 5 or 10 percent. A specimen assigned a value of 100 percent is assumed to be a perfect specimen of the species under consideration. It should be cautioned that very few specimens are perfect, and that assigning an appropriate condition rating requires a great deal of experience in plant appraisal. The CTLA (1992) recommends that the following factors be considered when assigning a condition rating to a specimen; growth rate, presence of decay, structural weakness, insects and diseases, survival conditions, and life expectancy (Table 5–3).

Location factor was added to the appraisal method in the 1975 edition to reflect the impact of a plant's location on its value. It is unrealistic to appraise a landscape plant at a value that represents a high percentage of the property value. A parcel of land worth $10,000 certainly should normally not contain a tree appraised at $7000. A tree in a group of trees will usually be worth less than a single landscape tree, and rural roadside trees will usually be worth less than boulevard trees in the city. Vegetation that provides other amenities, such as screening, noise reduction, climate modification, and air purification, or plants that have historic or

TABLE 5–3 GUIDE TO JUDGING PLANT CONDITION (From CTLA, 1992.)

Factor Ratings. Rate each of the five factors noted below from 1 to 5. High numbers indicate a high rating

Factors[1]			Condition Points
ROOTS			________
Root anchorage S[2]	*No problem[3]*	*5*	
Confined relative to top S	*No apparent problem(s)*	*4*	
Collar soundness S,H[2]	*Minor problem(s)*	*3*	
Mechanical injury S,H	*Major problem(s)*	*2*	
Girdling or kinked roots S,H	*Extreme problem(s)*	*0 or 1*	
Compaction or water-logged roots H[2]			
Toxic gases & chemical symptoms H			
Presence of insects or diseases H			
TRUNK			________
Sound bark & wood, no cavities S,H	*No problem[3]*	*5*	
Upright trunk (well tapered) S	*No apparent problem(s)*	*4*	
Mechanical or fire injury S,H	*Minor problem(s)*	*3*	
Cracks (frost, etc.) S,H	*Major problem(s)*	*2*	
Swollen or sunken areas S,H	*Extreme problem(s)*	*0 or 1*	
Presence of insects or diseases H			
SCAFFOLD BRANCHES			________
Strong attachments S	*No problem[3]*	*5*	
Smaller diameter than trunk	*No apparent problem(s)*	*4*	
Vertical branch distribution	*Minor problem(s)*	*3*	
Free of included bark	*Major problem(s)*	*2*	
Free of decay and cavities S,H	*Extreme problem(s)*	*0 or 1*	
Well-pruned, no severe heading back S,H			
Well-proportioned (tapered, laterals along branches) S			
Wound closure H			
Amount of dead wood or fire injury S,H			
Presence of decay, insects or diseases H			
SMALLER BRANCHES & TWIGS			________
Vigor of current shoots, compared to that of 3–5 previous years H	*No problem*	*5*	
Well-distributed through canopy H	*No apparent problem(s)*	*4*	
Normal appearance of buds (color, shape and size for species)	*Minor problem(s)*	*3*	
Presence of weak or dead twigs H	*Major problem(s)*	*2*	
Presence of insects or diseases H	*Extreme problem(s)*	*0 or 1*	

TABLE 5–3 CONTINUED

Factors[1]			Condition Points
FOLIAGE			________
Normal appearance (size and color) H	*No problem*	*5*	
Nutrient deficiencies H	*No apparent problem(s)*	*4*	
Herbicide, chemical or pollutant injury symptoms H	*Minor problem(s)*	*3*	
	Major problem(s)	*2*	
Wilted or dead leaves H	*Extreme problem(s)*	*0 or 1*	
Presence of insects or diseases H			
TOTAL POINTS			________

Total Points	Condition	Rating
23–25	Excellent	90–100
19–22	Good	70–89
15–18	Fair	50–69
11–14	Poor	25–49
05–10	Very Poor	05–24

[1]Give one rating for each **Factor.** The items listed under each Factor are to be considered in arriving at a rating for that **Factor.**
[2]S = item is primarily structural
H = item is primarily health
S,H = item may involve both structure and health
[3]A rating of "5" indicates no problems found after doing a root-collar inspection and/or climbing the tree to inspect the trunks and major limbs.

other significance, should be assigned higher location factors than those that do not. The location value is based on the average of values assigned for the following:

Site (10–100%)	Quality of the property and area
Contribution (0–100%)	Functional and aesthetic contribution to the landscape
Placement (0–100%)	Effectiveness in providing functional and aesthetic contribution

In summary, the CTLA appraisal method involves establishment of a base value for a landscape plant using either replacement cost or formula. The base value is a maximum value and is modified by multiplying by percentage factors for species, condition, and location. Table 5–4 provides an example of the calculation of a tree's value using this method.

TABLE 5–4 SAMPLE TREE VALUE CALCULATION USING THE CTLA VALUATION METHOD (From CTLA, 1992.)

Example: Obtain the **Appraised Value** of a 35-inch-diameter (90-cm) Northern Red Oak tree growing in an open landscape. The **Replacement Cost** and wholesale cost of a 4-inch (15-cm) tree are $1,040 and $415, and the **Species** and **Condition** ratings are 90% and 80% respectively. The **Site, Contribution,** and **Placement** ratings for **Location** are 90%, 80%, and 70% respectively.

$$\textit{Appraised Value} = \textit{Basic Value} \times \textit{Condition} \times \textit{Location}$$

$$\textit{Basic Value} = \textit{Replacement Cost} + (\textit{Basic Price} \times [TA_A - TA_R] \times \textit{Species})$$

1. **Replacement Cost:** largest transplantable tree — *$1,040*
2. **Basic Price** of replacement tree
   - A. *Wholesale, retail* or *installed* cost = *$ 415*
   - B. Replacement tree **Trunk Area (TA_R)** = *12.56* in^2
   - C. Divide cost by TA_R (2A ÷ 2B) = *$33.00* /in^2
3. Determine difference in trunk areas

   If d_A = 30" or less, determine TA_A
   - A. $TA_A = (0.785d^2 \text{ or } 0.080c^2)$ = ______ in^2
   - B. Replacement tree TA_R (2B) = ______ in^2
   - C. Subtract TA_R from TA_A (3A – 3B) = ______ in^2

   If diameter > 30", determine ATA_A from Table 4–1, 4–2, 4–3, or 4–4 or use the appropriate formula, such as (English units):

   $ATA_A = -0.335d^2 + 69.3d - 1{,}087$

   $= (-0.335 \times 35^2) + (69.3 \times 35) - 1{,}087$
   - AA. –410.38 + 2,425.5 – 1,087 = 928 in^2
   - BB. Replacement tree TA_R (2B) = 13 in^2
   - CC. Subtract TA_R from TA_A (3AA – 3BB) = 915 in^2
4. Multiply **Basic Price** by area differences

   $33.00/in^2 (2C) × 915 (3C or 3CC) = $30,195
5. Adjust step 4 by **Species** rating 90% = $27,175.50
6. **Basic Value** = $1,040 (Line 1) + $27,175.50 (line 5) = $28,215.50
7. Adjust Line 6 by **Condition** 80% = $22,572.40
8. Adjust for **Location:**

   Location = (Site + Contribution + Placement) ÷ 3

   = (90% + 80% + 70%) ÷ 3 = 80%

   Adjust Line 7 by **Location** 80% = $18,057.92
9. **Appraised Value** = Round to nearest $100 = $18,100

It should be emphasized that proper appraisal of landscape plants requires extensive knowledge of the species or cultivar being appraised, an understanding of local real estate values, and years of practical experience. It is recommended that those interested in landscape plant appraisal purchase the *Guide for Plant Appraisal* published by the International Society of Arboriculture, P.O. Box GG, Savoy, IL 61874.

Tree Council of the United Kingdom valuation. Helliwell (1990) describes a method of tree valuation he developed in 1966, subsequently published by the Tree Council of the United Kingdom in 1975. This method calculates tree value using a point system for seven parameters describing the tree and its surroundings (Table 5–5). The points for each parameter are multiplied together to arrive at a total score for the tree. The score is converted to currency values by multiplying it by an appropriate conversion factor. As of 1994 the conversion factor in the United Kingdom was set at 12 pounds sterling. If a tree fails to score on any of the first six parameters, it will receive a value of zero (Arboricultural Association, 1994).

Preservation costs. A different approach to tree valuation involves the cost of tree preservation during construction. When developing wooded parcels for housing, some, or all, trees are removed to make room for structures, driveways, and utilities. In a study by Lash (1977) of new home construction in Amherst, Massachusetts, builders reported an average cost of $1000 to clear the entire lot of trees prior to construction, and an average cost of $1700 to preserve residual trees. The cost of tree preservation in this case was $700 per lot. Builders reported that increased costs resulted from hiring consultants to advise them on which trees to preserve, and from the erection of protective barricades prior to construction. Builders felt that the investment was worth the cost based on how fast homes sold and higher prices for homes on wooded lots.

In similar studies in Athens and Atlanta, Georgia, builders reported lower costs associated with preserving trees compared to clearing the entire lot (Seila and Anderson, 1984). The authors feel that these results differ from the Amherst study because builders in Atlanta and Athens made no special effort to select trees for preservation or to protect them during construction. As in the Amherst study, builders felt that residual trees helped to sell their homes faster.

Property values. It has been well documented that trees contribute to property values. Payne and Strom (1975) created a model replica of undeveloped open land in Massachusetts and photographed it with varying amounts of tree cover. The photographs were shown to professional real-estate appraisers, and they were asked to assess the market values of the parcels. The assessors valued open parcels at an average of $1500 per acre and parcels two-thirds wooded at $2050 per acre, or 27 percent more for wooded parcels. In another study, Payne showed photographs of similar homes on parcels with varying amounts of trees present to realtors and persons

TABLE 5–5 METHOD FOR PLACING AN AMENITY VALUE ON INDIVIDUAL TREES (From Arboricultural Association, 1994.)

	Points[a]			
Factor	1	2	3	4
1. Size of tree	small	medium	large	very large
2. Useful life expectancy	2–5 yrs	5–40 yrs	40–100 yrs	100+ yrs
3. Importance of position in landscape	little	some	considerable	great
4. Presence of other trees	many	some	few	none
5. Relation to the setting	barely suitable	fairly suitable	very suitable	especially suitable
6. Form	poor	fair	good	esp. good
7. Special factors	none	one	two	three

[a]Points are multiplied together and the total score is multiplied by an agreed conversion factor (£12 as of 1994).

who had recently purchased homes. When asked to estimate market values, both groups valued homes on wooded lots by an average of 7 percent higher.

Unwooded house lots in an exclusive subdivision near Columbia, South Carolina, sold much faster when the builder started planting 2- to 3-in.-caliper pines (*Pinus* spp.) on the lots with a hydraulic tree spade. The builder reported that the trees increased the selling value of the parcel by more than $1500 per acre (Bickley, 1978).

Morales (1980) looked at fourteen valuables influencing the price of suburban homes in Manchester, Connecticut, including the presence or absence of tree cover. Analysis of these data by multiple regression found seven significant variables influencing the selling price of homes in the study area. Presence of mature tree cover ranked sixth in importance and contributed an average of $2686 (6 percent) to the selling price of the homes. In a similar study, Morales et al. (1983) used multiple regression to predict the selling prices of homes in Greece, New York. In this study, homes on lots with tree cover sold for an average of $9500 (15 percent) more than homes on untreed lots. For comparison, they used the CTLA method to appraise trees on the study parcels. They determined tree values to be an average of $6000 per lot using the CTLA method, or $3500 less than actual differences in real-estate values. They felt the difference was due to realization of the value that trees add to lots on the part of the contractor, location of the treed lots farther from the frontage road, and construction damage to residual trees not noticed when appraising the entire property but noticed when appraising trees separately.

Martin et al. (1989) conducted a similar study in the Austin, Texas, area, but rather than simply using the presence of trees, attempted to quantify them by number and CTLA condition using "the equivalent number of healthy trees (number of trees × average condition) multiplied by the average diameter of the trees on the lot." Home values were based on tax appraisals with the regression using 12 variables in addition to the tree variable to predict appraised value. Home values varied from $30,000 to $600,000 in the study. Predictive modeling determined trees contribute an average of 19 percent to residential property value. The CTLA formula was used to determine tree value for the same properties, estimating 13 percent of the value being attributed to trees.

Urban forest value. Nowak (1993) used the CTLA method of appraisal to calculate the compensatory value of the urban forest for the city of Oakland, California. The tree canopy cover was 21 percent with an estimated value of $385.7 million. Trees on residential properties accounted for $226 million, while street trees accounted for $9.9 million. The Tree Council of the United Kingdom valuation system was applied to amenity trees in the Borough of Tunbridge Wells, Kent, England (Dolwin and Goss, 1993). They estimated the value of the 15,000 trees in the community to be 18 million pounds sterling, or £1,200 per tree.

McPherson (1994) applied benefit-cost analysis to tree planting and care in the city of Chicago. Costs include planting, pruning, removal, waste disposal, infrastructure repair, litigation, inspection, and program administration. Benefits were calculated from energy savings, air quality improvement, sequestered and avoided carbon dioxide, avoided water runoff and use at power plants, and CTLA tree valuation. Since tree value as calculated using the CTLA method includes environmental benefits, the first five benefits were subtracted from the CTLA value to avoid double counting. Benefit-cost ratios calculated at a 7 percent discount rate found ratios ranging from 3.5 (residential plantings) to 2.1 (park plantings), with discounted payback periods from 9 to 15 years.

Urban Woodlands

Forestland in urban regions is held in a variety of public and private ownerships. Assigning economic values to these lands should consider the value of vegetation even though the value of the land will be much higher than its vegetative cover, especially if the land has alternative use potential. Miller and others (1978) studied changes in land and timber values from 1952 to 1971 in the vicinity of two urban centers in Massachusetts. They found that land values increased much more rapidly than timber values, and suggested that owning land for timber production in urban regions is a poor investment when analyzed using soil expectations or present-net-worth calculations. In other words, income from growing trees will not justify costs associated with owning forestland in urban areas. Yet forests and other plant communities do have value in cities, including forest products. The following discussion considers wood, alternative, and amenity values of urban forests.

Wood values. Even though land values are high in urban regions, it does not mean that trees have little or no value as forest products. In the same study of land and timber values in Massachusetts described above, the researchers report that some lands in the study areas supported high-value stands of timber, particularly white pine *(Pinus strobus)*. Many of these parcels are held by urban residents for speculation or recreation, and could be managed for timber production provided that foresters were sensitive to desires and values of landowners. Speculators might be advised to leave existing vegetation as potential landscape trees on future home sites, and developers assisted in marketing timber removed during construction. Recreation owners could be advised that timber harvest will provide income to help offset costs of ownership. Management plans for these owners should emphasize light cuttings, wildlife habitat improvement, and improved aesthetics.

Public urban forests in the United States and Canada are, for the most part, not managed for timber production. However, timber harvests are common in European urban forests, as domestic wood supplies are much more limited. For example, German forests grow only about 50 percent of domestic wood supplies, the remainder being imported from other countries. To meet a variety of demands, virtually all forestland is managed for multiple use, including timber production. Holscher (1971) reports that the Frankfurt, Germany, urban forest produces 1350 board feet per acre per year even though the dominant use of the forest has been recreation since 1927. Income from timber sales in German urban forests pays for all forest management activities. The urban forest in Zurich, Switzerland, produces 1.5 million board feet of wood per year, which provides an annual income of about $80,000 (Holscher, 1971).

Alternative value. Demand for land in urban regions is high, so numerous alternative uses exist for forestland in these regions, provided that the land can be physically developed. Land not developed for alternative uses when opportunities exist are subject to opportunity costs, which include foregone income and interest charges. In the case of public forestland, it is the desire of urban residents to bear the opportunity costs associated with park or open space in cities, rather than to dedicate the land to alternative uses. Public land sold for other uses will generate income to the community in the form of jobs, tax dollars, and increased economic development. This may be interpreted as meaning that land covered with trees is worth more to the community than the same land covered with buildings. A measure of the value of the land and its vegetation is the income foregone by not selling it for alternative uses.

Lynn Woods, a 2000-acre tract of forestland in the city of Lynn, Massachusetts, is used primarily as recreational land and the location of four municipal water supply reservoirs. In the late 1950s it was proposed to use a portion of Lynn Woods for an interstate highway corridor. Many residents of Lynn, concerned by the declining economic base of the city, saw immediate economic benefits to the community, including construction jobs, and better access to the city, which they hoped would make it more attractive to new industry. Other residents did not want to lose a portion of the land to highway development, and this sparked a decade-long struggle over the fate of Lynn Woods. In 1972, the governor of Massachusetts finally ruled out Lynn Woods as the location for the highway, as other alternatives existed to the taking of parkland. Lynn Woods was spared because it was more valuable to the community as forest than as a highway corridor with its accompanying economic benefits (Gordon and Lambrix, 1973).

Public open space is being included in many new developments, both as a selling feature by the developer or as mandated by local ordinance. In a survey of housing developments in which a substantial portion of the land was dedicated to public open space and managed by a homeowners' association, Strong (1982) reported that residents were willing to pay an average of $220 per year to preserve and maintain this open space.

Amenity values. Alternative values are a way of describing the amenity value placed on woodlands in a community. However, alternative values do not separate vegetation values from land values. Dwyer et al. (1989) attempted to determine the values of trees and forests in recreation areas in the Chicago area. Questionnaires were sent to users of the parks and forest preserves asking them to choose between paired descriptions of hypothetical recreation sites that varied in attributes including user fees. They found that respondents would be willing to pay an extra $1.60 per visit for a site that was mostly wooded with some open grassy areas compared to a site with few trees. The *Guide for Plant Appraisal* (CTLA, 1992), while primarily addressing individual landscape plants, includes forest trees under the location factor. This is to be used for trees in open woods and forests that serve primarily aesthetic or amenity functions, not for areas under timber management. Trees cannot be assessed as forest products and amenity trees at the same time, but as one or the other. For example, trees in rural woodlands owned for recreational purposes may be assessed using the CTLA method provided that they are not being managed for forest products.

Helliwell (1990) describes a system of amenity forest valuation in the United Kingdom that uses a point system similar to one he developed for individual trees (described previously). This system (Table 5–6) awards one to four points for each of seven factors, and these points are multiplied together to give value units. The value units are then to be multiplied by a monetary conversion factor to ascertain a forest's amenity value (Arboricultural Association, 1994).

LEGAL VALUES

Economic values of landscape plants are useful for a number of reasons; however, in most instances the appraisal of vegetation is done to ascertain economic values for legal purposes. These values are used for insurance claims, by the Internal Revenue Service, and in litigation.

TABLE 5–6 METHOD FOR PLACING AN AMENITY VALUE ON WOODLANDS (From Arboricultural Association, 1994.)

	Points[a]			
Factor	1	2	3	4
1. Size of woodland	very small	small	medium	large
2. Position in landscape	secluded	average	prominent	very prominent
3. Viewing population	few	average	many	very many
4. Presence of other trees and woodland	area more than 25% wooded	area 5–25% wooded	area 1–5% wooded	area less than 1% wooded
5. Composition and structure of the woodland	dense plantation or blatantly derelict woodland	even-aged pole-stage crops with mixed species	semi-mature or uneven-aged woodland with fairly large trees	mature or uneven-aged woodland with large trees
6. Compatibility	just acceptable	acceptable	good	excellent
7. Special factors	none	one	two	three

[a]Points are multiplied together and the total score is multiplied by an agreed-upon conversion factor (£30 as of 1994).

Insurance

Insurance policies for personal and business property will cover some loss or damage to vegetation, but this coverage is usually limited in extent. The Insurance Services Office of New York developed the Homeowners Broad Form Ed 7–77 that is used over much of the United States for homeowner insurance policies. Pertaining to landscape plants, the form states:

> Trees, shrubs, or lawns on the residence premises are covered for loss caused by the following perils; insured against fire or lightning, explosion, riot or civil commotion, aircraft, vehicles, not owned or operated by a resident of the resident premises, vandalism or malicious mischief or theft.

In other words, the loss must be a casualty loss due to an identifiable and sudden event. Homeowner policies typically exclude loss to insects or diseases, and losses due to wind. However, if a structure is damaged by a tree falling in high wind, the structure is covered, but not the tree. The extent of coverage in most states is limited to a maximum of $500 per plant, with the entire settlement not to exceed 5 percent of the appraised value of the property. Most insurance policies have a $100 deductible clause, so the homeowner assumes the first $100 of the loss. Homeowner policies do cover liability, and if a person's tree damages another's property, damages will be paid to the extent of liability limits of the tree owner's policy. A lawsuit may involve damage to another party's vegetation, and insurance companies pay what the courts direct them to pay. This amount is frequently based on an appraisal using the CTLA method, and sometimes includes payment for pain and suffering on the part of the plaintiff.

Internal Revenue Service

The Internal Revenue Service (IRS) recognizes the loss of landscape plants as a casualty loss that may be deducted from taxable income. Casualty loss is defined by the IRS as "the complete or partial destruction or loss of property from an identifiable event that is damaging to property, and is sudden, unexpected, or unusual in nature." A *sudden* event is swift in nature; an *unexpected* event is unanticipated; and an *unusual* event is extraordinary and nonrecurring. Casualty losses as they relate to landscape plants include damage by wind, flood, storm, fire, accident, vandalism, and pollution (provided that it is an unusual identifiable event). The loss of landscape plants to insect or disease attack may not be deducted if they are of a progressive nature. However, loss due to sudden, unexpected, or unusual infestations by insects may be deductible as a casualty los.

Personal property. Deductions may be made provided that each loss exceeds $100, and exceeds in aggregate 10 percent of the property owner's adjusted gross income for the year in which the loss occurs (Schedule A, Form 1040). The deduction is to be based on loss in fair market value, and this is to be determined by before and after appraisals of the property. The taxpayer must submit proof of loss in the form of appraisals, and it is recommended that photographs of the damage be taken.

Casualty deductions are determined by appraisal to determine loss in fair market value, less insurance recovery (if any), less the $100 limitation. Table 5–7 provides an example of a casualty deduction for tax purposes. If the damage is repaired, the cost of repair may be submitted as proof of loss provided that repair costs include only damage by the event, the property is restored to its condition just before the event, costs are not excessive, and repair does not make the property more valuable than it was immediately prior to the event. The fee for appraisal may be deducted under Miscellaneous Deductions as an expense in determining taxes on Schedule A (Form 1040), but not as a casualty loss.

Business property. The loss to property used for business is generally deductible regardless of the cause. The loss is determined by the appraisal of each item lost rather than on loss in fair market value, and this includes landscape plants. Gen-

TABLE 5–7 COMPUTATION OF CASUALTY LOSS FOR TWO SHADE TREES AS PRESCRIBED BY THE INTERNAL REVENUE SERVICE

Appraised value before casualty	$60,000
Appraised value after casualty	–55,000
Loss in value	5,000
Less insurance adjustment	– 1,000
Adjusted loss	4,000
Less $100 deductable	– 100
Casualty loss	$ 3,900

eral business properties are included in this category as well as nurseries, cemeteries, and public arboreta and gardens where admission is charged. Casualty losses for business property are determined based on the loss minus the amount of insurance settlement. The $100 limitation does not apply in the case of business properties (CTLA, 1992).

It should be remembered that the Internal Revenue Service is constantly revising tax rules based on legislation and rulings in tax court. The following are examples of rulings in the U.S. Tax Court as they apply to landscape plants:

1. *Nelson v. Commissioner,* 1968. A mass attack by southern pine beetles constituted a casualty, due to the suddenness of the event.
2. *Graham v. Commissioner,* 1960. Snow, ice, and sleet may be a casualty if of an unusually severe or violent nature.
3. *McKean v. Commissioner,* 1981. The loss of a century-old black oak to the two-lined chestnut borer constituted a casualty loss of $15,000, due to the suddenness of the event.
4. *Appleman v. Commissioner.* Casualty loss denied for phloem necrosis loss of an American elm due to the progressive deterioration of the tree.
5. *Miller v. Commissioner,* 1970. Casualty loss denied for the trees lost due to root suffocation caused by improper grading and spring-thaw weather, as these conditions were to be expected in the geographic area.

Litigation

In most instances litigation over trees and other vegetation involves common law or nonstatutory claims. The arborist or urban forester in these claims will be in court either as a defendant or as an expert witness in a lawsuit. The most common claims fall under the tort categories (civil wrongs in absence of a contract) of negligence, trespass, and conversion, or under breach of contract (DiSanto, 1982).

Torts. Negligence is damage resulting when a person fails to act reasonably or with care to another party. Three elements must exist before an award is made by the courts for negligence. First, a duty must exist on the part of the defendant to protect the plaintiff from injury; second, the defendant failed to perform such duty; and third, injury resulted from failure of the defendant (Borst, 1983).

Landowners are expected to exhibit a reasonable degree of care in regard to the safety of trees growing on their property. The courts have ruled that landowners are held to the duty of common prudence to prevent injury to their neighbor's person or property, even though they may not be aware of a hazardous condition. This means landowners are obligated to inspect their trees to determine if hazards exist (Merullo and Valentine, 1992). Rural landowners have historically been held to a lesser degree of responsibility regarding trees than urban landowners, but recent court decisions have increased the responsibility of rural landowners relative to the safety of trees along rural highways (Merullo, 1994).

Abutting property owners often come into conflict over trees growing on or near property lines. If the trunk of a tree is on the property line the tree belongs to both parties, and both parties must agree to any action that might affect the tree, including pruning and removal. Likewise both parties are responsible for any damages caused by the tree. A tree growing near a property line and extending over the line is the sole property of the landowner who owns the property upon which the trunk is located. However, the abutting property owner has the right to prune branches and roots that extend onto his or her property (Merullo and Valentine, 1992).

Litigation regarding trees often takes place when a person seeks compensation from another party for damage to, or loss of, a tree. There are four ways the courts determine the value of a tree for compensation (Merullo, 1994):

1. Depreciation of property value
2. Replacement costs
3. Value of the trees themselves
4. Aesthetic, comfort, or convenience loss to the owner

In most cases the loss is based either on the loss in market value of the property of appraisal of the individual plants using the CTLA method. Some states require punitive compensation of two to five times the appraised value for the lost trees and other plants.

Trespass is defined as the unauthorized entry on another's land, or the unauthorized use or interference with the property of another (DiSanto, 1982). Trespass involves not only physical entry by equipment and personnel, but includes nuisance complaints for pesticide drift, dust, and excess noise.

Conversion takes place when the personal property of another is appropriated (DiSanto, 1982). This could occur in a number of instances involving trees or other plants. Examples include the removal and use or sale of wood from a property when the owner expects it to be left behind, or removal of wood from the wrong property due to a faulty survey.

Breach of contract. Breach of contract occurs when one party fails to perform as agreed (DiSanto, 1982). This includes both services performed and payment for said services. A contract for services implies certain accepted standards for performance, and not meeting these standards constitutes a breach of contract.

Many states have statutes that pertain to tort or breach-of-contract cases, and provide additional penalties. If the action constitutes unfair or deceptive practices, a statute may prescribe additional penalties, such as doubling damages plus attorney's fees. Some states have special statutes dealing exclusively with ornamental, fruit, or timber trees, and these carry similar penalties (DiSanto, 1982).

Expert witness. Consulting arborists and urban foresters are often called upon to serve as expert witnesses during litigation. This usually involves appraisal of damage to vegetation, or giving testimony during trial proceedings. In court proceedings, a witness may only testify to the facts as he or she perceived them. The expert witness also testifies to the facts, but in addition may give opinions and draw inferences from the facts. Two criteria are necessary to serve as an expert witness. The expert witness must possess knowledge not held by the layperson and must have sufficient skill to be credible. The expert witness may not testify outside his or her area of expertise. The agreement to serve as an expert witness constitutes a contract and is subject to breach-of-contract proceedings should the expert witness fail to perform up to the standards of his or her profession.

Law and Municipal Forestry

Municipalities exist because state statutes permit their existence through issuance of state corporate charters. State governments have also passed laws that direct activities of municipalities, and these may be in the form of permissive statutes or mandatory obligations. In terms of urban trees, some states give communities permission to regulate public and private trees if they so choose, while other states require communities to so. This is important in litigation, for if the state law gives a community permission to care for public trees and the community chooses not to, the community is less likely to be found liable for damages caused by a tree on public property. On the other hand, if a state statute says that the community shall maintain public trees and it fails to do so, it is more likely that the community will be found liable if injury is caused by a public tree. However, in all cases, when a community accepts responsibility for public trees, it assumes liability for injury caused by the tree. State laws also require communities to maintain public rights-of-way free of obstructions and hazards, and this includes vegetation growing in or near the right-of-way. Even in the absence of statutes dealing specifically with public trees, communities must concern themselves with any vegetation growing in the right-of-way that may pose a public hazard (Borst, 1982).

Some states have "sovereign immunity" laws that prohibit bringing lawsuits against the state or municipalities. This immunity does not protect employees from being sued for alleged injury caused by them in performance of their duties. Many states have abolished sovereign immunity laws in recent years, and it is likely that more will do likewise in the future. However, it is important for municipal and state employees to understand that they may be held libel for injury to others in the performance of their responsibilities. Penalties are usually paid by the government, but it is the individual who is placed on trial.

Municipalities are most likely to be involved in tort action for negligence. For example, it is the duty of the municipality to exert reasonable care in keeping streets clear of hazards. If the city forester fails to notice a hazardous tree and an injury results, the city will be held liable provided that the plaintiff establishes that the city had a duty to notice the hazard and did not, or the city noticed the hazard and did nothing to correct it.

Many communities have ordinances concerned with the control of Dutch elm disease. Typically, these ordinances give the city forester permission to enter private property to inspect for the presence of the disease. However, entering private property to inspect for Dutch elm disease constitutes a search, and a search warrant is necessary if the property owner refuses to grant the forester admission without it (Borst, 1982).

Law and Utility Forestry

Utility companies manage vegetation that grows near or under their lines in the interest of reliable service and public safety. However, to prune or remove trees in the public right-of-way the company must have an agreement with the municipality. On private land utility companies need an easement to enter land for the purpose of constructing powerlines and clearing vegetation away from wires. The distance from the wires the company trims is based on safety standards relative to the voltage carried by the lines.

Utility companies are sometimes taken to court by property owners over cutting and trimming private trees. Merullo (1994), reviewing the liability of utility pruning, states:

> It would appear that certain steps must be followed to determine whether liability exists on the part of the utility company for the cutting or trimming of a landowner's tree. The first step is to determine whether the utility company even had the authority to trim or remove such trees. If such authority was present, the next determination to be made is whether or not an easement is present on the property owner's premises which would allow the public utility to enter upon the landowner's land. After finding that such an easement does, in fact, exist, the courts will then look to see if in the intended use of the land by the utility company, subject to the easement, is an added burden on the property owner's land. In making this determination, the courts will look to see if the proposed intrusion is unreasonable.

Law and Commercial Services

Commercial arborists and other private concerns that deal with urban vegetation are subject to various types of litigation, as described above. A frequent complaint brought against commercial arborists is for breach of contract. Any agreement to perform services, verbally or in writing, represents a legal contract. Standards exist for proper care of trees and other landscape plants, and improper care represents a breach of contract. DiSanto (1982) suggests the following defenses against a breach of contract claim.

1. No contract ever existed, so there was no violation.
2. The contract expired before the claim was made.
3. The terms of the contract were fulfilled.
4. The terms of the contract excluded the claim.
5. The performance expected was impracticable or impossible or the contract itself was unconscionable.

Negligence is another frequent complaint made against commercial arborists. Typically, this is for property damage or personal injury sustained in the performance of duties. DiSanto (1982) suggests the following basic defenses against a claim of negligence.

1. The claimant has named a defendant who is not responsible for the work or for actions that led to the injury.
2. The defendant is negligent, but the claimant is more negligent than the defendant.
3. The claimant knew the dangers either from express warning or from surrounding circumstances but assumed the risk of harm.
4. The defendant took all steps that a reasonably prudent person would take and either the harm occurred in spite of these efforts, or occurred in a manner that was not reasonably foreseeable.

LITERATURE CITED

ARBORICULTURAL ASSOCIATION. 1994. *Amenity Valuation of Trees and Woodlands.* Romsey, Hants, U.K.: Ampfield House.

BICKLEY, R. S. 1978. "Tree Value: A Case History." *Proc: Natl. Urban For. Conf.,* ESF Pub. 80–003. Syracuse: SUNY, pp. 192–193.

BORST, B. V. 1983 "Trees and the Law." *J. Arbor.* 8(10):271–276.

CHADWICK, L. C. 1980. "Review of Guide for Establishing Values of Trees and Other Plants." *J. Arbor.* 6(2):48–52.

COUNCIL OF TREE AND LANDSCAPE APPRAISERS, 1983. *Guide for Establishing Values for Trees and Other Plants.* Urbana, Ill.: Int. Soc. Arbor.

COUNCIL OF TREE AND LANDSCAPE APPRAISERS. 1992. *Guide for Plant Appraisal.* Savoy, Ill.: International Society of Arboriculture.

DARLING, J. S. 1989. "Landscape Waste: An Urban Problem or Community Resource?" *J. Arbor.* 15(8):198–200.

DEVOTO, D. F. 1979. "Paper from Municipal Trees." *Urban Waste Wood Utilization,* USDA-For. Serv., Gen. Tech. Rep. SE-16, pp. 52–61.

DISANTO, E. 1982. "Branches of the Law: Trees and Litigation." *J. Arbor.* 8(1):7–12.

DOLWIN, J. A., and C. L. GOSS. 1993. "Evaluation of Amenity Trees Mainly in Private Ownership within the Borough of Tunbridge Wells." *Arboric. Jour.* 17:301–308.

DRESSEL, K. 1963. "Street and Park Tree Evaluation." *Proc. Midwest Shade Tree Conf.* 18:105–112.

DWYER, J. F., H. W. SCHROEDER, J. J. LOUVIERE, and D. H. ANDERSON. 1989. "Urbanites' Willingness to Pay for Trees and Forests in Recreation Areas." *J. Arbor.* 15(10):247–252.

FOSTER, C. 1965. "Forestry in Megalopolis." *Proc. Soc. Am. For. Meet.,* pp. 65–67.

GORDON, B., and T. G. LAMBRIX. 1973. "The Battle of Lynn Woods." *Nat. Hist.* 82(9):76–81.

HELLIWELL, D. R. 1990. *Amenity Valuation of Trees and Woodlands.* Aboricultural Association. Ramsey, Hants, U.K.: Ampfield House.

HOLSCHER, C. E. 1971. "European Experience in Integrated Management of Urban and Suburban Woodlands." *Trees and Forests in an Urbanizing Environment,* Plann. Res. Dev. Series No. 17. Amherst: Univ. Mass., pp. 133–142.

HULL, R. B., IV. 1992. "How the Public Values Urban Forests." *J. Arbor.* 18(2):98–101.

KIELBASO, J. J. 1974. "Economic Values of Urban Trees." *Proc. Urban For. Conf.* Stevens Point: Univ. Wis., pp. 30–52.

LARUE, G.. 1982. *Utilization of the Urban Forestry Conf.* Washington, D.C.: Am. For. Assoc., pp. 179–181.

LASH, J. 1977. *The Cost of Preserving or Establishing Trees in Housing Development,* unpublished report. Amherst, Mass.: USDA-For. Serv.

LOGGENS, T. J. 1978. "Urban Waste Wood: Debit or Credit." *Proc. Natl. Urban For. Conf.,* ESF Pub. 80–003. Syracuse: SUNY, pp. 194–204.

MARTIN, C. W., R. C. MAGGIO, and D. N. APPEL. 1989. "The Contributory Value of Trees to Residential Property in the Austin, Texas Metropolitan Area." *J. Arbor.* 15(3):72–76.

MCPHERSON, E. F. 1994. "Benefits and Costs of Tree Planting and Care in Chicago." In E. G. McPherson, D. J. Nowak, and R. A. Rowntree, eds. *Chicago's Urban Forest Ecosystem: Results of the Chicago Urban Forest Climate Project.* Gen. Tech. Rep. NE-186. Radnor, Pa.: USDA Forest Service.

MERULLO, V. D. 1994. "Common Law Branches off into New Directions." *J. Arbor.* 20(6):341–343.

MERULLO, V. D., and M. J. VALENTINE. 1992. *Arboriculture and the Law.* Savoy, Ill.: International Society of Arboriculture.

MILLER, R. W., R. S. BOND, and B. R. PAYNE. 1978. "Land and Timber Values in an Urban Region." *J. For.* 76(3):165–166.

MORALES, D. J. 1980. "The Contribution of Trees to Residential Property Values." *J. Arbor.* 6(11):305–308.

MORALES, D. J., et al. 1983. "Two Methods of Valuating Trees on Residential Sites." *J. Arbor.* 9(1):21–24.

MURPHY, W. K., et al. 1980. "Converting Urban Tree Residue to Energy." *J. Arbor.* 6(4):85–88.

NOWAK, D. J. 1993. "Compensatory Value of an Urban Forest: An Application of the Tree-Value Formula." *J. Arbor.* 19(3):173–177.

PAYNE, B. R., and S. STROM. 1975. "The Contribution of Trees to the Appraised Value of Unimproved Residential Land." *Valuation* 22(2):36–45.

SEILA, A. F., and L. M. ANDERSON, 1984. *Estimation of Tree Preservation Costs.* Georgia For. Res. Paper No. 48, p. 7.

STRONG, A. 1982. "Making Urban Forestry a Part of the County and City Planning Process." Unpublished paper. *Sec. Natl. Urban For. Conf.*

WALKER, D. 1979. "Mulch from Limb and Trunk Debris." *Urban Waste Wood Utilization,* USDA-For. Serv. Gen., Tech. Rep: SE-16, pp. 62–65.

6

Street Tree Inventories

Municipal foresters have a number of responsibilities associated with the care and management of urban vegetation, but their primary responsibility in most communities is the care of city-owned street trees. A street tree is defined as the publicly owned tree growing in the public right-of-way, usually between the sidewalk and the curb. Exceptions to this are trees growing in boulevard medians, trees planted on streets without sidewalks and curbs, and trees planted on private property by special easement on streets with limited public right-of-way available for tree planting. Kielbaso (1989) reported in the results of a national survey of cities in the United States that city foresters spend an average of 61 percent of their budgets on the management of trees. The same survey found only 38 percent of 2787 respondents knew with any degree of accuracy the number of street trees under their care. A more recent survey found that by 1994, 78 percent of responding municipalities had spent money on inventories (Tschantz and Sacamano, 1994). This increase is primarily the result of federal and state community forestry assistance grants made available in the 1990s.

THE NEED FOR INVENTORIES

Management of any resource begins with an inventory of that resource, and urban forest management is no exception. Bassett (1978) feels that inventories are essential to provide a current record of resources being managed; to plan, schedule, and monitor maintenance tasks; and to assist in making management decisions, particularly when developing budgets. A street tree inventory need not be complex or expansive in parameters measured, but should at least provide some minimal level of information that will allow the manager to make intelligent management decisions.

The primary objectives of a good street management plan should be to maximize public benefits from street trees and to minimize public expense in achieving these benefits. Street tree management involves three primary functions: planting, maintenance, and removal. Planting includes the identification of a site in need of planting, selecting and planting the appropriate species on that site, and may include post-planting activities such as watering and fertilization. Pruning is the primary maintenance activity, while cabling and bracing, insect and disease control, damage repair, and watering and fertilization may also be done. Ultimately, all living organisms die, and for street trees this involves prompt removal before a hazardous situation develops. Scheduling of these various management activities should be on a priority basis to maximize community benefits and minimize public inconvenience or hazards. An inventory is essential to locate planting sites, identify management needs, and locate hazardous trees in need of repair or removal.

There are other reasons to conduct street tree inventory systems. Gerhold et al. (1987) suggest that a tree inventory can be used in public relations as a news item describing the public tree resource both in terms of the number and the value of trees. Smiley and Baker (1988) report that inventories can be used to increase work efficiency, and recommend that inventory information pertaining to the tree resource be used to educate the public and policy makers as to the value of community trees.

Inventory needs will vary with the size of the community, level of service desired, and potential vegetation problems. Some forestry programs will function well with simple estimates of a few tree parameters, while other communities will need sophisticated inventory systems that require the use of a computer. Each community is different and has different public tree management needs. The most effective inventory system is one that is tailored or adaptable to the specific needs of a community.

Existing community forestry programs function on some form of information descriptive of the street tree population. This information might be in the form of knowledge held by the city forester or other employees, based on periodic samples or surveys, or obtained from computer summaries that are updated as work is performed. As long as the information descriptive of the tree population is adequate to meet management goals and objectives, the method of obtaining this information need not be altered.

City foresters in communities initiating new forestry programs need to set preliminary management goals and objectives and then develop an information system to provide them with the necessary data to meet these goals and objectives. This system should be of a progressive nature, designed to provide inexpensive preliminary information in order to identify management priorities, including how much tree information will be needed to manage the tree population.

Ziesemer (1978) advises city foresters to ask themselves if an inventory is needed before deciding on implementation, and recommends a preliminary program analysis. His program analysis recommendations, with modifications by the author, follow.

1. How much work is currently needed?
2. Are there a large number of planting spaces available, or are the streets close to full stocking?
3. Is pruning done on scheduled basis, or by request?
4. How are tree and stump removal decisions reached?
5. Are priorities set based on tree needs or other considerations?
6. How is work scheduled and assigned?
7. How accurately are existing needs reflected in current priorities?
8. Do estimates of the volume of work correspond with actual needs of the tree population, or are there discrepancies?
9. Is the public satisfied with the level of service, or are there numerous complaints?
10. Is service performed routinely or by request; if by request, how long does it take to respond?

Once these questions have been answered, a city forester should have some idea as to the level of service provided in reference to management goals. If there are discrepancies, managers are receiving inaccurate or inadequate information on which to base decisions. These discrepancies indicate a need to improve the information system, but may not point to the need to overhaul the entire system. Ziesemer (1978) suggests that managers objectively analyze their program by flowchart to determine what information is needed for decision making at all levels. (Figure 6–1), and for determining the following information for each position level within the organization:

1. The function performed at each position
2. The decisions that must be made at each position to enable successful completion of each of those functions
3. The information required at each position to enable making those decisions rationally (Figure 6–2)

Once this list of information is obtained, tree information may be segregated from the list, and this is what needs to be known about the tree population to perform the various functions of the forestry department. This list may then be compared with the existing information system to determine if it is providing what managers need to perform their respective tasks. The following are typical outcomes from this process.

1. The existing information system is meeting the needs of the program; therefore, no changes are needed.

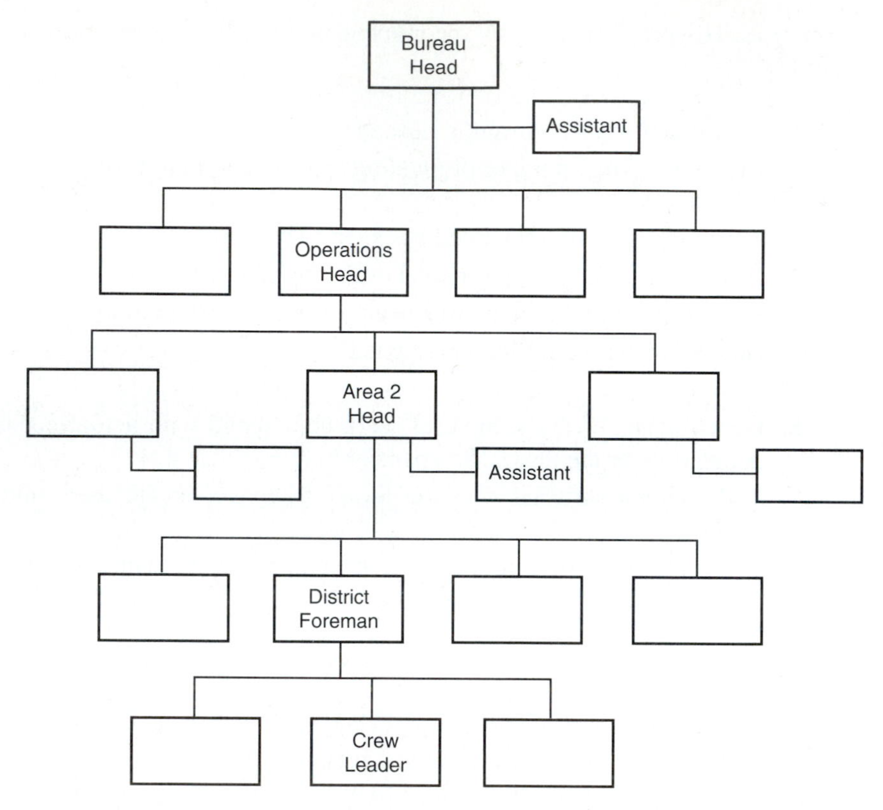

Figure 6–1 Sample municipal forestry organization. (From Ziesemer, 1978.)

2. The existing information system is partially inadequate and needs some overhauling.
3. The existing system is wholly inadequate and a new system needs to be implemented.

OBJECTIVES AND COMPONENTS OF INVENTORIES

Once the need for an inventory has been established, the next task is that of implementation. Selecting or designing an inventory system should be undertaken with the overall objective of providing essential information to assist managers in decision making. The analysis technique described above provides a guide to what information is needed by whom, when, and why (Ziesemer, 1978). The inventory should then be designed to provide this information on a timely basis at a reasonable cost.

Before entering into a discussion of what information should be collected in an inventory, it is important to distinguish between transitory information and informa-

POSITION ANALYSIS
(Bureau Head)

Functions	Decisions	Information
Directs development and initiation of programs	Are trees' physical needs being met?	Size, species, condition, and location of trees
Establishes program priorities	Do insects/diseases warrant control?	Location and number of planting sites
	Are ongoing programs on schedule?	Status of ongoing programs
	Are service requests handled promptly?	Number of requests for service, by category
		Turnover of requests for service, by category
Maintains awareness of status of all programs		
Maintains awareness of condition of trees		
Directs preparation of budget		
Etc.		

POSITION ANALYSIS
(Crew Leader)

Function	Decisions	Information
Schedule workers and equipment	Whether crew is at correct address	Addresses of jobs for day
	Whether work is done on correct tree	Identification of trees needing work at each address: size species, position
	Whether crew is doing correct work	Work to be done on each tree
	When to move to next job site	
	When to alter schedule	
Assure that crew does all work in accordance with established standards		
Enforce bureau rules and regulations		
Complete records and reports		

Figure 6–2 Partial three-step analysis of bureau head and crew leader positions to determined information needs from a street tree inventory. (From Ziesemer, 1978.)

tion of a more permanent nature. For example, identification of tree species provides information of a more durable nature than describing pruning needs of a particular tree. The species will remain unchanged as long as the tree is there, but pruning needs will change with the next scheduled maintenance. This distinction is related to the type of inventory system adopted. If an inventory is to be of an occasional or periodic nature, it might be helpful to ascertain transient information, such as where pruning needs are greatest, in order to schedule maintenance activities. If the inventory is to be used to keep maintenance and tree records and is to be continuously updated, the date and description of maintenance may be more important. The following discussion of inventory parameters and records is meant to be comprehensive in order to provide a list for the city forester to choose from when designing or selecting an inventory system, with the assumption that more sophisticated inventory systems will maintain more extensive records. These parameters fall into four general categories: time and personnel, site descriptors, tree descriptors, and maintenance records.

Time and Personnel

Time and personnel records include the date the record was made, who made the record, and who performed the maintenance activity. A record of the date may be used when reconstructing tree history or when evaluating a species or cultivar as to its longevity and/or adaptability to a particular set of site conditions. The date is also important to verify maintenance and response to complaints should damage caused by a tree be the subject of litigation. Identification of the people who made the record and the crews that performed the maintenance or service request may be used to verify maintenance, evaluate personnel performance, and evaluate contractual services.

Site Descriptors

Site descriptors include the location of the tree or planting site, width of the tree lawn (planting strip between the sidewalk and curb), presence and height of overhead wires or other utilities, sidewalk damage, soils, irrigation needs, land use, and pollution problems. The presence or absence of a tree, recommended species if the site is vacant, and planting priority may also be included as site descriptors.

Site location may be described by a number of variables. Sacksteder and Gerhold (1979) provide the following list of common variables used independently or in combination when describing a street tree location:

Area, division, or section number
Street name
Address, house number, or lot number
Block number
Distance from an intersection
Number of nearest utility pole

Sequence number, within addresses or within blocks
Coordinates of various kinds
Distances from the street or curb
Tag number
Map number
Geographic information systems

Many communities subdivide the street tree population into smaller management units along neighborhood, political district, or quarter section boundary lines (160 acres based on the U.S. Public Land Survey; Figure 6–3). Some communities use the management unit as the tree location, organizing information based on summaries of field count surveys. Communities with computer inventories often subdivide street segments into work units for data access and to generate data summaries by work unit.

The most common methods of tree location are based on block and sequential number, or address and sequential number. The main drawback with using an address system is how to inventory trees on corner lots where the address is on a different street. Various methods developed to solve this problem include assigning the tree to the next address, or assigning the tree to the street address around the corner (Figure 6–4). Some computerized inventory systems maintain a corner file by work unit to assist in locating an individual tree on a corner property for either street.

Some computerized tree inventory systems generate maps of tree locations. Many communities now maintain information describing the city in geographic information systems (GIS), a mapping system consisting of "layers" of information such as streets, publicly owned utilities, land use, census information, and so on. Each layer can generate an independent map, or layers combined to produce more complex maps. Tree location and information can be placed in a GIS system as a separate layer, or as part of an existing layer such as one describing city-owned infrastructure. Tree location information may be entered into a GIS file by street address or distance from a corner, from air photos, or using the global positioning system (GPS). The GPS uses a handheld unit that locates a point on the ground via a satellite system, and this information can be readily transferred to GIS.

Width of the tree lawn, height of overhead wires, and the presence and location of other utilities provide descriptive information of structural constraints for each potential or actual tree location (Figure 6–5). These data may be important when selecting a particular species for a given site, especially with reference to the ultimate size and growth rate. Maintenance costs may be estimated or anticipated from these data, such as expected pruning costs based on the height of wires and species present, or expected sidewalk and curb repairs based on the width of the tree lawn and growth rate and root habit of the tree present.

Sidewalk damage is information of a transitory nature, but may be important information to collect when conducting a tree inventory. These data may be correlated with species to determine which species are causing high sidewalk-maintenance

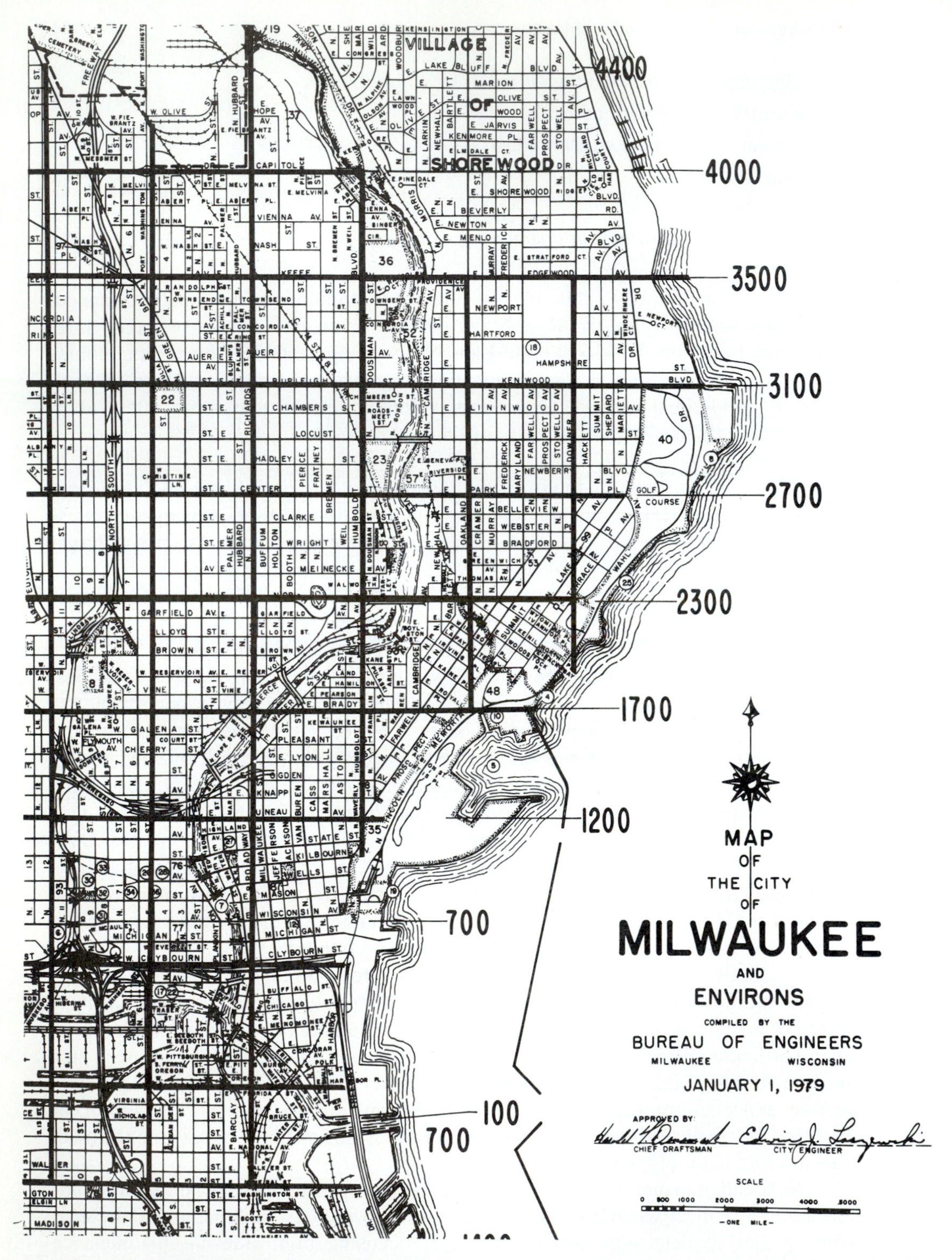

Figure 6–3 A portion of the city of Milwaukee, Wisconsin, subdivided into quarter-section urban forestry work units.

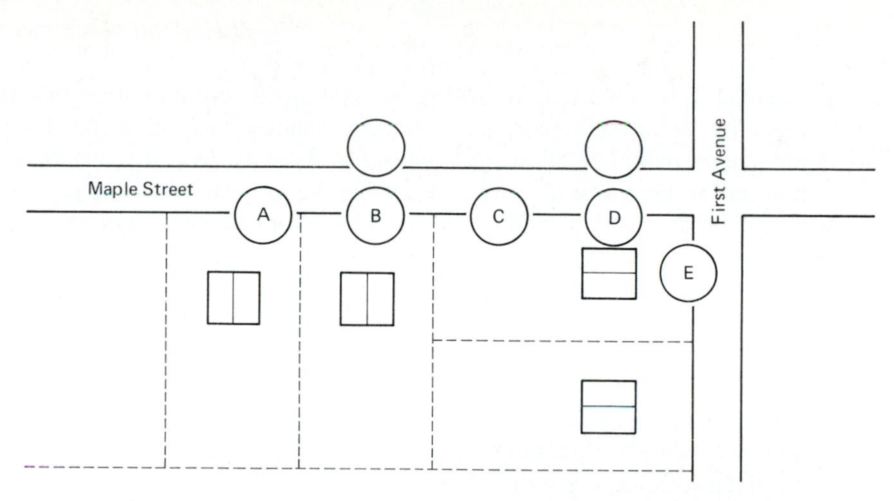

Corner lot trees assigned to adjacent address

Tree	Address	Tree Number
A	2142 Maple Street	1
B	2144 Maple Street	1
C	2144 Maple Street	2
D	2144 Maple Street	3
E	3601 First Avenue	1

Corner lot trees assigned to corner address

Address	Tree Number
2142 Maple Street	1
2144 Maple Street	1
3601 First Avenue	3
3601 First Avenue	2
3601 First Avenue	1

Figure 6–4 Two methods of assigning street addresses to street trees.

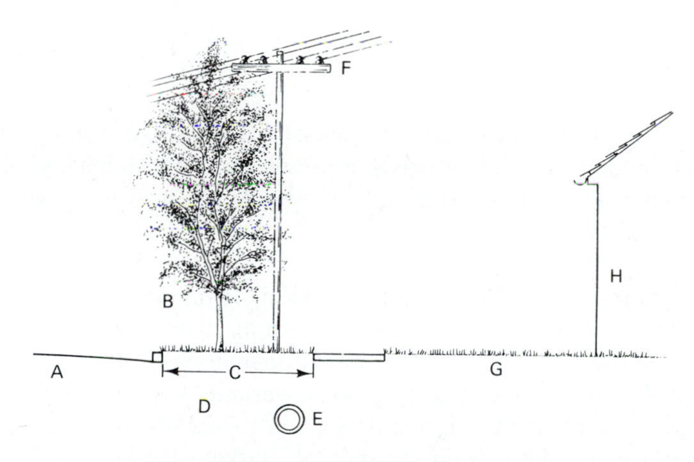

A Street width or traffic load
B Pollution problems (i.e., deicing salt, O_3, etc.)
C Width of tree lawn
D Soil type
E Underground utilities
F Presence and height of overhead wires
G Land use
H Structural constraints

Figure 6–5 Site descriptors that may be important in a street tree inventory.

problems. It is important to recognize that many communities bill the property owners for sidewalk repair, and excessive damage to sidewalks by public trees could cause public relations problems for the city forestry department. In most communities, however, the city engineering department will conduct periodic inventories of damage to sidewalks, and the tree inventory crew will not need to collect these data.

Soil data may be important to collect in some communities, especially if there are highly diverse conditions, or if an extensive planting program is to be undertaken. Along with soil data, irrigation needs may be identified in regions subject to drought problems or in parts of the city where soil or other site conditions dictate the need for irrigation. Land-use data can be useful when assigning management priorities such as where to concentrate planting, or when selecting the appropriate species or cultivar for a given location. Pollution data will assist in species selection for problem areas where chronic air pollution exists, or where heavy use of de-icing salt causes salinity problems in the soil.

An efficient means of determining planting needs and species recommendations is during the inventory. Many communities have tree ordinances that specify spacing between street trees, and these standards can be used to determine the location of vacant tree spaces. When a vacant tree site is encountered it should be identified as a potential tree site for future planting, and a species recommended for that site. Some sites may have a higher planting priority than others, and these priorities can be assigned during the inventory. For example, competing vegetation on private or public property may cause shading problems for a potential tree site, and such sites should be given a low-priority rating when considering planting needs.

Tree Descriptors

Tree descriptors generally fall into four categories: species or cultivar, size, condition, and management needs. Since management needs are transitory and ultimately result in maintenance activities, they will be discussed in the maintenance records section.

Tree species. Tree species may be identified by common or scientific name, and may be further described to the cultivar level. There is much discussion as to the desirability of using either scientific or common names in tree inventories. Scientific nomenclature will remove problems associated with regional variations in common names, but crew leaders and other personnel may be reluctant to use them. Since most inventories use some sort of coding system for species, Barker (1983) developed standardized mnemonic (memory-assisted) codes based on scientific names and suggests that use of these codes will eliminate the confusion sometimes caused by common names, and will allow easy merger of tree data files from various inventories to assess the performance of various species and cultivars under field conditions. These are sound arguments and should be considered when implementing an inventory system, but ultimately the decision must be made based on the city forester's understanding of local needs and conditions.

Tree size. Diameter, height, crown spread, and biomass are the parameters most used to describe the size of street trees in inventories. The first three parameters are measured directly, with biomass estimated based on diameter or diameter and height measurements. Tree diameter is the most frequently used measurement as management costs are often associated with tree diameter (i.e., pruning and removal bids are usually based on tree diameter). Diameter is measured in inches (cm) at 4.5 ft (1.4 m) above the ground on large trees, but may be measured lower on smaller trees. The Council of Tree and Landscape Appraisers recommends that trees under 12 in. (30 cm) in diameter be measured at 12 in. (30 cm) above the ground and trees less than 4 in. (10 cm) measured at 6 in. (15 cm) above the ground when making tree appraisals (see Chapter 5). Diameter classes are frequently used in inventory systems, ranging from 1-in. classes to grouping the entire range of tree diameters into three or four classes. Selection of the appropriate diameter class range will depend on the management needs of agency using the data. Tree diameter measurements may be made by diameter tape, ocular estimation, calipers, and cruiser sticks (Figure 6–6). Existing inventories rely primarily on tree diameter for the size measurement, with other measurements being made to address particular management problems.

Tree height is sometimes measured, particularly when the tree is growing beneath utility wires. In this instance, height measurement may be a means of estimating future pruning needs, but in most cases line clearance is maintained by the appropriate utility rather than the city. When used, this measurement is estimated from the ground ocularly, or by using an instrument such as a clinometer, Abney level, altimeter, or hypsometer (Figure 6–7). As in diameter measurements, tree heights are recorded in height class intervals.

Measurement of crown spread may be made if there is an interest in estimating the amount of canopy cover in the community. This measurement is usually made by pacing the crown radius from the trunk in one or more directions. Crown spread data appear to have marginal value, and if an estimate of tree canopy is desired, it may be more efficient to conduct a tree canopy analysis using aerial photographs, a technique described in Chapter 7.

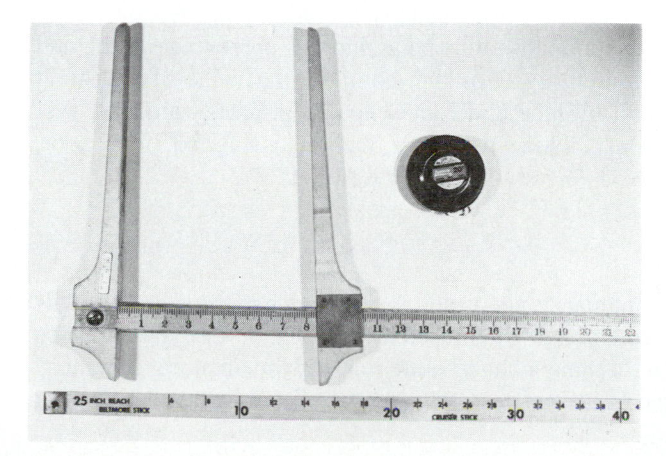

Figure 6–6 Instruments used to measure tree diameters. Bottom to top: 25-in.-reach Biltmore stick, caliper, and diameter tape.

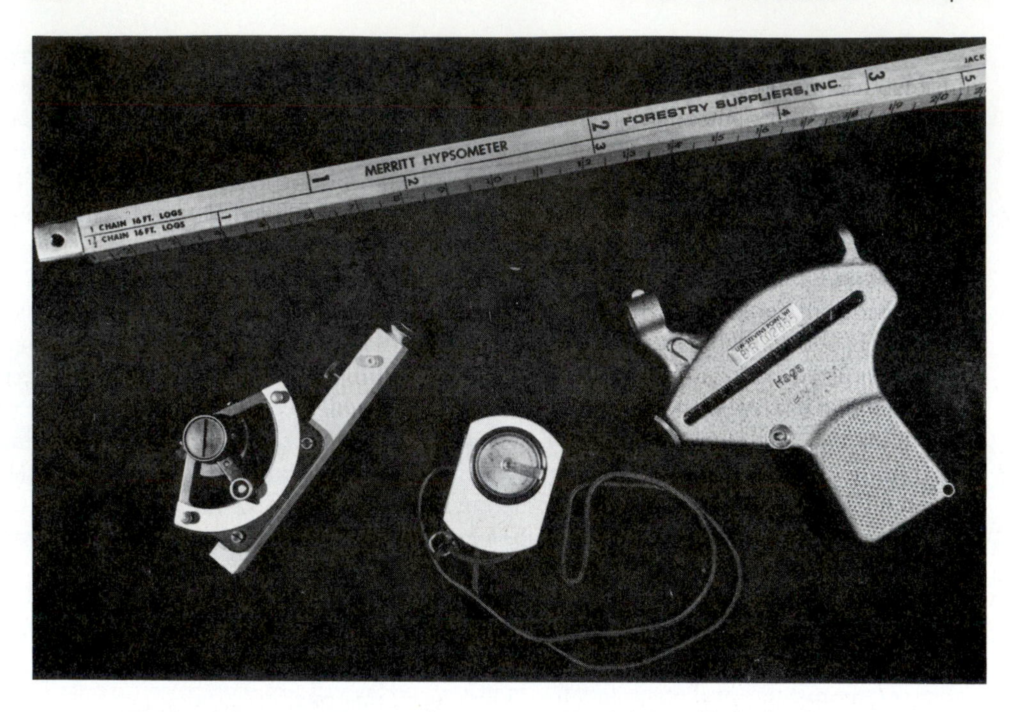

Figure 6–7 Instruments used to measure tree heights. Clockwise from top: Merritt hypsometer, altimeter, clinometer, and Abney level.

Tree diameter or tree diameter and tree height may be used to determine tree biomass. Biomass estimations are important when attempting to determine the volume of waste wood that will need to be disposed of, or when a market exists for wood products. These products can vary from chips for energy and mulch to an occasional veneer-quality log. Volume estimates based on diameter and height may be made from standard volume tables available from the appropriate state forestry agency.

Tree condition. Tree condition values usually include some description of tree vigor, health, or damage. Tree condition may be assigned a percentage rating and used as a condition class in inventory systems that compute the value of the street tree population using the Council of Tree and Landscape Appraisers valuation system. Other condition descriptions include health or vigor class (good, fair, poor, dead), injury (insects, disease, mechanical), or structural soundness.

Maintenance Records

Tree descriptors in terms of maintenance needs and maintenance records are two different ways of looking at the same thing. When used as tree descriptors, maintenance needs are transient information that may change quite readily, thus making inventory data obsolete quickly. This type of information should be used only when conducting periodic inventories to ascertain specific management needs. On the other hand, when using an inventory system that is continuously updated, then recording the maintenance activity as a permanent record with the inventory may provide important information about an individual tree, species or cultivar, or site conditions. In general, tree maintenance records will include date and sometimes cost or time information for planting, fertilization, watering, pruning, cabling and bracing, insect and disease control, service request, complaints, and ultimately, removal.

When maintenance needs are recorded as transient information in periodic inventories, some estimation of when and how much work is needed should be provided. A tree may be described as needing pruning immediately, in the near future, or in the distant future. The need to repair damage could be described in a similar fashion, as might recommendations for removal. Insect or disease problems may be identified and appropriate controls recommended.

Updating

An inventory may be periodic, or updated as part of the regular work routine. A periodic inventory consists of a reinventory of either part of or the entire community. Reinventory of a portion of the community usually takes place on an annual basis, with each management unit being reinventoried at a predetermined interval. This approach allows inventory to be a regular item in the budget, and maintains an even annual expenditure. However, at any given point in time a portion of the inventory will be as old as the inventory cycle. Periodic inventory of the entire city will provide current information on the entire tree population on a periodic basis, but the information will become progressively obsolete until the next inventory. This method will also put a high periodic expense in the department budget, and such items are often subject to elimination in budget hearings.

Inventories updated on a continual basis will provide the most current information about the tree population, but this assumes that all work is done by city crews and crew leaders are trained to record accurate data reflecting the status of each tree serviced. This also assumes that office staff are trained and have time to update inventory files. Continuous updating becomes a part of the regular work routine and inventory is a part of the annual budget. As a budget item, inventory is then an expense to maintain with an existing clerical system, while data collection becomes part of tree maintenance. However, when first adopting such a system it should be understood that startup costs will be high and must include training so that the personnel will use the system accurately and efficiently.

TYPES OF INVENTORIES

The foregoing discussion of parameters that may be included in a street inventory is intended to provide a menu for the city forester to select from in developing or adapting a system to local needs. Collecting marginally useful or useless information is a waste of time and money on the part of forestry department personnel. By following procedures outlined earlier in this chapter a manager may ascertain with a degree of confidence those parameters essential to manage the street tree population effectively, and how to go about collecting data describing these parameters. According to Ziesemer (1978), a street tree inventory consists of three phases: planning, implementation, and operational. Planning involves selecting those population attributes necessary for management, and deciding how to obtain data descriptive of those attributes. Implementation is the initial collection and reduction of data by either city employees or consultants. The inventory becomes operational when the information gathered and reduced is used to assist in planning and management activities, and when inventory maintenance and updating become part of the overall street tree management program.

The types of inventory systems available fall into four general categories: street tree surveys, sampling, tree data files, and computer inventories. Sacksteder and Gerhold (1979) define a street tree inventory system as "a method of obtaining data about urban trees and organizing it into usable information." They further distinguish between data and information, pointing out that data consist of individual observations that may not be very useful until aggregated as totals, averages, percentages, graphs, or cross tabulations to provide management information. The following discussions will describe each general type of inventory in terms of their use and capability to provide useful information through data reduction.

Street Tree Surveys

A street tree survey is a count of trees in a community that can provide a surprising amount of useful information. The survey can be citywide in a small community, or by work unit in larger cities. Figure 6–8 illustrates a tally sheet designed to provide information about species, size classes by species, condition by species and size class, and the number of available planting spaces. Species diversity can be determined, diameter distributions summed across species, average diameters calculated for individual species, and species recommendations made for planting. Numeric values can be assigned to, or be used in place of, condition class—such as dead = 0, poor = .25, fair = .5, and good = .75. Average conditions can be calculated to compare species performance, and the value of the street population calculated using condition class in the Council of Tree and Landscape Appraisers formula. If collected by work unit, work units can be compared and management priorities set by work unit. Street tree surveys can be collected as a separate task, or can be collected by city crews as they perform tree maintenance in individual work units.

The Community Forestry Program in the state of Kansas is designed to assist small rural communities with limited budgets. A state-employed forester aids these communities in creating a tree board, developing a street tree ordinance, and in conducting a windshield survey of the tree population. Minimal data that are required by the survey must provide summaries of the number of trees by species, diameter, tree condition, and planting and removal needs. All trees are tallied in communities with

INVENTORY DATA SHEET

Crew Names Smith/Jones: Work Unit 3

Date October 16

	DIAMETER				CONDITION				VACANT
Species	0–6	7–12	13–18	19 +	*Good	Fair	Poor	Dead	Space*
Live Oak	☒ ☒	☒ ☒	☒ ⁛	☒ ·	☒ ☒	☒ ☒ ⁘	☒ :	☐	☒ ⁛
Cypress									
Chinese Elm									
Red Maple									

64 Live Oaks

20 good, 24 fair, 12 poor, 8 dead; can plant 13 more in work unit 3. Average diameter is 9".

**Determined by tree spacing standards*

Figure 6–8 Street tree tally sheet with data for one species. (From Miller, 1994.)

fewer than 2500 people, while every other street is surveyed in communities with populations between 2500 and 10,000. Larger cities usually survey between 25 and 50 percent of their streets (Atchison, 1978). The average cost of the survey is about $0.04 per tree, with summary data costing about $50 per town (Sacksteder and Gerhold, 1979).

The initial inventory is conducted by three to five people from a moving automobile. Inventory personnel are trained by the forester to estimate parameters used in the inventory. Usually, the driver reports tree data from the left side of the street and a person in the right front seat reports data from his or her side of the street. One or two people sit in the back and tally data as reported. Additional personnel may accompany the inventory crew to record specialized data and to record progress on a city street map. Once the data are gathered, the forester completes the process by reducing them to a usable form. Since an estimate of tree condition is made during the inventory, the value of the street tree population may be estimated using the Council of Tree and Landscape Appraisers (1992) method (Atchison, 1978). It should be cautioned that many inventory systems are used to compute tree value, and these values should only be used as an estimate of the overall street tree population value and are not to be used as an appraised value of an individual tree. As discussed in Chapter 5, tree appraisal is a difficult task and should be undertaken only by an experienced arborist. Table 6–1 provides a summary of the street tree population in Waterville, Kansas, following a windshield survey.

Sample Survey

Sampling as an inventory procedure has been applied to rural forest lands for many years, but only recently has the method been used to describe street tree populations. As a means of data reduction, sampling by definition provides estimated summaries and means of parameters measured. Geiger (1977) describes a system developed for the city of Chicago designed to estimate wood volume for potential marketing. At the time of inventory, removals averaged one tree per city block, for a total of 28,000 trees per year. City forestry officials faced with disposal problems of this magnitude elected to attempt to market the wood fiber and needed an estimate of potential volumes. The inventory system assumed a homogeneous population and that distinct boundaries could be defined. Three management areas were defined, and sampling was random, using subgrids of 10 acres each (two city blocks or their equivalent) as depicted in Figure 6–9. Only treed subgrids were measured (defined as containing a minimum of three trees each); thus the city was found to be 67 percent "forested." Geiger (1977) concluded that reliable estimates could be made of wood fiber in Chicago based on 172 field plots throughout the city with an overall sampling error of 5 percent at 1 standard deviation, provided that only treed subgrids were measured. Additionally, this system provided some information as to the general condition of the street tree population (Figure 6–10) and species desirability distribution.

Valentine et al. (1978) and Mohai et al. (1976) developed a street tree inventory system based on sampling parameters that have more important management implications. They recommend a combination of cluster and systematic sampling using

TABLE 6–1 SUMMARY TABLE FOR STREET TREES INVENTORIED USING THE WINDSHIELD SURVEY TECHNIQUE (From Atchison, 1978.)

Species	Number of Trees	Average Age	Average Diameter (in.)	Percent of Trees That Are: Good	Fair	Poor	Dead or Dying	Percent of Total Trees	Value
Hackberry	403	80	16	77	15	8	0	30	$306,280
American elm	244	80	18	56	23	11	10	18	141,240
Chinese elm	167	40	13	53	30	16	1	12	47,762
Silver maple	124	10	4	87	7	3	3	9	6,696
Green ash	65	40	12	65	22	13	0	5	37,050
Hard maple	38	15	4	97	3	0	0	0	4,294
Lombardy poplar	35	15	10	97	0	0	3	3	4,318
Black walnut	34	30	8	50	26	20	4	2	6,154
Honeylocust	24	30	9	79	15	6	0	2	10,880
Hybrid elm	28	15	5	100	0	0	0	2	1,960
Fruit	25	10	4	76	12	12	0	2	1,975
Sycamore	21	25	8	81	10	9	0	2	4,557
Pin oak	18	35	10	83	11	6	0	1	10,170
Pine	14	60	12	64	36	0	0	1	6,846
Red cedar	14	60	10	36	21	43	0	1	3,948
Cottonwood	15	5	3	93	0	7	0	1	450
Mulberry	12	60	17	25	33	42	0	1	3,924
Miscellaneous[a]	67							5	20,227
Total	1362								$618,731

[a]Redbud, apricot, Russian olive, Kentucky coffeetree, English oak, bur oak, basswood, catalpa, flowering crab, persimmon, silver poplar, red oak, buckeye, hickory, goldenrain, sweetgum, tree-of-heaven, black locust, willow.

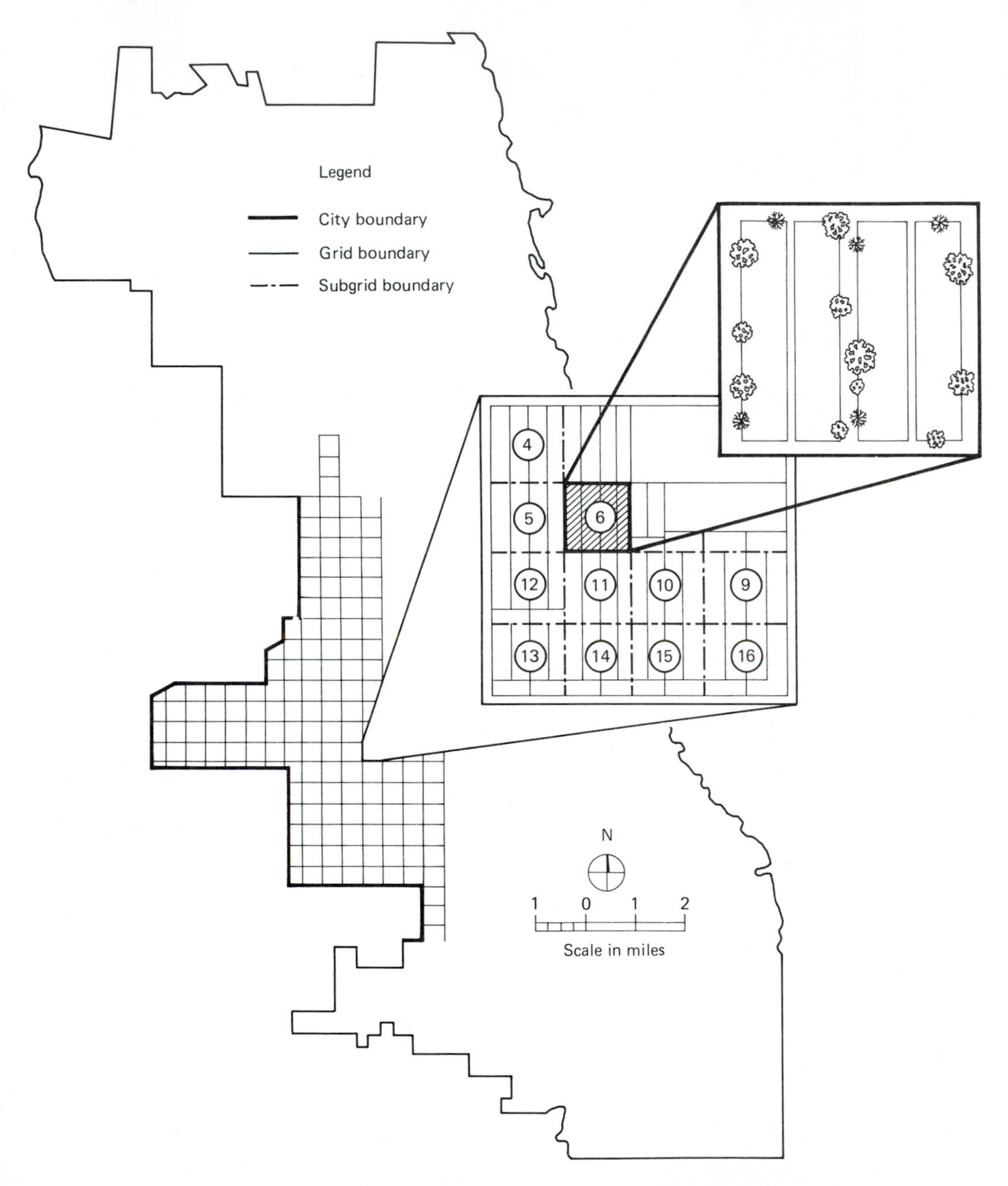

Figure 6–9 Subgrid sampling system used in Chicago, Illinois, to inventory the street tree population. (From Geiger, 1977.)

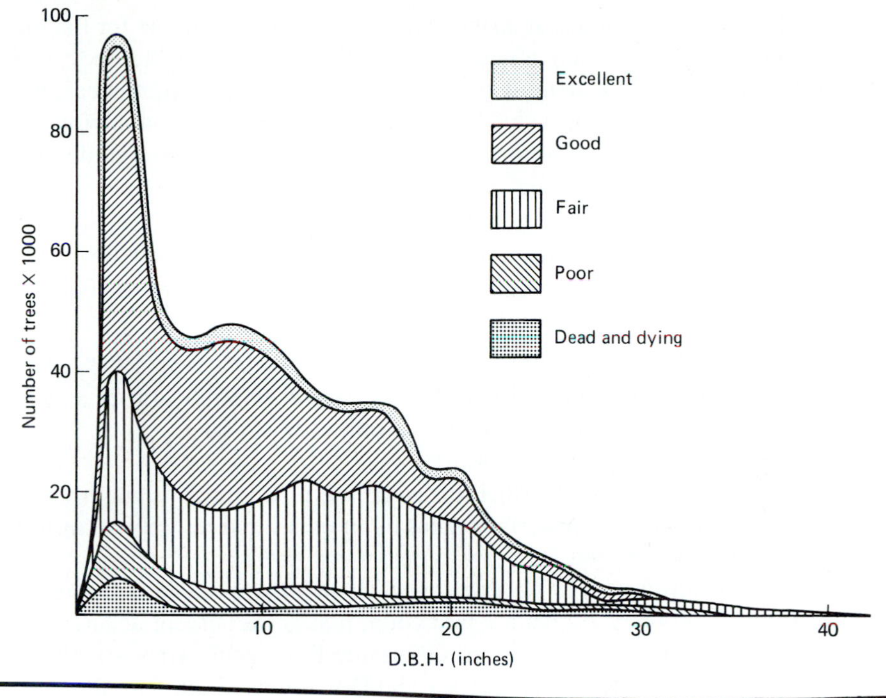

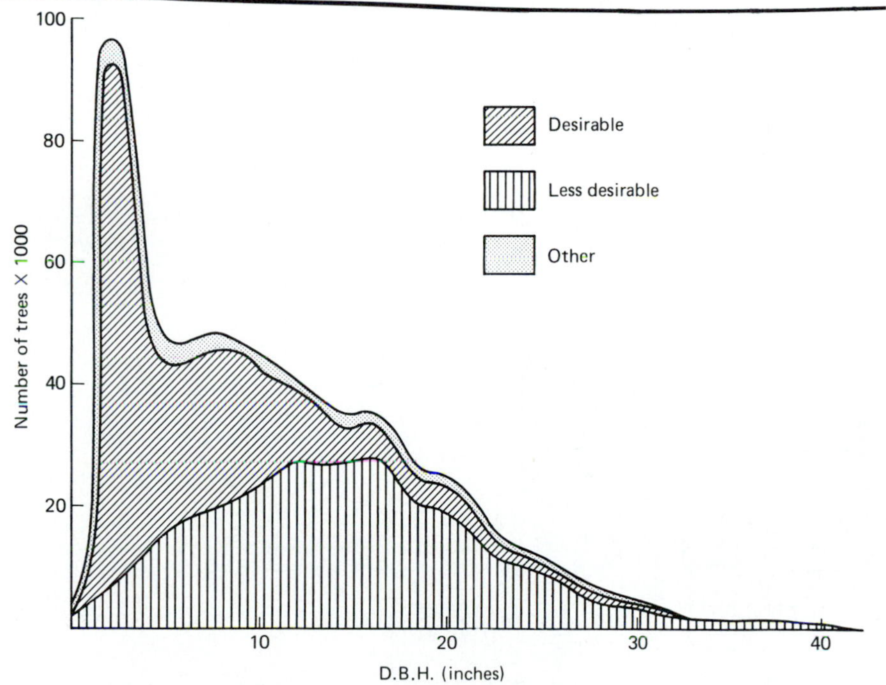

Figure 6–10 Condition and species desirability graphs generated from Chicago, Illinois, street tree inventory data. (From Geiger, 1977.)

clusters of streets and systematic samples on selected streets for the most commonly occurring species. Using tree diameter as the sample variable, they were able to estimate accurately tree stem diameter variances by species (not exceeding a 50 percent coefficient of variation value). Based on a number of combinations of clusters and systematic samples, they recommend the following specific guidelines for conducting a street inventory based on sampling.

1. The sample population size should be about 100 per species.
2. The number of randomly selected streets to be surveyed should be at least 50, but should not exceed 100.
3. A length of twc or three city blocks should be surveyed on each street.
4. The sampling interval per species should be based on its incidence, so that a sample population size of 100 is achieved.

They suggest further that this sampling procedure has another management potential, such as to estimate the extent of an existing tree problem or to predict potential insect infestations.

Jaenson et al. (1992) developed an inventory method that uses a sample of 2000 to 2300 trees to obtain a tree count that is 90 percent accurate to two standard deviations. The community is stratified into three zone types: Rectilinear Residential (RR), Curvilinear Residential (CR), and Downtown (DT). RR zones contain standard city blocks, CR zones have curved streets and cul-de-sacs, and DT is industrial and commercial land. Since the sample is based on the average number of trees per sample unit, city blocks need not be the same size. Each zone is then subdivided into homogeneous subareas for sampling (Figure 6–11) ranging from 20 to 500 sampling

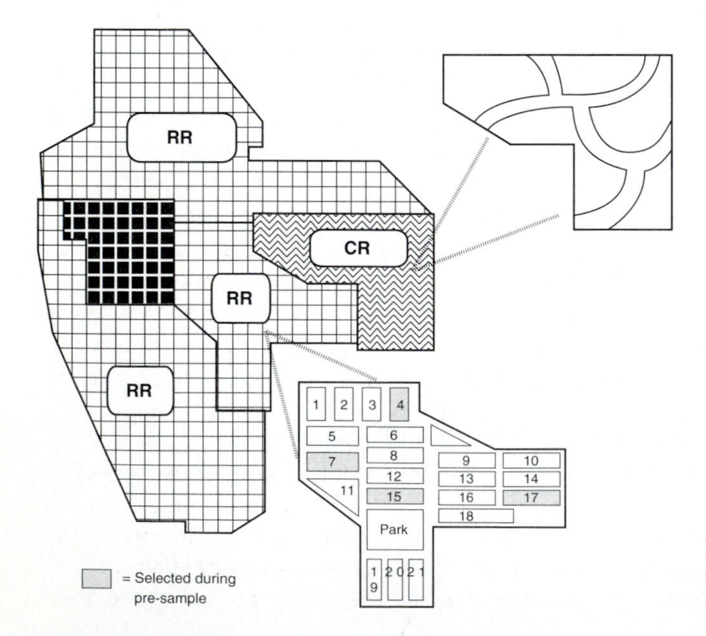

Figure 6–11 Example of zone types subdivided into subareas for sampling. (From Jaenson et al., 1992.)

units (blocks in RR and DT, and street segments in CR). Street segment lengths in CR zones are determined from the mean block perimeter in RR zones. All blocks and segments are numbered by subarea and the presample uses 20 percent (up to 10 sampling units) randomly selected samples per subarea. Data collection can be on foot or by automobile. The mean number of trees per block (street segment) is multiplied by the number of blocks (street segments) to estimate the number of trees per subarea. The sample of 2000 to 2300 trees is then distributed by subarea tree density, and blocks (street segments) are again randomly selected. Tree data are collected, expanded by subarea, and summarized by total number of trees and by number per species. This method gave consistent results in four communities with street tree populations ranging from 5571 to 113,000 trees (Jaenson et al., 1992).

Urban tree inventory by sampling has a number of applications, such as estimating parameters in a windshield survey or as a sampling technique to provide special data, such as wood volume or average tree diameter. It is possible to estimate other parameters, such as tree conditions, frequency, and management needs for the entire or portions of the entire population. For example, a sample of street trees in Mexico City revealed that species diversity was low, many plantings were in inappropriate locations, and the general level of maintenance was too low (Chacalo et al., 1994). Sampling does not require continuous updating, but rather is done on a periodic basis when the need for new inventory information has been determined.

Tree Data Files

Tree data files differ from windshield surveys and sampling in that they record the location of individual trees and maintain a file system describing each tree in the community. Typically, these systems record all street trees, which are maintained in a card file by address, with one tree per data card. Data file systems are good inventory tools in small communities with few street trees (less than 1000). Updating is done by hand as trees are added to or deleted from the population, but a complete reinventory on a periodic basis is essential to provide current information for management. Maintenance records may be kept on tree data cards, but continuous updating is time consuming when each maintenance activity is recorded. However, it may be advisable to record complaints and homeowner service requests on tree cards should the tree ever become the subject of litigation. Tree data files are simple and easy to use when there are not too many trees, but data reduction is very time consuming when summaries of tree population are desired for management purposes. Figure 6–12 illustrates an example of a tree data file card used for inventory purposes.

The Florida Division of Forestry devised a street tree inventory system that utilizes tracings made of city streets from aerial photographs (representative fraction 1/2400) as cruise maps and for tree location. Data are collected to provide information on the existing vegetative theme for a street, to identify areas of the community with few trees, and to determine species diversity and adaptation. Three data forms were used to collect tree data, with data reduction on each form providing the sum-

Index No. *1846*

1	Location	*1274 Main*	11	Land use	*Residential*
2	No. trees	*1*	12	Lawn width	*8'*
3	Species	*Norway Maple*	13	Wires	*None*
4	Plant date	*10/72*	14	Replantable	
5	DBH	*8"*	15	Replaced	
6	Condition	*Good*	16	Removed	
7	Vigor	*Good*	17	Rem. reason	
8	Damage	*Basal damage*	18	Value	*$600*
9	Deadwood	*5%*	19	Remarks	
10	Pruning rec.	*4-years*			

Figure 6–12 Sample of a card file inventory record.

mary information described above. Updating this type of inventory is a matter of periodic reinventory (Theobald, 1978).

Computer Systems

Street tree inventory systems in many cases are best maintained on a computer. This does not mean that all communities should use a computer; rather, a computer should be used when the costs associated with such a system can be justified in terms of better tree care and/or more efficient management. It is important to remember that use of a computer involves not only the initial cost of inventory and data processing, but also involves updating and recording procedures that can be expensive and time consuming. Before deciding on a computerized system, a community should carefully analyze its management needs to determine what data are essential, how these data should be reduced, if the computer will provide this information at a reasonable cost, and who will be responsible for the system in terms of report generation and updating.

Computers were first used as a means of street tree inventory during the 1970s with the introduction of a number of systems designed for use on mainframe computers. Computers were available, and city foresters saw them as an opportunity to finally devise an inventory that was accessible, usable, and would provide summaries of specific tree parameters for management. Prior to using computers, city foresters, particularly in large cities, managed street trees based on the knowledge of their crews and a general understanding of the tree population and management needs, or by using tree tallies from work units. However, maintenance of inventories on mainframe computers proved expensive and time on the computer had to be shared with other city departments, many of whom had higher priority for use.

Microcomputers. The development of inexpensive microcomputers increased access to computerized inventories by the 1980s. As the technology developed, the storage capacity and processing capability increased, while at the same time the cost went down. Initially some city foresters developed their own inventory programs using commercial database software, but this can prove time consuming and expensive. There are a variety of street tree inventory programs now available commercially. Smiley (1989) describes thirteen programs and suggests that if a commercial program suits your needs, it is cheaper to purchase an existing program than develop a custom program. He further lists six functions that the thirteen inventory programs have in common and that should be assessed by the buyer.

1. Tree data files manage the tree data and should be readily accessible so tree information can be updated, new trees added, and trees removed.
2. Work history files store work records by activity, time spent doing the work, equipment, date, and crew.
3. Service request files store requests from citizens for tree service, date of response, and activity performed in response to request.
4. Data summaries summarize the three data files above for planning and management, and for preparation of reports and budget requests.
5. Tree lists are generated of trees in need of service to be used as work orders, such as trees identified for immediate removal or cabling.
6. Computer mapping is available on some systems, which have the capability to generate maps showing tree locations.

Wagar and Smiley (1990) report that as newer inventory systems are developed, they have increasing capacities to assist with tree management. They describe the following capabilities in order of increasing power.

1. Retrieving, displaying, and reviewing records. This should be by address as it is particularly useful when responding to homeowner requests, and for work records on specific trees.
2. Creating work orders. These provide listings of work needs for a specific area, or to respond to a homeowner request or complaint.
3. Computing tree values. Tree diameter and condition can be used by the computer to compute tree values for the city, information useful during budget requests.
4. Summarizing records. This is the ability to describe the tree population in terms of number of trees, average diameters, species mix, condition, value, and work summaries for the entire city and for individual work units.

5. Mapping tree locations. This is the ability to display maps of tree locations using systems such as Map Info or AutoCAD, or the inclusion of tree locations in a GIS.
6. Creating graphs. A graphic display of information describes tree population parameters such as species mix, diameter distributions, tree condition, and so on in pie charts, bar graphs, or other displays.
7. Tracking costs. Work records can be summarized by costs to compare species, crews, equipment, management alternatives, and so forth.
8. Forecasting future work loads. Records of current costs by species and diameters can be used to forecast future workloads for planning purposes.

The increasing capabilities of computer systems carry with them increasing costs both in initial purchase of software and hardware, data collection and entry, and future maintenance of information in that system. When selecting a program for a community, it is important to decide just what information is really needed and to get a system that meets those needs. It is also a good idea to purchase a system that can be expanded as the forestry program grows. Wagar and Smiley (1990) suggest selecting the software that will meet your needs, and then purchasing a computer to run that software. Appendix A contains a guide to thirteen inventory programs available in the United States.

Implementing a computer inventory. Just as there are a number of tree inventory programs commercially available, there are a number of ways a community can implement a system. Commercial packages vary from the user purchasing the software and installing the system to packages that includes software installation, data collection, entry and initial processing by the vendor. Consulting arborists and urban foresters provide inventory services for communities, as well as management plans based on the inventory. Whichever approach a community selects, future budget allocations will be necessary to maintain the system, including training of personnel who will use the system.

Data collecting and updating. Data collection for street tree inventories has traditionally involved collection of information on field data sheets and transferring that data by hand into the computer. However, the development of handheld data collection devices eliminates manual transfer of data into the computer, a considerable cost savings. These units are cheap, lightweight, durable, and programmable. They can be programmed to validate information, and they allow direct or modem data transfer into the computer, further reducing the risk of error. They also reduce field data collection time by 30 percent. Some data collectors use a wand to read bar codes rather than typing in entries, an innovation that further speeds up the data collection process (de Vries 1993).

Updating the inventory on a microcomputer may be done by periodic reinventory or as part of normal work procedures. If the city maintains street trees on a regular basis with city crews, updating becomes a matter of the crew leader recording tree information each time the tree receives maintenance service. These data may be recorded in a field computer or manually, and the inventory updated following each workday. Service requests and complaints relative to street trees may be recorded via data entry screens by office personnel and become a part of the tree history file.

Computer inventory costs. The cost of a computerized inventory system will include the design or purchase of the inventory program, field data collection, data transfer to the data base, data verification, data processing, report generation, and file updating. These costs will vary based on the type of computer and software, amount of data collected per tree, and how the data are collected (work sheets versus handheld computer, and labor costs). Overall inventory costs in 1979 were reported to vary from $0.31 per tree to $1.25 per tree, while data collection costs ranged from $0.12 to $0.72 per tree (Sacksteder and Gerhold, 1979).

Once the appropriate computer hardware and software have been selected, data processing costs will remain somewhat fixed. However, data collection costs are highly variable, depending on the quantity of data collected, tree density, and the organization, knowledge, and training of inventory personnel. While quantity of data per tree and tree density are set prior to data collection, crew organization, knowledge, and training are the responsibility of the city forester and will have a real impact on data collection costs. In many cases, the initial inventory or reinventory is conducted by a crew specifically hired for that purpose. The best personnel affordable should be hired for this task, and students majoring in arboriculture, urban forestry, or related fields are a good choice, as they have an expressed interest in the profession. Crews should be organized in the most efficient manner for local conditions, with one person given responsibility for the operation. This responsibility will include planning, organization, work assignment, quality control, and maintenance of maps to see that all streets are covered in an efficient manner. Data summaries and reports are only as good as their database, and it is essential that crews be well trained and motivated to do the job as correctly as possible.

Sacksteder and Gerhold (1979) report inventory crew size for various systems they reviewed range in number from one to four. Experience by the author conducting inventories in a number of Midwestern cities indicates that the most efficient means of data collection for an initial inventory or reinventory is using one-person crews. Individuals working alone are more productive than when working as part of a crew, as there is a tendency to discuss tree features when two or more work together. Even when inventorying opposite sides of the street, the faster member has a tendency to wait for the slower, a situation that can easily arise when there are more trees on one side of the street than the other. Larger and slower-moving crews are more likely to attract the attention of homeowners, who often wish to discuss their personal landscapes with the crew, thus slowing the progress of the inventory considerably. The size of the inventory crew should be tempered with judgment, as in some sections of cities personnel safety may be a problem dictating the necessity for more than one person per crew.

OTHER USES OF INVENTORIES

General management uses of inventories were discussed at the beginning of this chapter and will be elaborated upon in Chapter 10. There are a number of specialized uses that inventories have been put to in recent years, a discussion of which follows.

A number of tree inventory systems calculate tree value, a figure that may be useful in obtaining funding for the department responsible for tree care. Most people will admit that they place some value on trees, but are reluctant to put this value in monetary terms. If these persons are shown the value of street trees and are made aware of the contribution that these trees make to property values, they will be more likely to support the efforts of the forestry department. Most operational funds used by city government come from property taxes, and higher-value property will generate more revenues for municipal operations, a fact of which elected officials should be made aware. You can be sure other municipal departments know the monetary value of streets, sewers, water mains, street lights, and so on, and make sure that elected officials are aware of what it costs to maintain the value of these improvements. All other municipal improvements depreciate in value over time, with the exception of trees, which appreciate in value.

Chan and Cartwright (1979) developed an inventory system for the city of Sacramento, California, that noted general tree health and specifically categorized the severity of dwarf mistletoe infestations. They report use of the inventory system in developing integrated pest management strategies by identification of species to avoid, analysis of specific problems, location of trouble spots, and pest monitering during inventory. Miller and Sylvester (1979) used inventory data gathered for UWSP/URBAN FOREST inventory system to determine the location of residual American elm (*Ulmus americana*) populations in three Wisconsin communities, and to ascertain what species were being used to replace them. Bassett (1978) reports on an inventory used in Ann Arbor, Michigan, to study girdling roots and to evaluate species and cultivar performance.

Miller and Marano (1986) developed URBAN FOREST, a street tree computer simulation designed to utilize tree inventory data for management simulation and training purposes. Inventory data from three communities in Wisconsin were used to simulate four levels of Dutch elm disease *(Ceratocystis ulmi)* control over twenty years (Miller and Schuman, 1981). Results indicate that when control costs are compared to tree value over time, the cost is exceeded in all cases by increased tree value, and the more intensive the control effort, the higher the future values of the street tree population.

A street tree inventory in and of itself represents static information when it describes the current status of a street tree population. When the inventory includes those sites that are vacant and assigns a planting priority, the inventory begins to take on a more dynamic character as it describes not only the present population, but describes what is needed in terms of future management. Perhaps the most important aspect of a street tree inventory is when it is carried out on a continuous or periodic basis to establish a baseline of where the population has been, where it is now, and where it is going in the future. Whether the city forester elects to conduct periodic inventories by windshield or sample, or uses a computerized inventory that tallies all trees and is updated on a continual basis, it is the dynamic analysis of the population over time that will form the basis of sound management.

LITERATURE CITED

ATCHISON, F. 1978. "Community Forestry Inventories in Kansas." *Proc. Natl. Urban For. Conf.,* ESF Pub. 80–003. Syracuse: SUNY, pp. 767–773.

BARKER, P. A. 1983. "Microcomputer Data Bases for Data Management in Urban Forestry." *J. Arbor.* 9(11):298–300.

BASSETT, J. B. 1978. "Vegetation Inventories: Need and Uses." *Proc. Natl. Urban For. Conf.,* ESF Pub. 80–003. Syracuse: SUNY, pp. 632–644.

CHACALO, A., A. ALDAMA, and J. GRABINSKY. 1994. "Street Tree Inventory in Mexico City." *J. Arbor.* 20(4):222–226.

CHAN, F. J., and G. CARTWRIGHT. 1979. "Tree Management Aided by Computer." *J. Arbor.* 5(1):16–20.

COUNCIL OF TREE AND LANDSCAPE APPRAISERS. 1992. *Guide for Plant Appraisal.* Savoy, Ill.: International Society of Arboriculture, 103 pp.

DE VRIES, R. 1993. "Advanced Technology for the Urban Forester." *Proc. 6th Natl. Urban Forestry Conf.,* Washington, D.C., American Forests, pp. 74–76.

GEIGER, J. R. 1977. "A Sampling Technique to Inventory the Urban Forest," *Proc. Urban For. Workshop.* Stevens Point; Coll. Nat. Resour., Univ. Wis., pp. 50–62.

GERHOLD, H. D., K. C. STEINER, and C. J. SACKSTEDER. 1987. "Management Information Systems for Urban Trees." *J. Arbor.* 13(10):243–249.

JAENSON, R., N. BASSUK, S. SCHWAGER, and D. HEADLEY. 1992. "A Statistical Method for the Accurate and Rapid Sampling of Urban Street Tree Populations." *J. Arbor.* 18(4):171–183.

KIELBASO, J. J. 1989. *City Tree Care Programs.* In G. Moll and S. Ebenreck, eds., *Shading Our Cities.* Washington, D.C.: Island Press, pp. 35–46.

MILLER, R. W. 1994. *Introduction to Urban and Community Forestry: An Independent Study Course by Correspondence.* Gainesville: Univ. Fla., Dept. of Ind. Study.

MILLER, R. W., and M. S. MARANO. 1986. "URBAN FOREST: A Street Tree Management Simulation." *Proc. For. Microcomputer Software Symp.,* June 30–July 2, 1986. Morgantown, W. Va., pp. 659–670.

MILLER, R. W., and S. P. SCHUMAN. 1981. "Economic Impact of Dutch Elm Disease Control as Determined by Computer Simulation." *Proc. Dutch Elm Dis. Symp. Workshop.* October 5–9, 1981, Winnepeg: Manitoba Dept. of Nat. Res., pp. 325–344.

MILLER, R. W., and W. A. SYLVESTER. 1979. "Report on the Use of UW/SP URBAN FOREST Computer Inventory Program as Part of the Dutch Elm Disease Demonstration Project in Wisconsin." *Wisconsin Dutch Elm Disease Demonstration Project 1979 Accomplishment Report.,* Madison: Wis. Dept. Nat. Resour., pp. 65–67.

MOHAI, P., et al. 1976. "Structure of Urban Street Tree Populations and Sampling Designs for Estimating Their Parameters." *METRIA: 1 Proc. First Conf. of the Metropolitan Tree Improvement Alliance,* pp. 28–43.

SACKSTEDER, C. J., and H. D. GERHOLD. 1979. *A Guide to Urban Tree Inventory Systems.* Sch. For. Resour. Res. Paper No. 43.

SMILEY, E. T. 1989. *Computer Software for Urban Forest Management: A Buyers Guide.* In G. Moll, and S. Ebenreck, eds., *Shading Our Cities.* Washington, D.C.: Island Press, pp. 288–293.

SMILEY, E. T., and F. A. BAKER. 1988. "Options in Street Tree Inventories." *J. Arbor.* 14(2):36–42.

THEOBALD, W. F. 1978. "Urban Forest Inventories in South Florida: Need for and Uses." *Proc. Natl. Urban For. Conf.,* ESF Pub. 80–003. Syracuse: SUNY, pp. 757–766.

TSCHANTZ, B. A., and P. L. SACAMANO. 1994. *Municipal Tree Management in the United States.* Kent, Ohio: Davey Tree Expert Company.

VALENTINE, F. A., et al. 1978. "Street Tree Assessment by a Survey Sampling Procedure." *J. Arbor.* 4(3):49–57.

WAGAR, J. A., and E. T. SMILEY. 1990. "Computer Assisted Management of Urban Trees." *J. Arbor.* 16(8):209–215.

ZIESEMER, D. A. 1978. "Determining Needs for Street Tree Inventories." *J. Arbor.* 4(9):208–213.

7

Park and Other Urban Natural Resource Inventories

While street tree inventories are of primary interest to city foresters, there is often a necessity to inventory other natural resources in urban areas. Responsibilities of the city forester will vary from community to community and may range from street trees only, to broad responsibilities for both public and private vegetation. In some communities, the city forester may be the only public employee with a background in natural resources or biology and may be called upon to assist with or conduct inventories beyond those specifically related to tree management. Other vegetation managers, such as commercial arborists, private and public grounds supervisors, utility arborists, and landscape maintenance contractors, may also need inventories to serve as a basis for work planning and scheduling.

As in street tree inventories, the vegetation manager must carefully select those parameters most important to include in the database for management decision making. Once the inventory parameters are selected, the method of data collection should be determined. This could consist of a sampling or a 100 percent inventory, depending on management needs, variability of the population, and available funds for inventory. A wide variety of techniques are available, and some of these will be explored in this chapter with specific examples presented. Vegetation and other resource inventory procedures form a large body of information in the literature. Coverage of the topic in this text is meant to introduce the reader to a number of these techniques, but is not intended to be all-inclusive.

As with street tree inventories, management planning and decision making require not only the present status of the resource, but an understanding of the resource from a temporal perspective. Periodic inventories establish baseline data from which the manager can draw inferences as to the future of the ecosystem. It is this anticipation of the future of the system that allows the manager to make the best management decisions in the present, thereby shaping the future. In this chapter we explore vegetation inventories from a general perspective, followed by specific applications to the urban environment. The discussion of other natural resource inventories will describe what is already available for most municipalities, and discuss techniques available for specific information.

GENERAL COVER TYPE MAPPING

Cover type mapping involves a number of techniques designed to describe the dominant vegetative communities of a selected area. In most instances, the process begins with the preliminary identification of dominant plant communities from aerial photographs, followed by ground checks to ascertain the reliability of the assessment. Ground surveys may be used to gather additional information, such as overstory species mix, or to describe forest strata not visible on aerial photographs. In the absence of recent aerial photographs, vegetation maps may be made by means of several methods of ground surveys. More elaborate systems of ecologic land classification are available, including landform, vegetation, soils, and aquatics. In general, these techniques are used to survey residual vegetation, and vegetation on public and private open-space lands. Specific inventory situations described later in this chapter refer to techniques described in this section.

Photogrammetry

Photogrammetry is defined as the measurement of aerial photographs to construct maps and describe features of topography, vegetation, and cultural activity. Aerial photographs are frequently used for natural resource inventories because they provide a vantage point, permanent record, and a stereo perspective of the landscape. For vegetation surveys they are particularly useful, especially when special film types are used. Films that record visible black-and-white or color wavelengths will provide a "normal"-appearing landscape with little distinctions between different cover types. Infrared black-and-white or color film will provide strong contrasts between certain vegetation types, particularly conifers and hardwoods. Black-and-white infrared film will also record water and wetlands darker than visible spectrum film. Timing of the photography can provide distinctly different views of the landscape. Photographs made when deciduous trees are in full leaf will emphasize vegetation contrasts, while photographs without leaves provide the best information concerning land form, topography, and infrastructure.

The U.S. Geological Survey contracts for periodic aerial photographic surveys over many parts of the country for other agencies. These photographs are available to the public for a fee. Additionally, various public agencies and private organizations contract aerial photographic work, and these photos can usually be purchased. Municipal governments often contract periodic photographic work for planning and engineering purposes, while state agricultural departments fly annual photographic surveys to estimate crop patterns and yields, often leaving cameras on while passing over communities, thus providing an annual source of urban aerial photographs.

Stereoscopic photographs. Aerial photographs are normally taken in flight patterns that overlap prints by at least 60 percent. This overlap allows the interpreter to view two photographs at once through a stereoscope and provides an exaggerated vertical perspective of the landscape. This exaggerated perspective allows for easy identification of important landforms and assists in separating vegetation types based on size and age.

Scale. The scale of an aerial photograph is determined by the height of the camera over the terrain and is usually given as the representative fraction (RF):

$$\text{RF} = \frac{\text{distance on photo}}{\text{distance on the ground}}$$

Both the numerator and denominator must be in the same units and the fraction must be reduced so that the numerator is 1.

Since the aircraft cannot maintain a precise altitude during flight and variations in land elevation exist, the representative fractions given with the photographs are approximate and the true scale must be determined using a quadrangle map or other map with a known scale. This is done by selecting two points that are evident on both the map and photograph and precisely measuring the distance on each. The true photo representative fraction may then be determined by entering the ground distance between the two points on the map (determined using the map RF) as the denominator, entering the photo distance as the numerator, and reducing to a unitary fraction (see Table 7–1 for an example). Once the true representative fraction of the photograph has been determined, the information on the photo may be equated to map scales for vegetation mapping purposes.

Cover types. Cover type interpretation on aerial photographs will vary from region to region, but there are general patterns common to most areas. These vegetation patterns fall into two general categories: nonforested, including upland clearings, upland brush, and wetlands; and forested, including hardwoods, conifers, and conifer bogs. The following description of these cover types are for photographs made with black-and-white infrared film.

A. Nonforest

1. Upland clearings: fine texture, shadows from surrounding trees, distinctly lower; rock outcrops sometimes present

TABLE 7–1 CALCULATION OF THE REPRESENTATIVE FRACTION FOR AN AERIAL PHOTOGRAPH

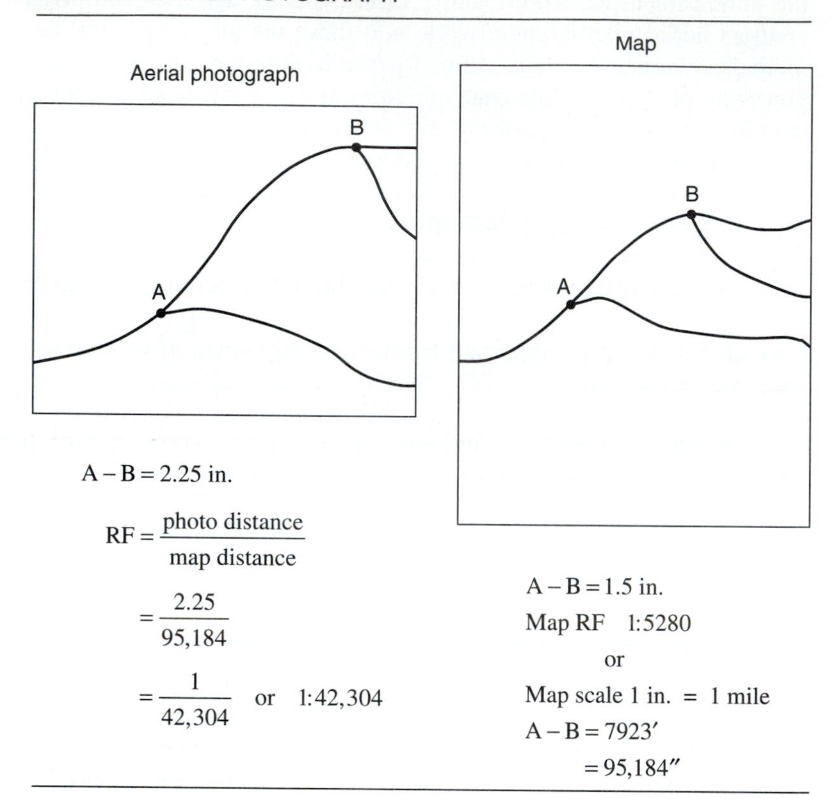

2. **Tilled fields:** distinct furrows or row crops
3. **Upland brush:** coarse texture, shadows from surrounding trees, lower than surrounding trees
4. **Open water:** black and smooth
5. **Bogs and swamps:** smooth and flat, presence of surface water makes them dark

B. Forest

1. Hardwoods: light shade
   - **a.** Even age: smooth/uniform, few or no shadows present
   - **b.** Uneven age: coarse texture, shadows from larger crowns present
2. Conifers: dark shade (usually even-aged)
   - **a.** Conifer plantation: rows often evident, especially in young plantations
   - **b.** Conifer bogs: dark, flat, often sparse cover (becoming sparse toward center of bog)

Figure 7–1 illustrates a number of cover types readily distinguishable on an aerial photograph taken in the north temperate region of the United States. Aerial

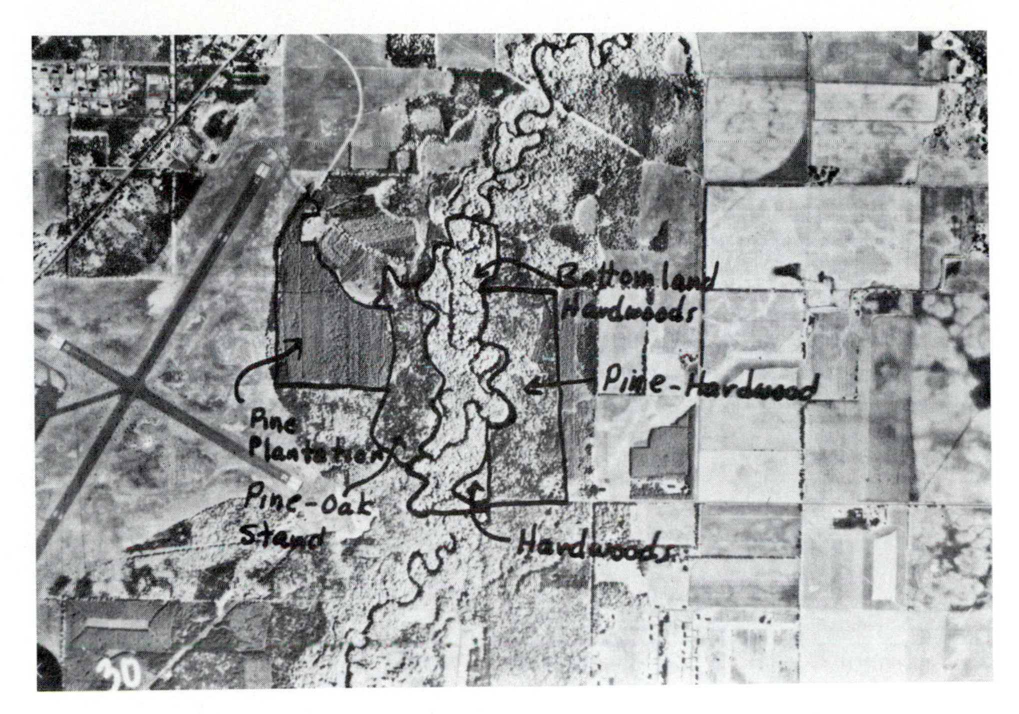

Figure 7–1 Cover types identified on an aerial photograph in the northeastern United States.

photographs can be used to identify more specific cover types within local areas. Descriptions of these cover types may be obtained from state or regional forests and planning offices.

Area estimation. Once preliminary cover type maps have been made, the areas of various communities can be determined if this information is desired for management. The simplest method of obtaining this information is through the use of a dot grid, as illustrated in Figure 7–2. Many dot grids have area conversion tables for common representative fractions found on most maps and aerial photographs. However, when using maps with different scales or when using dot grids without conversion tables, it is an easy matter to calculate area from the map or photo scale. Simply determine the area of a square on the dot grid using the representative fraction of the map or photo and divide this area by the number of dots in that area. Each dot counted will then represent that area of land on the map or photo.

To use the dot grid properly, it is dropped on the map or photo three times, the area estimated each time, and then the average area computed. Every other dot falling on a cover type boundary line should be counted for the most accurate use of the system. Figure 7–2 illustrates a dot grid used to determine the area of a cover type on an aerial photograph.

Other methods of determining the area of cover types include average width

Figure 7–2 On the photo each dot represents 1.11 acres. The pine plantation contains 47 dots, or covers an area of 52.17 acres.

times length and use of the polar planimeter. Use of the average width times length will provide a rough estimate of the area for preliminary surveys, but it is not accurate enough for intensively managed land. Planimeters measure the periphery of the area and convert the measurement to area on the machine scale (Figure 7–3). It is an accurate method of determining areas when properly adjusted and used, but like the dot grid, an average of three measurements should be used.

Ground Inventory

Ground surveys serve three general functions: to verify maps made from aerial photographs, to gather additional information for these maps, or to serve as the inventory tool in the absence of recent photographs. Aerial photographs record an accurate picture of ground information at a given point in time, but many changes can take place within plant communities over time due to cultural and natural events, especially in urban areas. If the photographs are a few years old, ground verification is necessary to record these changes. General vegetation maps may be made from aerial photographs, but ground surveys are necessary to provide specific descriptions of species present, and in forested communities, descriptions of understory strata. In the absence of aerial photographs, ground inventories are the only means to obtain vegetation inventories.

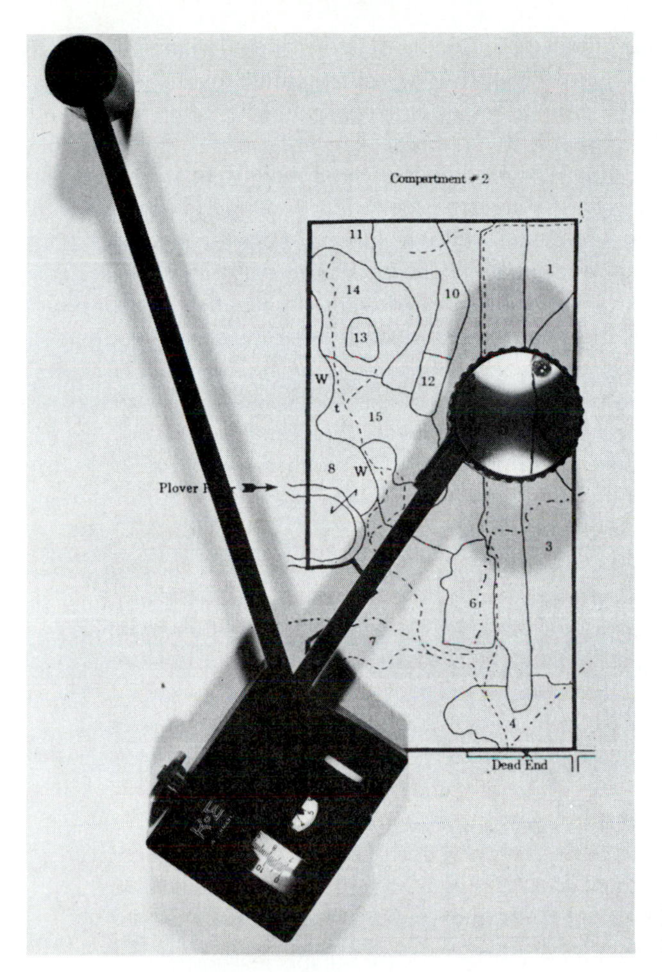

Figure 7–3 Planimeter being used to measure areas of forest cover types.

Selecting the cover types to describe in a survey will range from general distinctions between major differences in vegetation to a precise inventory of each vegetative community, depending on management needs. Classification and descriptions of specific forest cover types for North America are available from the Society of American Foresters (1980). The U.S. Department of the Interior (Cowardin et al., 1979) has developed a classification system for wetlands and deep-water habitats, as have a number of state agencies involved in wetland protection projects. In arid regions, classifications of various prairie, savanna, and desert communities may be obtained from local resource agencies and from universities.

Common techniques for vegetation inventories are presented below. Several of these methods may be subjected to statistical analyses to determine sample size or the reliability of the inventory. Specific statistical tests will not be described, and the reader is advised to refer to various statistical and biometric texts for additional information.

Graham cover type mapping. The Graham system of cover type mapping is easy to use and will provide a reasonably accurate vegetation inventory. The system is flexible and adaptable to a wide variety of geographic regions. When used in conjunction with preliminary aerial photographic mapping, data are collected by traversing the communities and recording the relative frequency of species present in the various strata and recording evidence of past disturbance. Physiographic conditions may be observed in the field or obtained from soil and geologic surveys. Data are collected along transect lines or at specific points that are selected either randomly or systematically. When using this method without the aid of aerial photographs, the survey lines or points may be used to locate transitions from one cover type to another, thus establishing data for construction of a cover type map. Species lists and appropriate abbreviations may be developed for whatever region the inventory is used. Although this system is not quantitative in providing statistical estimates of species frequency or biomass, it does provide a description of community stratification, the most important species present, successional trends, and site conditions. Table 7–2 provides an example of the Graham system applied to northeastern United States (Graham, 1945).

Line transect. The line transect method is used primarily to determine the plant coverage or percentage of area a species occupies in a community. The sampling unit may be visualized as a vertical plane that has length and height only. All plants with part of their structure intersecting this plane are tallied, and the length of intersection is recorded (Figure 7–4). Randomness in sampling may be introduced in line location and length. Cover and relative cover for a species are then calculated by the formulas

$$\text{cover} = \frac{\text{total intercept lengths for species}}{\text{total length of intercept}}$$

$$\text{relative cover} = \frac{\text{total intercept lengths for one species}}{\text{total intercept lengths for all species}} \times 100$$

If the relative frequency of species is all that is needed, the number of intercepts per species is counted, not the length of intercept. The line transect method is used primarily to estimate the cover of the shrub and herbaceous layers in forest cover types and to inventory grassland, wetlands, and other nonforested ecosystems. Reliability of the inventory may be tested statistically through analysis of the coefficient of variation for the frequency of a given species or group of species on each transect.

Quadrat sampling. The quadrat sampling method tallies all individuals by species within an area of known size. This method may be used to sample a forest overstory, shrub layer, or herbaceous growth on the forest floor, as well as plant communities on nonforested sites. The methodology is the same, with plot size varying based on the strata being sampled. In the case of forest ecosystems, combination surveys are made at randomly or systematically preselected locations, with the

TABLE 7–2 COVER TYPE MAPPING SYSTEM DEVELOPED FOR USE AT THE UNIVERSITY OF WISCONSIN-STEVENS POINT SUMMER CAMP. THIS SYSTEM IS A MODIFICATION OF THE GRAHAM SYSTEM. (From Graham, 1945.)

		EDAPHIC CONDITIONS									
		HYDRIC			INTERMEDIATE				XERIC		
		B	M	S	W	L	F	C	P	D	R
		Acid bog	Marsh	Seepage or basic bog	Spring wet/Fall dry		Low permeability		Permeable upland	Sandy and droughty	Rock outcrop droughty
					Depression	Flood plain	Fragipan	Fine textured			
0	NO PLANT COVER	Stagnant open water		Flowing open water							Bare rock
1	FIRST PIONEER SPECIES	Submerged aquatics									Lichens
2	PIONEERS OR BARE SUBSTRATE	Floating-leaf plants			BARE SURFACE						Mosses Lichens
3	EMERGENTS OR PIONEERS	Sphagnum and sedges	Sedges		ANNUAL PLANTS						Crevice plants
4	PERENNIALS LOW DIVERSITY	Sphagnum and heaths	Sedges and cattails	Sedges and grasses	PERENNIAL HERBS					Herbs, mosses and lichens	As above more cover
5	PERENNIALS HIGH DIVERSITY	Heaths	Hydrophytic perennials		PERENNIAL HERBS MORE SPECIES					Perennials Dwarf shrubs	More cover, a few shrubs
6	SHRUBS	Heaths, very small trees	Hydrophytic shrubs	Shrubs, esp. alders	SHRUBS						Shrubs, a few trees
7	INTOLERANT TREES	TREES									
8	MID-TOLERANT TREES	TREES									
9	TOLERANT TREES EVEN-AGED	TREES									
10	TOLERANT TREES UNEVEN-AGED	TREES									

(continued)

TABLE 7–2 CONTINUED

PHYSIOGRAPHIC CONDITIONS:

Glacial deposits:

o – outwash
k – outwash with kettleholes
m – unsorted moraine

Wind and water deposits:

r – river alluvium
l – lake bed
w – wind (loess)
z – dunes

Bedrock:

c – calcareous (limestone)
g – crystalline
s – sedimentary, excluding limestone

DISTURBANCE:

n – built up (fill)
h – farmed
f – flooded
t – thinned
x – burned
e – eroded
p – pastured
v – plantation
d – ditched
y – cut over
b – blowdown
j – strip mined
a – browsed by wild animals

COVER/ABUNDANCE SCALE:

	Trees	Shrubs/herbs
Dense	///	≡
Medium	//	=
Scattered	/	–
Present	+	+

COVER/ABUNDANCE GUIDELINES:

	/// & ≡	// & =	/ & –
Tree/shrub/ sapling	over 50% cover	10–50% cover	plentiful but below 10% cover
Herb/seedling	over 25% cover	5–25% cover or numerous	plentiful but below 5% cover

EXAMPLE: Pmp 7 TA///12–18 WB/14–18
Ha– Rd– WP///1–3
bb≡ m= sa= bf= ly– SM/

This means: Permeable soils (P), unsorted glacial till (m), pastured (p).
Dominated by shade-intolerant trees (7).
Specifically with the following vegetational composition:
Canopy—dense stocking of 12–18" DBH quaking aspen; scattered stocking of 14–18" white birch (TA///12–18,WB14–18).
Shrub and understory tree layer—hazelnut, scattered density; white pine saplings, dense stocking of 1–3" DBH; red-ozier dogwood, scattered density.
Herb layer—Bluebead, dense; Canada mayflower, wild sarsaparilla, and bracken fern, all medium density; club moss (*Lycopodium*), scattered; sugar maple seedlings, scattered.

TA = *Populus tremuloides*
WB = *Betula papyrifera*
WP = *Pinus strobus*
SM = *Acer saccharum*
Ha = *Corylus americana*
Rd = *Cornus stolonifera*

cm = *Maianthemum canadense*
bb = *Clintonia borealis*
sa = *Aralia nudicaulis*
bf = *Pteridium aquilinum*
ly = *Lycopodium* sp.

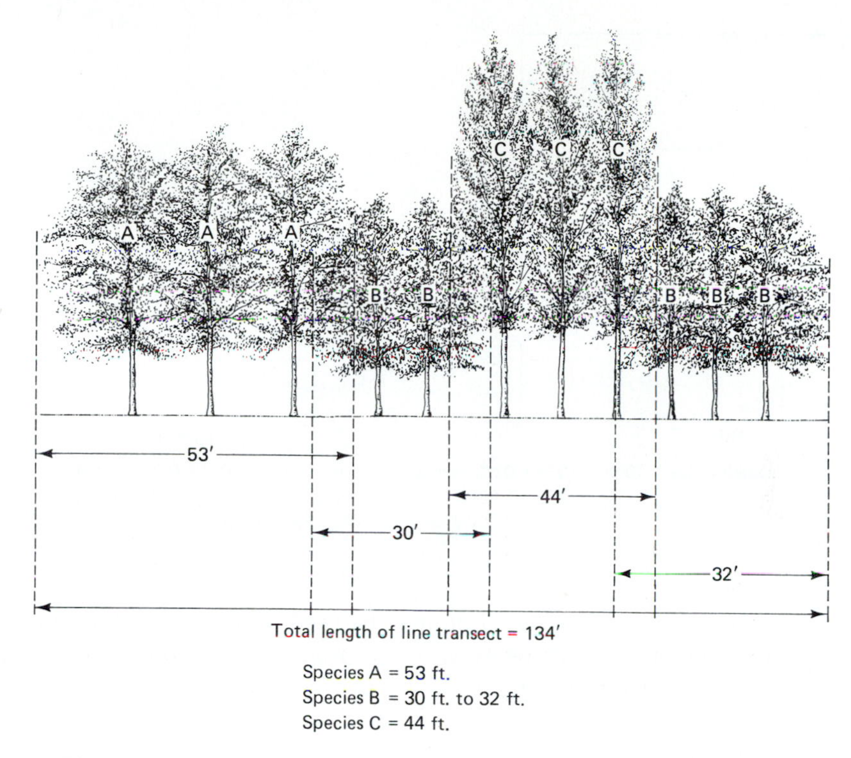

Figure 7–4 Three species of trees (A, B, and C) intersecting a transect line.

overstory, shrub layer, and ground cover sampled in progressively smaller quadrats. Commonly used quadrat sizes in the United States are 1/40 acre (33 by 33 ft.) for trees, 1/1000 acre (6.6 by 6.6 ft.) for saplings and shrubs, and 1/4000 acre (3.3 by 3.3 ft.—a square meter) for herbaceous growth (Figure 7–5). Once the survey has been made, the density, relative density, frequency, relative frequency, and total number of organisms can be calculated using the following formulas:

$$\text{density} = \frac{\text{no. individuals of the species}}{\text{total no. quadrats}}$$

$$\text{relative density } (\%) = \frac{\text{density of the species}}{\text{total density of all species}} \times 100$$

$$\text{frequency} = \frac{\text{no. quadrats containing the species}}{\text{total no. quadrats}}$$

$$\text{relative frequency } (\%) = \frac{\text{frequency for the species}}{\text{total frequency of all species}} \times 100$$

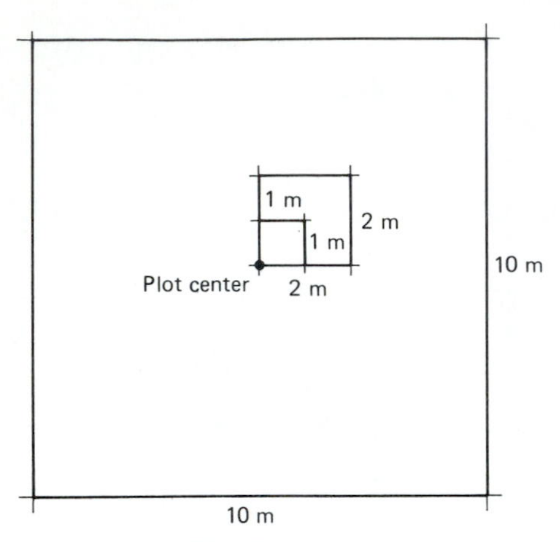

Quadrat size 10 m X 10 m = 0.01 ha or 0.025 acre
2 m X 2 m = 0.0004 ha or 0.001 acre
1 m X 1 m = 0.0001 ha or 0.00025 acre

Figure 7–5 Quadrat sampling for overstory vegetation (10 in. × 10 in.), shrub layer (2 m × 2 m), and herbaceous layer (1 in. × 1 in.).

The coefficient of variation may be used to determine the statistical reliability of the inventory through analysis of the frequency variability for a given species or group of species for the sample.

Point sampling. Point sampling is a technique commonly used by foresters in timber cruising to determine the volume of standing timber. Although timber volumes may not be of interest when managing or mapping urban forests, the technique will provide a quick and reliable estimate of stand density and relative species biomass for the predominant overstory species.

Point sampling uses a wedge prism or an angle gauge to determine the relative density of the stand in square feet of wood at breast height (4.5 ft. or 1.4 m) on a per acre basis. Prisms are ground to a precise angle designed to offset the line of vision a known amount. The angle of the prism provides the basal area factor (BAF), and it is this number that converts the field observation to an estimate of basal area. When viewing the trunk of a tree through the prism at 4.5 ft. (1.4 m) above the ground from a sampling point, the prism will offset a segment of the trunk. If the offset is greater than the diameter of the trunk, the tree is not tallied. However, if the offset is less than the trunk diameter, the tree is tallied as an "in" tree (Figure 7–6). The number of "in" trees tallied per plot is multiplied by the BAF (typically 10), and this estimates the basal area per acre based on the point where the sample is taken.

Sample points are selected systematically or randomly, and the average basal area per acre is calculated using the formula

$$\text{average basal area} = \frac{\text{total “in” trees}}{\text{no. sample points}} \times \text{BAF}$$

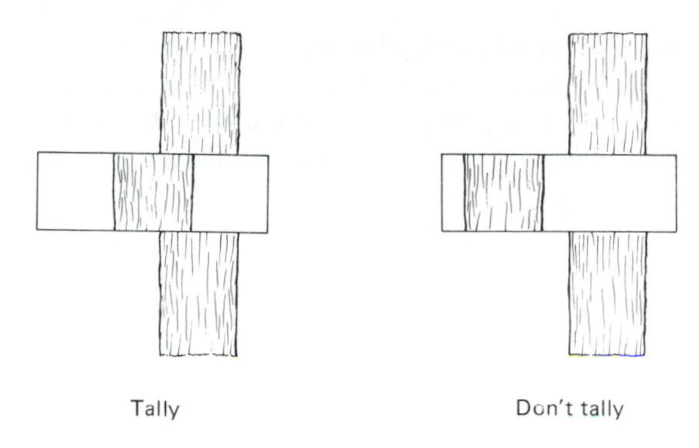

Figure 7–6 Prism view of trees when determining basal area of forest stands. Borderline trees should be measured from plot center to determine if they are to be tallied.

Basal area may also be used to compare the relative density of trees in the overstory when "in" trees are tallied by species and expressed by their basal area per acre and as a percent of the total basal area (Table 7–3). It should be noted that larger trees are more likely to be sampled using the point sampling technique. A species that has a high frequency but small average diameter will be underrepresented when basal area is used to describe species relative density. The coefficient of variation for the variability of basal area by plot may be tested to determine the reliability of the overall basal area per acre estimate.

TABLE 7–3 BASAL AREA SUMMARY TABLE FROM A PRISM SURVEY IN AN URBAN GREENBELT FOREST

Plot	Red Oak	White Oak	Red Maple	Shagbark Hickory	Trees Tallied	BA/Acre per Point
1	• •	•	• • | • •	• • | • •	13	130
2	• • •	• •	• • | • •	• • | • •	15	150
3	•	• • • •	• •	• • •	10	100
4	• • •	•—• | • •	• • • •	• •	15	150
5	• • • •	• • •	•—• | • •	• • •	16	160
6	• • •	• •	• • • •	•—• | • •	15	150
Trees tallied	16	18	27	24	84	
X (BA/Acre)	26.6	30	45	40	140	

Perhaps the most important use of basal area as a stand descriptor in urban forest management is its application to determining when a stand of trees needs to be thinned. Excessive basal area in young intolerant stands is a sure sign of stagnation and susceptibility to other problems, such as insect or disease outbreaks. Timely thinning of these stands will invigorate them and concentrate growth on remaining stems. While the amount of basal area a young stand will support and still maintain high vigor will vary by region and species, a rough guideline is that basal area in excess of 140 ft^2 per acre is getting high, and 160 is definitely stagnating and in need of immediate attention.

Ecologic land classification. A number of ecologic or habitat classification systems have been proposed over the years, ranging from the simple to the highly complex. When developed for a specific region these systems have been accurate and useful, but when applied to other geographic regions, weaknesses became apparent. In discussing the concept of a biological classification system for forest landuse allocation, Daubenmire (1980) described the three following ecologic principles necessary to develop such a system.

1. The probable outcome of successional trends, as indicated by population structure in the arboreal stratum, indicates a certain breadth of environmental potentialities.
2. The character of the herbaceous and shrubby undergrowth reflects different environmental potentialities.
3. Each combination of climax overstory and undergrowth defines a specific and narrow range of plant-growth conditions wherever it appears.

He further defined a unit of land area so described as a habitat type, and such a system has been broadly applied throughout the Rocky Mountain region of the United States. Systems such as habitat type are available for other regions, and may be useful for inventorying and mapping urban vegetation.

An ecological land classification adopted by the United States Forest Service and other federal agencies defines broad ecoregions based on climate, which are then subdivided into progressively smaller areas of increasingly uniform ecological attributes (Avers et al., 1993). The system is hierarchical with variations in landform, microclimate, and soils increasingly defining smaller subunits. The system is used at the ecoregion level for broad strategic planning, and at the land unit of less than 100 acres (45 ha) for project management, planning, and analysis (Table 7–4).

At the subregional level, systems of ecologic classification systems have been developed based on classification specific habitats. In Wisconsin a system has been developed that classifies landform and soil types, and uses associations of understory plants to define specific habitat types based on the potential climax or late successional community (Kotar et al., 1988). Understory and forest floor plants are used for classification because late successional forests are often not available for classification due to past disturbance, and because understory plants progress much more

TABLE 7–4 USDA FOREST SERVICE NATIONAL HIERARCHY OF ECOLOGICAL UNITS. (From Avers 1993.)

Planning and analysis scale	Ecological unit	Purpose, objectives, and general use	General size range
Ecoregion			
Global Continental Regional	Domain Division Province	Broad applicability for modeling and sampling. Strategic planning and assessment. International planning.	1,000,000s to 10,000s of square miles.
Subregion	Section Subsection	Strategic, multiforest, statewide, and multiagency analysis and assessment.	1,000s to 10s of miles.
Landscape	Landtype association	Forest or areawide planning, and watershed analysis.	1,000s to 100s of acres.
Land unit	Landtype Landtype phase	Project and management area planning and analysis.	100s to less than 10 acres.

rapidly to late successional associations than the forest overstory. For each habitat type successional tendencies are then described and management options discussed. Figure 7–7 is an example of successional tendencies in three specific habitat types. Systems such as this can be readily used to classify natural forest communities in urban areas.

URBAN COVER TYPE INVENTORIES

Inventories of urban vegetation cover types include both developed and undeveloped land and may be applied to public and private ownerships. Two broad categories are discussed, tree canopy analysis and landscape inventories, both of which utilize techniques described in the preceding section.

Tree Canopy Analysis

The analysis of urban tree canopies uses aerial photographs to provide a description of tree cover in developed areas and for specialized functions such as determining the distribution of certain tree species or location of tree problems. To be most useful, aerial photographs should be recent and should be taken at the right time of year and day, on the appropriate film, and from a height that provides a useful scale. Commercial firms may be hired to perform this service at given specifications, but the process is expensive and it is not likely that such a project would receive funds under normal operating circumstances. However, a 35mm camera may be used from a rented

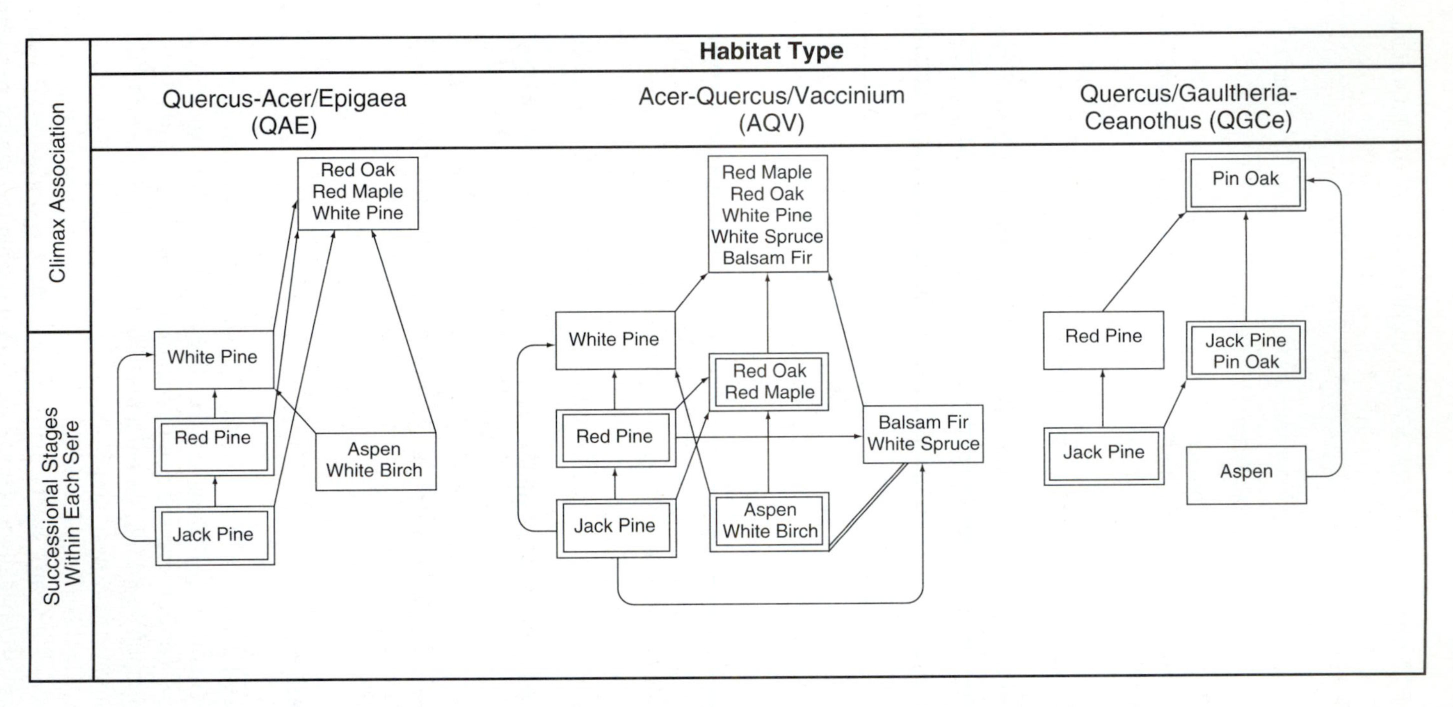

Figure 7–7 Common successional stages and probable successional pathways on dry/nutrient-poor habitat types in northern Wisconsin. (From Kotar et al., 1988.)

aircraft to provide this information for a preliminary canopy analysis, provided that rigid vertical, horizontal, and elevational control are not necessary. Photographs taken for other purposes are also available, as described earlier in this chapter.

The assessment of overall tree canopy in urban areas was developed by the Florida Division of Forestry to serve as a preliminary planning tool for communities embarking on an urban forestry program (Theobald, 1978). The technique involves the use of a dot grid on existing photographs (Figure 7–8). Dots that fall on a tree crown are tallied, and percent crown cover for a community or neighborhood is calculated using the formula

$$\% \text{ crown cover} = \frac{\text{dots on tree crowns}}{\text{total no. dots}} \times 100$$

When using 35mm slides, percent crown cover may be determined by projecting the slides on a sheet of paper on which a dot grid has been established, and then following the same procedure. In addition, canopy cover may be transposed from slides to city maps by projecting slides on the map and adjusting the distance of the projector from the map to obtain the proper scale.

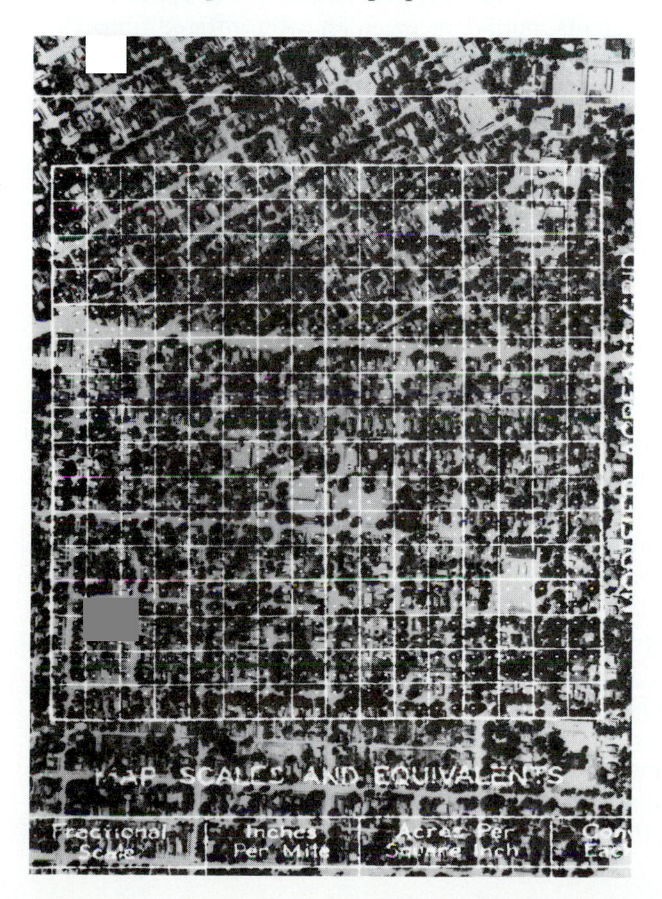

Figure 7–8 Urban tree canopy analysis using a dot grid. Percent canopy cover is determined by dividing the number of dots on tree crown for a predetermined area by the total number of dots in that area.

Tree canopy analysis may be used to determine the overall percent canopy cover for a community and the distribution of tree cover (Theobald, 1978). Caution should be exercised before deeming a portion of a community as being in need of immediate attention, as the area may have a lot of young trees not yet providing much canopy cover. Ground verification following canopy analysis is essential before drawing conclusions.

Under certain circumstances remote sensing may be used to inventory a particular species. Rodgers and Harris (1983) used color infrared film to survey the canopy coverage of pecan *(Carya illinoensis)* trees in five Texas cities, a valuable and widely planted ornamental tree in the southern United States. Color infrared film is sensitive to infrared wavelengths, with active foliage appearing various shades of red. In this study photographs were taken in April, when most other hardwood trees were in full foliage and pecans were just beginning to leaf out. Other species were red in color while the pecans had a distinct pink appearance. A grid system was used to index each hectare, and tree population data recorded by percent coverage of all species and percent coverage by pecan trees. The data files were stored in a computer and reproduced in diagrams and tables. The accuracy of the pecan survey was verified by field checking a sample of the inventory grids.

Healthy foliage reflects more infrared energy, with healthy plants appearing red while unhealthy plants appear magenta, purple, or green on color infrared film. This color difference has been much used to detect insect and disease problems in crops and commercial forests. Howe and Harker (1977) used this difference in tree vigor appearance to detect shade tree decline in Downers Grove, Illinois. Color infrared 35mm photographs were taken at various heights from a helicopter on a late morning in July. They report that chlorotic white oaks *(Quercus alba)* were distinctly pink on the prints compared to the strong red color of healthy oaks. Declining trees were detectable on color infrared film, but they were not able to clearly identify stressed trees prior to normal visible symptoms of decline.

Landscape Inventories

General landscape inventories, conducted on both private and public lands, range from cover type maps of natural ecosystems (residual plant communities or early successional communities on disturbed areas) to locating, mapping, and describing individual plants. General cover type mapping will utilize techniques described previously, while more intense inventories will use grid systems, aerial photographs, plane-table maps, or computers to locate specific plants. The following discussion considers these various tools when used to inventory public parks and forests, private landscapes, and utility rights-of-way.

Parks and forests. It is difficult to distinguish between parks and forests on the basis of ecosystem type, as the vegetation exists on a continuum ranging from carefully arranged and maintained landscape plants to true forest ecosystems. For this discussion, parkland is defined as open land with a predominant sod cover and scattered woody landscape plants. Parkland is further defined as requiring intensive man-

agement. Forestland is a natural vegetation association, including nonforest cover types, and receives more extensive management.

Inventories of forest and other natural ecosystems are most efficiently made by preliminary cover type mapping on aerial photographs followed by a more intensive ground survey employing techniques described above. Of particular usefulness in urban settings is periodic reinventory of these ecosystems to provide baseline data for continuous monitoring. Air and other pollution problems in urban regions, as well as overuse, may have a profound impact on natural ecosystems. Monitoring species frequency and relative frequency in the various strata will often provide early detection of impending problems, thus allowing time for corrective actions.

Park inventories are usually more intensive than forest inventories, often with all woody plants being recorded. In most cases a system of plant location is included in the inventory, and this will range from plotting locations on a map to computer mapping by a global positioning system (GPS). Aerial photographs may be useful to locate large specimens on maps, followed by ground surveys to identify them by species, and to collect other management information. Landscape plants may also be located on maps using general mapping techniques such as compass and pacing, transit and chaining, or mapping by intersect on a plane table. Grid systems have been used to create a series of cells located by horizontal and vertical coordinates. Each cell is then inventoried, plants cataloged, and data stored by coordinates describing the cell.

Parameters describing plant attributes and cultural needs are selected based on management objectives. Green (1984) describes a system that locates and sequentially numbers trees on a map, and these numbers are affixed to the tree in the field with aluminum tags. Each tree is identified by species or cultivar, height and diameter measured, and condition class evaluated. The following condition classes, modified by Green since publication in 1984, are used to characterize the size and relative health of each tree.

Class 1.0: Newly planted; not established but expected to live a long time.

Class 1.5: Newly planted; not established but has a problem(s) that may require treatment(s) to prevent premature death.

Class 2.0: Established tree; not mature size but expected to live a long time.

Class 2.5: Established tree; not mature size but has a problem(s) that may require treatment(s) to prevent premature death.

Class 3.0: Mature tree; expected to live at least 20 years.

Class 3.5: Mature tree; but has a problem(s) that may require treatment(s) to prevent premature death. Any key or specimen tree that rates a 3.5 should strongly be considered for fertilizing, pruning, cabling, mulching, or other arboricultural treatments.

Class 4.0: Mature tree with such a problem(s) that tree death is expected within 20 years regardless of any treatments. Under certain circumstances treatment is warranted to prolong tree life.

Class 4.5: Mature tree that is so near death that treatment is not recommended

and is considered a waste of time and money. Immediate removal is not usually necessary for safety reasons. Young trees rated 4.5 are considered inferior species (e.g. Siberian elm, Mulberry, Boxelder maple), and their removal is recommended before they become too large.

Class 5.0: Mature tree that is dead or nearly dead and immediate removal is recommended for safety reasons.

Kelsey and Hootman (1988) used Green's inventory system to inventory groves of oaks *(Quercus* spp.) on the Northwestern University campus in Evanston, Illinois. The groves are remnants of a presettlement oak savanna with many trees appearing to be in a state of decline. The inventory revealed that while the overstory was still dominated by oaks, the understory was dominated by maples (*Acer* spp.). However, 60 percent of the declining overstory trees were oaks suffering chlorosis, leading to the conclusion that without intervention the groves would no longer be dominated by oaks within fifteen years. An accompanying soil inventory revealed increasing alkalinity and disturbance were contributing factors. A program of soil acidification and understory oak planting was recommended as a result of the inventory.

Computerized park inventory systems have been developed that locate trees on maps, store management information for retrieval, and provide data summaries. Myers and McCurdy (1976) describe a system they developed that draws park boundaries and locates landscape features by symbols. The boundary survey and landscape features (located from fixed points) are measured by azimuths and bearings. Four characteristics can be measured or recorded for each feature by measurement or code. These data are entered into the computer and the program draws and locates features on the map. When used for a tree inventory, species, size, age, and condition are measured, and trees are located on the map by a species symbol. The system prints maps with a coordinate scale, and this scale may be used to add or change landscape features for planning purposes.

Management of Central Park in New York City is aided by a computerized inventory system that describes tree characteristics such as size, species, and condition; includes environmental characteristics such as soil conditions; and locates trees in a 50- by 50-ft. grid system. A computer program provides locations of all trees and their management needs. Data reduction produces species summaries, profiles of tree parameters, maintenance needs and condition, and listings of trees with specific insect and disease problems (Weinstein, 1983).

McPherson et al. (1985) developed a method to locate park trees with a coordinate grid system and to inventory these trees using a program developed for a microcomputer (Park Tree Inventory System). Trees are located in predetermined cells described by grid coordinates. These location data are then used to describe tree location when storing inventory data in the computer. Twenty-one tree parameters are described in the inventory, with computer-generated reports describing species summaries, species composition of the park, condition and value of trees, overall maintenance needs, maintenance needs by diameter class, individual records, species records, and tree history records.

Urban region vegetation surveys. Nowak (1994) used a sampling system to inventory the urban forest of the Chicago region (Cook and DuPage counties), an area covering 1,292 square miles (3,350 square km) and containing nearly six million people. The area was stratified by estimated tree cover and 652 randomly located sample plots were allocated proportionally based the estimate within each strata. He found an estimated 50.8 million trees within the two county area, with the city of Chicago accounting for 4.1 million of these trees. In the city of Chicago street trees account for 10.1 percent of the tree population, while in the two counties street trees account for only 2.0 percent of total population. However, street trees are larger on average than the total population average, accounting for 9.5 percent of the total leaf surface area regionwide.

Private landscapes. Vegetation inventories of private land are conducted by both government agencies and private contractors. Local and regional planning agencies inventory undeveloped lands at the urban periphery to determine those lands suitable for development, ascertain where development will pose clear hazards such as on flood plains or unstable underlying strata, and lands suitable for parks, open space, or environmental protection. City foresters may be involved in landuse planning surveys as resource specialists, or they may find it necessary to conduct inventories on private land in special situations such as locating private elms as part of a Dutch elm disease control program.

Surveys of selected tree species to determine density and spatial arrangements are often an important component of insect and disease control programs. Control efforts on public trees will be ineffective unless control is also directed to private vegetation. Inventories of susceptible species can usually be conducted from public property, but it may be necessary to obtain permission to enter private property. Susceptible species should be recorded by location and concentrations of host species mapped to serve as the basis of control efforts.

Tree canopy analysis, as described previously, will provide an assessment of general tree cover, but cannot normally distinguish between private and public trees or distinguish between species except in special circumstances. However, ground surveys in conjunction with canopy analysis can provide this information. Many urban neighborhoods contain predictable overstory cover based on neighborhood age, type of development, and sometimes, ethnic characteristics. One neighborhood may have a mature elm population while another is dominated by maples or oaks. New subdivisions in wooded areas often retain an overstory of native species while a subdivision placed in an open field represents what is available from local nurseries. A survey of tree cover in Stevens Point, Wisconsin, by the author revealed three distinct cover types in developed areas. The oldest neighborhoods in the community have a canopy of American elms *(Ulmus americana).* Subdivisions developed from 1940 to 1960 are predominantly maple (*Acer* spp.) covered and developments since 1960 are covered with a canopy of residual oaks (*Quercus* spp.) and pines (*Pinus* spp.), the predominant woodland cover in the region.

Predevelopment surveys are an important component of land-use planning, and aid in the preservation of residual vegetation and other landscape amenities. In his book *Design with Nature,* McHarg (1969) describes a technique of resource inventory prior to development that identifies critical resources affecting the development process and assigns relative values to descriptive attributes for each resource. These values are then put on a map overlay for each resource, with increasing restrictions to development indicated by increasing intensities of shading. The accumulative shading effect on the map of the combined overlays will identify those areas that should be protected and those areas best suited to development.

An adaptation of this method is used to assist large-scale developers by urban foresters employed by the state of Maryland. Natural features are put on transparent map overlays to determine logical neighborhood layouts and to reduce the impact on vegetation and natural drainage patterns (Moll, 1978). Theobald (1978) describes a similar application in Florida that inventories soil associations, aquifer recharge areas, hydrology, tidal flood prone areas, and other resources. Rapidly urbanizing counties are inventoried in their entirety (both public and private lands) to provide resource information for preliminary land-use planning. This information is then used to guide development, preserve unique ecosystems, and to improve management of public and private properties.

A computerized tree inventory system was developed for use on private land prior to development. Allen (1978) describes this system as being used to inventory all trees greater than 5 in. (13 cm) in diameter on a defined area. Trees are located on aerial photographs taken from 2000 ft. (600 m) and inventories made on the ground to record tree identification number, species, diameter, and biological-aesthetic tree quality. Biological-aesthetic tree quality is a rating of expected survival to construction disturbance and is ranked as high, medium, or low. Tree location coordinates are digitized and data stored in a computer. These data can be plotted on a map at a desired scale with a plotting code describing survival chances. These maps are used when designing the development to determine which trees are preserved and which trees should be removed. Allen reports that the system is best applied in open woodland situations where individual trees are readily located on aerial photographs. The availability of GPS can make such systems considerably less expensive than using field coordinates.

Commercial arborists and landscape maintenance firms will find inventories of customer properties useful, especially if they provide continuous contractual services. Large-scale industrial and commercial properties may be inventoried using a number of the techniques described above. Smaller parcels can easily be mapped using preprinted grids adaptable to different property sizes and shapes by drawing boundaries and locating structures, landscape features, and individual plants in an appropriate scale (Figure 7–9). Computer software for commercial arborists is now available that will allow a residential property inventory map and field information to be entered into a handheld field computer and be transferred to a database at the office. The database can then be used to generate reports for the property owner and work orders for field crews.

Utility corridors, although narrow, add up to a significant area of land in urban regions. Utility right-of-way vegetation management is enhanced by inventory systems, and these systems generally fall into two categories, plant communities and individual trees. Plant community inventories are usually made beneath transmission lines, where maintenance of nonforested ecosystems is the primary management goal. A number of the systems described above have application for this type of

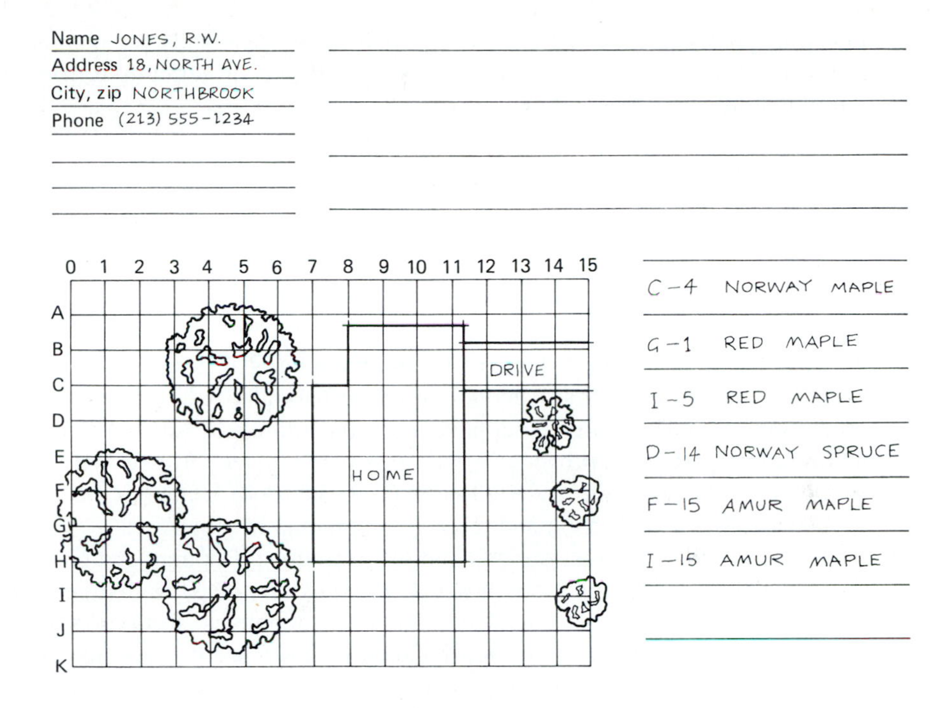

Figure 7–9 Inventory grid system developed by a commercial arborist. (Buckley Tree Service, Waukesha, Wisc.)

inventory, particularly those that employ the use of aerial photographs. Distribution-line clearance is maintained by pruning of trees in the right-of-way corridor, and these trees can be inventoried individually or collectively in management units. These data will be useful in accepting bids from contractors and in estimating clearing costs when using utility crews.

OTHER URBAN NATURAL RESOURCE INVENTORIES

A number of inventory systems described above incorporate resources other than vegetation. Frequently, information describing these resources is readily available from planning offices, other agencies, or may be inventoried as needed. The following discussion briefly describes some of these sources and/or techniques.

Soil Surveys

Perhaps one of the most useful sources of local resource information is the soil survey. Not only are local soils mapped, but descriptions of the soils contain information on slope, water table, and engineering capacities of each soil type. Recent soil surveys also contain information relating to natural plant communities found on various soil types and lists of tree species adapted to these soils for planting in urban areas. Potential uses for each soil type are described for agriculture, forestry, recreation, wildlife, and construction.

Floodplains

Floodplains exist throughout most cities and are defined as those lands adjacent to streams and rivers subject to periodic flooding. The U.S. Army Corps of Engineers has identified and mapped most floodplains according to their statistical frequency of flooding (i.e., 20-year floodplain, 50-year floodplain, 100-year floodplain, etc.). These maps are avai'able at local and state planning offices and provide an excellent source of information when conducting urban natural resource inventories. In most communities, undeveloped floodplains are subject to restrictive zoning, thus forming an important part of the urban park and open space system. Morris (1977) defines the following five floodplain site conditions in northeastern United States as they relate to their wildlife management and recreation potential in urban areas.

Site I. Poorly drained wetland subject to frequent flooding and covered primarily with willow (*Salix* spp.), cottonwood *(Populus deltiodes),* and silver maple *(Acer saccharinum).* Wildlife management is the most suitable use for these sites.

Site II. Well-drained silty bottomlands with similar tree species as Site I plus sycamore *(Platanus occidentales).* Stands are not aesthetically pleasing, with dense undergrowth containing poison ivy *(Rhus radicans)* and stinging nettles (*Utrica* spp.), limiting recreation use.

Site III. Frequently flooded, extremely unstable, and on point bars of coarse textured deposits or on first bottoms. Flood-tolerant species are most common, such as cottonwood, willow, and elm (*Ulmus* spp.). Little recreation potential due to flooding, unstable soils and stinging nettles (*Utrica* spp.).

Site IV. Old stable point bar deposits that support mature hickory (*Carya* spp.) and silver maple forests. Nettles and flood debris are absent and flooding is minor, making these areas valuable recreation sites for camping or water-related recreation.

Site V. Terraces covered with upland forests that are flooded only occasionally (100-year flood probability). These areas have the least limitations for use and are aesthetically attractive for recreation.

Riparian Surveys

Riparian areas are of special concern in urban areas. They provide diverse habitat for wildlife, corridors linking fragmented forests and other habitats, protection of water quality and aquatic habitat, and provide recreational opportunities (Henson-Jones, 1993). The first step is to delineate the watershed on the appropriate topographic map, and then determine the various cover types within the watershed. In urban areas much of the watershed will be covered with urban development, but riparian vegetation often remains on flood planes, lakeshores, and stream banks. This vegetation can be mapped and categorized according to cover type, and managed to protect water quality. In peripheral areas, riparian surveys are essential to protect water quality during development, and to provide wildlife habitat and recreation after development is completed.

Aesthetic Surveys

Aesthetics are becoming increasingly important in making land-use allocation and urban design decisions. These surveys are usually made by landscape architects to obtain data for input in land-use planning. The *Urban Forestry Handbook* (1973) developed by the Florida Division of Forestry describes a two-part survey to inventory landscape aesthetic qualities.

The first part of this system involves a description of three-dimensional site characteristics, including buildings and other structures. The site characteristics are categorized into six landforms: (1) level or gently sloping or rolling sites, (2) sloping sites backed by hills are steeper slopes, (3) valley or gorge sites, (4) amphitheater or fan-shaped sites, (5) bowl-shaped sites, and (6) ridged or hilltop sites. Buildings and other cultural features are added to the sites in five basic forms: (1) urban textures, (2) green areas, (3) circulation facilities, (4) paved open space, and (5) individually significant architectural masses.

The second part of the aesthetic survey identifies and records significant paths and vantage points for viewing. These locations are then described in terms of how the city is perceived, such as panoramas, skyline, vista, urban open space, and through experience of individuals in motion. Aesthetically pleasing and unsightly views should be identified and these data used to guide development, identify open space resources, and enhance or screen urban scenes.

Gobster (1992) devised a method for visual quality assessment for urban parks. Lincoln Park, Chicago, was divided into five landscape zones: shore/water, developed, open, formal, and treed. Photo points were established within each zone, and park users and other groups asked to evaluate slides based on their scenic beauty using a ten-point scale. Scenic beauty values were then mapped using a scale of four values ranging from low to very high. The method identified areas of highest scenic value and areas in need of visual mitigation, and suggested visual management strategies to enhance the park experience for users.

Wildlife Habitat

Wildlife populations are usually a response to available habitats. Techniques are available to estimate populations of individual wildlife species, but these techniques will not be discussed in this text. The reader is referred to the *Wildlife Management Techniques Manual* edited by Schemnitz (1980) and *Ecology and Field Biology* by Smith (1974) for detailed descriptions of these methods.

A number of techniques are available to inventory wildlife habitats and are similar to methods of vegetation cover type mapping discussed earlier in this chapter. Matthews and Miller (1980) developed a system that is being used to inventory New York State's urban wildlife habitat. Their method starts with census tract information from the U.S. Department of Commerce, a land-use and natural resource inventory

New York State Department of Environmental Conservation
Division of Fish and Wildlife
Bureau of Wildlife
Ecological Zone ________________

________________ Region
________________ County
________________ City
________________ Census Tract No.
________________ NYTM Coordinator

I. Census Information

________ 1970 population (1)
________ 1960 population (2)
________ Total Area in acres (T)
________ Land area in acres (L)
________ Water area in acres (W)
________ Nonresidential area in acres (N)
________ Gross density (1/L)
________ Net residential density (1/L–N)

Count of housing units by occupancy/vacancy status

________ Owner occupied
________ Renter occupied
________ Other

Count of all housing units

________ 1 unit or structure
________ 2 or more units/structure
________ Mobile home or trailer

Population by race

________ White
________ Black
________ Other

Population by sex and age

Male	Female	Age
		0–4
		5–14
		15–24
		25–34
		35–44
		45–54
		55–64
		65+

Year structure built total occupied and vacant year-round

________ 1965–March 1970
________ 1960–1964
________ 1950–1959
________ 1940–1949
________ 1939 or earlier

________ Average income of families
________ Average income of unrelated individuals
________ Count of families below poverty status

II. Air Photo Information

________________ Photograph Source
________________ Date
________________ Scale

Number	Area	Cover Type
	01.	Wet meadow
	02.	Flooded deciduous trees
	03.	Flooded dead trees
	04.	Flooded shrubs
	05.	Emergents
	06.	Drained muckland
	07.	Reverted drained muckland
	08.	Floating vegetation
	09.	Open water
	10.	Upland body
	11.	
	12.	Flooded conifers
	13.	Submergents
	14.	Mudflats
	15.	Streams and creeks
	16.	Rivers
	17.	Tidal
	18.	Rock area
	19.	Sandy area
	20.	Lawns
	21.	Pasture
	22.	Surfaced
	23.	Cultivated
	24.	Disturbed
	25.	Old field
	26.	Orchards
	27.	Brush
	28.	Deciduous trees
	29.	Coniferous trees
	30.	Burned area
	31.	Shrubs and trees

Number	Area		Land Use III
		A.	Agriculture – 32
		B.	Campground – 68
		C.	Commercial area – 75
		D.	Dump – 51
		Da.	Discharge (air) – 56*
		Dw.	Discharge (water) – 55*
		Dj.	Junk yard – 54
		Di.	Land fill – 52
		Ds.	Sewage treatment plant – 53
		E.	Extractive excavation – 76
		F.	Forest
		G.	Golf course – 66
		H.	Institutional – 60
		Ih.	Industry heavy – 78
		IL.	Industry light – 77
		J.	Cemetery – 61
		K.	Canal – 88
		L.	Construction – 74
		M.	Marina – 69
		N.	Airports – 91
		O.	Race tracks
		P.	Parks – 67
		Qa.	Railroad active – 86
		Qi.	Railroad inactive – 87
		Rd.	Residential development – 47
		Rm.	Residential multiple units – 46
		Rs.	Residential single homes – 43
		S.	Water engineered feature – 97
		V.	Vacant land
			*Not Used

Number	Area	Centroid	Significant Habitat IV
			1. Significant plants
			2. Significant wildlife
			3. Significant plants and wildlife
			4. Potentially significant plants
			5. Potentially significant wildlife
			6. Potentially significant plants and wildlife
			7. Known deer concentration areas
			8. Known deer concentration areas not in use
			9. Aerial survey yards—not field checked
			10. Other—such as unique geological formations

Present	Absent	Structural Wildlife Habitat V
		1. Bridges and/or overpasses
		2. Highways
		3. Slanted roofs
		4. Flat roofs
		5. Gables
		6. Porches
		7. Steeples
		8. Fences
		9. Street lights
		10. Telephone and power pole lines
		11.
		12.
		13.

Number	Animal Nuisance Complaints VI
	1. Small mammals
	2. Large mammals
	3. Insects
	4. Birds
	5. Reptiles

Centroid	Wetland Type VII
	1.
	2.
	3.
	4.
	5.
	6.
	7.

Number	Area	Natural Area—2 VIII
	Within the CT	
		Centroid
		Centroid
		Centroid
	In adjoining CT	
		Centroid
		Centroid
		Centroid
	Juxtaposed to CT	
		Centroid
		Centroid
		Centroid

Figure 7–10 Urban Wildlife Habitat Data Sheet, New York Department of Environmental Conservation. (From Matthews and Miller, 1980.)

made by the state, reports on significant habitats, the state wetland inventory, tidal wetland maps, and recent aerial photographs. This information is integrated and used to develop an inventory of urban wildlife habitats with the following procedure:

1. Each census tract is outlined and transposed from the census tract maps to base maps.
2. Data sheets are prepared for each census tract.
3. Aerial photographs of each census tract are interpreted for vegetative cover types, natural areas, and urban land use by block. A block is a well-defined piece of land bounded by roads, railroads, streams, or other landscape features.
4. Cover types are identified according to vegetative characteristics and land surrounding structures within each block. Each block is assigned an individual or mixed cover type on the base map.
5. Land use is identified from aerial photographs in twenty-seven categories and placed on the base map.

The data collection sheet used for this inventory is presented in Figure 7–10. Data collected and base maps made for this inventory provide the groundwork for an urban wildlife program by identifying habitat availability. This information may be further used by urban planners, zoning boards, and land developers, and urban residents desiring to include wildlife in the overall urban design (Matthews and Miller, 1980).

GEOGRAPHIC INFORMATION SYSTEMS

Historically, geographic information was generally kept and maintained on cloth or paper maps. Increasingly this information is electronically stored, maintained, and manipulated by geographic information systems (GIS). These systems not only allow us to enter and store geographic information, we can also rapidly manipulate, analyze, and display geographic or spatial data. Geographic information in GIS is typically stored in layers; such information includes roads, utilities, land use, topography, soils, hydrology, lakes and streams, vegetation, and census tract information. The user can combine layers to generate new maps, or extract specific information from a data layer to add it to another layer. GIS is used to model and predict changes over time, such as vegetation changes through plant succession. Updating information on GIS maps can be done much more quickly than updating more traditional maps (Congalton and Green, 1992).

Information for GIS can come from a variety of sources. For example, the United States Geological Survey provides not only hard-copy maps, but also map information in digital format on computer tapes that can readily be transferred to GIS. The United Census Bureau has digital information available on roads, railroads, census tracts and blocks, political areas, latitude and longitude, feature names, and classification codes through the Topologically Integrated Geographic Encoding and Referencing (TIGER) system (Congalton and Green, 1992). Remote sensing information from aerial photographs can be digitally scanned and georeferenced into GIS, as can satellite imagery and airborne video (Lachowski et al., 1992).

GIS Applications in Urban Forestry

Many communities have, or soon will have, information such as census tracts, land use, zoning, utilities, floodplanes, and so forth on GIS. When this is the case, city foresters have the opportunity to develop or enter natural resource data layers in the system, including vegetation (street trees, canopy density, and cover types), soils, wetlands, and critical habitat.

The city of Ann Arbor, Michigan, used a ground-based survey to inventory street trees and color infrared aerial photography to map wetlands, woodlots, and canopy cover, and to establish continuous urban forest inventory plots. Wetlands and woodlots were mapped and scanned into separate GIS layers, and urban tree crowns were evaluated for energy savings and placed in another GIS layer. Wetland and woodlot layers serve as the basis for future wetland and woodlot protection ordinances, and the tree crown evaluation layer is used to identify those areas of the community where more tree planting will conserve energy (Laverne, 1993).

The city of Baltimore, Maryland, uses a GIS that relates biophysical and socioeconomic information in layers that include land use, vegetative cover, streams, watersheds, soils, roads, and census tracts. Maps generated from the system are put to a variety of uses, including managing and conserving stream and greenway corridors, identifying critical bird habitat threatened by forest fragmentation, identifying neighborhoods where tree survival is low, and selecting appropriate species for planting based on soils and slope (Grove and Hohmann, 1992).

A GIS layer was prepared for the city of Stevens Point, Wisconsin, containing all vegetation in and around the city. Developed areas were mapped by percent canopy cover, while vegetation in undeveloped areas was mapped by cover type using aerial photographs. These data were transferred to the GIS and are used for areawide land-use and open-space planning (Dwyer, 1996).

Urban ecological analysis. A system of using GIS information to inventory and place economic values on urban forest resources was developed by American Forests, Washington, DC (Moll et al., 1995). Natural system values have usually not been included in the land-use planning and development processes, primarily because we have either regarded nature as an impediment to development or because we have had difficulty placing quantifiable values on the function of these systems. However, just as we regard forests and other natural systems as ecosystems, urban areas are ecosystems we have developed for human habitation. Natural processes such as material cycling and energy flows occur in urban ecosystems just as they do in all ecosystems. Trees, forests, and other natural ecosystems systems perform very important ecological functions around and in urban areas.

The urban heat island as discussed in Chapter 4 is a function of replacing vegetation with built objects, resulting in higher temperatures and increased air pollution associated with higher temperatures. The urban heat island results in higher air conditioning costs, higher air pollution control costs, and the cost of increased air pollution on public health. Likewise, stormwater runoff increases along a gradient from rural land to the central city, the result of increasing amounts of pavement and roofs, and decreasing amounts of vegetation. Increased runoff carries with it increased pollution loads that end up polluting aquatic systems, or being treated at a high cost to the community. Urban areas interrupt much natural nutrient cycling, with nutrients becoming a part of the urban waste load that is costly to deal with. A significant part of these costs can be mitigated through increased use of urban vegetation and land-use planning that retains as many natural system functions as possible (Moll et al., 1996).

Planners have long recognized the function and value of natural systems, but have not been able to place an economic value on these functions that justifies their inclusion in the development process. Urban ecological analysis is a method to identify and quantify these values to insure their consideration in urban development. Planners use GIS for land-use planning, but traditionally have used satellite images for vegetation analysis. These images work well for undeveloped land, but the resolution is not sensitive enough to recognize trees in developed landscapes. However, low-level aerial photographs do recognize trees in developed landscapes. Thematic Mapper satellite data detected a 1 percent tree cover in a neighborhood of Atlanta, Georgia, where low-level aerial photographs detected a 16 percent tree cover (Moll et al., 1996).

Urban ecological analysis uses low-level aerial photographs to stratify communities into nine ecostructures for analysis:

1. Low-density single-family residential with light canopy
2. Low-density single-family residential with medium canopy
3. Low-density single-family residential with dense canopy
4. Medium-density single-family residential with light canopy
5. Medium-density single-family residential with medium canopy
6. Medium-density single-family residential with dense canopy
7. High-density single-family residential with light canopy
8. High-density multi-family residential with light canopy
9. High-density commercial with light canopy

Each ecostructure is sampled and field measurements made describing the quantity of pavement, grass and tree canopy. The impact each ecostructure has on energy conservation, water retention, air quality, and nutrient cycling can then be calculated. Table 7–5 is an example of energy savings derived using the urban ecological analysis method. Energy savings in the United States from urban trees has been estimated to be $4 billion annually. The urban ecological analysis system predicts planting more urban trees in strategic locations is likely to double this value. As this

TABLE 7–5 ECONOMIC BENEFITS IN ENERGY SAVINGS FROM URBAN FORESTS IN FREDERICK, MARYLAND. ECOLOGICAL TYPES REPRESENT FOUR ECOSTRUCTURES IDENTIFIED IN THE COMMUNITY. (From American Forests, 1995.)

City Classified According to Ecological Features

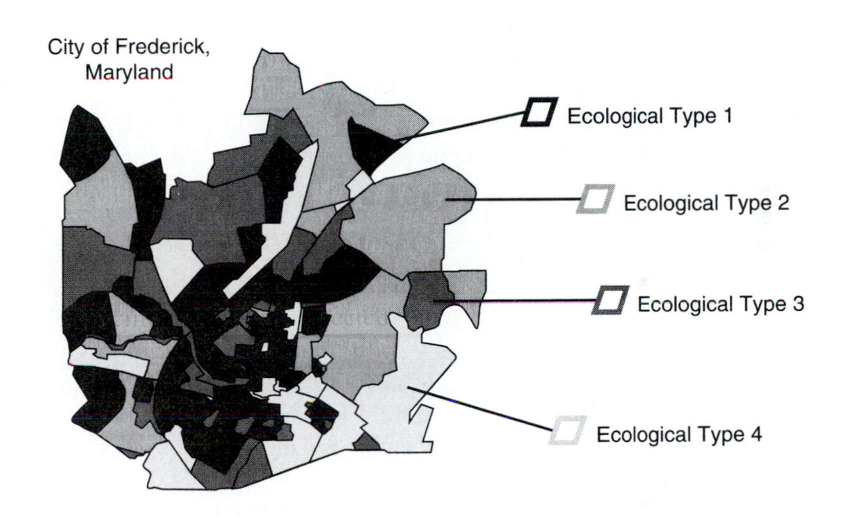

Benefits of Urban Forests: Frederick Maryland

Ecological Type	Energy Savings	Erosion Control	Storm Water	Air Pollution	Total Benefits
One	$503,600	+	+	+	=______
Two	$422,300	+	+	+	=______
Three	$62,600	+	+	+	=______
Four	$400	+	+	+	=______
Total	$988,900	+	+	+	GRAND TOTAL

technique develops further, values associated with storm water management, erosion control, wildlife, carbon reduction, air quality, and others can be added to benefits associated with energy conservation (Moll et al., 1995).

Using these kinds of calculations, the full range of urban forest values can be inferred by comparing ecosystem values of land developed at different intensities with different amounts of vegetation to similar undeveloped land. As development intensities increase, the natural benefits of ecosystems decrease. Benefit/cost analyses of removing or retaining trees and natural ecosystems can be calculated, and will provide a powerful incentive to protect and retain trees and natural ecosystems for their ecologic functions in developed landscapes. Planners will be able to use the economic benefits of ecosystems to argue strongly for using natural ecological processes in urban development through the retention and inclusion of trees and other plants, and natural ecosystems in the overall landscape (Moll et al., 1995).

LITERATURE CITED

ALLEN, M. S. 1978. "Creating an Urban Tree Environment in New Residential Communities: An Adaptation of the Computer to Suburban Tree Management." *Soc. Am. For. Proc.,* pp. 477–483.

AVERS, P. E., D. T. CLELAND, W. H. MCNAB, M. E. JENSEN, R. G. BAILEY, T. K. KING, and M. E. GOUDEY. 1993. *Summary, National Hierarchical Framework of Ecological Units.* Washington, D.C.: Ecomap, USDA Forest Service.

CONGALTON, R. G., and K. GREEN. 1992. "The ABCs of GIS." *J. Forestry* 90(11):13–20.

COWARDIN, L. M., ET AL. 1979. *Classification of Wetlands and Deepwater Habitats of the United States,* FWX/OBS–79/31. Washington, D.C.: USDI.

DAUBENMIRE, R. 1980. "The Scientific Basis for a Classification System in Land-Use Allocation." *The Scientific and Technical Basis for Land Classifications.* Soc. Am. For. Proc., pp. 7–10.

DRISCOLL, R. S., ET AL. 1984. *An Ecological Land Classification Framework for the United States.* USDA For. Serv. Misc. Pub. No. 1439.

DWYER, M. C. 1996. "Ecosystem Based Urban Forest Planning for the Stevens Point, Wisconsin Area." M.S. thesis, Univ. Wisc., Stevens Point.

FLORIDA DIVISION OF FORESTRY. 1973. *Urban Forestry Handbook.* Tallahassee: Fla. Div. For.

FYRE, F. H., ed. 1980. *Forest Cover Types of the United States and Canada.* Washington, D.C.: Soc. Am. Foresters.

GOBSTER, P. H. 1992. "Managing Visual Quality in Big Diverse Urban Parks: A Case Study of Chicago's Lincoln Park." *Managing Urban and High-use Recreation Settings.* Minneapolis: USDA Forest Service Gen. Tech. Rep. NC-163, pp. 33–40.

GRAHAM, S. A. 1945. "Ecological Classification of Cover Types." *J. Wildl. Manage.* 9:182–190.

GREEN, T. L. 1984. "Maintaining and Preserving Wooded Parks." *J. Arbor.* 10(7):193–197.

GROVE, M., and M. HOHMANN. 1992. "Social Forestry and GIS." *J. Forestry* 90(12):10–15.

HENSON-JONES, L. 1993. "Riparian Reforestation: Restoration Goals." *Proc. Sixth Nat. Urban For. Conf.* Washington, D.C.: Am. For. Assoc., pp. 217–220.

HOWE, V. K., and G. R. HARKER. 1977. "Infrared Photography in the Detection of Shade Tree Decline." *Proc. Midwest Chap. Int. Shade Tree Conf.,* pp. 40–44.

KELSEY, P. D., and R. G. HOOTMAN. 1988. "Soil and Tree Resource Inventories for Campus Landscapes. *J. Arbor.* 14(10):243–249.

KOTAR, J. J., A. KOVACH, and C. T. LOCEY. 1988. *Field Guide to Habitat Types of Northern Wisconsin.* Dept. of Forestry, Univ. Wisc.–Madison, and Wisc. Dept. of Nat. Res., Madison.

LACHOWSKI, H., P. MAUS, and B. PLATT. 1992. "Integrating Remote Sensing with GIS." *J. Forestry* 90(12):16–21.

LAVERNE, R. J. 1993. "Mapping the Complete Urban Forest." *Sixth Nat. Urban For. Conf.* Washington, D.C.: Am. For, Assoc., pp. 172–175.

MATTHEWS, M. J., and R. L. MILLER. 1980. "New York State's Urban Wildlife Habitat Inventory." *Trans. Northeast Sec. Wildl. Soc.,* pp. 11–18.

MCHARG, I. L. 1969. *Design with Nature.* Garden City, N.Y.: Doubleday & Company.

MCPHERSON, E. G., ET AL. 1985. "A Microcomputer-Based Park Tree Inventory System." *J. Arbor.* 11(6):177–181.

MOLL, G. A. 1978. "Urban Forest Planning." *J. Arbor.* 4:213–215.

MOLL, G., J. MAHON, and L. MALLET. 1996. "Urban Ecological Analysis: A: A New Public Policy Tool." *Urban Ecosystems* 1(1).

MORRIS, L. 1977. "Evaluation, Classification, and Management of the Floodplain Forest of South-Central New York." M. S. thesis. Syracuse: SUNY.

MYERS, C. C., and D. R. MCCURDY. 1976. "Computer Mapping in the Urban Forest." *Trees and Forests for Human Settlements,* IUFRO P1.05 U. Toronto, Ontario, Canada, pp. 15–26.

NOWAK, D. J. 1994. *Urban Forest Structure: The State of Chicago's Urban Forest.* In E. G. McPherson, D. J. Nowak, and R. A. Rowntree, eds. *Chicago's Urban Forest Ecosystem: Results of the Chicago Urban Forest Climate Project.* Gen. Tech. Rep. NE-186, Radnor, PS.: USDA Forest Service.

RODGERS, L. C., and M. HARRIS. 1983. "Remote Sensing of Pecan Tree in Five Texas Cities." *J. Arbor.* 9(8):208–213.

SCHEMNITZ, S. D., ed. 1980. *Wildlife Management Techniques Manual.* Washington, D.C.: Wildl. Soc.

SMITH, R. L. 1974. *Ecology and Field Biology,* 2nd ed. New York: Harper & Row Publishers.

THEOBALD, W. F. 1978. "Urban Forestry Inventories in South Florida: Need for and Uses." *Proc. Natl. Urban For. Conf.,* ESF Pub. 80–003. Syracuse: SUNY, pp. 756–766.

WEINSTEIN, G. 1983. "The Central Park Tree Inventory: A Management Model." *J. Arbor.* 9(10):259–262.

PART 3
Planning for and Management of Urban Vegetation

8

Planning and Urban Forestry

Planning is nothing more than thinking out a course of action in anticipation of the future. Anticipation of the future is based on analyses of current trends and projection of these trends beyond the present. Predicting the future is an uncertain task at best, and the risk of error increases with the length of the projection. In spite of the risk associated with erroneous projections, we cannot afford not to plan. Many of our current problems are the direct result of not attempting to anticipate these problems in the past.

Planning is an interdisciplinary effort drawing on the expertise of individuals representing many different professions. The planner identifies those disciplines needed to develop the plan and coordinates their efforts. Past trends are analyzed and a series of future scenarios developed based on different plans. The course of future action is then decided upon by managers in the private sector, or by managers or elected officials in the public sector.

Municipal foresters are involved in planning at two levels, as resource experts in overall community planning and as managers developing plans for the community forestry program. At the community level, municipal foresters provide expertise for land-use planning based on their knowledge of vegetation, plant communities, and overall natural resources. As urban forest managers, municipal foresters must develop management plans to deal with urban forest resources on a day-to-day basis and develop long-term management plans to ensure continued urban forest amenities for the community. In this chapter we describe the basic planning process, relate this process to land-use planning, and discuss the urban forester's input in this process. The planning process is then discussed as a tool in developing short- and long-term management plans for community forestry.

THE PLANNING PROCESS

To be an effective tool, planning must be an ongoing process that is continuously open to new data and changes in values. The planning model described below is presented in the context of community planning, but certainly has application as a planning tool for the private sector. The basic planning model asks three questions and includes a feedback loop.

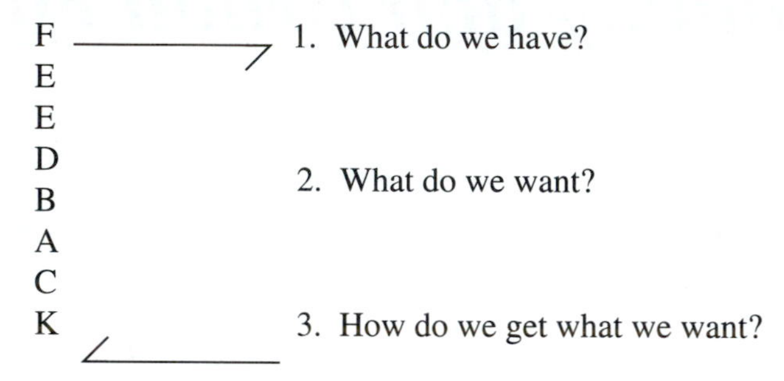

What do we have? The first step in planning is the establishment of baseline data that describe the recent past and the current situation, and project the future if the status quo is maintained. These data are collected in an inventory descriptive of the present situation and are compared to past inventories to draw inferences regarding the future. An example of this is community population growth. If the population is 40,000 and for the past decade has grown at an average rate of 2 percent per year, then anticipating the population size at points in the future is a simple task. However, this assumes that all other factors influencing community population growth remain constant, and this becomes an increasingly risky assumption the further we project our figures into the future. For example, the location of a large new industry in the community will certainly make population projections in error. However, it is far more risky not to anticipate community growth and end up with overtaxed public services such as schools, water, and sewage treatment.

What do we want? In community planning this becomes a rather complex task. Different interest groups view the future through different eyes, and this often results in conflict. Consider the community growth example discussed above. The Chamber of Commerce may think that 2 percent annual population growth is too low and ask public officials to entice new industry to locate in the community. On the other hand, the local conservation organization may be of the opinion that 2 percent annual population growth is too high and propose public policies that encourage slower growth. Ultimately, public policy decisions must be made that consider opposing viewpoints and develop a consensus for future planning.

On the other hand, what we have may in fact be what we want, and if so, the process is complete, as there is no need to change direction by public policy. This does not mean that planning is to be abandoned, as it is an ongoing process designed to establish a system of gathering information and assessing community values.

How do we get what we want? Once a consensus has been reached concerning future goals, procedures must be enacted to ensure that these goals are reached. In public planning reaching these goals involves decision making, public information and education, and legislation. Legislation may invoke the three powers of government, power of policies, power to tax, and the right to take by eminent domain. These three powers as they relate to planning will be discussed in detail later in this chapter.

Feedback. Planning is a process, and being a process must have a feedback system to be self-correcting. For example, to evaluate whether public policies enacted as part of the planning process are effective, we must again assess what we have. If the new inventory ascertains that the policies are effective, what we want must be re-evaluated. Community values may change, the outcome of our policies may not be what we wanted after all, or a new factor may have entered the picture that changes everything.

LAND-USE PLANNING

Planning at the public level falls under the general umbrella of comprehensive community planning. Comprehensive planning is a process designed to fulfill objectives of social, physical, and economic well-being for the public in the present and in the future. Comprehensive planning integrates transportation, economic, health care, sanitation, water supply, and land-use planning efforts.

Land-use planning is concerned with the location and amount of land developed for various uses. The land-use planning process attempts to minimize land-use conflicts and maximize human benefits through allocation of land to uses such as industrial, commercial, residential, conservancy, and agricultural. Prior to land-use planning, land use was determined primarily by the interaction of economic and social determinants (Florida Division of Forestry, 1973).

Economic Determinants

Land-use allocations are determined primarily by economic determinants or their market value. Historically, productive land was farmed, less productive land grazed, and even less productive land used for wood supplies or not at all. Community location was determined by access to transportation (water) or by its defendability.

Within the community, homes were built near places of employment and shops opened near the homes. The "highest and best use" of land was that use that generated the highest income. City land is worth more than farmland, and farmland is worth more than forestland. Changes in land uses generally went from a lower to a higher use; thus conversion of farms to housing and housing to commercial or industrial sites was viewed as a good thing.

Social Determinants

Social determinants interact with economic determinants and, to a lesser extent, influence land-use patterns. People tend to aggregate in neighborhoods with others of like interests and backgrounds. In the United States each wave of immigrants settled in communities or neighborhoods with the people of the same ethnic background. Sometimes the same neighborhoods served successive waves of immigrants from different parts of the world as they came, improved their lot, and moved to more affluent neighborhoods. Segregation imposed from without also influenced the ethnic character of some neighborhoods, as these areas were the only places housing was available. Perceived social values and norms influence land-use patterns. The urban exodus to the suburbs in the United States following World War II was based not only on increased income, but the notion that life was somehow better in the suburbs.

As long as population density is low and land abundant, economic and social determinants will generally allocate land to appropriate uses. It is when conflict occurs that the need to plan for land use becomes important. Earlier cities were small and people needed to walk to their jobs and to stores. However, when industrialization invaded cities, conflicts arose over the allocation of land to various uses. Furthermore, social determinants that created voluntary ethnic neighborhoods provided diversity and interest in cities, but enforced segregation violated the civil rights of segments of the population.

In rural areas conflicts arose over the invasion of prime agricultural lands by shopping centers and subdivisions. Modern farms and rural housing are not necessarily good neighbors, as new residents complain of farm noises and smells, while farmers are driven to sell their land because of inflated property values and subsequent property taxes.

Public Interest

When conflicts arise over land-use allocation and other matters affecting the general health, safety, and welfare of the public, the government becomes involved in the public interest. Land-use planning in the public interest had its start in the industrial city to resolve land-use conflicts, and has since spread to rural areas following the suburban sprawl of the past several decades. Initially, the public interest was concerned with controls such as zoning and public health codes, but now includes maximizing the livability of communities. In general, the public interest as a land-use determinant includes safety, health, convenience, economy, and amenity. The urban forester can contribute to land-use planning efforts in all of these areas, through involvement in density controls, convenience, hazards, pollution abatement, and amenities (Florida Division of Forestry, 1973).

Density control is placed on housing as a control on the number of people allowed per unit area. Detached housing on large lots represents the low density end of the spectrum, with high-rise apartments having the highest density. Density also considers the ratio of people in a given area to available open space. A planned community might have high-density housing but include ample public open space throughout the development, while a subdivision of detached houses may have no public open space.

Convenience as a factor in land use involves many things in planning, such as the location of stores and transportation corridors. For the urban forester this is related to density, and considers the location of parks and other open space in relation to residences. Identification of appropriate sites for recreation and open space in new developments is an important function within the overall urban vegetation management plan. In addition, the location, identification, and protection of rare and endangered species and ecosystems prior to development may be an important function of the urban forester.

Public hazards associated with the development of housing in floodplains and other areas prone to flooding can be averted through controls on land use. Ecosystems and soils associated with periodic flooding can be identified in the absence of flood hazard maps by analysis of vegetative communities. Other hazards include soils unsuitable for building and/or septic tanks, and geologic strata unstable for construction.

Hazards from various sources of pollution pose a public health threat in most cities. Land-use planning can allocate land use in spatial arrangements to minimize the impact of various types of pollution. As discussed in Chapter 4, urban vegetation can make a substantial contribution to reducing some of these hazards. Proper location of forest belts will reduce air and noise pollution, and exurban forests may be used to neutralize urban wastewater. Water pollution can be further reduced by the use of urban vegetation to control erosion and to act as sinks for surface runoff.

Amenity is defined as the quality of the environment in which we work, live, and recreate. The contribution to the quality of life of trees and other vegetation on streets, in parks and open spaces, and throughout the community is well documented. Planning for the inclusion and protection of this vegetation is an important contribution to the land-use planning process both in the absolute quantity of vegetation and in its spatial arrangement.

Planners are increasingly using ecological concepts in community development (see Chapter 7). Natural elements serve broad ecological functions in human communities, and inclusion of these elements in the land-use planning process not only makes cities more amenable to live in, but conserves resources, reduces pollution, and reduces the cost of operating our communities.

McPherson (1989) recommends that not only should trees and other vegetation be included in planning efforts, but that urban landscapes be compatible with the local environment. Tucson, Arizona, is located in a desert, and like many cities in deserts most early landscaping efforts included turf, shade trees, and other water-demanding plants. In recent years concern for water conservation has led to a very definite shift to xeric landscaping using drought-tolerant trees and other plants. However, using desert plants that require little water for urban landscapes is not the same as using gravel, stone, and other nonliving materials as the primary landscaped materials. Urban forest benefits as described in Chapter 4 are not obtained with nonliving materials.

Powers of Government

To achieve planning goals identified in the planning process, public policies are enacted that utilize the three powers of government: (1) the power of police, (2) the power to tax, and (3) the power to take by eminent domain.

Power of police. Police power is the use of regulation for the protection of society or individuals, including the taking of rights without compensation. In land-use planning, subdivision ordinances and zoning utilize police power.

Subdivision ordinances exercise power over the design of subdivisions, including physical layout, street standards, utility service, and the inclusion of public open space. In some cases all the land may be developed and funds given to the community by the developer to purchase other land for park or open space.

Subdivision ordinances can also be used to provide adequate space for public street trees because these ordinances regulate the width of the space between the sidewalk and the curb. A width of four feet is enough to grow a tree with a small mature size, but large trees that will make substantial ecological contributions to the community should have a minimum of an eight-foot planting strip or greater if possible. Wide planting spaces will also reduce future costs associated with repairing sidewalks and curbs damaged by tree roots.

Zoning regulations control land use by dividing the land into different use districts and setting standards for development. Typically, zoning ordinances regulate parcel use, lot size, density, street and property line setbacks, and structure size. The concept of zoning is that while landowners have restrictions placed on the use of their land, they are protected by having these same restrictions on adjoining parcels.

Zoning can be used to identify development zones, and areas such as forests or farmland we wish to protect from development. Development rights can be purchased outright from property owners as easements by government in protection zones. Development rights can also be purchased by developers for the right to develop land in areas zoned for development, a process known as transferable development rights. For example, Montgomery County, Maryland, uses transferable development rights to protect farmland from development (Schwab, 1993).

Overlay zones are areas that have been identified as environmentally sensitive for reasons ranging from groundwater protection and flood protection to wildlife habitat protection. The overlay zone may cover several different zoning districts and impose restrictions on development that are designed to protect environmental resources. Overlay zones can also be used to protect urban forest resources while meeting other needs. For example, groundwater protection zones used to protect water supplies by restricting housing densities and agricultural practices over aquifer recharge areas may also protect natural areas from development (Schwab, 1993).

The town plan for the community of Oakville, Ontario, recognizes five environmental land-use designations or zones. These consist of (1) major creeks and ravines, (2) high-quality upland woodlands, (3) poor-quality woodlands, (4) regional biological sensitive areas, and (5) shoreline protection areas. Acquisition of these areas is primarily through parkland dedication by land developers. Where it is not feasible to obtain the entire environmental land-use designation zone through park dedication, the developer is required to retain and protect tree cover in the zone during development (McNiel, 1991).

Zoning is well accepted in cities but is sometimes controversial in exurban and rural areas. Zoning to prohibit development of land in the public interest for health and safety purposes (i.e., floodplains, unsuitable soils, and so on) has strong support in the courts. However, zoning to preserve land for open space and aesthetic reasons is often interpreted as a taking by courts of the United States, with the community directed to compensate the landowner. Zoning for open space and aesthetics has been upheld in other countries, particularly where land ownership is interpreted as carrying social responsibilities.

Many communities now have zoning ordinances that permit cluster housing. Smaller-than-standard lots are permitted, provided that the overall density within the development does not exceed community standards and provided that land so saved is placed into public open space (Figure 8–1). Cluster housing still only permits a

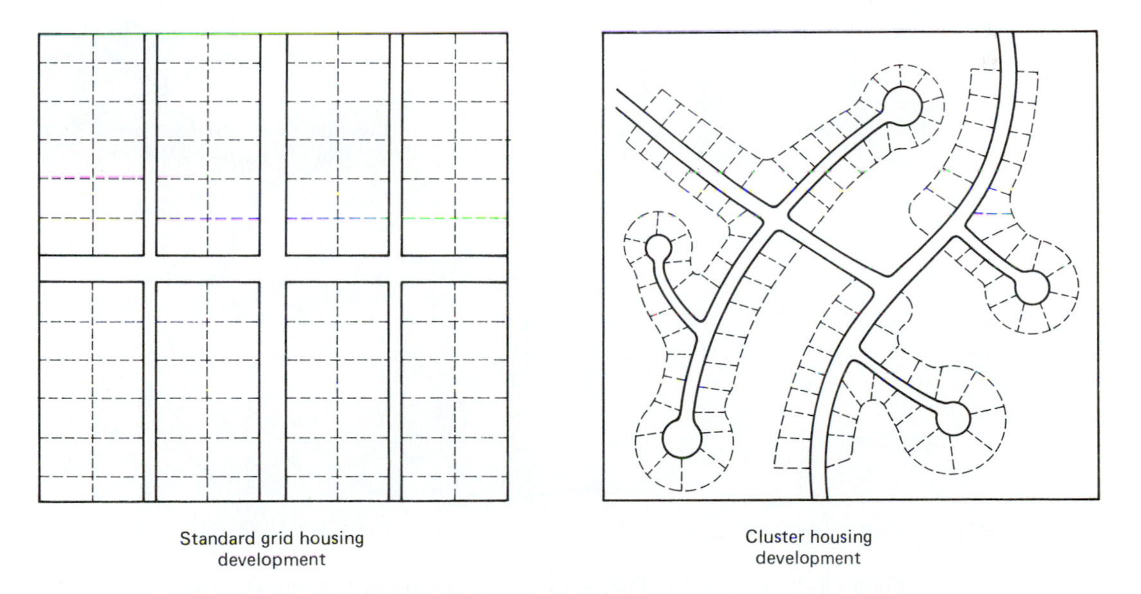

Figure 8–1 Examples of grid and cluster housing developments. Both areas are the same size and have 96 house lots. The cluster housing site retains 50 percent open space.

specified land use, usually single-family dwelling units. Many developers desire to provide a variety of housing in large developments but are restricted by zoning to provide only the type specified. Some communities now have a planned community development (PCD) amendment on their zoning ordinance that allows the developer greater freedom to provide different housing options, commercial areas, more open space, and greater sensitivity to the landscape. In PCDs all zoning is waived except the total number of dwellings for the entire parcel, with the plan being accepted on the basis of performance (Figure 8–2).

American suburban developments and sprawl have been criticized for their waste of land and resources, and their lack of community focus. In an article summarizing recent trends in land use planning and development, Adler (1995) suggests that subdivision ordinances mandate smaller lots, narrower streets, mixed housing types, and large trees planted curbside. Also recommended are more public open space, town centers, corner stores, mass transit, lower-intensity street lamps, and definite boundaries to limit sprawl. Planners for the Portland, Oregon, urban area drew a line around 325 square miles of land that encompasses 24 municipalities and designated the included landscape to absorb all population growth. Twenty years after implementation average detached house lot size has shrunk from 13,000 square feet to

Figure 8–2 Reston, Virginia, is a planned community that allows for a variety of housing options, including apartments, townhouses, and detached housing. (Courtesy of Reston Land Corporation.)

8,500 square feet. Between the years 1990 and 2040 population in the Portland area is expected to increase 77 percent, but land used for residential development is expected to increase by only 6 percent (Adler, 1995).

Power to tax. Land use is influenced by taxation, both as an incentive and as a penalty. Property taxes are levied based on the appraised or market value of land. This has caused problems in preserving prime agricultural lands near cities, as land values and property taxes inflate when nearby parcels are developed, forcing farmers to sell. Some states have enacted legislation that allows land to be taxed at its present use provided that the owner agrees not to develop it, and includes penalties should the owner break the agreement. Other states provide income tax credits to offset high property taxes on prime agricultural land near urban areas. Real-estate taxes are sometimes used to raise capital for the purchase of open space or the purchase of development rights to the land.

Power of eminent domain. The government has the power to take private property for public use, but must compensate the owner for the fair market value. If the property owner refuses to sell, condemnation proceedings may be used to obtain the property, with the courts deciding the fair market value. Eminent domain may also be used to obtain development rights or easements to property for the public good. Public parks are sometimes obtained by outright purchase, while open space and greenbelts are preserved by easements or the purchase of development rights. Eminent domain may also be used in the private sector for the public good; for example, power companies may condemn land to obtain easements for transmission lines.

PLANNING FOR OPEN SPACE

Land-use planning in the past was concerned primarily with allocation and distribution of land for various uses. Open space and park land were identified by two criteria: location and availability. The first criterion was, and is, an important consideration in planning for urban green space. Open space land should be in a location where most of the population have access to it. Kürsten (1993) recommends that all residents in a community be within a fifteen-minute walk from contact with nature, and suggests planning park and open space location based on this distance. However, land values in urban regions are determined largely by location. Land located in an inconvenient place for industry, commerce, or housing will probably be inconvenient for recreation. Location is related to the second criterion, availability. Availability is obviously important in obtaining open space land, but has often led to the inclusion of unsuitable land for recreational use. Land in urban areas that has not yet been built upon is probably unsuitable for construction due to site problems, and these limitations may also affect the development of recreation facilities.

Snyder (1973) states that until recently land-use planners dealt with forestry and open space casually. These lands were included in plans as open space based on the fact that they were not yet built upon. Data from recent aerial photographs and topographic maps were used to identify open space, with little or no site analysis done to determine if these lands were suitable for this use. Snyder (1973) further remarks: "The planner is desperately in need of the technical expertise of the urban forester if he is to balance the demands of man and nature. The forester on the other hand will find the planner's frame of reference helpful in making his work relevant in the long run."

Land included in an open space and park system should be selected on the basis of its location, availability, and suitability. Not all open space land will be subject to intensive recreational use, nor should it. Land that is unsuitable for development may have high value as wildlife habitat, flood buffers, pollution sinks, or aesthetics, and should be identified and protected for these reasons. Other land that has high potential for development may also have high recreational or environmental values, and the use of these lands should be determined based on the long-term public benefit. Criteria for land-use decisions should consider economic, social, and environmental factors. Urban foresters and other natural resource professionals can make important contributions to the land-use planning process by their expertise in the analysis of natural systems.

Urban foresters have a strong role to play in urban and regional planning. Schafer and Moeller (1979), in discussing the integration of urban forestry with urban planning and development, suggest that urban foresters become involved in the planning process and develop the following:

1. Strategies for incorporating urban forest management and protection procedures into a more comprehensive urban planning process
2. Information exchange systems and methods to ensure public involvement in urban forestry management decisions
3. Systems of monitoring technology and social change to evaluate their impacts on future urban forestry programs
4. Ways to integrate urban forest management technology into community planning systems that emphasize natural ecological processes

Involvement of urban foresters in land-use planning takes place on many levels. Municipal foresters often serve in an advisory capacity to urban and regional planning agencies, providing expertise on natural resource issues. Some county and state forestry agencies employ urban foresters to serve as advisors to local governments and the private sector. Consulting urban foresters offer land-use planning assistance for site and ecosystem analysis to both the public and private sectors. Consultants also offer services to communities to develop urban forest management plans.

An example of urban forestry in land-use planning comes from the state of Maryland. Counties south of the Baltimore-Washington metropolitan area were receiving strong development pressure from an expanding urban population. The Maryland Department of Natural Resources, the Forest Service, and other agencies concerned by the impact of rapid urbanization in these counties developed the concept of urban forest planning to retain valuable forest vegetation. This program is concerned with tree preservation on home sites, preservation of forest screens and buffer strips, preservation of forests on recreation areas, and preserving prime forestland for future wood production. State urban foresters assist county planners in the development of county master plans, zoning ordinances, and subdivision regulations. They also provide assistance to developers in site analysis and with preservation of trees and open space. Information-education is an important aspect of the urban forestry assistance program. The public and public officials are made aware of the consequences of uncontrolled sprawl and advised of the planning tools available to control the process. Literature is available concerning the preservation of trees during development, and advising appropriate species for urban planting (Moll, 1978).

From their very conception, planned communities have the option of including open space. Analysis can be made of all sites prior to development, and land-use plans then include those areas best suited to open space. Reston, Virginia, is a totally planned community founded in 1960 (Figure 8–3). The community, with a population in excess of 40,000, now has over 850 acres of open space, 50 miles of

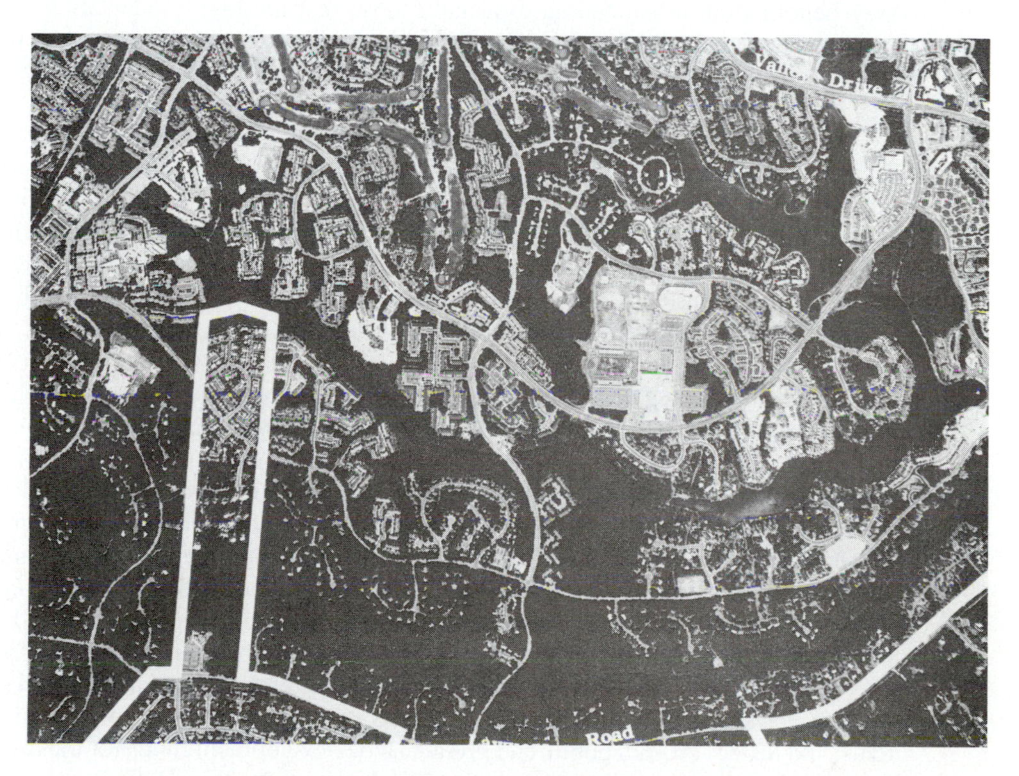

Figure 8–3 Reston, Virginia, combines a planned community with abundant greenbelts, parks, and other open space. (Courtesy of Reston Land Corporation and Air Survey Corporation.)

pathways, athletic fields, garden plots, and a 70-acre nature center. This land is managed by the Reston Homeowners Association (RHDA), a private nonprofit service corporation. The goal of the RHDA is to provide the best living environment for residents, "best" being defined as "a healthy ecosystem that provides, in the long run, the most productive growth of vegetation and wildlife compatible with surrounding land use" (Ziminiski, 1978).

Diversity in Urban Open Space Planning

Much of the effort for protection of biological diversity is focused on rural lands, but land in urban greenbelts and other open spaces can make a contribution to the protection and enhancement of biodiversity. Urbanization typically leads to a decline in native species richness as critical habitats are lost, natural systems are fragmented, and introduced exotics compete with native species for space. As landscapes are developed for urban use, some loss of species is inevitable because some species such as wolves or large cats are incompatible with humans. Likewise populations of other species will be reduced in number as habitat is lost to development. Some species may benefit from urbanization, especially those that are highly adaptable and have more generalized habitat needs. However, much diversity can be protected through careful planning of open space, and an understanding of the principals of conservation biology.

Conservation biology is broadly defined as the science of preserving biological diversity. Anderson (1993) recommends large blocks of native habitats to protect diversity rather than dispersed smaller areas. Smaller parcels will not meet the home range needs of some species, and the effect of the edge will reduce the habitat for other species. Fragmented habitats can isolate some species genetically, leading to the eventual decline of a viable population. Blocks of native ecosystems can be made more effective for protecting diversity by connecting them with corridors of vegetation. Connecting corridors also makes them more useful to humans.

Corridors of vegetation are frequently called greenways, and are often used to connect parks and other open spaces. Greenways typically contain trails for human use, but will be used by wildlife as well. Rails to trails programs purchase rail lines as they are abandoned, and develop them as trails for a variety of uses. Utility corridors also provide habitat corridors between ecosystems and between rural and urban lands. By far the most important corridors from biological and recreational perspectives follow streams, rivers, and other aquatic systems. Communities are restoring stream banks and waterfronts, protecting floodplanes from developments, and acquiring land and restoring habitat along many urban waterways (Brown, 1989). A group of citizens from the Stevens Point, Wisconsin, area planned and developed a twenty-four mile trail system that encircles four communities. The trail uses and connects parks, follows river systems, and resides on both public and private land. Development funds for the trail came from local governments, private donations, and state grants (Figure 8–4).

Schaefer and Brown (1992) identify river corridors as among the most important lands to protect for wildlife. It is generally recognized that to be most effective

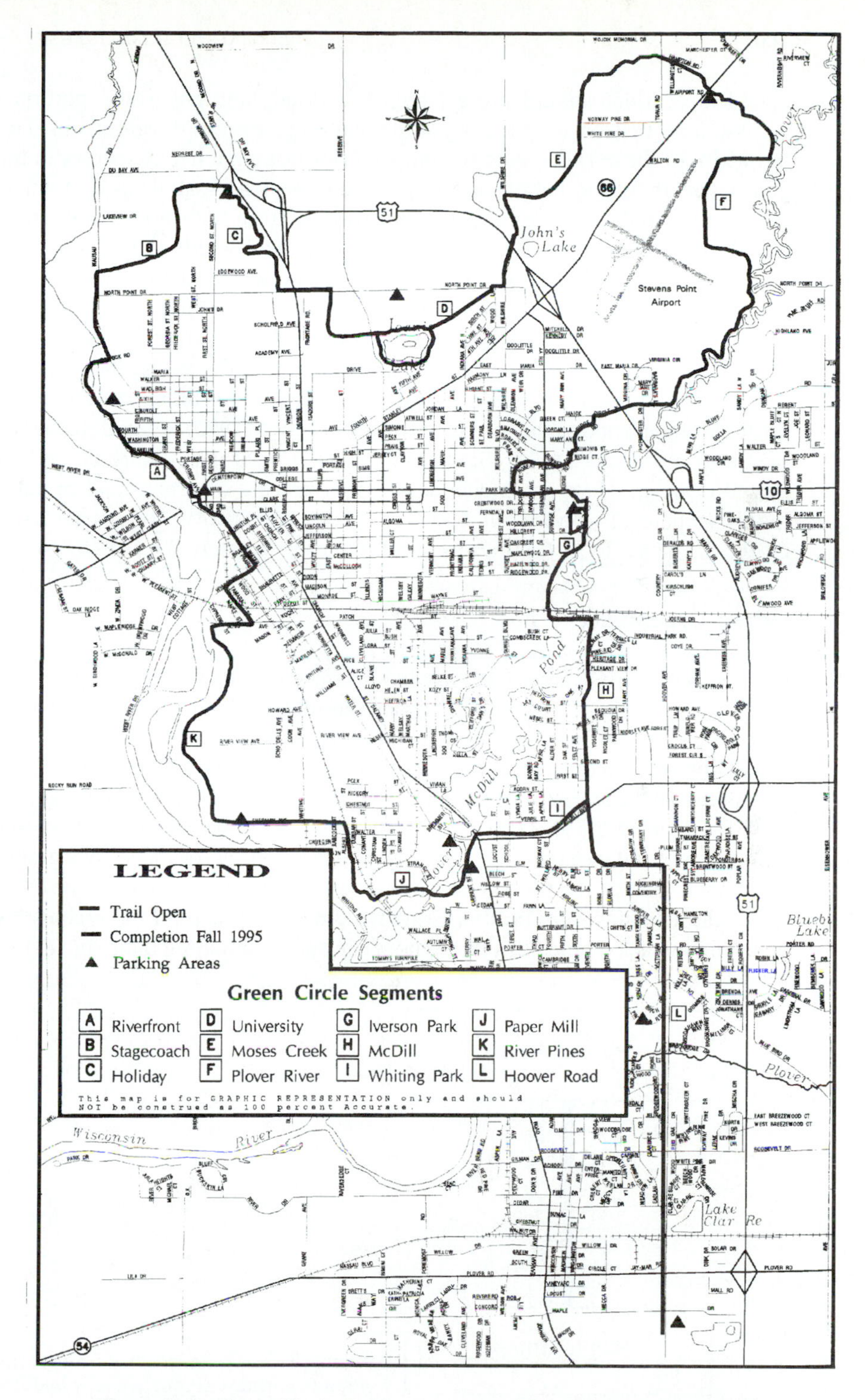

Figure 8–4 Green Circle Trail system around Stevens Point, Wisconsin.

the ideal corridor should protect both the floodplane and some portion of upland habitat, a process best carried out at the planning stage of development. To determine the width of the corridor they recommend developing a list of species in the corridor, and determining the habitat needs for each species. Spatial needs for each species that resides in the corridor are then determined from their home range requirements, and these requirements used to determine the ideal width of the corridor. Corridors that connect blocks of larger habitat provide the best situation where species within the corridor have adequate habitat, while species with larger habitat requirements use the corridor to enlarge their home range (Figure 8–5). Buffers along the corridor edge provide additional protection for species negatively affected by the edge effect.

Some species of wildlife can be included in more developed landscapes through the planned inclusion of open space. Stout (1995) studied the habitat requirements of red-tailed hawks (*Buteo jamaicensis*) in urban, suburban, and rural southeast Wisconsin. He recommends leaving 16 percent of urban land in natural habitat that is 40 percent wooded and 60 percent herbaceous cover to provide adequate habitat for the hawk. Ideal nesting sites require approximately 9 ha of woods.

Urban Greenbelts

Open space in urban areas can vary in size from tot-lots to large tracts of undeveloped land. The concept of a greenbelt carries the idea of open space further by suggesting that cities be confined by belts of undeveloped land at their periphery. Ebenezer Howard promoted the concept of garden cities in Great Britain late in the last century as a response to the excesses of the Industrial Revolution. He proposed that new towns be built for populations of 30,000 to 50,000 people, and that each town contain industry to support the local population. These towns were to be surrounded by a greenbelt of undeveloped land in the ratio 5 hectares of greenbelt to each hectare of city. Although some garden cities were built in Great Britain, the concept of the greenbelt drew wide acceptance, with many cities throughout Europe establishing them on their peripheries.

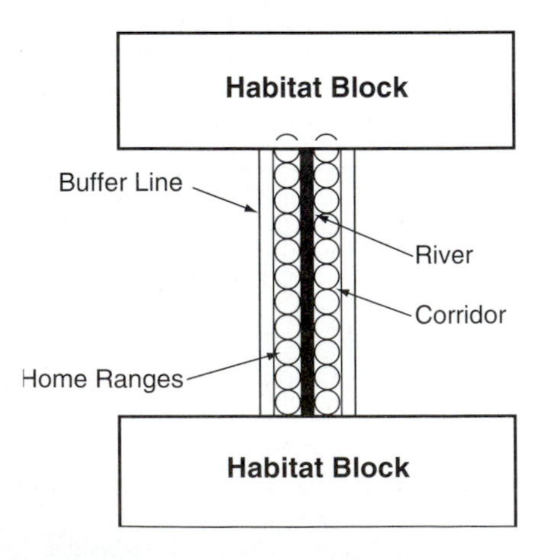

Figure 8–5 Components of an ideal riparian wildlife corridor. (From Schaefer and Brown, 1992.)

Germany has a number of major cities with peripheral greenbelts and forests, including Frankfurt, Stuttgart, Munich, Nuremberg, and Cologne (Schabel, 1980). The national government passed the Federal Nature Protection Act in 1975 to ensure the protection of natural areas and landscape amenities throughout the country. This law has been particularly useful in protecting private lands in urban greenbelts and forests from development. Several key provisions of the act follow (Bundesminister, 1976).

1. In order to achieve the objectives of nature protection and landscape management, landscape programs, guidelines, and plans must be worked out and the applicable portions utilized in state, regional, and local land-use planning.
2. Unnecessary encroachment on nature and landscape is to be avoided and unavoidable damage must be offset by the responsible party.
3. Property owners or those authorized to use properties can under certain circumstances be required to tolerate nature protection and landscape management, that is, with respect to maintenance and management of their property.
4. Certain areas may be designated as nature protection areas, nature parks, natural monuments, or protected features of a landscape and are to be accordingly protected, managed, or developed.
5. Wild plants and animals and their habitats are to be protected and preserved, through international cooperation if necessary.
6. Public access to agricultural, forest, and idle land via paths and public roads is a basic right.

Of the many provisions of the act, landscape protection is the one most used to protect urban greenbelts. Land is included in landscape protection areas to assure or restore the health of the environment, because of the uniqueness or beauty of the landscape, or because of its significance for recreation. In landscape protection zones, any activity that alters the character of the area is forbidden. No compensation is given to property owners in landscape protection areas for land-use restrictions (Bundesminister, 1976).

Europeans have a long history of government involvement in land-use allocations, and have generally accepted development restrictions on private land included in the greenbelts. However, North Americans have a strong tradition of land ownership, with few restrictions placed on their land by government. Economic determinants were, and are, the primary factor in allocating land use. Although city planners have long promoted greenbelts around North American cities, few cities have them because the high cost of land at the urban periphery prohibits purchase. Restrictions on development of private land have been historically ruled by the courts to represent a taking of land value. The only exception to restriction of development permitted so far is if a public hazard results from development. This is changing, however, as recent court decisions reflect increasingly more liberal interpretations of public hazards, and stronger court support of good land-use planning.

The development of a greenbelt around the city of Ottawa, Canada, reflects the growing acceptance of this concept in North America, but also serves to illustrate problems associated with attempts to restrict land use. The Plan for the National Capital, prepared in 1950, recommended the establishment of a 2.5-mile (4-km)-wide greenbelt around the southern periphery of the city to define an area large enough to contain an ultimate population of 500,000 people (Figure 8–6). Initially, zoning was to be used to restrict development and compensation was not to be paid for financial loss since the greenbelt was for the good of the community. However, public resistance to the use of zoning without compensation was so strong that in 1958 the Canadian government decided to acquire all remaining land not in public ownership through purchase, using condemnation if necessary. At the same time the following purposes of the greenbelt were described (McDonald and Cole, 1978):

1. To prevent further haphazard urban sprawl around the Capital, including ribbon development along approach highways, and so to protect adjacent farming areas from being swallowed up by uncontrolled development

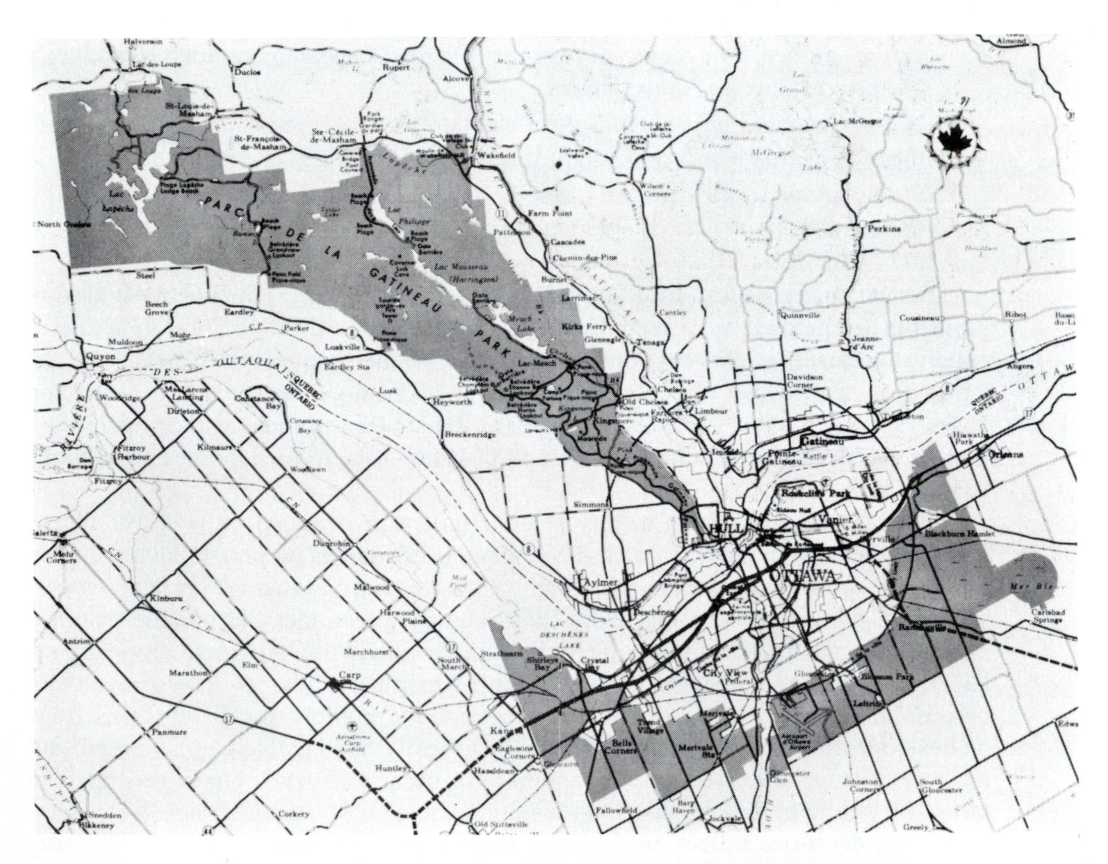

Figure 8–6 Greenbelt around Ottawa, Ontario, Canada.

2. To meet long term National Capital planning needs by ensuring that when the central area was built up, an adequate reserve of sites for future buildings, for government and public institutional purposes, was retained
3. To place a practical and economic limit on the growth of the capital by confining intensive building development to an area which could be provided with municipal services at a reasonable cost

URBAN FOREST MANAGEMENT PLANNING

The planning process is an integral part of management. Management involves an inventory of the resource base, development of goals and objectives, specific programs to meet goals and objectives, and a periodic assessment of the entire management plan (feedback). McPherson and Johnson (1982) describe a similar approach to community forestry planning that involves four phases: foundation building, data collection, analysis and evaluation, and implementatioin—with feedback throughout the process.

Management planning takes place in two time horizons, short and long term. Short-term planning is the development of methods to deal with day-to-day management activities in such a manner as to make the most efficient use of personnel and equipment. This involves scheduling work activities, equipment maintenance, supervision, reporting, record keeping, and action plans to deal with crisis situations. Long-term management planning is the establishment of overall goals and objectives for the organization, prioritizing of these goals, and action plans to achieve these goals. Management planning should provide contingency plans should goals not be met, or should goals be redefined by feedback.

In the public sector, management planning will involve the development of management plans for forested greenbelts, parklike vegetation on recreation areas, the development of master street tree plans, and plans to deal with private vegetation as needed. In the private sector, consulting urban foresters and arborists develop similar management plans for public vegetation on a contractual basis, and management plans for privately owned vegetation when employed as agents of the owners.

A problem sometimes encountered in development of management goals for urban vegetation is the diversity encountered in natural and human ecosystems. Byrne (1978) recommends dealing with diversity problems through stratification of the population. Obviously, different management plans will be needed for parks, forests, and street trees, but even the street tree population may need to be stratified based on cover, species, age structure, and neighborhood characteristics. Each strata will have different management needs and subsequent management plans. The overall management plan will integrate these strata through the allocation of resources for management.

The remaining chapters of this book deal with different aspects of managing the urban forest resource. Once inventories have been conducted and goals set, ordinances are used in public forestry to provide the tools for management. In the following chapter we discuss various ordinances that deal with urban vegetation and how these ordinances fit into urban vegetation management planning. Subsequent chapters deal with the specifics of planning and managing street tree populations, managing park and open space vegetation, commercial and utility arboriculture, and overall program administration and management. Personnel training and supervision are discussed, and the need for and techniques of public relations outlined.

LITERATURE CITED

Adler, J. 1995. "Bye-Bye Suburban Dream." *Newsweek,* May 15, pp. 40–53.

Anderson, E. M. 1993. "Conservation Biology and the Urban Forest." *Sixth Natl. Urban For. Conf.* Washington, D.C.: Am. For. Assoc., pp. 234–237.

Brown, C. N. 1989. "Greenways: Urban Forestry's Best Friend." *Sixth Natl. Urban For. Conf.* Washington, D.C.: Am. For. Assoc., pp. 181–186.

Bundesminister fur Ernahrung Landwirtschaft und Forsten. 1976. *Bundesnaturschutzgesetz.* Coburg, Germany: Neue Presse GmbH.

Byrne, J. G. 1978. "A Planners Viewpoint on the Urban Forestry Effort." *Proc. Natl. Urban For. Conf.,* ESF Pub. 80-003. Syracuse: SUNY, pp. 672–675.

Florida Division of Forestry. 1973. *Urban Forestry Handbook.* Tallahassee: Fla. Div. For.

Kürsten, E. 1993. "Landscape Ecology of Urban Forest Corridors." *Proc. Sixth Natl. Urban For. Conf.* Washington, D.C.: Am. For. Assoc., pp. 242–243.

McDonald, D. L., and J. W. P. Cole. 1978. *Urban Greenbelts.* Ottawa: Natl. Cap. Comm.

McNiel, J. 1991. "Sustainable Development in the Urban Forest." *J. Arbor.* 17(4):97–97.

McPherson, E. G., and C. W. Johnson. 1982. "A Community Forestry Manual and Case Study." *Proc. Sec. Natl. Urban For. Conf.,* Washington, D.C.: Am. For. Assoc., pp. 325–333.

McPherson, G. 1989. "Creating an Ecological Landscape." *Proc. Fourth Natl. Urban For. Conf.* Washington, D.C.: Am. For. Assoc., pp. 63–67.

Moll, G. A. 1978. "Urban Forest Planning." *J. Arbor.* 4(9):213–215.

Schabel, H. G. 1980. "Urban Forestry in Germany." *J. Arbor.* 6(11):281–286.

Schaefer, J. M., and M. T. Brown. 1992. "Designing and Protecting River Corridors for Wildlife." *Rivers* 3(1):14–26.

Schafer, E. L., and G. H. Moeller. 1979. "Urban Forestry: Its Scope and Complexity." *J. Arbor.* 5(9):206–209.

Schwab, J. 1993. "Planning for the Urban Forest." *Proc. Sixth Natl. Urban For. Conf.* Washington, D.C.: Am. For. Assoc., pp. 254–256.

Snyder, R. 1973. "Urban Forestry and Land Use Planning." *Proc. Natl. Urban For. Conf.* Syracuse: SUNY, pp. 26–27.

STOUT, W. E. 1995. "An Urban, Suburban, Rural Red-Tailed Hawk Nesting Habitat Comparison in Southeast Wisconsin." M.S. thesis, Univ. Wisc., Stevens Point.

ZIMINSKI, A. 1982. "Reston Homeowners Association Open Space Management and Development." *Proc. Sec. Natl. Urban For. Conf.* Washington, D.C.: Am. For. Assoc., pp. 71–77.

9

Vegetation Ordinances

Urban areas contain a highly complex mosaic of public and privately owned vegetation. Establishment, protection, and management of this vegetation is influenced by decisions of private landowners and by public regulations. This chapter deals with ordinances and laws as they specifically relate to the establishment and management of vegetation on public land, the protection, preservation, and management of vegetation on private land, and requirements for landscaping and screening. Vegetation regulation is carried out as specified by municipal ordinances, country ordinances, and state or national laws that deal directly with vegetation, or indirectly by ordinances and laws that control land use, development, and construction.

MUNICIPAL FORESTRY PROGRAMS

Municipal forestry programs exist as a reflection of community interest in urban forestry and operate as specified by ordinances. However, the passage of a community forestry ordinance can be a complex issue. Weber (1982) notes that community forestry ordinances could affect the jobs or property of "arborists, realtors, developers, the mayor, homeowners, building contractors, nurserymen, horticulture teachers, utilities engineers, the city attorney, city council members, an urban forester, a public works manager, and many others." With such a diverse group having a stake in forestry ordinances, he recommends a broad base of community support be developed prior to the introduction of an ordinance. All communities are different, and they differ in four ways: nature (climate, soils, and the like), ethnic tradition, political-economic climate, and legal framework. The forestry ordinance best suited to, and most likely to be approved in, a community is written with a thorough understanding of these four factors (Weber, 1982).

Forestry ordinances to be effective must do three things: provide authority, define responsibility, and establish minimum standards for management. The authority to carry out the forestry program must be assumed by the city or other unit of local government. The ordinance must then assign authority to an individual or agency to carry out provisions of the ordinance, and it must set management standards to provide for the public safety and convenience (Grey, 1978).

Ordinances that deal with urban vegetation are not new. The evolution of these ordinances started by addressing certain plants as public nuisances and prohibiting their planting in the community. As cities embarked on beautification efforts following the Industrial Revolution, tree ordinances were passed calling for the planting and maintenance of trees on public rights-of-way and in parks. In North America, this was followed by the introduction of Dutch elm disease (*Ceratocystis ulmi*), which devastated city tree populations and brought about enactment of ordinances designed to deal with the problem. Other communities, experiencing rapid urban development and concerned with the rapid loss of native vegetation, enacted ordinances to protect vegetation during development and to require appropriate landscaping following construction. Screening ordinances have been approved and are being used to soften the visual impact of unsightly land uses, such as parking lots, industrial activities, and transportation corridors.

Today a great variety of municipal ordinances exist to deal with urban vegetation. Perry (1978) requested copies of tree ordinances from sixty American cities. Fifty-one cities responded and their ordinances classified based on provisions and scope of authority (Table 9–1). He found that 20 percent provide some authority to control insects and diseases on public land, to address public safety issues on private

TABLE 9–1 FREQUENCY OF MUNICIPAL TREE ORDINANCE TYPES IN THE UNITED STATES BASED ON A SURVEY OF 51 CITIES (From Perry, 1978.)

Ordinance to:	Number of Cities with Ordinance
Protect public lands	48
Protect trees and vegetation on private lands	10
Landscape parking lots	10
Provide minimum green space for new developments	3
Protect environment during development	1

trees, and to set landscaping standards on parking lots. In addition, 7 percent require green space dedication in new developments.

The following discussion of vegetation ordinances describes them in general terms. For an ordinance to work in a particular community it must be tailored to meet local needs. Each type of ordinance is presented independently, but in practice many communities will have an ordinance that incorporates aspects from several of those described below. The ordinances are arranged in approximate order of their historical development, although the exact evolution of vegetation ordinances on a local level may be different.

Undesirable Vegetation

The earliest ordinances dealing with urban vegetation dealt with their undesirable elements. Trees cause property damage, interrupt utility service, pose public safety hazards, block intersections, create a public nuisance from fruits and other tree parts, and attract undesirable insects or wildlife. Trees, shrubs, and herbaceous plants can have a negative impact on aesthetic and property values, cause allergic reactions in some people, and pose serious fire hazards. Urban foresters are so accustomed to thinking of the positive aspects of urban vegetation that it is easy to forget that this same vegetation can be classified as a pest. Nosse (1982) defines pest as anything that causes trouble, annoyance, or discomfort, and provides the following assessment of the potential pestilent nature of urban trees.

1. *Traffic safety.* At age 10 branches on street trees will begin to extend into the public right-of-way. By age 15 the tree will begin to block traffic signs; at age 25, street lighting will be reduced; and by age 40 the tree will interfere with traffic lights.
2. *Structural integrity of sidewalks and curbs.* The tree's root system may lift and break the sidewalk and curb. Even assuming the species planted won't do this, 10 percent of trees uprooted in a 100-year storm will likely lift up sidewalks.
3. *Drainage systems.* Leaf litter plugs rain gutters and drainage systems, fills catch basins, and may reduce storm drainage capacity by as much as 10 percent.
4. *Property damage.* Once trees exceed 25 years of age they pose a threat of considerable property damage should they fail structurally.
5. *Attractive nuisance.* Trees attract insects and diseases, and attract children to climb if branches are low enough.
6. *Electric and communication systems.* By 15 years of age trees will normally extend into power and communications lines. Fast growing species may need annual pruning to reduce chances of service interruption.
7. *Energy influence.* Trees block solar energy, reduce snow melt, and accelerate wind velocities when arranged in a manner as to cause a vortex.

8. *Tree removal.* All trees eventually die and must be removed if they present a hazard. Removal of mature trees growing over wires and homes requires trained personnel, special equipment, and is expensive.

Legal values of urban trees and other vegetation in terms of liabilities were discussed in Chapter 5. Due to potential problems associated with urban vegetation, states and communities have enacted statutory controls on trees and other plants to protect the health, safety, and welfare of citizens. Public health laws controlling plants causing dermatitis or allergic reactions to pollen have been accepted by the courts, as have public safety ordinances in relation to fire hazards and right-of-way obstructions (Widrlechner, 1981).

General welfare involves a number of issues in relation to urban vegetation. Statutes to control noxious plants such as Canadian thistle (*Cirsium arvense*) or cottonwood (*Populus deltoides*) are common in communities. Ordinances protecting property values by requiring certain standards of weed control, mowing, and landscape maintenance are also widespread, but are sometimes not supported in the courts. This is particularly true where alternative forms of landscaping are used, such as the establishment of natural ecosystems. Some communities now provide exceptions in their noxious weed ordinances to allow alternative landscaping (Widrlechner, 1981).

General welfare statutes to control the spread of plant diseases fall into three categories. The first declares the disease to be a nuisance and directs the destruction of diseased plants or alternate hosts of the disease. The second group of statutes confers powers to an agency to discover and control plant diseases in general. The third type of statute enables local governments to create special disease control districts. The control of Dutch elm disease generally involves the first two categories: sanitation or the removal of diseased trees, and the right of the community to discover and control the disease (Widrlechner, 1981).

Dutch elm disease. The control of Dutch elm disease in communities is carried out primarily through inspection and sanitation programs. Municipal employees are given jurisdiction over private elms, and trespass is permitted to search for diseased trees. Public trees are removed when the disease is diagnosed and private trees are condemned and removal ordered. A typical Dutch elm disease ordinance will contain all or some of the provisions listed below.

> Section 1: Public nuisance declared. Any living or standing elm tree or part thereof infected with the Dutch elm disease fungus or which harbors any of the elm bark beetles; any dead elm tree or part thereof, including logs, branches, stumps, firewood, or other elm material from which the bark has not been removed and burned or sprayed with an effective elm bark beetle destroying insecticide; is hereby declared a public nuisance.
>
> Section 2: Nuisances prohibited. No person, firm or corporation shall permit any public nuisance as defined in Section 1 of this ordinance to remain on any premises owned or controlled by him within the city of __________ .

Section 3: Inspection. The Shade Tree Board shall inspect or cause to be inspected all premises and places within the City at least twice each year to determine whether any public nuisance as defined in Section 1 of this Ordinance exists thereon, and shall inspect or cause to be inspected any elm tree reported or suspected to be infected with the Dutch elm disease fungus or any elm bark bearing material reported or suspected to be infested with the elm bark beetle.

The Shade Tree Board shall have the authority to enter upon the premises at all reasonable times for the purpose of carrying out any of the provisions of this ordinance.

Section 4: Abatement of Dutch elm disease nuisances. Any public nuisance as defined by this ordinance found on public property shall be sprayed, removed, burned, or otherwise disposed of in such a manner as to prevent the spread of the Dutch elm disease fungus. Whenever any public nuisance as defined by this ordinance is found on private property the Shade Tree Board shall notify the property owner in writing, if he can be found, that the nuisance must be abated as directed in the notice within thirty (30) days from the date of the notice. If the property owner fails to comply with the notice within the time limit, the Shade Tree Board shall cause such nuisance to be abated and bill the property owner for the costs incurred.

Section 5: Transporting of elm wood prohibited. No person, firm, or corporation shall transport within the City of __________ any bark bearing elm wood or material without first securing written permission from the Shade Tree Board.

Section 6: Interference with the Shade Tree Board or its agents prohibited. No person, firm, or corporation shall prevent, delay, or interfere with the Shade Tree Board or any of its agents or city employees while they are engaged in the performance of duties imposed by this ordinance.

Section 7: Penalty. Any person, firm, or corporation which shall violate any of the provisions of this ordinance shall, upon conviction thereof, forfeit not less than $ __________ nor more than $ __________ together with the cost of prosecution, and in default of payment of such forfeiture and costs shall be imprisoned in the county jail until said forfeiture and costs are paid, but not exceeding 90 days.

Municipal Tree Ordinances

The purpose of a municipal tree ordinance is to express municipal authority over public trees, to assign responsibility to a municipal employee, to set maintenance and management standards, and to define nuisance conditions on private trees. A typical municipal ordinance will contain some or all of the headings below. Each heading is followed by a brief description of what they normally include. A complete municipal tree ordinance may be found in Appendix C.

1. *Purpose:* includes a brief description of the ordinance and why it is necessary.
2. *Definitions:* provides definitions of legal and technical terminology in the ordinance.
3. *Establishment of a Shade Tree Board:* either creates a Shade Tree Board and

describes its membership, or assigns responsibility for the program to the Park Board.

4. *Municipal arborist or forester:* describes the educational and experience qualifications of the city forester.
5. *Duties:* outlines the duties of the city forester in managing public vegetation, in dealing with private nuisances, and in enforcing the provisions of the ordinance.
6. *Authority:* gives the city forester the authority to supervise all work done by permit as described in the ordinance, and to affix reasonable conditions to the granting of permits. The city forester is also given the authority and responsibility to develop a master street tree plan for the community.
7. *Permits:* prohibits all planting, maintenance, and removal of public vegetation by anyone other than the city forester or agent of the city forester except by permit.
8. *Maintenance:* sets standards for planting, maintenance, and removal of public trees, and establishes a replacement policy. May also describe spacing standards, corner setbacks, and mature size restriction based on width of the tree lawn and other site restrictions.
9. *Obstructions:* requires owners of private trees to keep sidewalks clear to a height of 10 ft. (3 m), streets clear to a height of 12 ft. (3.6 m), truck routes clear to a height of 16 ft. (4.8 m), and to keep street signs and lights clear.
10. *Nuisance and condemnation:* defines a nuisance tree on the basis of insects or disease problems, undesirable characteristics for urban use, or being a threat to public safety; and gives authority to the city forester to condemn a nuisance tree. May include a list of trees prohibited in the community.
11. *Abuse of public trees:* prohibits mutilation or other abuse of public trees.
12. *Protection of trees:* requires protection of public trees during construction or other activities which may harm them.
13. *Interference:* prohibits interference with the city forester in performance of duties.
14. *Penalties:* provides penalties for failure to comply with the provisions of the ordinance.

The preparation of a municipal tree ordinance must be undertaken with great care. An ordinance that does not adequately define the position of city forester or arborist, and does not provide the authority to carry out its provisions, will be ineffective. The ordinance must set minimum standards for maintenance of public trees and reflect community support for an urban forestry program. On the other hand, the ordinance must be flexible enough to allow the city forester latitude in making management decisions. For example, the ordinance may list species acceptable for planting on streets, but should be open ended to allow for modification of the list. A Standard Municipal Tree Ordinance is available from the International Society of Arboriculture, P.O. Box GG, Savoy, IL 61874.

Certification and licensing. The International Society of Arboriculture (ISA) certifies arborists on the basis of candidates passing a written exam concerned with the fundamentals of tree biology and tree care. Some communities have a provision in their municipal tree ordinance requiring that all contractors performing work on city trees be Certified Arborists as designated by the ISA.

A number of communities license arborists to do commercial tree work within the city limits. The city of Bismarck, North Dakota issues an arborist's certificate based on passing an exam after paying a fee of $50 to take the exam. The fee is $50 per year thereafter and the license is good for up to five years.

VEGETATION PROTECTION

Many communities have enacted ordinances to protect trees, forests, rare plants, and ecosystems. These ordinances range from requirements for land developers and builders to protect trees during development and construction, to ordinances designed to prohibit the removal of any trees in the community without a permit. Duerksen and Richman (1993) regard tree and woodland protection ordinances as a major environmental and planning issue, reporting that in 1984 there were fewer than 100 tree protection ordinances in the entire United States, but by 1989 there were more than 80 in California alone. Communities experiencing rapid growth are especially prone to enact vegetation protection ordinances due to the strong visual impact of such growth. Vegetation protection can be specifically directed by an ordinance enacted for that purpose, or can be a provision of an ordinance controlling other land-use activities.

Tree Protection Ordinances

Tree protection ordinances are designed primarily to protect existing trees and other vegetation during land development. These ordinances fall under the police power of county and local governments, and regulate the removal of trees by establishing definitions, procedures, penalties, and appeals necessary for enforcement (Florida Division of Forestry, 1973). A tree protection ordinance will contain some or all of the provisions described below. (See Appendix D for an example of a tree protection ordinance.)

1. *Findings of fact:* describes the importance and function of trees in providing healthy environments and other amenities in communities, and defines the purpose of the ordinance
2. *Definitions:* defines terms unique to the ordinance
3. *Enforcement:* identifies the political subdivision in which the ordinance applies, and assigns authority for enforcement of the ordinance
4. *Applicability:* defines the types of property to which the ordinance applies and may specify parcel size

5. *Permits:* requires tree removal permits, and defines permit requirements, fees, site plan, and application review procedure
6. *Tree protection:* outlines tree protection procedures to be followed during construction
7. *Relocation or replacement:* requires the relocation of trees during development where feasible, or the replacement of trees removed based on lot-size specifications
8. *Exceptions:* granted for agricultural clearing, nurseries, silviculture, nuisance trees or species, and emergencies
9. *Appeals:* provides a vehicle through which the landowner or developer may appeal the provisions of the ordinance
10. *Penalties:* specifies penalties for failure to comply with provisions of the ordinance

Other provisions may be added to tree protection ordinances to provide additional protection during development. Robson (1983) describes a tree preservation ordinance adopted in Lake Forest, Illinois, which restricts construction activity to a "building envelope." The building envelope (Figure 9–1) is defined as "that area of the lot needed for the movement of equipment and materials to be used in construction . . . and includes the area where the house will be built, plus a 20 foot (6 m) wide area around the house." The ordinance requires the building envelope to be roped off and cleared by chain saw and stump grinder only (no bulldozing permitted). Foundation soil may be stored and backfilled in the building envelope only, with excess fill

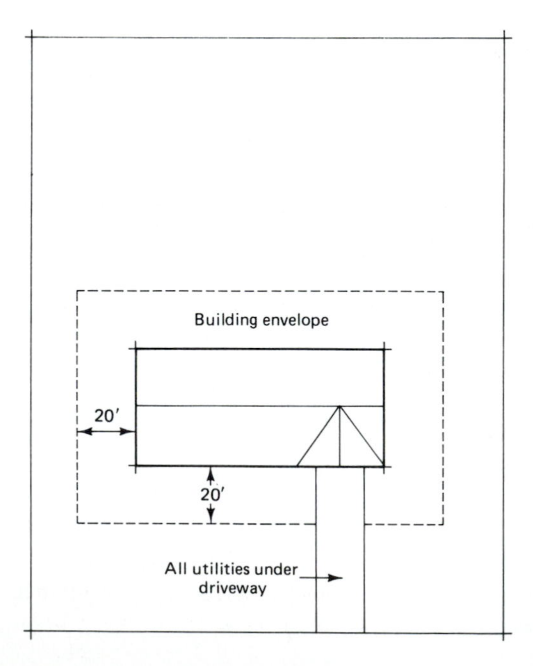

Figure 9–1 Building envelope as described in the Lake Forest, Illinois, tree protection ordinance. (From Robson, 1983.)

removed from the site. Heavy equipment is not permitted outside the building envelope. Utility installations are restricted to beneath the driveway.

While most tree protection ordinances are directed to protect trees during development, increasing numbers of communities are protecting all trees, including trees on private property. Under these circumstances all trees in the community are regarded as public assets by performing both social and environmental functions within the community as described in previous chapters. Highland Park, Illinois, enacted such an ordinance in 1991 in order for the city to exercise some control over the "indiscriminate destruction of trees on both public and private property." The ordinance requires a permit to remove any tree in the community larger than 8 in. (20 cm) in diameter at 4.5 ft. (1.4 m) above the ground for a single stem tree, or 15 in. (38 cm) aggregate diameter for multistemmed trees. Permits may be issued by the city when the tree is dead or dying, diseased, hazardous, or removal is constant with good forestry practices (City of Highland Park, 1994).

Special Trees

Special or specimen trees are protected for size, species, historic, or other special characteristics. Beijing, China, prohibits the removal of any tree that is taller than a building or is over 100 years old without notification of the Beijing Institute of Landscape and Gardening, and approval of the Beijing Forestry Bureau (Profous, 1992). Austin, Texas, requires a permit for removal of any tree greater than 60 in. (150 cm) circumference, while the community of Thousand Oaks, California, concentrates protection on oaks (*Quercus* spp.). Tampa, Florida, combines species and size characteristics to protect "grand" trees. For example, bald cypress (*Taxodium distichum*) must score 200 points using one point per inch (2.5 cm) of circumference, one point per foot (.3 m) of height, and one point per foot of average crown spread to be considered a grand tree. Sanibel Island, Florida, protects all native vegetation outside development zones and requires that new landscaping use only native plants. Alexandria, Virginia, protects "any tree with notable historical association" (Duerksen and Richman, 1993).

The Town and Country Planning Act of 1971 contains provisions to protect amenity trees in the United Kingdom. To be eligible for protection by a Tree Preservation Order, an amenity tree "should be of clear and substantial amenity value and a reasonable degree of public benefit should accrue. The tree(s) must be visible in whole or in part from a public place such as a roadway or footpath, etc., although other exceptional trees can be included within an order" (O'Callaghan, 1991).

Rare Species and Ecosystems

Rare and endangered species and their habitats are identified and protected by U.S. law. Some state and local governments have similar legislation designed to protect locally endangered species and ecosystems. The state of Florida passed enabling legislation permitting communities to adopt procedures for the protection of "environmentally endangered" lands. Dade County, Florida, adopted such an ordinance for the protection of areas containing mangrove forests, hardwood hammocks and tree islands, pinelands, wetlands, and cypress forests (Figure 9–2) (Dade County, 1979).

Owners of 5 or more acres of endangered land in Dade county may enter into voluntary covenant with the county for a period of 10 years and have their land taxed at its present use rather than market value. The owner must maintain the land in its present state for the contract period. If the landowner violates the terms of the agreement, all deferred taxes plus 6 percent interest must be paid to the county.

Tree Abuse

Communities have provisions in their tree ordinance that regulate the maintenance and pruning of public trees, and some communities regulate the treatment of some or all private trees. Tree abuse ordinances can deal with a variety of improper tree care practices such as leaving pruning stubs, using climbing spurs on trees not being removed, flush pruning, excessive removal of branches and foliage, and topping trees. Tree abuse ordinances usually target topping as the most serious mistreatment of trees. Topping is cutting the tops or main leaders back on trees to drastically reduce the size of the tree, especially mature trees. Cuts are usually made between nodes. The practice causes the tree to regrow a structurally weak crown, and introduces a

Figure 9–2 Endangered southern Florida slash pine (*Pinus ellottii*) ecosystem.

great deal of decay into the tree. Over time these trees can become serious hazards in the community, and this provides the reason to regulate the practice. Topping should not be confused with pollarding, an acceptable practice of annual pruning to maintain a tree at a specified size by removing current annual growth.

Alachua County, Florida, requires that all public agencies and public utilities follow the American National Standards Institute (ANSI A300, 1995) standards for tree care operations when pruning both public and private trees. A San Juan Capistrano, California, ordinance prohibits anyone in the Tourist Commercial Zone or any residential zone within 500 feet of a scenic highway or drive from topping trees, while Thousand Oaks, California, prohibits pruning any living tissue on oaks (*Quercus* spp.) without a permit (Duerksen and Richman, 1993). Trees protected by a Tree Preservation Order in the United Kingdom may not be abused, and violators are subject to a substantial fine should they do so (O'Callaghan, 1991).

Augering

Underground installation of utilities such as water, sewer, electricity, phone, and television cables can do considerable damage to tree roots. Lines and pipes are usually placed under sidewalks and tree lawns because of their public ownership. The standard and cheapest method of installation is to trench and backfill, thus cutting tree roots. In many instances following trenching near trees, decline sets in, frequently resulting in tree mortality several years after the operation.

Morell (1984) conducted a survey of tree dieback and mortality approximately 10 years after trenching in tree lawns for new water mains in Park Ridge, Illinois. One installation lost 44 percent of the trees, while another lost 25 percent. Although trenching is cheaper, Morell found that the costs of removing dead trees and replanting nearly offset savings incurred by not augering. In conjunction with other foresters in northeastern Illinois, Morell developed specifications for an augering ordinance that was subsequently adopted by a number of communities. The ordinance establishes a minimum augering depth of 24 in. (0.6 m) and prohibits trenching near the tree, as specified in Table 9–2 and Figure 9–3.

TABLE 9–2 AUGERING ORDINANCE SPECIFICATIONS DEVELOPED IN NORTHEAST ILLINOIS (From Morell, 1984.)

Tree Diameter (dbh) (in.)	Auger Distance from Face of Tree (ft)
0–2	1
3–4	2
5–9	5
10–14	10
15–19	12
Over 19	15

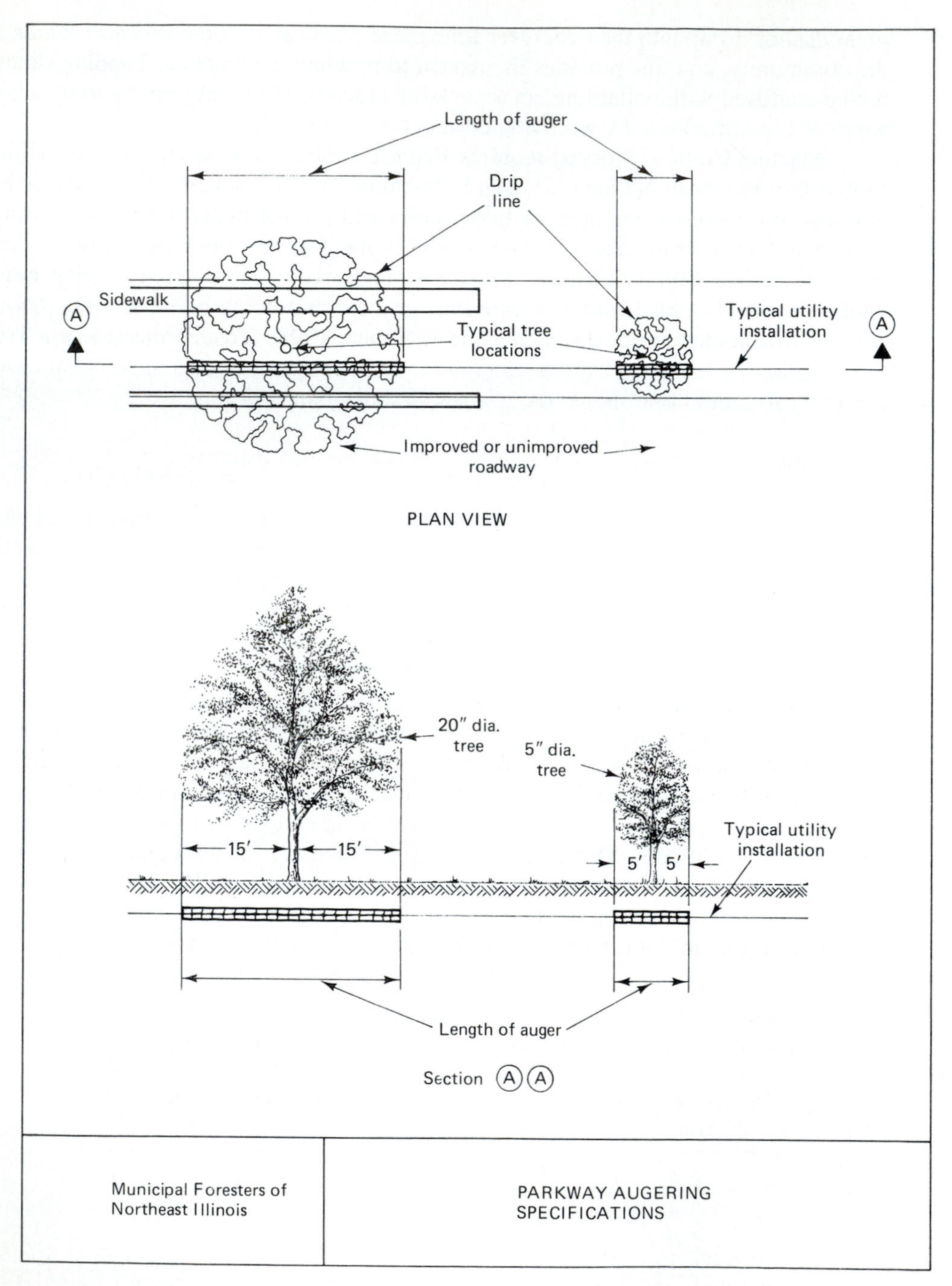

Figure 9–3 Parkway augering specifications used in northeastern Illinois communities. (From Morell, 1984.)

Zoning and Subdivision Regulations

Planning ordinances as they relate to preserving open space and parkland are discussed in Chapter 8. However, some zoning and subdivision ordinances contain sections on vegetation preservation and landscape planning. Many levels of government now have zoning districts designed to protect agricultural land, wetlands, and other environmentally sensitive areas from development. These ordinances can make a significant contribution to an overall urban greenspace plan.

Shorelines and associated vegetation are protected in some states by special zoning districts. Local governments in the state of Wisconsin are required to meet minimum specifications in regulating development of shorelands. Among other requirements, these ordinances specify minimum lot widths of 100 ft. (30 m), water setbacks of 75 ft. (22.5 m), and prohibits removal of more than 30 percent of the shoreline vegetation by the builder or homeowner.

Subdivision regulation may specify vegetation protection and replacement during development. The subdivision ordinance for the city of Green Bay, Wisconsin, specifies:

> Street, Block, and Lot Layouts shall be so designed as to: least disturb the existing terrain, flora, fauna, and water regimen; and meet all the use, site, sanitary, floodland and shoreland regulations contained in the City Zoning, Sanitary, and Building Ordinances.
>
> Shorelands. Due consideration shall be given to the reservation of a suitable area along ponds, streams, lakes, flowages, and wetlands for the purpose of providing public access to such waters and creating a buffer between private shoreland owners and public water users.

With reference to tree protection, the ordinance further specifies:

> Street Trees and Landscaping. All trees in excess of 6 in. (15 cm) in diameter which are removed during site grading, road and structural construction shall be replaced by the developer with trees of the same species with a diameter of not less than 3 in. (7.5 cm).

In subdivision review proceedings developers can be required to alter site plans to protect vegetation and other resources. Landscape plans may also be required before permits are issued for the construction of commercial or industrial buildings.

Woodland Protection

Entire woodlands are increasingly given protection in land development. Lake County, Illinois, requires 70 percent of mature woodlands on a site be protected as open space, and 40 percent of young woodlands protected. Mature woodlands must cover an area larger than 1 acre (.4 ha) and have at least 50 percent of the canopy cover trees exceed 10 in. (25 cm) in diameter or can be any smaller grove where eight or more trees occupy at least 50 percent of the canopy and are at least 12 in. (30 cm) in diameter. Young woodlands are defined as not meeting the above specifications (Duerksen and Richman, 1993).

Concern over an annual loss 10,000 acres (4,000 ha) of forest land to development from 1985 to 1990 led to the state of Maryland's 1991 Forest Conservation Act, which among other provisions requires retention of trees and planting of trees during land development. The law establishes six Forest Conservation Thresholds (FCT) defined by land use and forest cover:

Agricultural/resource areas	50% forest
Medium density residential	25% forest
Institutional developments	25% forest
High density residential	15% forest
Commercial and industrial use	15% forest

Tracts that exceed the FCT must plant ¼ acre of forest for every acre cleared, and areas below the FCT must plant 2 acres of forest for every acre cleared. Additionally, agricultural/resource and medium density residential land with less than 20 percent forest cover, must be afforested up to 20 percent of the land area when developed. The other land use areas must be afforested to 15 percent tree cover when developed (Piotrowski, 1991).

Prince George's County, Maryland, uses a woodland conservation ordinance tied to development density. Low-density residential zones require up to 50 percent woodland retention, medium-density 20 percent, and high-density 10 percent. Fulton County, Georgia, requires 15 "units per acre" of trees in housing developments. Large trees are given higher unit values than small trees. If the developer cannot meet the minimum requirement through preservation of existing trees, he or she must plant more trees (Duerksen and Richman, 1993).

The Federal Nature Protection Act in Germany is discussed in Chapter 8 as it applies to land-use planning. The act provides for the protection of vegetation by setting aside Nature Protection Areas, Landscape Protection Areas, National Parks, and Nature Parks (Bundesminister, 1976). In the United Kingdom trees are protected in designated Conservation Areas, which are defined as having specific architectural or historic interest, and this includes trees within these areas greater than 150 mm (6 in.) diameter. Permission must be granted by the local planning authorities before any protected tree can be removed or altered in any way (O'Callaghan, 1991).

Corridors

Vegetation along highway and river corridors can be protected by ordinance. Austin, Texas, enacted the Hill Country Roadway Ordinance that prohibits the clearing of vegetation within 100 feet of roadways in the hill country, except for site and utility access, or if the buffer exceeds 20 percent of the adjacent property. Fulton County, Georgia, requires protection of all trees within 35 ft. (10.6 m) of the banks of the Chattahoochee River and its tributaries (Duerksen and Richman, 1993).

LANDSCAPE ORDINANCES

Some communities have specific landscape requirements for developers and builders. The village of Arlington Heights, Illinois, requires that a landscape plan be submitted with all construction in the village. This plan must be reviewed and approved by the Forestry Department before a building permit is issued (Page, 1983). A landscape ordinance in Lake Forest, Illinois, requires "a tree location plan for each new subdivision . . . (for) existing trees 4 inches (10 cm) or larger . . . according to location, size, species and condition." The ordinance further requires that trees to be removed be so noted and the location of all new plant materials in the subdivision be on the plan. The subdivider is required to plant street trees according to city specification and to guarantee them with replacement for two years (Robson, 1983).

The city of Chicago requires the planting of parkway (street) trees when new buildings are constructed, when old buildings receive additions or are rehabilitated, or when parking lots that hold more than four spaces are built or rehabilitated. One 2.5-in. (6 cm) shade tree must be planted for every 25 ft. (7.6 m) of street frontage. One-, two-, and three-family homes are exempted from the ordinance, as are streets where less than a 9-ft. wide (3 m) planting strip exists (City of Chicago, 1991).

SCREENING ORDINANCES

Screening ordinances apply primarily to off-street parking. They generally set standards for structural and/or vegetation screening on the periphery of the lot and may require islands of vegetation within the parking area. Screening ordinances usually apply to off-street parking in all land-use categories except single- and two-family dwelling unit districts.

The Park Ridge, Illinois, ordinance requires that all parking lots be screened on "each side and along the rear lot line by a fence or wall not less than 4 feet (1.2 m) high, plus a planting strip of 4 feet minimum width." For every 1500 ft.^{2} of paved area, up to 6000 ft.^{2}, a tree greater than 3 in. (7.5 cm) must be planted. For each additional 3000 ft.^{2} over the original 6000 ft.^{2}, a similar-sized tree must also be planted. The builder must submit a plan to the city forester for approval that describes tree species and ground covers to be used. The property owner is then required to maintain the fence and vegetation, and to replace dead trees following the builder guarantee period (Morell, 1983).

The city of Lynnwood, Washington, has similar parking lot screening requirements, and further requires screening between certain zoning districts. The ordinance describes five types of landscape screens and recommends the type to be used between designated land uses. Appendix E contains the complete ordinance for the city of Lynnwood.

ENERGY CONSERVATION

Chapters 4, 7, and 8 discuss the contributions that urban trees, forests, and other plant materials make to conserve energy by reducing wind speed in winter and mitigating the urban heat island in summer. Patterson (1992) suggests that communities revise tree ordinances to include energy conservation provisions as they relate to reducing the effects of the urban heat island. Provisions can encourage private property owners to plant trees where they shade windows, air conditioners, roofs, walls, and other key areas. Tree canopy cover requirements over parking lots and other paved surfaces will reduce heat buildup and community energy costs. Developers can be required to maintain a specified amount of canopy when land is developed, to use light-colored building materials, and to plant trees where they will provide maximum energy conservation. Landscape and development plan approval is then based, in part, on optimizing energy conservation. Tree planting on city streets should use trees with broad mature crowns where adequate space exists for these species both above and below ground. An energy conservation ordinance must also recognize that trees and other plants can work against energy conservation by shading solar collectors. Solar access provisions are a necessary part of energy conservation ordinances.

LITERATURE CITED

AMERICAN NATIONAL STANDARDS INSTITUTE (ANSI A300). 1995. *For Tree Care Operations: Tree, Shrub and Other Woody Plant Maintenance Standard Practices.* Washington, D.C.: American National Standards Institute.

BUNDESMINISTER FUR ERNAHRUNG, LANDWIRTSCHAFT UND FORESTEN. 1976. *Bundesnaturschutzgesetz.* Coburg, West Germany: Neue Presse GmbH, 32 pp.

CITY OF CHICAGO. 1991. *Guide to the Chicago Landscape Ordinance.*

CITY OF HIGHLAND PARK. 1994. *Tree Preservation in the City of Highland Park.*

DADE COUNTY, FLORIDA. 1979. *Ordinance No. 79-105.*

DUERKSEN, C. J., and S. RICHMAN. 1993. *Tree Conservation Ordinances.* Chicago: American Planning Association, Rep. No. 446.

Florida Division of Forestry. 1973. *Urban Forestry Handbook.* Tallahassee: Fla. Div. For.

GREY, G. W. 1978. "Tree Ordinances and Related Policy." *Proc. Natl. Urban For. Conf.,* ESF Pub. 80-003. Syracuse: SUNY, pp. 627–631.

MORELL, J. D. 1983. "Municipal Ordinances' Relation to Trees: Screening and Landscape Ordinance." *J. Arbor.* 9(5):120–136.

MORELL, J. D. 1984. "Parkway Tree Augering Specifications." *J. Arbor.* 10(5):129–132.

NOSSE, R. A. 1982. "Is Management Recognizing That the Urban Forest Can Be a Pest?" *Urban and Suburban Trees: Pest Problems, Needs, Prospects, and Solutions.* East Lansing: Mich. State Univ., pp. 241–244.

O'CALLAGHAN, D. P. 1991. "Legal Protection for Trees in Britain and Ireland." *J. Arbor.* 17(11):306–312.

PAGE, E. 1983. "Municipal Ordinances' Relation to Trees: The Tree Ordinance in Arlington Heights." *J. Arbor.* 9(5):128–136.

PATTERSON, F. 1992. *Ordinances.* In L. Akbari, S. Davis, S. Dorsano, J. Huang, and S. Winnett, eds. *Cooling Our Communities: A Guidebook on Tree Planting and Light-Colored Surfacing.* Washington, D.C.: U.S. Environmental Protection Agency, pp. 111–128.

PERRY, T. O. 1978. "Constraints to the Effectiveness of Urban Forestry Programs." *Proc. Natl. Urban For. Conf.,* ESF Pub. 80-003. Syracuse: SUNY, pp. 652–660.

PIOTROWSKI, G. 1991. "Maryland's Forest Conservation Act of 1991." *Proc. Fifth Natl. Urban For. Conf.* Washington, D.C.: Am. For. Assoc., pp. 114–117.

PROFOUS, G. V. 1992. "Trees and Urban Forestry in Beijing, China." *J. Arbor.* 18(3):145–153.

ROBSON, H. 1983. "Municipal Ordinances; Relation to Trees: Tree Preservation." *J. Arbor.* 9(5):128–136.

WEBER, C. C. 1982. "Developing a Successful Urban Tree Ordinance." *Proc. Sec. Natl. Urban For. Conf.,* Washington, D.C.: Am. For. Assoc., pp. 227–232.

WIDRLECHNER, M. 1981. "Legal Controls for Undesirable Vegetation." *J. Arbor.* 7(6):145–151.

10

Management of Street Trees: Planning

Municipal forestry programs vary greatly in terms of their responsibilities. These responsibilities can involve the care of municipal street trees, the management of park and greenbelt vegetation, and regulation of private of private vegetation in the interest of the public safety and welfare. Most municipal forestry programs have as their primary responsibility the management of street trees planted in the public right-of-way. A survey of municipal forestry programs in the United States by Tschantz and Sacamano (1994) found that 57 percent of funds allocated for tree management activities were spent on street trees, with most of the remaining tree management funds being spent on park trees and trees on other public grounds (Table 10–1).

A number of terms are commonly used to describe the location of street trees, the most common being tree lawn, lawn extensions, tree terrace, planting strips, and boulevard. These strips of land lie primarily between the sidewalk and street, but sometimes are located between the sidewalk and private property or in median strips between traffic lanes (Figure 10–1). In many cities the initial function of the strip of land between the sidewalk and street was to provide a buffer for the pedestrian from traffic on unpaved and often muddy streets. Trees were added later by abutting property owners or by the city, with city government often assuming responsibility for these trees because of their location in the public right-of-way. Although streets have long since been paved, these strips are still considered essential for snow storage dur-

TABLE 10–1 MEAN AMOUNT SPENT ON TREE-RELATED ACTIVITIES BY POPULATION SIZE IN 1993. (From Tschantz and Sacamano, 1994.)

Population	Park Trees	Street Trees	Public Grounds	Cemeteries	Nursery Maintenance
Entire Population	$61,966	$226,361	$89,106	$6,287	$15,667
Over 1,000,000	$454,000	$791,667	$27,500	$2,000[a]	$75,000[a]
500,000 to 1,000,000	$437,813	$1,043,643	$48,333	$20,000[a]	N/A
250,000 to 499,999	$262,962	$1,276,867	$861,722	$24,000	$104,071
100,000 to 249,999	$63,014	$397,622	$111,783	$18,389	$10,400
50,000 to 99,999	$55,137	$129,350	$22,150	$3,844	$2,344
25,000 to 49,999	$32,070	$73,723	$49,500	$4,083	$4,594
10,000 to 24,999	$12,213	$25,026	$14,759	$3,367	$4,278
5,000 to 9,999	$4,018	$20,650	$2,738	$1,909	$4,500
2,500 to 4,999	$1,500	$8,719	$2,714	$1,000	$1,100

[a]One community responded to this question.

ing snow removal operations in cold climates, because of their aesthetic value when planted with trees, and as a place to locate underground utilities.

Kielbaso and Cotrone (1990) sampled 320 cities in 1989 to describe the status of urban forestry programs in the United States. Based on their survey there are 60 million street trees in the United States, or an average of 102 trees per mile (1.6 km). The number of potential street trees in the United States has been estimated at 100 million by Barker (1976), assuming a spacing of 58 feet (17.6 m) per tree and 620,000 miles (387,500 km) of municipal streets, or 182 trees per mile (114 per km). Normal recommendations for spacing on city streets is 40 to 50 feet (12 to 15 m), but this does not allow for intersections, unplantable sites, streets where there is no room to plant trees, or for variability between communities. The city of Milwaukee, Wis-

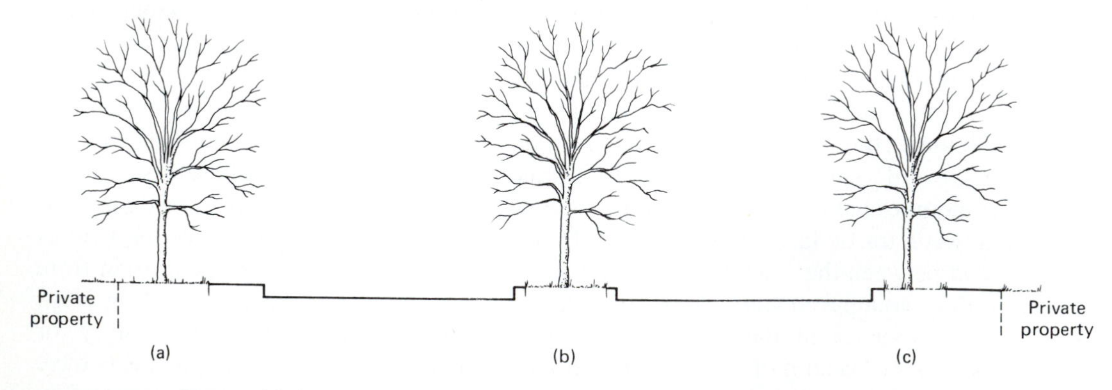

Figure 10–1 Trees located in the public right-of-way. (a) Between the sidewalk and private property. (b) In a median strip. (c) Between the curb and the sidewalk.

consin, is very close to full stocking with 200,000 street trees on 1,400 miles (2,240 km) of streets, or 143 trees per mile (Ottman, 1993). Minneapolis, Minnesota, is 92.6 percent stocked with 150,000 street trees on 1000 miles (1600 km) of streets for a stocking of 150 trees per mile (Hermann, 1993). Based on percent stocking, Minneapolis has room for 162 trees per mile (1.6 km). Full stocking of street trees in a community is likely to be approximately 140 to 180 trees per mile of street (87 to 112 per km), depending on the number of sites available for trees.

SUCCESSFUL MUNICIPAL FORESTRY PROGRAMS

Municipal forestry programs in the United States thrived in many cities following the city beautification era. House lots were small, many people lived in apartments, and one of the few places available for trees was the tree lawn. Communities prided themselves on their appearance and a key attribute was the street tree population. Municipal forestry programs enjoyed strong public support during the first half of the twentieth century. For example, the city of Milwaukee, Wisconsin, appointed its first city forester under the jurisdiction of the park commissioner in 1918, one year after enabling legislation was passed by the state legislature. In 1937 Milwaukee's park system was turned over to the County Park Commission. The Bureau of Forestry was created under the Commissioner of Public Works and assigned responsibility for street trees, boulevard median plantings, small parks, and other city-owned land.

Many cities, both large and small, followed a similar pattern in the development of forestry programs. However, Bartenstein (1982) describes two events that eventually diluted the constituency for urban forestry. Air conditioning reduced the importance of trees in modifying urban microclimates, and the automobile provided city dwellers with a means of escaping to suburban environments where larger house lots provided private green space. Other activities contributing to the erosion of city tree populations include urban renewal, street widening, absentee ownership of housing, budget constraints, safety concerns by municipal government, and Dutch elm disease. Eventually many municipal officials came to regard street trees as more of a liability than an asset.

Further compounding the problem for city tree programs, it appears that new subdivisions in many communities are leaving even less space for street trees in spite of larger house lots. Barker (1976) reports that the width of the tree lawn as required by ordinance in Utah communities is generally 3 ft. (1 m), while these same communities specified up to 12 ft. (3.5 m) prior to 1900. To avoid problems with sidewalk and curb lifting, only slow-growing species that reach a small mature size should be planted in 3-ft. (1-m) tree lawns. However, Kalmbach and Kielbaso (1979) report that residents of five cities surveyed prefer larger to smaller trees.

With all these forces working against street trees, it is no wonder that many city forestry programs suffered disproportionately during urban fiscal problems of recent decades. Some programs experienced severe budget cuts and some programs were eliminated entirely, but other programs continued to thrive in the face of a great deal of adversity. The ability of these programs to do well during recent years is related to three factors: the severity of fiscal problems, good management, and the maintenance of a supportive constituency. In some communities facing severe budget emergencies, no amount of good management could have protected city forestry budgets. Citizens in these communities placed a higher priority on essential services such as police and fire protection, sanitation, and street maintenance, and this left little financial support for forestry.

Communities where forestry continued to do well were those communities where good management was supported by a long-term program of maintaining public support through information and education programs. Nighswonger (1982) describes a successful community forestry program as one that is cost-effective, involves people in the community, stimulates community pride, is well planned, and is educational. The ensuing discussion of municipal tree management draws on the lessons learned from successful programs and describes techniques used to plan and execute a successful street tree management program. This involves development of a master street tree plan, selecting and planting trees, tree maintenance, tree removal and replacement, task scheduling, and sources of funding. Public involvement should be a part of all street tree management programs. Public relations techniques are discussed in Chapter 13.

PLANNING A STREET TREE PROGRAM

Good management involves the setting of goals and objectives, prioritizing them, and developing specific management strategies to achieve them. All too often managers are encountered that manage on a crisis basis only, with each day spent reacting to the job rather than working toward long-term goals. To be sure, crisis managers get the job done on a day-to-day basis, but over the long term it is questionable whether they are meeting the needs of their employer. In urban forest management the crisis approach to management is easy to slip into by reacting to tree problems as they arise. Obviously, the city forester cannot ignore a dangerous street tree in the interest of working toward long-term goals. Crises are demanding and immediate and can require a lot of a manager's time. However, no matter how much time is spent dealing with day-to-day problems, some effort must be made to formulate and address long-range management objectives of the organization. Too much time spent dealing with crisis situations may be an indication that a manager is not performing his or her job correctly (i.e., not delegating enough responsibility to subordinates).

Effective management should operate in at least two general time frames; short term, or day-to-day activities including crisis management; and long term, or the setting and prioritizing of long-term organizational goals. A planned system of day-to-day management activities such as task assignments and equipment allocations should be provided to each crew. Crisis situations should be prioritized in a manner to minimize hazards, as should regularly scheduled maintenance activities. At higher levels of the organization, managers must be aware of community needs, the political arena, and must maintain a system of communication with the public. The goals and objectives of the city forester cannot be different than those of the community. If they are different, the forester needs either to influence community goals through an information-education program, or accept community goals as program goals. In any event, good long-term management requires a thorough understanding of community values, the development of goals and objectives based on these values, and the use of these goals on a day-to-day basis to assist in guiding short-term management decisions.

Crisis Management

In an attempt to determine the economic costs associated with crisis management of street trees, Callahan and Bunger (1976) conducted a study of an urban forestry program in a large midwestern city (population 1,000,000). The city has four established management objectives:

1. To systematize the management of park and street trees
2. To improve the quality of city trees and other woody plants
3. To assist in the creation and maintenance of a pleasing urban environment within the constraints imposed by silvical, ecological, budgetary, and urban planning considerations
4. To reduce and/or stabilize the ultimate costs of the city's urban forestry program

To meet these objectives, the current management system consists of servicing street trees on the basis of requests by abutting property owners. When a request is made, the site is visited by forestry personnel and the work given a priority rating of A, B, or C for servicing. Daily work schedules are built around "A" or top-priority jobs. "B" jobs are serviced if near an "A" job, and "C" jobs take from several months to two years for servicing. Callahan and Bunger (1976) describe this system as efficient and responsive for crisis management, but inefficient and expensive when used as an overall management system. They recommend: "The alternative to a reactive response system is one of instituting management principles and regulating the urban forest by removing potential trouble-maker trees, by periodic preventative maintenance of the better trees, and by the replacement of monocultures with an aesthetically pleasing mixture of environmentally tolerant trees." Regulating the urban forest in this fashion would meet the first three goals of the forestry program, but at an additional cost of $5 million over a fifteen-year period. The reactive system in the city studied is the result of chronic underfunding of the forestry program, and can be corrected only by convincing local voters that the added expense is justifiable.

Funding problems plague many urban forestry programs. In a survey of municipal tree management in New Jersey, Tate (1984) found that three-fourths of the city foresters surveyed felt that funding was the major problem in providing adequate tree care in their communities. Many communities suffering substantial budget cuts were forced to shift from systematic management to reactive responses or crisis management systems. From 1986 to 1994 the percentage of the mean municipal budget for tree management in the United States decreased from 0.49 percent to 0.32 percent (Tschantz and Sacamano, 1994). However, in six of the nine population size categories the percent of the mean municipal budget for tree management increased (Table 10–2).

Public Attitudes and Management

In spite of setbacks in many municipal forestry programs during recent decades, there is strong public interest in street tree management programs. Social needs as they apply to urban vegetation are discussed in Chapter 2, and these needs directly relate to street tree management programs. Kalmbach and Kielbaso (1979) used the semantic differential technique to determine personal attitudes toward photos of urban scenes with and without trees in two Michigan communities. They found that the presence of trees does enhance urban scenes in the minds of the viewers (Figure 10–2). In addition, people were found to favor large trees over 25 ft. (7.5 m) tall in residential neighborhoods, as opposed to small trees.

In a similar study made in Ohio, Schroeder and Cannon (1982) measured the visual prominence of features in photographs of street scenes by the area occupied by the feature (e.g., vegetation) or by the frequency of the feature (e.g., automobiles).

TABLE 10–2 PERCENT OF MEAN MUNICIPAL BUDGET FOR TREE MANAGEMENT BY COMMUNITY POPULATION SIZE. (From Tschantz and Sacamano, 1994.)

Population Size	Mean % of Total Municipal Budget Allocated to Tree Management in 1986	Mean % Total Municipal Budget Allocated to Tree Management in 1994
Entire survey population	0.49	0.31
Over 1,000,000	0.09	0.13
500,000 to 1,000,000	0.02	0.28
250,000 to 499,999	0.06	0.41
100,000 to 249,999	0.06	0.40
50,000 to 99,999	0.28	0.38
25,000 to 49,999	0.62	0.35
10,000 to 24,999	0.70	0.30
5,000 to 9,999	0.12	0.22
2,500 to 4,999	1.91	0.30

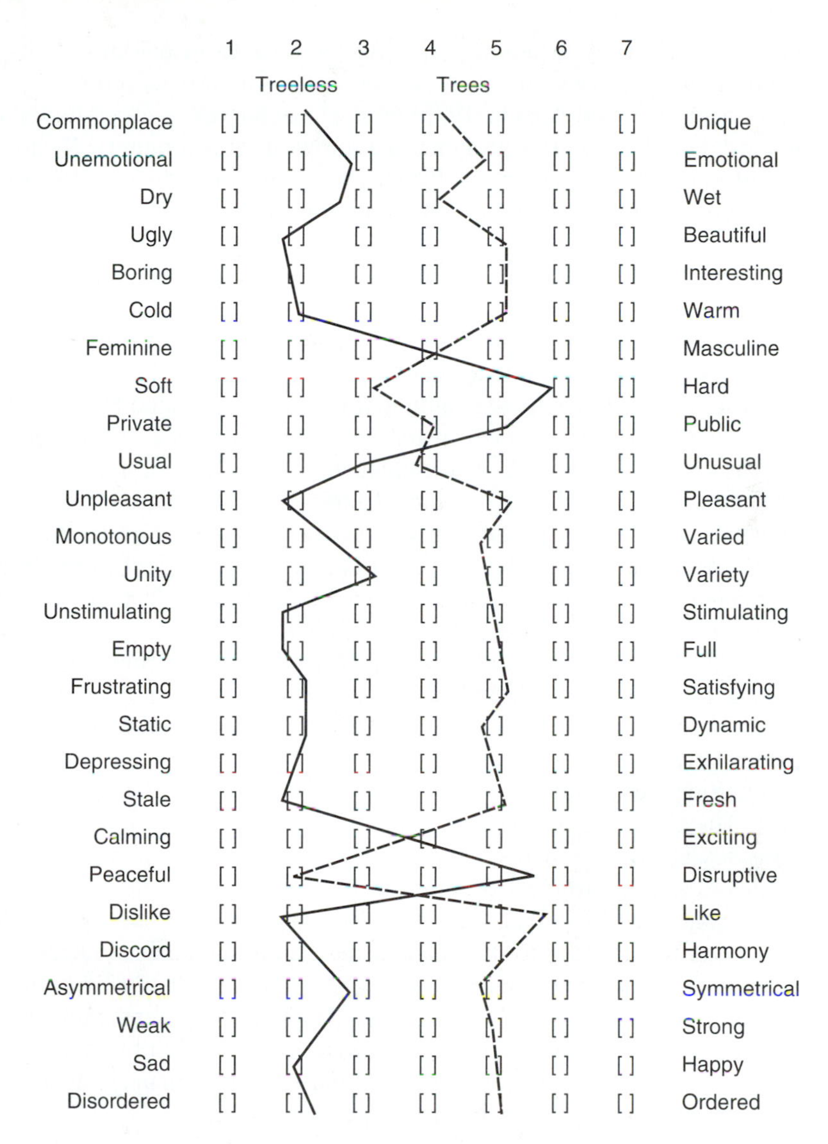

Figure 10–2 Composite semantic differential comparisons of photos with and without trees for two Michigan cities. (From Kalmbach and Kielbaso, 1979.)

University students and local residents were asked to rate the photographs on a 10-point scale as to attractiveness of the scenes. Regression analysis (accounting for 60 percent of the variation in responses) indicated street trees to be the most important influence on the attractiveness of the street to the individuals surveyed. In a later study Schroeder and Cannon (1987) found that private trees in front of homes also enhanced street views, and that street trees are most important where there are few or no yard trees. They recommend that for the greatest visual impact city foresters concentrate planting where there are few or no trees on private property.

Schroeder and Appelt (1985) used a mail survey to ascertain the public attitude toward the urban forestry program in a midwestern community. Results of the survey found that most respondents felt that trees were important to the community, were satisfied with the forestry program, and placed high priority on public tree services, especially planting and removal. The authors felt that this information would be very useful to argue against budget cuts that would require the forestry department to reduce services.

Fourteen towns in Ohio ranging in population from 13,000 to 82,000 were studied by Hager et al. (1980) to assess their street tree policies and programs. Public concern for street trees in the communities surveyed ranged from apathy to strong interest, and this was reflected in ordinances and the quality of the tree population. Towns with long-established and adequately funded street tree programs were found to have a greater variety of species, healthier trees, more consistent and higher tree densities, and an uneven age distribution of size classes. Towns with a new or nonexistent street tree program tended toward fewer species, less healthy trees, fewer trees, less consistent tree density, and even-aged tree populations that were becoming decadent, an indication of no replacement planting.

The key to a successful urban forestry program is strong public support of good management. The city of Milwaukee, Wisconsin, has an exemplary urban forestry program that has stood the tests of Dutch elm disease and fiscal emergencies during recent decades. The program maintains a highly visible posture in the community and participates in numerous programs and projects that elicit strong community support. These include a city-wide Arbor Day program, cooperative planting programs with downtown merchants, champion tree contests, maintenance of flower beds on boulevards, and participation in citywide festivals and celebrations. In a short-term time frame the Milwaukee Bureau of Forestry is responsive to crisis situations, service requests, and day-to-day management and maintenance needs. On a long-term basis, the program is managed to meet goals of (1) full stocking with a tree population diverse in species and age classes, (2) adequate maintenance through scheduled pruning and other services, and (3) prompt removal of problem trees followed by replacement. Table 10–3 provides a summary of 1983 management activities for the forestry program in the city of Milwaukee (Skiera, 1984).

Urban Forest Planning Model

The basic planning process consists of four steps as described in Chapter 8. These are:

1. What do we have? (Inventories)
2. What do we want? (Goals)
3. How do we get what we want? (Management plans)
4. Feedback

TABLE 10–3 SUMMARY OF TREE MANAGEMENT ACTIVITIES FOR MILWAUKEE, WISCONSIN, 1983 (From Skiera, 1984.)

Management Activity	Number of Trees
Pruned	56,053
Planted	8,485
Removals	4,733
Tree surgery	2,678
Storm damage service	1,777
Service requests	6,923
Root cutting	163
Dutch elm disease control	63
Insect control	15,041

This process is adapted to serve as the basis of a planning model designed to specifically address the overall management of urban vegetation resources (Figure 10–3). The ensuing discussion describes the Urban Forest Planning Model and is followed by its specific application to street tree management. The model is also used in subsequent chapters as the basis for the development of management plans for park, greenbelt, private, and other urban vegetation.

Inventories. The items to be considered in the inventory portion of the planning model include identification of the planning unit and inventories of vegetation, public attitudes, and agents of change within the unit. The planning unit consists of the vegetation resource base being planned for, such as a street tree population, a specific park or landscape, an entire park system, or the total vegetation resource of a community.

Once the planning unit has been identified, a vegetation inventory should be conducted using a system such as those described in Chapter 7. This usually takes the form of a preliminary survey used to develop initial management plans, followed by a detailed inventory if deemed necessary by the management plan. The inventory should include a review of past management activities and inventories (if available) to develop baseline information descriptive of the dynamic nature of the resource. Some very basic information essential for preliminary planning includes:

1. How many trees are there?
2. How big are they?
3. How many vacant spaces are there?
4. How many do we remove each year?
5. How many do we plant each year?

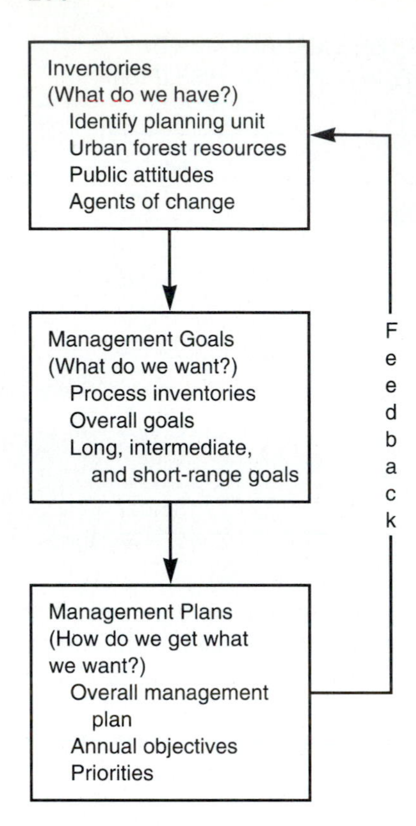

Figure 10–3 Urban Forest Planning Model.

An inventory that provides this information will allow us to develop preliminary plans and budgets for planting, pruning, and removals. The number of vacant spaces and annual removals provides a measure of the current planting program relative to whether the population of trees is increasing or decreasing, and when or if the community can expect to have all vacant spaces filled. Pruning costs are related to tree size, and tree diameter summaries can be used to estimate pruning costs relative to the number of trees scheduled for pruning. Information on the size and species of trees being removed will give an indication of problem species and can be used to anticipate future removal costs.

Some understanding of public attitudes, perception, and knowledge is essential in the establishment of management goals. This information may be obtained through discussions with members of the shade tree or park board, elected officials, community leaders, service organization, or by direct survey of public attitudes, perceptions, and knowledge. This survey can take the form of a questionnaire designed to determine the following:

1. Public perception of the quality of community vegetation on streets, and in parks and recreation areas
2. Public attitudes toward public vegetation

3. Public knowledge of proper vegetation maintenance
4. Public satisfaction with public vegetation management

In the case of private vegetation management, a thorough understanding of the desires of the client or organization is essential when setting goals.

The vegetation resource of a community is not static but a dynamic system that will improve, remain the same, or deteriorate over time. The agents that influence vegetation should be identified and actual or potential impact on vegetation by these agents be described. Agents of change are categorized as environmental, public, and private.

1. *Environmental.* Existing and potential biotic agents, such as insects and diseases, should be identified and potential hosts evaluated. This also includes abiotic agents such as air pollution, de-icing salt, storms, and so on.
2. *Public.* Public agents of change should be identified and potential impact on vegetation determined. These include existing vegetation ordinances, elected officials, public utilities, streets and sanitation departments, park or tree board, educational institutions, planning departments, state and federal agencies, and so on.
3. *Private.* Private agents of change should be identified and potential impact on vegetation evaluated. These include the news media, public service organizations, private utilities, chamber of commerce, land developers, builders, business associations, the green industry, and so on.

Management goals and objectives. Management goals describe what you hope to achieve and objectives set specific points in time when you anticipate meeting those goals. For example, a goal of a community forestry program might be to fill all vacant tree spaces on city streets, the objective being to accomplish this goal in ten years. Management goals should be defined based on an understanding of public attitudes, perceptions, and knowledge, a review of the agents of change, and the expressed needs and concerns of the community (client). These goals should be compared to a dynamic or temporal description of the resource base as ascertained from the inventory, to develop specific management objectives. Mudrack et al. (1980) suggest that urban vegetation management goals be developed in the following time frames:

1. Overall or comprehensive goals
2. Long-range goals (10 or more years)
3. Intermediate-range goals (5 to 10 years)
4. Short-range goals (3 to 5 years)
5. Annual objectives

Lobel (1983) recommends public urban foresters extend long-range goals out well beyond ten years to cover the life span of trees being managed, as is done in rural forest management.

Management plans. Management plans are developed to meet the goals and objectives of the organization in the time frames identified. Mudrack et al. (1980) suggest that "short- and intermediate-range goals should lead toward fulfillment of longer range and overall goals. Conversely, overall goals should be able to be broken down into shorter goals and annual objectives." Objectives should be prioritized within each time frame, and agents of change needed to accomplish these goals identified.

Feedback. Feedback makes management planning an ongoing process that is responsive to community or client needs and changes. This involves a reinventory of the resource base to determine the impact of the management plan, a reassessment of public needs, and reevaluation of the plan to determine if the objectives are being met (Lobel, 1983). Feedback keeps the management system open and sensitive to changes in public or client values, new management techniques, and changes in the resource base.

Contracting versus In-house Services

Community tree management consists of planning, planting, maintaining, removing, and replacing community trees. These services can be carried out by community employees, or can be done wholly or in part by consultants and contractors. Three types of services are generally available: management services, establishment services, and maintenance services. Management services include conducting inventories, writing tree ordinances, and developing complete management plans for the community. Establishment services include design of plantings, providing nursery stock, and tree planting. Maintenance services cover tree pruning, repair, protection, tree removal, stump removal, and disposal of waste wood. The quantity of services contracted varies from none to virtually all tree management activities, depending on local attitudes and conditions. Which approach is best for a community is certainly a matter for debate, but a community should look at a number of factors before making a decision.

Cost is an important consideration when deciding whether or not to contract arboricultural services. If a contractor can provide high-quality professional services for a lower cost than municipal crews, good management may best be served by a contractor. The key is quality of work, and the city forester must determine if there are good commercial firms in the community interested in a city contract and capable of providing the desired service. Awarding a city contract to the lowest bidder may not result in a satisfactory job on public trees. Minimum standards of performance must be established and described in the contract, with the city forester having the authority to reject a bid from a firm with a reputation for substandard performance.

Tate (1993) feels contracting is cheaper than using in-house crews for many tree care services because contractors do not provide the same level of fringe benefits for employees, and because of the competitive nature of contract bidding. City employees are often unionized, and this can also drive up costs through higher wages. Contracts are administered by city employees, and this requires time and skills different than city crew supervision. The contract manager must focus on overall performance rather than on the day-to-day activities of the contractor (Tate, 1993).

Three additional factors influencing whether to contract or perform services inhouse relate to community size, specific problems, and equipment costs. Large communities have larger work loads, and local commercial firms may not have the personnel or equipment to handle a contract for hundreds of trees. A catastrophic event such as a wind or ice storm may exceed the capacity of city personnel to deal with the problem, and emergency services must be contracted on an immediate basis. Equipment costs relate to community size and back to cost-effectiveness. It is unlikely that a small city with a few thousand trees can justify purchasing bucket trucks, chippers, and stump grinders for pruning and removal work, yet this equipment is necessary for operations. There must be enough work on an annual basis to justify equipment costs, and this is more likely to occur in large versus small communities.

A final factor to consider goes beyond tree management, and that is the overall management of the city work force. Some communities shift crews from department to department in the course of the year to gain maximum productivity from employees. The same crew may plant trees in the spring or fall, maintain park facilities in the summer, and remove snow and service city equipment in the winter months. The city forester must look at overall municipal services before deciding whether to recommend contractual tree services.

The quantity and type of work contracted varies from community to community. Robson (1984) conducted a study of urban forestry programs in eight suburban Chicago communities, and compared contractual versus in-house services in six management activity areas (Table 10–4). He found that all eight communities provided a full range of street tree services, with removals and planting most likely to be contracted. In a similar survey of city forestry programs in New Jersey, Tate (1984) reports contract services being used by 70 percent of the respondents; the top three contract items being removal (69 percent), planting (63 percent), and pruning (53 percent). In 1994, 21 percent of tree management expenditures in the United States were spent on contract services, with more than half of respondents stating they prefer in-house work to contracts (Tschantz and Sacamano, 1994). It is also important to note that contracts are tempting targets during times of budget reductions, as it is much less painful for the city not to advertise for bids than it is to lay off community employees.

The legal implications of tree contracts are discussed in the legal values section of Chapter 5. City foresters and commercial arborists should be aware of, and follow, the Uniform Commercial Code as described by individual state statutes before negotiating a contract. Hoefer (1982) recommends that contractors interested in bidding on city tree work be given a notice to bidders containing the following information:

TABLE 10–4 PROGRAM VARIATION IN CONTRACT VERSUS IN-HOUSE CREWS IN EIGHT CHICAGO SUBURBAN COMMUNITIES (From Robson, 1984.)

Service	Contract	In-house Crew	Combination
Trimming	1	3	4
Removal	2	2	4
Insect and disease	0	5	2
Planting	5	3	0
General maintenance	0	8	0
Wood disposal	3	2	3

1. Who is requesting bids
2. Where to send the bids
3. Deadline time
4. "Earnest" money or bid deposit
5. Bonding requirements, if any
6. Arborist license or certification requirements, if any
7. Insurance requirements
8. How to bid (i.e., by tree, by diameters, by total, etc.)
9. Penalties for failure to perform
10. Liability disclaimer for the contractee
11. A statement that gives the right to reject any or all bids
12. Date, time, and location of prebid meeting
13. Equal opportunity and affirmative action requirements
14. Any other statements legally required by the community

A contract itself contains five general parts: document name, introductory remarks, recital, operative provisions, and closing (Hoefer, 1982).

Document name. A typical contract name begins with "Agreement for . . ." followed by a description of the work to be performed. An example of a municipal tree contract title is "Agreement for Pruning City Trees," or "Agreement for the Removal of City Trees."

Introductory remarks. This section names the parties entering into contract, the date, and an agreement statement.

Recital. Recitals are the "whereas" statements, and are used to insure the validity of the contract.

Operational provisions. This section of the contract describes what is to be done, when it is to start and expiration date, inspection process, operational procedures, safety procedures, standards of performance, billing procedures, and penalties for failure to perform. The American National Standards Institute has developed standards for a number of tree maintenance operations, and it is recommended these standards be used in contracts.

Closing. The closing is where the document is signed providing the parties agree upon the contents of the contract.

Types of contracts. Three types of contracts are normally used when bidding for tree work; unit, time and material, and lump sum (Tate, 1986). Unit contracts are based on a price per unit such as cost per tree in planting contracts, or cost by diameter class for pruning or removal contracts. Time and material contracts involve the cost per hour plus the cost of materials used. A lump sum contract is a single bid to do an entire job such as pruning or planting a given number of trees. Tate (1986) feels time and material is the best system for city forestry programs because the city has greater control over the contractor, and the contractor will be more competitive in bidding. Municipal forestry contract work in New York City was first done on unit pricing, but later changed to time and materials because of lower cost, greater efficiency, and greater flexibility for management (Lough, 1991).

Master Street Tree Planning

Many street tree ordinances direct the municipal forester to develop a Master Street Tree Plan. Unfortunately, this has often taken the form of a listing by street of those species to be planted throughout the community. However, a Master Street Tree Plan when developed through use of the Urban Forest Planning Model becomes a much more comprehensive document. It becomes a management plan that is process oriented, sets long- and short-term goals and objectives, and is responsive to community needs. King (1979) lists the following advantages of a properly developed Master Street Tree Plan:

1. Less interference with buildings
2. Fewer disease problems
3. Lower tree maintenance from the standpoint of trimming, removal, and spraying
4. Less or no expense when streets are widened
5. Less damage to sidewalks and curbs
6. Aesthetically more pleasing
7. Safer to the public using the roads and sidewalks
8. Lower trimming and removal costs to utilities
9. Fewer outages of electrical and telephone service

A Master Street Tree Plan should be based on a dynamic street tree inventory system, reflect community values, establish short- and long-term goals and priorities for tree planting and maintenance activities, and establish removal and replacement policies. Street tree planning and management involves:

1. Identification of planting sites and inventory of existing trees
2. Selecting appropriate species and planting vacant sites
3. Scheduled fertilization and watering during the establishment period
4. Periodic inspections for problems and hazards
5. Regularly scheduled pruning
6. Insect and disease control as warranted
7. Repair as needed
8. Removal and replacement when the tree becomes a liability

The tree popul ıtion should be continuously monitored for problems, and this information fed into the management process to alter the management activities listed above. Lobel (1983) suggests that urban forest managers consider the additional criteria of age structure and rotation age when formulating street management programs. Even-aged tree populations by block are easier to maintain and aesthetically more pleasing. However, a city-wide diversity of age classes, even with even-aged populations on individual blocks, will provide a more stable population by reducing the chance of deforestation should an unexpected event cause heavy losses in a specific age class.

Tree attrition. Planting blocks or entire sections of a community at the same time will result in even-aged populations of trees in those areas, at least for a time. However, tree losses occur continuously throughout the community for a variety of reasons, and replacing lost trees tends to develop a wide variety of age classes over time. An understanding of tree attrition is essential to developing a comprehensive Master Street Tree Plan and subsequent management plan.

Foster and Blaine (1978) in a study of street tree attrition in Boston concluded that the average tree in sidewalk cutouts needed to be replaced every ten years. They further reported that Washington, D.C., is on an eighteen-year rotation based on the number of trees planted annually. Based on a survey of city foresters, Skiera and Moll (1992) report that the average life span of downtown trees is thirteen years while the average suburban tree lasts thirty-seven years.

Tree attrition can be described by average life span or by median life span. Using average life span requires knowing how long each tree in a population of trees lives, and then finding the average life span for that population. Polanin (1991) examined contract removal records and planting records for London plane (*Plantanus x acerifolia*) and Norway maple (*Acer plantanoides*) trees in Jersey City, New Jersey. These species represent 55 percent of the trees in the city. London plane trees attained an average age of thirty-nine years and Norway maples an average age of forty-eight years. However, the author did not include planting mortality in these calculations, and this would reduce the average life span.

Median life span describes how long one-half of the trees in a population of trees last. For example, a median life span of twenty-five years for a population of 200 trees means that 100 of them will be lost in that period and 100 will remain. The median life span can be readily calculated from removal rates, assuming the city is replacing trees as they are removed. For example, the City of Milwaukee has a population of 200,000 street trees, and removes approximately 3,500 per year (Ottman, 1993). Based on removals and assuming removed trees are replaced, it will take fifty-seven years to replace the entire tree population. The median life span for street trees in Milwaukee is 28.5 years (57 divided by 2). Of course, some trees will live longer than fifty-seven years just as other trees will be replaced more than one time. Median life span will usually be different than the average life span for a population of trees because tree mortality rates change as the population of trees gets older.

It is important that records of all tree removals be included in life span calculations, including planting mortality. Nowak et al. (1990) reported 34 percent mortality after two years of newly planted trees on a boulevard through two California communities. They also report that a review of the literature reveals that planting losses range from 3 percent after one year to 99.5 percent after six to ten years. Planting mortality was studied in three Wisconsin communities that have well-established and -funded forestry programs (Miller and Miller, 1991). Mortality rates were highest in all three communities the first year, and declined until year four, when normal attrition rates of less than 1 percent per year were attained. The survival rate after four years was 75 percent for all three communities, with 50 percent of the planting attrition occurring the first year (Figure 10–4). Based on this study, doubling the mortality

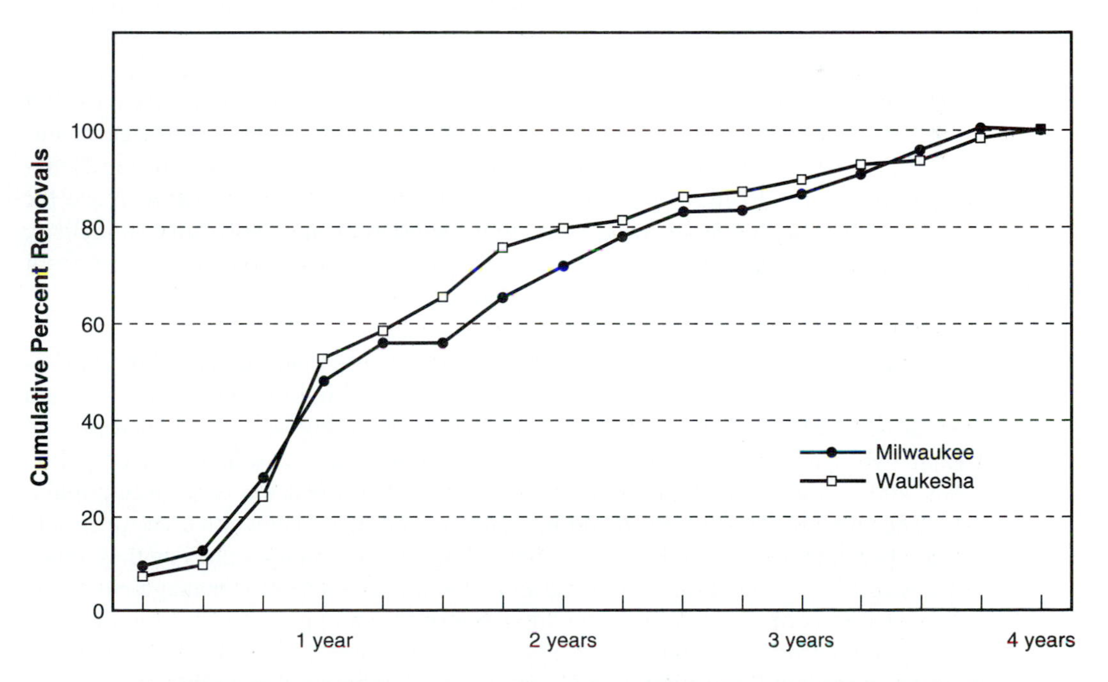

Figure 10–4 Cumulative percent removal over a four-year period for trees planted between 1980 and 1982 in Milwaukee and Waukesha. (From Miller and Miller, 1991.)

after one year may provide a reasonably accurate estimate of total mortality that can be attributed to planting during the establishment period.

The computer simulation described in Appendix B was used to estimate the median life span of a population of 9,000 street trees drawn from two street tree inventories in Wisconsin. Annual mortality rates by diameter class were based on actual mortality rates from five cities in Wisconsin and ranged from a low of 0.8 percent for trees 4 in. (10 cm) in diameter and increased linearly to 3.2 percent for trees 24 in. (60 cm) in diameter. Trees larger than 24 in. in diameter were given accelerated mortality as very few street trees in Wisconsin attain diameters in excess of 30 in. (75 cm). Planting mortality was set at 13 percent at planting and at 3 percent per year for the first four years, values that produced a total 25 percent planting mortality. The simulated diameter growth rate was 0.5 inches (1.25 cm) per year (Churack et al., 1994). The simulation yielded a median life span of 21 years for the tree population, meaning that every twenty-one years one-half of the population of trees is replaced.

The above calculations and simulations show that tree attrition in communities results in a fairly rapid turnover in community trees, certainly higher than we think of in conventional forest tree rotations. However, rotation age of forest trees for forest products should not be compared to the rotation of trees on city streets. A newly regenerated forest stand contains far more trees than will be around when the final harvest is made. The median life span of forest trees in naturally regenerated stands is likely much less than the median life span of a city tree due to intense competition for growing space.

Most urban trees have ample room to grow above ground, but are often limited by the amount of suitable soil available. Street trees in residential areas often exhibit rapid growth when young, but growth rate tends to decline as trees get larger. Churack et al. (1994) found that street tree diameter growth in Milwaukee began to decline in trees over 12 in. (30 cm) in diameter. Lack of sufficient growing space for roots combined with stresses from a variety of sources such as pollution, vandalism, higher urban temperatures, de-icing salts, and so on prevent most street trees from reaching their biologic potential in size and age.

Inventories. Actual development of a street tree management plan consists of taking inventories, developing management goals, writing management plans, and generating feedback (Figure 10–5). The inventory phase not only includes trees and vacant tree spaces, but an assessment of community values, solicitation of public input, and a description of those agents of change that affect the street tree population. The street tree inventory should be dynamic and be updated either continuously or on a periodic basis. A thorough understanding of community values as they relate to the street tree population is essential in formulating policies for management. The assessment of community values should also provide an opportunity for public input in the planning process, and should be an ongoing process throughout the management program. Environmental, public, and private agents of change that affect street trees should be assessed and their existing or potential impacts described.

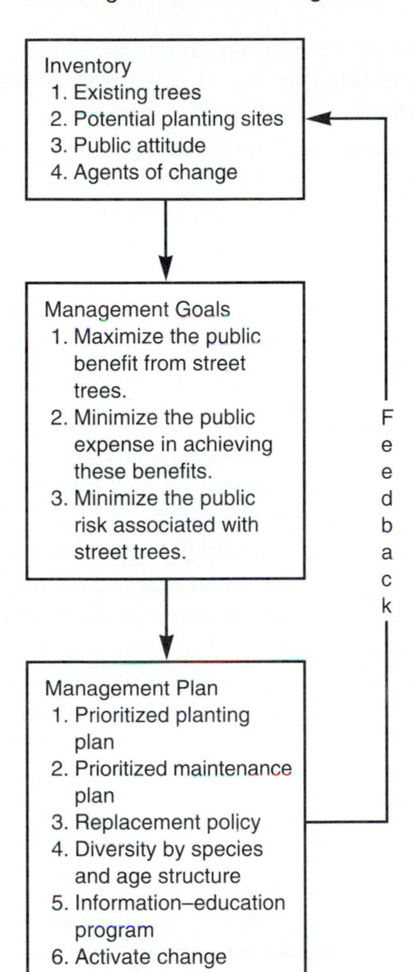

Figure 10–5 Master Street Tree Planning Model.

Management goals. The life of a street tree involves planning for that tree, planting it, maintaining it, and eventually replacing it with another tree when it becomes a liability rather than an asset to the community. To accomplish this, a Master Street Tree Plan should be designed to meet the following overall goals:

1. Maximize the public benefit from street trees.
2. Minimize the public expense in achieving these benefits.
3. Minimize the public risk associated with street trees.

Within these broad goals inventory information is processed and used to develop specific management goals to provide the following:

1. *Maximum stocking of street trees.* To provide an optimum-value tree population, the plan should attempt to achieve full stocking of street trees in the community at some realistic point in the future.
2. *Low maintenance costs and public safety.* Maintenance costs should be balanced against needs for public safety, a high-value tree population, and budgetary constraints. In the short run, cutting maintenance expenses will save money, but in the long run this will result in public hazards and trees that are a liability rather than an asset to the community.
3. *Stability.* A stable street tree population is one that is not threatened by catastrophic losses due to poor management. Poor management includes overreliance on a few species, and minimal or no scheduled tree maintenance.

Management plan. A management plan is designed to address management goals by priorities within the planning time frames selected. Overall comprehensive plans to meet long-term goals should extend into the future at least as long as the median life span for the tree population. Based on estimates of life span discussed earlier in this chapter, the minimum planning horizon for a street tree population should be twenty to thirty years, and longer if feasible. Intermediate- and short-term planning periods, and annual work plans may then be developed to meet long-term goals for the tree population. For example, a goal of full stocking of street trees may be fifteen years in the future. To meet this goal annual work plans need to allocate sufficient funding for tree planting. Likewise, establishing a maintenance (pruning) cycle of a fixed number of years at some point in the future will require annual adjustments in workloads and budgets for a period of years to meet that goal.

Within the time periods specified in the Master Street Tree Plan the following should be included in the management plan:

1. *Prioritized planting plan.* Develop planting priorities by work unit, land use, street, and site to provide the optimum tree cover for the community.

2. *Prioritized maintenance plan.* Determine in advance those management activities that will receive the highest priority. Tree planting is expensive, and follow-up watering, corrective pruning, and fertilization may be cheaper than replacement. Periodic pruning is necessary to maintain a high-value population and to protect the public from hazards. Elected officials usually prefer planting to maintenance because of the immediate visual impact of planting, but it is better to plant fewer trees and maintain what is there than to plant more trees and not have enough funds to care for them adequately. In addition, controlling epidemics of insects and diseases will be a high priority, as will be the establishment of acceptable levels of damage from these factors.

3. *Replacement policy.* Tree removal is based on two factors: the potential hazard, and maintenance costs in comparison to value. Serious liability problems are associated with damages caused by trees (Chapter 5), and removal of hazards must be a top priority of any tree management program. Removal of trees when their value relative to maintenance declines is a more difficult decision, and one that demands careful

analysis and good public relations. Replacement policies should be based on expected survival chances, reasons for removal, and the overall planting plan.

4. *Diversity by species and age structure.* Species diversity throughout the community will help to ensure against catastrophic losses such as those experienced in many communities due to Dutch elm disease. Age diversity provides temporal continuity of tree cover as different age classes are more or less susceptible to external disturbances. Normal tree attrition followed by replacement will produce a wide array of tree ages that will meet the criteria of diversity in age classes. The age profile of a street tree population can be determined by graphing frequency over diameter class. Since each age cohort steadily and increasingly loses trees over time, the result will be a J-shaped curve similar to Figure 10–6A, which describes an all-age population. On the other hand an age profile of street trees that has a bell shape such as Figure 10–6B is indicative of trees mostly the same age, a sign of a declining population where removals are not being replaced.

5. *Information-education program.* Throughout the entire management process public input and public education must remain a high priority. The public should know what the city forester is doing and why, and the city forester must keep informed as to the desires and values of the community. A successful municipal forestry program must have the support of the community.

6. *Activate change.* For each management goal, existing and potential agents of change influencing that goal have been identified and described by the inventory. To accomplish management goals, agents of change must be activated in a positive manner by the municipal forester.

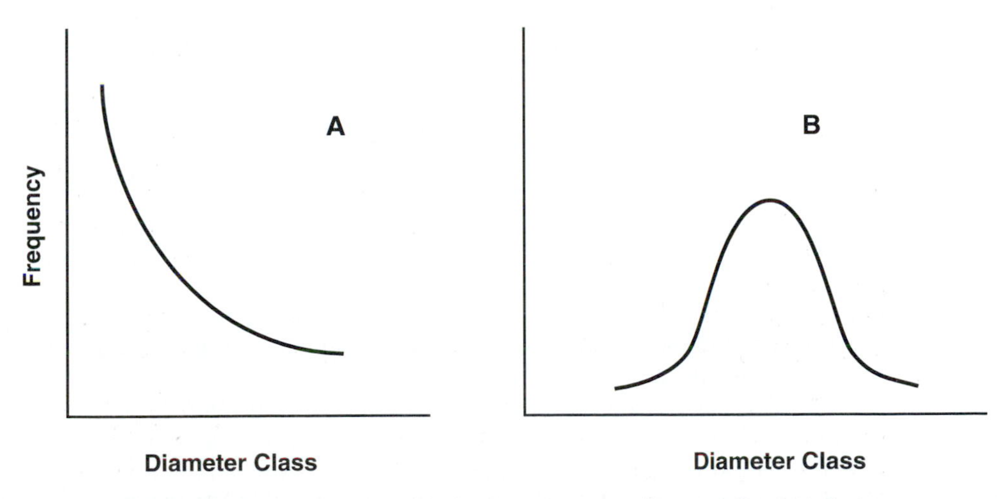

Figure 10–6 Street tree profiles by frequency over diameter. Graph A depicts an all-aged population, while Graph B depicts an even-aged population.

Feedback. Feedback provides a method of continuously evaluating the street tree management program. Feedback also allows the manager to evaluate community attitudes and values continuously, and provides community input into the management program. Management planning for periods of twenty or more years is uncertain at best. The most important function of feedback is that it makes the management plan responsive to the uncertainties of the future.

In summary, management planning for a street tree population involves an inventory of trees and community values, using the inventory to develop management goals, developing a management plan to achieve these goals, and feedback to monitor the entire process. Specifically, street tree management is the establishment and maintenance of a population of trees in the public right-of-way to provide community amenities. This consists of developing a Master Street Tree Plan to select and plant trees, to maintain these trees, and to replace them as needed.

PLANNING ACROSS COMMUNITY LINES

In many urban areas community boundaries abut one another, creating continuous urban regions. Opportunities exist for intercommunity cooperation in urban forestry as much urban and regional planning crosses community lines when attempting to treat urban areas and regions as a whole.

The Little Calumet Watershed (LCW) consortium was formed in 1987 to promote unified urban forestry programs. The LCW is in northern Illinois and consists of thirty-seven municipal governments, covers an area of 132,957 acres (53,183 ha), and has a population of 541,200 (Duntemann et al., 1989). In cooperation with the Illinois Department of Conservation, the University of Illinois, and the United States Forest Service the consortium has developed and is advocating five endeavors:

1. Promote Tree City USA programs.
2. Implement comprehensive tree ordinances.
3. Develop standard planting specifications.
4. Apply for Federal Land and Water Conservation and Department of Community Affairs grants.
5. Develop and coordinate forestry intern programs.

The long-term goal of the project is to have each municipality develop a comprehensive urban forest management plan, and to coordinate planning for greenspace and overall regionwide tree cover.

On a smaller scale, four small communities in Wisconsin joined together to apply for a Community Forestry Assistance grant to hire a consultant to inventory their tree population. Had they made independent applications it is unlikely a consultant would have found the grants sufficient to be cost-effective (Rideout, 1991).

LITERATURE CITED

BARKER, P. A. 1976. "Planting Strips in Street Rights-of-Way: A Key Public Resource." *Trees and Forests for Human Settlements.* Toronto: Univ. Toronto, pp. 263–274.

BARTENSTEIN, F. 1982. "Meeting Urban and Community Needs through Urban Forestry." *Proc. Sec. Natl. Urban For. Conf.,* Washington, D.C.: Am. For. Assoc., pp. 21–26.

CALLAHAN, J. C., and T. P. BUNGER. 1976. "Economic Costs of Managing Street Trees on a Crisis Basis." *Trees and Forests for Human Settlements.* Toronto: Univ. Toronto, pp. 245–262.

CHURACK, P. L., R. W. MILLER, K. A. OTTMAN, and C. KOVAL. 1994. "Relationship Between Street Tree Diameter Growth and Projected Pruning and Waste Wood Management Costs." *J. Arbor* 20(4):231–326.

DUNTEMANN, M. T., T. GARGRAVE, and J. W. ANDRESEN. 1988. "Community Forestry Initiatives." *J. Arbor.* 14(4):90–93.

FOSTER, R. S., and J. BLAINE. 1978. "Urban Tree Survival: Trees in the Sidewalk." *J. Arbor.* 4(1):14–17.

HAGER, B. C., W. N. CANNON, and D. P. WORLEY. 1980. "Street Tree Policies in Ohio Towns." *J. Arbor.* 6(7):185–191.

HERMANN, J. G. 1993. Personal communication. Minneapolis Park District.

HOEFER, P. 1982. "Negotiating Successful Contracts and Agreements." *Proc. Sec. Natl. Urban For. Conf.,* Washington, D.C.: Am. For. Assoc., pp. 233–237.

KALMBACH, K. L., and J. J. KIELBASO. 1979. "Resident Attitudes toward Selected Characteristics of Street Tree Planting." *J. Arbor.* 5(6):124–129.

KIELBASO, J. J., and V. COTRONE. 1989. "The State of the Urban Forest." *Proc. Fourth Urban For. Conf.* Washington, D.C.: Am. For. Assoc., pp. 11–18.

KING, G. S. 1979. "Proper Landscaping Can Minimize Tree Problems in the Urban Environment." *J. Arbor.* 5(3):62–64.

LOBEL, D. F. 1983. "Managing Urban Forests Using Forestry Concepts." *J. Arbor.* 9(3):75–78.

LOUGH, W. B. 1991. "Contracting for Urban Tree Maintenance." *J. Arbor.* 17(1):16–17.

MILLER, R. H., and R. W. MILLER. 1991. "Planting Survival of Selected Street Tree Taxa." *J. Arbor.* 17(7):185–191.

MUDRACK, L., ET AL. 1980. *Urban Vegetation: A Reference for New York Communities.* New York Dept. Environ. Conserv.

NIGHSWONGER, J. 1982. "Urban-Community Forestry, Any Which Way You Can." *Proc. Sec. Natl. Urban For. Conf.,* Washington, D.C.: Am. For. Assoc., pp. 317–321.

NOWAK, D. J., J. R. MCBRIDE, and B. A. BEATTY. 1990. "Newly Planted Street Tree Growth and Mortality." *J. Arbor.* 16(5):124–129.

OTTMAN, K. A. 1993. Personal communication. Bureau of Forestry, Milwaukee.

POLANIN, N. 1991. "Removal History and Longevity of Two Street Tree Species in Jersey City, New Jersey." *J. Arbor.* 17(11):303–305.

RIDEOUT, R. 1993. "Reaching out to Small Communities." *Proc. Sixth Urban For. Conf.* Washington, D.C.: Am. For. Assoc., pp. 100–102.

ROBSON, H. L. 1984. "Urban Forestry in the Chicago Suburbs." *J. Arbor.* 10(4):113–116.

SCHROEDER, H., and P. APPELT. 1985. "Public Attitudes toward a Municipal Forestry Program." *J. Arbor.* 11(1):18–21.

SCHROEDER, H. W., and W. N. CANNON, JR. 1982. "The Contribution of Trees to Residential Landscapes in Ohio." *Proc. 1982 Annu. Meet. Soc. Am. For.,* pp. 333–335.

SCHROEDER, H. W., and W. N. CANNON, JR. 1987. "Visual Quality of Residential Streets: Both Street and Yard Trees Make a Difference." *J. Arbor.* 13(10):236–239.

SKIERA, B., and G. MOLL. 1992. "Trees in the Red." *Urban Forests* 12(1):9–11.

SKIERA, R. W. 1984. *Bureau of Forestry, Annual Report.* Milwaukee.

TATE, R. L. 1984. "Municipal Tree Management in New Jersey." *J. Arbor.* 10(8):229–233.

TATE, R. L. 1986. "Contracting for City Tree Maintenance Needs." *J. Arbor.* 12(4):97–100.

TATE, R. L. 1993. "How to Compete for Budget Dollars by Privatizing the Tree Care Operation." *J. Arbor.* 19(1):44–47.

TSCHANTZ, B. A., and P. L. SACAMANO. 1994. *Municipal Tree Management in the United States.* Kent, Ohio: Davey Tree Expert Co.

11

Management of Street Trees: Planting

Street tree management is defined in Chapter 10 as the establishment and maintenance of a population of trees in the public right-of-way to provide community amenities. The overall goal of a street tree management program is to provide the highest benefits from the population for the costs incurred. Tree value is an important concept in street tree management. Tree value can be used to assist in making management decisions and to justify management expenses. The Council of Tree and Landscape Appraisers' (1992) method of determining the monetary value of a landscape tree (Chapter 5) considers tree location, species, and condition. These factors as they relate to street trees are under considerable management control. The city forester selects the location and species planted, and influences tree condition through management.

Good management will develop a street tree population of high monetary and amenity value; conversely, a tree population of lower value can result from poor management. A number of street tree inventory systems now being used compute tree value (Chapter 6), and these values can be used to evaluate different management strategies and techniques. However, it is important to remember that proper tree appraisal is time consuming, and under normal inventory circumstances condition class is quickly estimated. Tree value obtained during street tree inventories should only be used as an estimate of the value of the entire population, not as appraised values for individual trees.

Using monetary tree value in management decision making assumes that the community values the street tree population, and that is why an urban forestry program exists in the first place. However, this is not always the case. Communities that have recently incurred high costs due to public trees (i.e., Dutch elm disease, ice or wind storms, etc.) may have low or negative feelings toward street trees, and this is where an understanding of community needs and values as discussed in Chapter 10 is important. Discussion of tree value in this and Chapter 12 assumes that the community values street trees and that elected officials and policy boards have been made aware of the monetary, ecologic, and amenity value of the tree population through an effective information-education program.

Once goals have been established for the street tree program, the management plan should describe how these goals are to be reached. An overall street tree management program consists of the establishment of trees, including species selection, acquisition, and actual planting. This is followed by scheduled and unscheduled maintenance, such as pruning, fertilization, and protection. All street trees ultimately must be removed, disposed of, and replaced if warranted. These topics are the subject of this and Chapters 12 and 13.

STREET TREE ESTABLISHMENT

Street trees are planted in the tree lawn by municipal forestry departments, abutting property owners, service organizations, or through cooperative efforts of these groups and/or individuals. Tree planting activities vary from communities where all trees are planted by private efforts, to communities where all planting is done by the municipal forestry department. However, most communities rely on some combination of public and private efforts for street tree establishment. Whatever the program, the tree ordinance should establish municipal control of all planting on tree lawns, and a permit system established to enforce that control. From a legal standpoint, the tree lawn is in the public right-of-way and in most communities the local government is responsible for all trees present, regardless of who does the planting.

Species Selection

The first phase in a street tree planting program consists of selecting the appropriate species for the community and individual planting sites. This begins with a list of approved species that is open-ended should the city forester wish to add or delete species. The approved species list also includes species that are prohibited by the street tree ordinance. Working within the street tree species list, each planting site should be analyzed and the appropriate species selected for planting. The process for selecting species for the list and for individual sites can be facilitated through use of the Species Selection Model in Figure 11–1. Three main factors influence species selection: site, social, and economic.

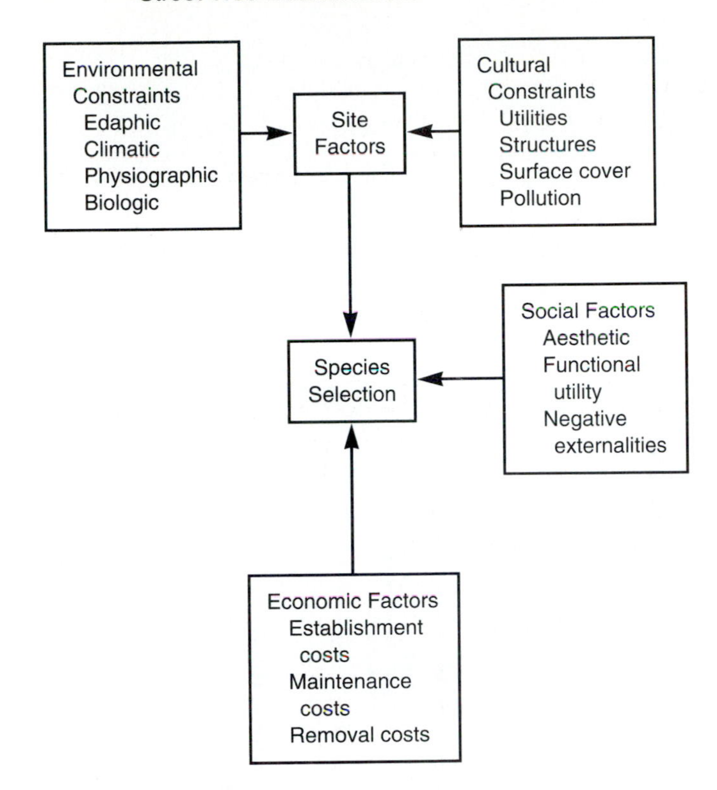

Figure 11–1 Species selection model.

Site factors. Site factors consist of cultural and environmental constraints. Cultural constraints refer to physical limitations of the site caused by human structures and activity. This includes above- and below-ground utilities; the size, style, and proximity of buildings; width of the tree lawn including the presence, type, and amount of hard surface; and pollutants. Above-ground utilities can limit crown development, while below-ground utilities limit placement of trees on planting sites. A survey of street trees in New Orleans found a strong relationship between the presence of overhead wires and poor tree condition due to repeated heavy pruning (Talarchek, 1987). Bloniarz and Ryan (1993) recommend designing street tree planting plans near above-ground utilities with tree planting setbacks, slow-growing small mature-size species, or columnar varieties to avoid tree/wire conflicts and provide a higher-quality tree population (Figure 11–2).

Neighborhood architecture and building setback influences selection relative to crown development. Tall species with spreading crowns will provide the greatest aesthetic and environmental effects in neighborhoods of two- and three-story homes, while shorter species may be better adapted to streets with one-story homes. Narrow streets with small or no building setbacks provide limited space where columnar varieties might be best adapted.

Clark and Kjelgren (1989) recommend rigorous site analysis when making species selections, especially relative to radiation load, temperature, water, and soils.

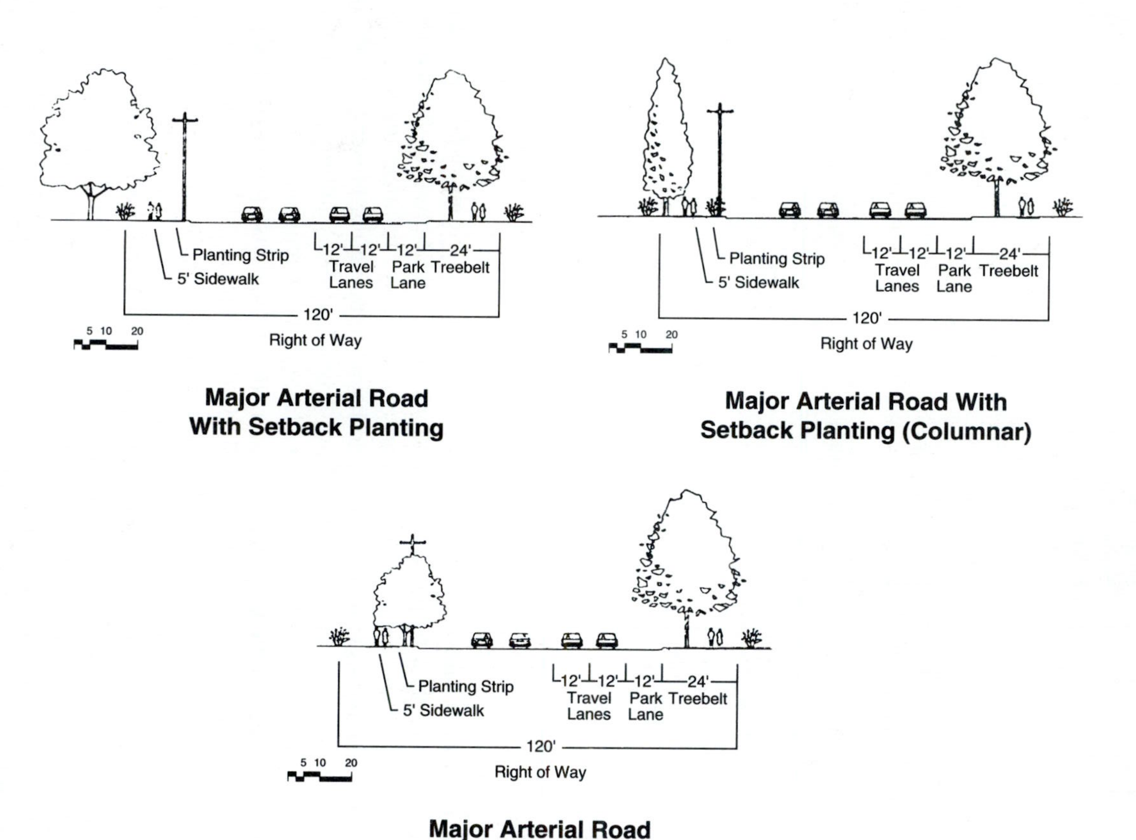

Figure 11–2 Design alternatives to accommodate trees and utility lines. (From Bloniarz and Ryan, 1993.)

A species that does well on a suburban street may be ill adapted to a sidewalk cutout in the central business district where microclimate-induced stresses are extreme. Figure 11–3 illustrates the general crown shapes common in shade trees.

Tree lawn width provides mature size constraints on the species planted. Considerable damage to sidewalks and curbs results from fast-growing, shallow-rooted

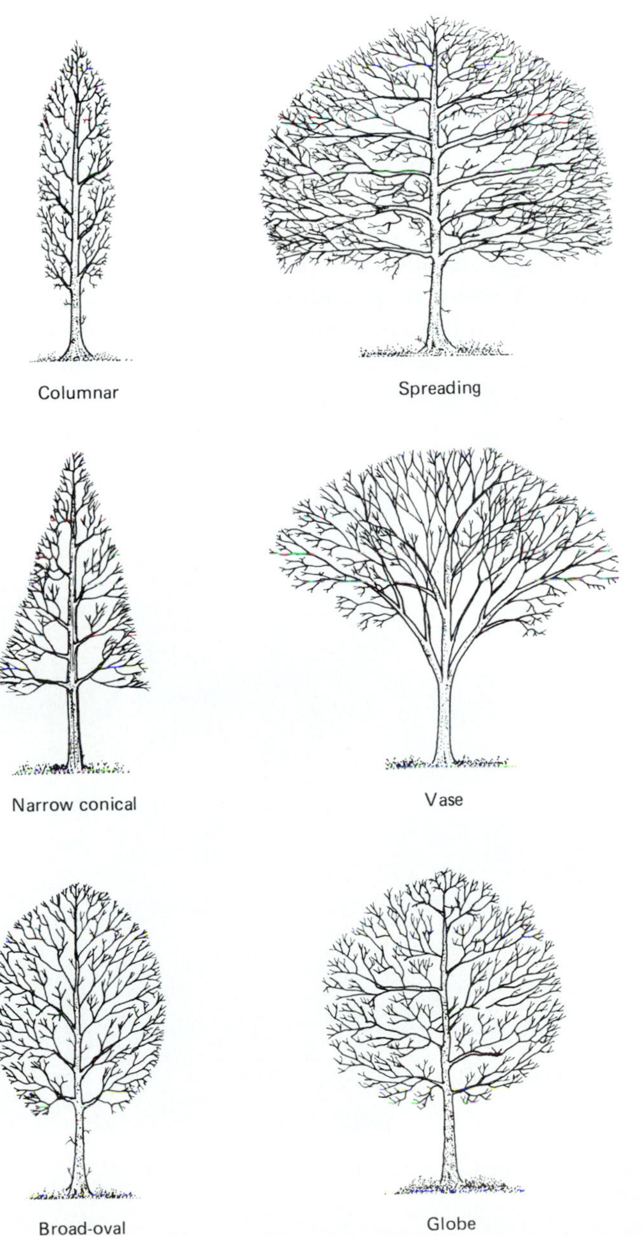

Figure 11–3 Typical shade tree shapes. (New Jersey Federation of Shade Tree Commissions.)

species planted in narrow tree lawns (Figure 11–4). Wagar and Barker (1983) suggest planting slow-growing species with a smaller mature size in narrow tree lawns. They further report variation in damage within species and recommend propagation of cultivars with this characteristic.

Air pollution is chronic in most larger cities. Types and levels of pollutants should be identified and measured throughout the community and resistant species selected for planting. Air pollutants are mapped by area and intensity in many larger cities, and this information is available from local pollution-monitoring agencies. Deicing salts cause soil salinity problems, and salt-sensitive species should be avoided on streets where heavy salting is standard maintenance. Above- and below-ground structures interfere with normal tree development. If chosen properly, species planted near structures should be able to develop a normal crown without requiring frequent and costly pruning.

Environmental constraints as site factors consist of insects, diseases, climate and microclimate, and soils. Urban trees are subjected to stress far in excess of rural trees. Insect and disease problems of minor importance in the forest can become major problems for trees stressed by the urban environment. Before extensive planting of a newly marketed species or cultivar, field tests should be conducted to determine if insect and disease problems are going to develop.

Figure 11–4 Sidewalk lifting caused by tree roots. (Richard Rideout, Milwaukee Bureau of Forestry.)

The species selected should be hardy for the climatic zone in which it is planted. There is a tendency to plant trees in hardiness zones colder than recommended in cities because of the urban heat island. However, it is not the normal winter that kills a specimen in a colder hardiness zone, but the periodic severe winter with below-normal temperatures that results in widespread losses. Microclimate is also important on a site-by-site basis, as urban areas contain frost pockets and have areas of high evapotranspiration due to light reflection from buildings and other reflective surfaces and high temperatures.

Urban soils are unlike rural soils. They are often compacted, lack topsoil, contain building debris, and are covered over with concrete or asphalt. Although complete soil analysis may be too expensive when planting street trees, an understanding of overall soil conditions within the community will assist in making species selection decisions. Soil pH is simple and inexpensive to determine, and this is far cheaper than attempting to correct pH-related nutrient deficiencies in established trees. Poor drainage is frequently a problem in compacted urban soils. A simple percolation test prior to planting may avoid excessive loss of newly planted trees.

Social factors. Social factors include neighborhood and community values, functional utility, species aesthetics, public safety, and negative social externalities. Attitudes toward public trees can vary by neighborhood, community, region, and ethnic background. Being aware of neighborhood, local, and regional values is extremely important in selecting species for city streets. Different ethnic groups will sometimes have preferences for different species. In one Milwaukee neighborhood only lindens (*Tilia* spp.) are planted because of an ethnic custom of making wine from linden flowers.

Functional utility includes ecologic and engineering functions such as reduced heating and cooling costs, stormwater interception, and interception of pollutants. It also includes modification of microclimate relative to outdoor comfort, increased property values, and improved wildlife habitat and diversity.

Tree aesthetics is a subjective evaluation and will vary from person to person. As discussed in previous chapters, urban residents have a preference for streets lined with large trees. Mature size is constrained by growing space, but species that will fully utilize that space without demanding excessive maintenance should receive high priority. Pruning standards for pedestrian and vehicular traffic are included in tree ordinances and influence both the appearance of street trees and community values as to what constitutes an attractive tree. Other aesthetic considerations in species selection include seasonal foliage color, leaf size, branching habit, flowers, bark, twig color, crown shape, and mature size.

Public safety is extremely important in street tree selection. In addition to proper and scheduled pruning, angle of branch attachment, wood strength, and resistance to decay determines resistance to breakage in high winds and ice storms (Figure 11–5). Species with large or toxic fruits, thorns, a low branching habit, and shallow roots that lift sidewalks all present public hazards easily avoided through species selection.

Figure 11–5 Storm-damaged street tree. (Richard Rideout, Milwaukee Bureau of Forestry.)

Negative social externalities are those factors that do not threaten public safety, but do result in public dissatisfaction with the tree. Excessive fruiting, prolonged heavy leaf fall, shedding bark, root suckering, and bad odors all make a species undesirable as a street tree. Species that attract nuisance wildlife and other pests—such as aphids, which drip honeydew on automobiles—should be avoided, as should species with dense, leafy crowns that shade out and kill lawns.

Economic factors. Economic factors include establishment, maintenance, and removal costs. Establishment of a street tree includes not only the cost of purchasing and planting, but must also include survival for a given establishment period. It may be more economical in the long run to purchase an expensive species with a higher survival rate when planting costs are evaluated based on surviving trees. Similarly, inexpensive species that demand frequent maintenance during the establishment period may cost more in the long run (Table 11–1). An appropriate establishment period should be used to determine the success of planting. The establishment period should be the amount of time needed for mortality associated with transplanting to drop to normal tree attrition levels. As discussed in Chapter 10, research has shown that it takes four years for street trees to become fully established in the northern United States.

Tree condition and growth rate as they influence tree value should also be consid-

TABLE 11–1 COMPARISON OF THE ULTIMATE ESTABLISHMENT COST PER TREE FOR TWO SPECIES[a]

	Species A	Species B
Purchase price/tree	$ 35.00	$ 50.00
Planting cost/tree	15.00	15.00
Other maintenance during 5-yr. establishment period	12.00	10.00
Total establishment cost per tree	$ 62.00	$ 75.00
Number of trees planted	100	100
Total planting cost	$6,200.00	$7,500.00
Percent survival after 5 years	60	80
Number of surviving trees	60	80
Cost per surviving tree	$ 103.33	$93.75

[a]Species A is less expensive to start with, but when evaluated on the basis of cost per surviving tree, species B is a better choice.

ered when evaluating a species. If a high-value tree population is a management objective, a species that has a high condition value and good growth rate at the end of the establishment period will have a higher net value than will a species with a lower condition value and size (Table 11–2). Areas where vandalism of newly planted trees is a problem should be planted with strong-wooded species. Much vandalism is by children, and a thornless honeylocust (*Gleditsia triacanthos inermis*) is much more difficult to break than a linden. Also, larger planting stock is more resistant to vandalism.

Maintenance costs can vary considerably for different species. Street trees need to be pruned on a regular cycle to maintain their highest value in relation to pruning costs (Miller and Marano, 1986). Fast-growing species that demand more frequent or heavier pruning are more expensive to maintain, resulting in significantly higher

TABLE 11–2 COMPARISON OF THE NET VALUE OF TWO SPECIES AFTER FOUR YEARS[a]

	Species A	Species B
Tree caliper	4 in. (10 cm)	3 in. (8 cm)
Base value determined by replacement (CTLA, 1983)	$430.00	$290.00
Species class (%)	90	90
Condition class (%)	80	60
Location class (%)	70	70
Tree value	$217.00	$109.00

[a]Both species cost the same to establish, but species A is better adapted to site conditions, is faster growing, and has a higher condition class. This, of course, assumes that both species were in the same initial condition at establishment.

pruning budgets in the future. Insect and disease problems can be expensive to control and may preclude some species for use on city streets. Species with high ascending branches are subject to splitting and may require cabling and bracing. Species that lift up sidewalks and curbs must be root pruned when the concrete is replaced or before replacement is necessary.

Removal costs depend on several factors: tree height, crown spread, and nearby structures and utilities. Species that attain a large mature size will be more expensive to remove, and this must be weighed against community preferences for big trees. Kalmbach and Kielbaso (1979) reported that residents in three Michigan cities preferred trees over 25 ft. tall, but a tree of this size is not that large when compared to the potential mature size for many species. Crown spread and nearby structures and utilities as they affect removal costs are related. Species with widespreading crowns are more likely to be difficult to remove due to their proximity to structures and overhead utility lines. Conversely, species with wide crowns provide more shade and subsequent modification of microclimate.

Species selection involves an understanding of the ecological characteristics of species considered for street planting. It also involves keeping accurate records of important parameters for species being planted in order to evaluate their adaptability to urban conditions. Unfortunately, many municipal foresters maintain only partial records. Gerhold and Steiner (1976), in a survey of municipal and highway arborists, found that 91 percent of the respondents kept planting location records, but only 21 percent recorded site characteristics (Table 11–3). Other questions dealing with record keeping showed similar responses.

Diversity and Stability

There was a historic tendency on the part of municipal foresters to select a few species that seemed most appropriate for planting on city streets. That is just what happened decades ago when American elm (*Ulmus americana*) was the tree of choice for many communities. The species was tolerant of urban site conditions, had few insect and disease problems, was highly regarded by urban residents, and was economical to plant and maintain. Overplanting of American elm set the stage for Dutch elm disease, and many communities are still trying to recover from its loss.

We try to learn by our mistakes, so diversity became the key to successful street tree management. Planting a variety of species helps to ensure against future catastrophic losses; however, attempting too much diversity can create problems. Communities in Wisconsin report difficulty in achieving diversity in street tree planting due to limited availability of species adapted to street conditions (Miller and Bate, 1978). From an architectural perspective, a single species of the same age provides aesthetic unity to a neighborhood. In addition, maintenance activities are easier to plan and more efficient when dealing with uniform population of trees. Richards (1982–83) cautions: "Undue emphasis on species diversity in replacement plantings may further threaten stability by causing inadequate replacement of the proven adapted species in the older population."

TABLE 11–3 TYPE OF RECORDS KEPT BY MUNICIPAL FORESTERS AND HIGHWAY TREE PLANTERS IN NORTHEASTERN UNITED STATES (From Gerhold and Steiner, 1976.)

Type of Record	Kept by Respondents (%)
Planting location	91
Planting-site characteristics	21
Species	83
Cultivar, clone	44
Number of trees	77
Size of trees	76
Survival after 1 year	47
Survival after 2 years	33
Survival beyond 2 years	29
Causes of mortality	32
Pest occurrence	30
Cultural treatments	20
Maintenance operations	63

Diversity is important in street tree management, from both a species and an age perspective. Architectural unity and maintenance efficiency can be achieved by planting a single species on a block or several blocks. Species diversity is introduced by changing species in a regular fashion by street and by block (Figure 11–6). From the standpoint of disease risk, Guries and Smalley (1985) state: "Planting each of ten boulevards with one of ten cultivars may be little different from planting ten boulevards with a mix of ten cultivars." However, all one species or cultivar on a block or series of blocks does carry the risk of total loss of trees on those blocks in the event of a new disease, insect, or other taxon-specific problem. Some communities group species by similar crown structure, growth rate, and appearance, and assign two or more species to be planted in specific blocks or segments in their Master Street Tree Plan.

Age diversity results from two sources: the planting plan and replacements (Figure 11–6). In many communities entire blocks are planted at once, either in new subdivisions or in replanting following Dutch elm disease. New plantings create age classes by block or neighborhood, while individual tree replacement introduces age diversity within neighborhoods. Based on street tree longevity as discussed in Chapter 10, age diversity should be less of a concern than species diversity in most communities.

Species diversity or richness is more than a long list of species and/or cultivars being planted on city streets. Thirty varieties may seem like good diversity, but if most of these are cultivars of three or four species, the interest of diversity is not

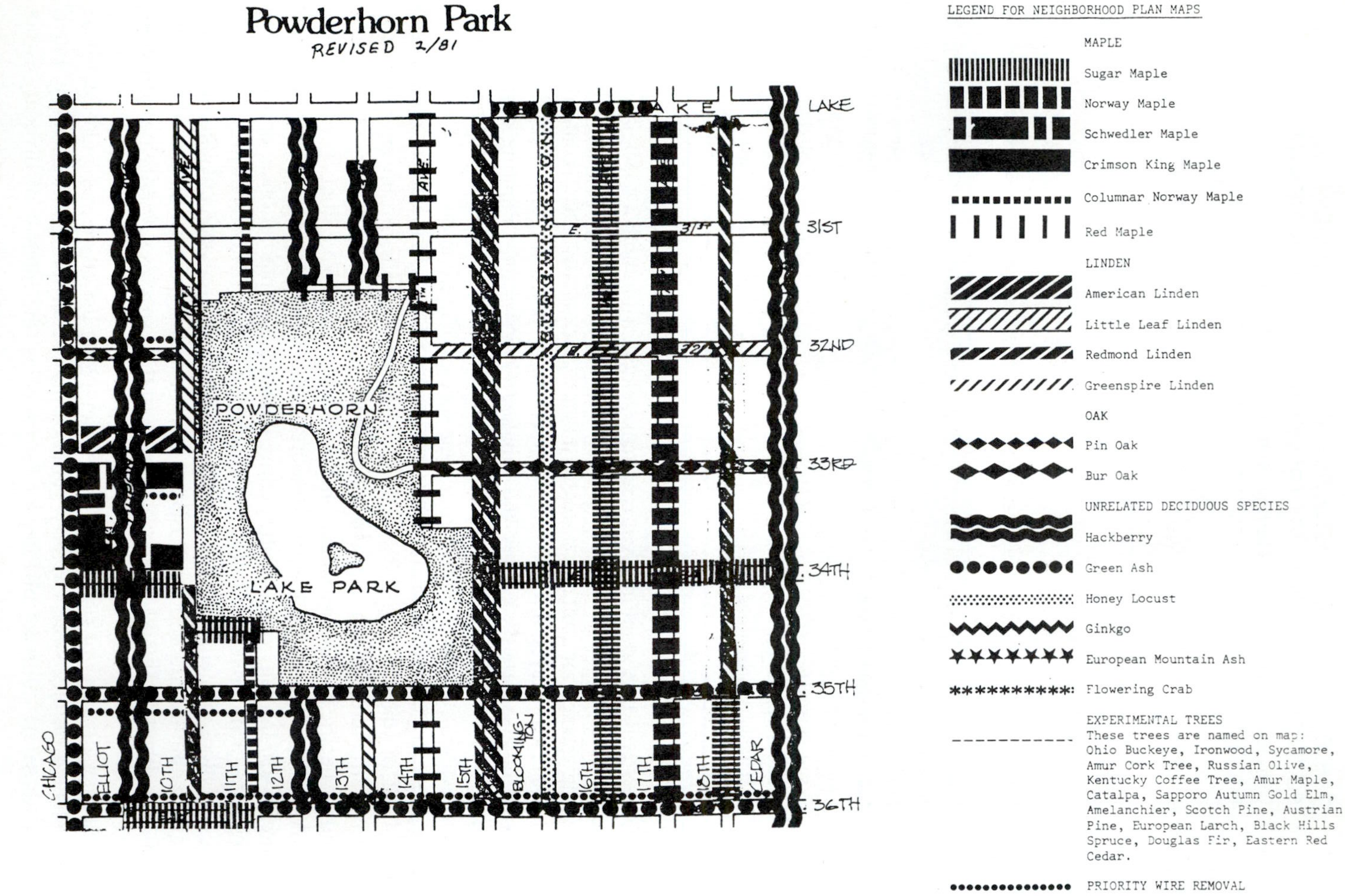

Figure 11–6 Planting plan for a neighborhood in Minneapolis, Minnesota, which specifies species diversity by block and street. (Courtesy of Minneapolis Park and Recreation Board.)

served. Likewise, if there are thirty separate species being planted, but a few species make up the bulk of the planting, diversity is also not served. Heavy dependence of just a few cultivars presents a greater diversity problem, since all individuals of a cultivar or clone are genetically identical and may all be susceptible to the same perturbation. Miller (1989) examined planting records in three Wisconsin cities and found that twenty-three cultivars were commonly planted, but these cultivars were drawn from only ten species. In two of the three cities just four species accounted for 84 and 88 percent of the species planted during a four-year period.

Barker (1975) recommends that communities establish maximum population densities for each species as a percent of the entire street tree population, and suggests that no more than 5 percent of any one species be used. Miller and Miller (1991) make more liberal recommendations, suggesting that proven species not exceed 10 percent, with more limited use of untested species. Richards (1993), on the other hand, feels it is illogical to set numerical limits on species use if it encourages planting unproven or ill-adapted species in place of proven species. He suggests that a species is overused if it is planted when other proven varieties are likely better suited. However, overuse of one or a few species can create conditions for high tree loss in the event of a taxon-specific event. Whether a city forester chooses to use numerical limits or just good judgment, diversity is important in selecting species for street planting. Some states provide lists of recommended species and cultivars through extension offices for use by homeowners and city foresters (Appendix F).

Concern is sometimes expressed over the use of introduced or exotic species in urban planting, and justifiably so. Some introduced species escape cultivation and invade native habitats, altering their structure and driving indigenous species out. This concern has sometimes led to blanket recommendations for use of native species only. Ware (1994) observes that locally indigenous species generally do well in open parks and other favorable sites, but may not survive long in more demanding urban sites along streets and in parking lots. He suggests selections for harsh sites be made from more difficult habitats for trees such as in river bottoms and savannas for both native and non-native species.

Exotic species that have proven adaptable to urban conditions and have been used long enough to establish that they are not invasive do not constitute a threat to the integrity of native ecosystems. In some circumstances they may be the only selection available for some planting sites. However, if the effect of an introduced species on local ecosystems is not known, it may be wise to avoid its use until it has been adequately tested.

Sources of Trees

Street trees are either grown in municipal nurseries, purchased from commercial growers, or transplanted from the wild. Historically, whips and seedlings were transplanted from the wild to small municipal nurseries for additional growth, and then planted on city streets. Seedlings were free, labor costs low, and commercial nurseries could not compete in price with municipal nurseries. Eastern and midwestern cities in the United States were initially planted in this fashion, thus the predominance of American elm and sugar maple (*Acer saccharum*) on city streets early in the century (Tate, 1977).

Disease problems and increasing environmental stress in cities brought the demise of this system of producing trees. As cultivars of native species and exotics were introduced to replace these species and to cope with urban stress, wildlings virtually ceased to be used, especially as a source of street trees. Today, most communities purchase either trees ready for street planting or whips for municipal nurseries from commercial nurseries (Tate, 1977).

Municipal nurseries. A study of municipal forestry programs in Wisconsin revealed that 11 (19.6 percent) of 56 communities surveyed have municipal nurseries. Three communities produced most of their planting stock from seed or seedlings, while the remainder used their nurseries to hold stock purchased from commercial nurseries (Miller and Bate, 1978). Tate (1984) reports a smaller percentage of communities with municipal nurseries in the northeastern United States. Of 233 communities sampled, 24 (10.3 percent) reported the presence of a municipal nursery. These communities buy all or most of their nursery stock from commercial nurseries. Fifty-four percent of the communities with nurseries used them to supply over one-half of their planting stock, with only three growing all their trees. Two communities report collecting a portion of their nursery stock from the wild, and one community used seed for a small percentage of its production. A survey of municipal forestry programs in the entire United States in 1994 revealed 23 percent of responding communities had a municipal tree nursery (Tschantz and Sacamano, 1994).

Tate (1977) reported that the cost of planting stock raised by a municipal tree nursery in Michigan was 84 percent of the cost of purchasing similar stock from commercial nurseries. The city of Milwaukee purchases whips or liners from commercial producers and grows them for four to six years before planting on city streets (Griffith and Associates, 1993). An analysis of their operation found the city produced trees at a lower cost than purchasing trees from commercial sources (Table 11–4).

City foresters generally attempt to plant a diversity of trees in their community, but often the desired species are not available from commercial nurseries, or are not available in sufficient quantity. Cities with their own nurseries can grow these species, or the city can contract with commercial nurseries to grow them. Contract growing allows the city access to species that are difficult to get and insures the nursery a buyer for species they might not otherwise grow in quantity. Tate (1977) suggests that contract growing will also result in savings for the city by not having to go to bid just prior to planting.

Commercial nurseries. Commercial nurseries provide most trees being directly planted on streets, and provide most of the lining-out stock planted in municipal nurseries. Advantages to purchasing trees ready for field planting from commercial nurseries include (Tate, 1977):

TABLE 11–4 MILWAUKEE CITY NURSERY VERSUS PRIVATE NURSERY COST COMPARISON. (From Griffith and Associates, 1993.)

Growth Term	Nursery's Direct Cost/Tree	Full Cost Multiple of Direct Cost[1]	City Nursery's Full Cost/Tree	Private Nursery's Cost/Tree	City Percentage of Private
4 year	$50.77	2.2	$111.69	$135.00	82.7%
5 year	$53.11	2.2	$116.84	$178.00	65.6%
6 year	$55.45	2.2	$121.99	$234.00	52.1%

[1]Includes overhead, annualized land value, and building use charge.

1. A commercial tree is ready for planting upon arrival with little root or branch pruning necessary.
2. If the tree doesn't meet contract specifications, it is rejected at no cost to the city.
3. Trees grown in commercial nurseries usually receive more intensive care due to the competitive nature of the industry and other demands placed on municipal personnel.

Purchasing good stock from a commercial nursery involves writing a carefully worded contract specifically describing what is being purchased, actual cost, guarantees, and the right to reject the stock upon delivery should contract specifications not be met. These specifications should include (Harris, 1983):

1. Species and/or cultivars by scientific name, with each plant labeled
2. Transplanting method (i.e., balled and burlapped, bare root, or container-grown)
3. Root condition, characteristics, and unacceptable defects (especially in container-grown)
4. Root spread in relation to tree height by transplanting method
5. Height-to-caliper ratio and taper
6. Crown configuration and branching pattern
7. Vigor
8. Unacceptable damage and pests
9. Delivery date

The American National Standards Institute (ANSI Z60.1) (1986) has established standards for nursery stock. These standards should be used in requesting bids and writing contracts. Many cities contract with nurseries well in advance to produce street-ready trees and liners. This results in savings to the city by not requesting bids when competition for nursery stock is highest during the planting season.

Planting

Under normal circumstances a street tree planting program will have a goal of full stocking of all available sites at some point in the future. The inventory should have identified existing vacancies, and these, plus normal attrition, should be filled in an orderly fashion. Additional sites may become available through development or redevelopment projects, removal of competing private vegetation, or annexation by the city.

Communities utilize four strategies in dealing with their street tree population:

1. *No planting or planting below replacement levels.* Trees are removed as they die or become a hazard, and the population dwindles over time due to no or minimal planting.
2. *Replacement.* Trees are replaced as they are removed, resulting in no net increase in the population. Communities at or near full stocking maintain the status quo in the tree population; communities experiencing population growth move toward a lower per capita public tree population; and communities with a low population of street trees initially do not gain anything.
3. *Full stocking goals.* Communities desiring to develop and maintain a maximum stocking of street trees allow for planting mortality and plant in excess of replacement and community expansion levels.
4. *Optimum stocking.* Optimum stocking can be defined by maintenance or by percent stocking levels. Optimum maintenance stocking considers present and projected budgets for tree care. Street trees reach their highest value and make their greatest contribution to the community when they receive regularly scheduled pruning and other maintenance (see Chapter 12). Increasing the number of street trees without increasing maintenance budgets will lead to a tree population at less than optimum value, and may create negative community attitudes toward trees due to hazards and nuisance problems. Richards (1993) suggests that optimum stocking consider tree spacing and size because standard spacing guidelines can result in some streets that are overstocked. When large trees mature, up to one-third of the trees can be missing without compromising most tree values. Replacing individual trees under these circumstances would be impractical, as shade from existing trees would make establishment difficult. He recommends that a street not be considered understocked until losses approach two-thirds because tree values would then be declining, and there would be room to plant in the gaps. A comparison of two inventories taken fifty years apart in Urbana, Illinois, supports this contention, as the number of trees in two neighborhoods declined by 41 percent but the total basal area of the trees declined by only 12 percent (Dawson and Khawaja, 1985).

If tree value to the community is used as a management goal, the street tree population will increase in value over time as the trees increase in size. On the other hand, tree planting is expensive, and reducing planting will save money for the community. If planting costs are subtracted from tree value over time, the net value of the street tree population as it relates to planting can be determined.

Using a street tree management simulation (Appendix B) that computes net tree values (tree value – management costs), three tree planting scenarios were tested over a forty-year period to determine their effect on net tree value: no planting, replace removals, and full stocking in ten years (Miller and Marano, 1986). Net tree values for all three scenarios increased over time, primarily due to tree growth. Results of the simulation indicate that the net value of the tree population increased most rapidly early in the no-planting simulation, due to savings on planting costs (Figure 11–7). The full-stocking-in-ten-years scenario yielded a net decrease in the value of the population initially, but in the long run produced the highest net value street tree population, even though other management costs of pruning and removal increased due to a larger tree population.

Overall tree survival on city streets varies greatly from community to community and within a community (Chapter 10). Most trees are lost the first few years after planting, particularly during the first growing season. Growing or purchasing a tree represents a sizable investment for a community, and it is an extreme waste of money not to ensure the highest possible chance for its survival. Much of this loss can be avoided through proper handling of planting stock followed by additional care during

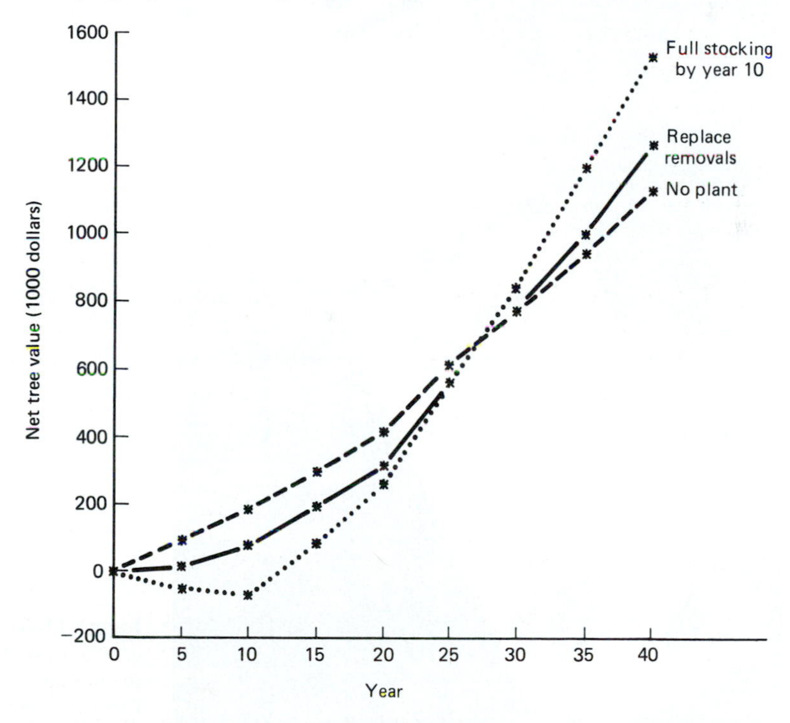

Figure 11–7 Net tree values (annual tree value – annual management costs) over 40 years for tree planting scenarios using a computer simulation. (From Miller and Marano, 1986.)

the establishment period. Trees should be planted at the proper depth, correctively pruned, watered, mulched, and fertilized following one growing season. Staking should be done only if absolutely necessary to protect the tree or to anchor the roots until the tree is securely established. Stakes should be removed after one growing season if possible, with one year being the maximum time allowed. Foster and Blaine (1978) report that stakes caused more tree damage in Boston than autos or vandalism, 84 percent in one section of the city.

Newly planted street trees need follow-up care during the first growing season. This includes watering during dry weather, checking for stake damage, and removal of stakes at the end of the season. Budgets may put constraints on intensive follow-up maintenance, but an effective public information effort can get adjacent property owners to water newly planted trees (Figure 11–8). However, the city should take responsibility for stake removal. During the establishment period the tree should

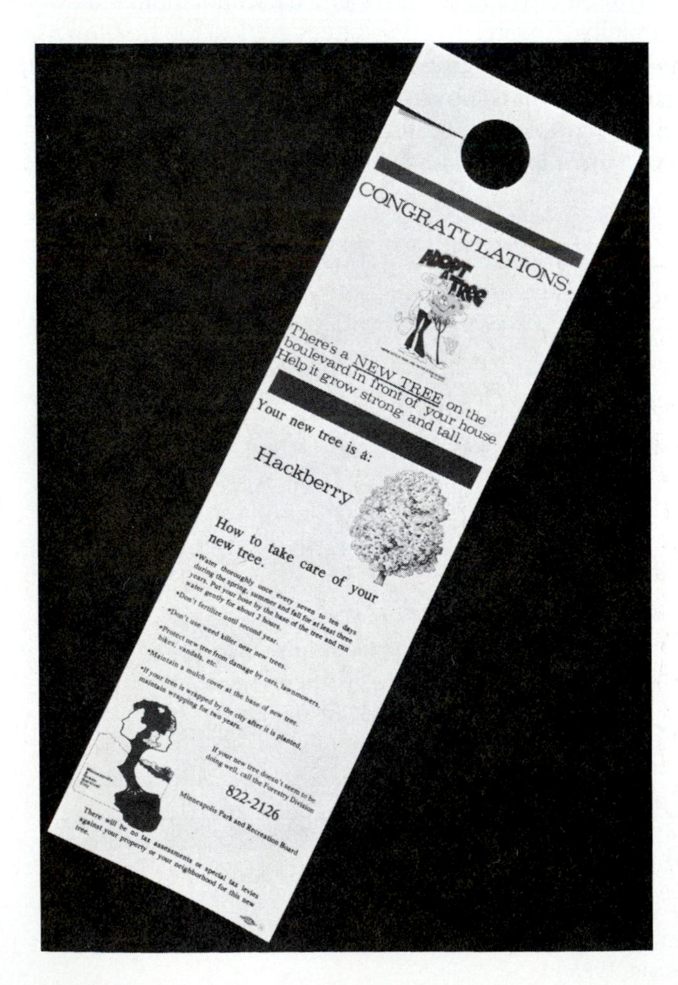

Figure 11–8 Doorknob hanger used to inform homeowners of the tree species planted and maintenance needs. (Courtesy of Minneapolis Park and Recreation Board.)

receive a minimum of one additional corrective pruning, and supplemental fertilization if budgets permit. In the long run it is better to plant fewer trees and give them more intense care if the goal is high survival. In many instances the cost per surviving tree will be lower. Proper planting and post-planting practices are described in detail in the texts *Arboriculture* (Harris, 1992) or *Modern Arboriculture* (Shigo, 1991).

Vandalism is often cited as a major cause of tree mortality for newly planted trees. Black (1978) describes three steps taken in Seattle, Washington, that have cut vandalism losses in that city by 50 percent. These steps are physical changes, managerial changes, and public involvement. Physical changes include planting larger stock or stronger-wooded species in areas noted for high vandalism. In addition, the tree staking method was changed from two wooden stakes that provided a fulcrum for breaking at the top tie, to a single steel reinforcing bar at least 5 ft. tall when installed. The tree is attached to the single stake for one year with wire and hose in a figure-eight configuration. Managerial changes involve identifying vandal-prone locations and either not planting or changing physical planting methods. Other managerial changes include being sensitive to neighborhood wishes and providing high-quality planting and maintenance of street trees. Finally, Black recommends a strong public involvement program that stresses pride of ownership so that the community senses that the planting is their own to take pride in.

Spacing. There is a tendency on the part of many individuals to plant young trees close together for an immediate visual impact. Many early street tree planting programs emphasized close spacing, with as little as twenty to thirty feet between species such as American elm, sugar maple, and silver maple (*Acer saccharinum*). The obvious problem is disease transmission through root grafts, such as in the case of Dutch elm disease. Not so obvious is the impact on future management costs. Trees planted close together will become more expensive to prune as they mature. Mutual shading will create more deadwood, and live branches from adjacent trees interfere with each other, necessitating additional pruning. Overcrowding causes stress, making trees more susceptible to insect and disease problems, thus demanding control efforts. There will also be more trees on city streets in need of maintenance, at little additional aesthetic premium.

Tree ordinances described community standards for corner, driveway, and alley setbacks; hydrant and utility pole setbacks; and spacing. Corner, driveway, and alley setbacks are based on vehicular traffic safety standards and may vary based on traffic loads for specific streets. Ordinances typically establish minimum setbacks of 30 ft. (9.1 m) from intersections, and 15 ft. (4.5 m) from driveways and alleys. Setbacks for hydrants and utility poles are usually 10 ft. (3 m).

Spacing standards are either community-wide, such as 50 ft. (15 m) for all streets, or relate to the mature size of the species planted. For example, trees attaining a small mature size are planted a minimum of 25 ft. (7.6 m) apart; medium, 35 ft. (10.6 m) apart; and large, 55 ft. (16.5 m) apart. Spacing also relates to the width of the tree lawn, as the species selected will be influenced by available growing space for its mature size. Additionally, some communities, by policy or by ordinance, prohibit planting in tree lawns less than 3 or 4 ft. (0.9 or 1.2 m) in width.

Spacing standards should be based on the characteristics and needs of each community. Width of the tree lawn and other constraints on available growing space influences the size of species chosen for planting. Lot width is also important in setting spacing standards, as each property owner may feel entitled to a tree in front of his or her property. Kalmbach and Kielbaso (1979), in a survey of resident attitudes toward street tree planting in five midwestern cities, report that a planting density of one per home satisfied the majority of respondents. Once a community spacing standard is described in the ordinance, it should be enforced throughout the community. If multiple spacing categories are permitted in the community, the Master Street Tree plan should describe the spacing for each street based on the foregoing considerations.

Priorities. If unlimited funds were available for municipal forestry, there would be no need to prioritize planting. All vacant sites identified could be immediately filed and full stocking achieved. However, funds for tree planting are limited, so priorities are established to provide the highest benefit to the community for funds expended. A thorough understanding of community values is necessary to set planting priorities. Two levels of planting priorities should be established: area specific and site specific. Areas may be described by individual neighborhoods and/or land-use districts. For example, a neighborhood could receive a low planting priority if the residents are elderly and not interested in raking leaves, or if it is scheduled for new water or sewer lines to be placed in or close to the tree lawn in the near future. Streets with abundant private trees in front yards can receive lower planting priority than streets with few private trees. The visual impact of street trees for urban residents is much greater where there are few private trees (Schroeder and Cannon, 1987). Shade from private trees may also hinder the growth and development of street trees, warranting planting where they will do the most public good.

Land-use priorities as they relate to street trees can be described based on three general land uses in communities: residential, commercial, and industrial. Residential districts are frequently assigned the highest priority, followed by commercial and industrial districts. Getz et al. (1982) surveyed inner-city residents in Detroit, Michigan, and found that their highest priority for public tree programs was for street trees in residential neighborhoods, which limited support for trees in industrial areas and in parking lots. Planting priorities by land use will be different for each community and will change over time within communities. A large redevelopment project in a commercial district, for example, may be a short-term high-priority project that consumes a large share of the planting budget for one or several years.

Site priorities relate to specific characteristics of each site available for planting. Planting priority by site is a local decision made for specific community characteristics. A number of site-priority descriptions are listed below, but not necessarily in order of priority:

1. New residential street
2. Replacement residential
3. Business
4. School/church
5. Existing overstory created by adjacent trees
6. Tavern
7. Homeowner refusal
8. Known high mortality rate for site
9. Not plantable at present (low overhead wires or other obstructions)

In many cases site priorities can be assigned for entire blocks or longer street segments. Priority rankings for sites, and areas, should be used only as general guides, with flexibility for exceptions.

Full stocking. A street tree management plan should have as a long-term objective the full stocking of all available tree sites. This goal, however, may prove to be elusive, due to city expansion, budgetary constraints, and unexpected tree mortality (i.e., storms, insects, or diseases). Even if funds are available, there will always be unplantable sites for a variety of reasons, and annual mortality to contend with. In spite of potential problems in reaching a goal of full stocking, it should remain an important management goal.

A goal of full stocking is used in management planning by selecting a realistic time frame in which to achieve full stocking, knowing how many vacant spaces are available for planting, and anticipated planting survival. The time frame is based on estimating how long it will take to achieve full stocking at current funding levels, and in consultation with the tree or park board. This frame is then used to determine how many trees will need to be planted each year to reach full stocking by the target date using the formula

$$N = \frac{R + V/G}{S}$$

where N = annual planting
R = annual removals
V = existing vacant sites
G = years left to achieve full-stocking goal
S = anticipated planting survival (used only if annual removals [R] do not include planting mortality)

Annual removals are based on past average annual removals, vacant sites taken from the inventory, and anticipated survival based on historic records of planting success. Values used in the formula are subject to change due to unanticipated mortality or city expansion. The overall full-stocking goal in the specified time frame should be kept constant if at all possible, with changes in planting frequency being used to adjust to external influences. Table 11–5 provides an example of an annual planting plan to achieve full stocking in eight years.

The goal of full stocking is related to spacing standards as described by city ordinance or policy. Planting sites should be identified in accordance with these standards, and trees scheduled for removal should have a yes-or-no replacement recommendation made based on the same. It is possible to have a stocking goal lower than the existing tree population once the Master Street Tree Plan is developed. This situation will arise if the community is overstocked based on current spacing standards.

Planting analysis. According to Kielbaso (1988), tree planting consumes 14 percent of municipal forestry budgets in the United States. Yet less than one-half of municipal foresters and highway tree planters surveyed by Gerhold and Steiner (1976) kept survival records beyond one year. A variety of cultivars and grades are available from nurseries, and the city forester should make selections based on their long-term benefits to the city. For example, it may be desirable to plant large trees for an instant visual impact, but Litzow and Pellett (1982) report that smaller grades grow faster and may catch up to larger grades after a few years. Good record keeping over time is essential to a good tree planting program.

Record keeping begins with selecting a tree establishment period of three or four years, depending on local conditions. The location of each newly planted tree is recorded, and the source of that tree identified (municipal nursery, commercial nursery A, commercial nursery B, etc.). Nursery bids vary and survival by nursery may also vary. Once the location and source of each tree has been recorded the site should be visited on an annual basis throughout the establishment period, and survival or cause of mortality noted. At the end of the establishment period survival of each species by transplant method and nursery can be established and causes of mortality analyzed (Table 11–6). Results of the survival analysis are then used to determine the cost per surviving tree and the net value added to the city tree population.

TABLE 11–5 ANNUAL PLANTING CALCULATION FOR A FULL STOCKING GOAL OF EIGHT YEARS

Vacant sites (V)	1624
Last year's removals (R)	120
Anticipated planting survival (S)	80%
Years to full-stocking goal (G)	8

$$N = \frac{R + V/G}{S}$$

$$= \frac{120 + 1624/8}{0.80}$$

$$= 404 \text{ trees}$$

TABLE 11–6 PLANTING MORTALITY ANALYSIS AT END OF ESTABLISHMENT PERIOD[a]

Cause of Mortality	Species 1						Species 2					
	Nursery A			Nursery B			Nursery A			Nursery B		
	1-in. BR[b]	1½-in. BR	2-in. BB[c]	1-in. BR[b]	1½-in. BR	2-in. BB[c]	1-in. BR[b]	1½-in. BR	2-in. BB[c]	1-in. BR[b]	1½-in. BR	2-in. BB[c]
Drought	1	2		1	2		3			1	3	
Insects								1				
Disease					1	2				1	2	
Auto	1		1		2	1		1	1	3	3	
Vandalism	11	2	1	14		1	1		1	5	2	
Basal injury		8	3		2		2		5		1	
Construction			2		1			4	1		2	
Unknown	2		1	1		2		20	10		10	3
Number planted	30	50	50	40	60	60	25	100	100	60	75	40
Number lost	15	12	8	16	9	6	5	26	18	10	23	3
% Survival	50	76	84	60	85	90	80	74	82	83	69	92

[a]Species 1 appears to be better adapted to street conditions, while Nursery B seems to have better survival, especially BB stock.
[b]BR, bare-root stock.
[c]BB, balled-and-burlapped stock.

The initial cost of tree procurement by nursery and transplant type is recorded along with the initial costs of planting, fertilization, corrective pruning, and staking. Additional costs during the establishment period are recorded for activities such as pruning, spraying, fertilizing, and stake removal. The cost per surviving tree may then be determined by the number of surviving trees at the end of the establishment period (Table 11–7). Surviving the tranplant period is not enough to evaluate the success of planting, as tree condition and value determines the ultimate value of the tree to the community. If a management goal is to develop the highest-value population of street trees for costs incurred, additional analyses can be made to determine the net value per tree by subtracting establishment costs from the appraised value of surviving trees (Table 11–7).

It is obvious from data presented in Tables 11–6 and 11–7 that 1-in.-caliper bare-root stock of species 1 is being vandalized to the point that its lower cost is offset by its low survival rate. However, 1-in. bare-root stock for species 2 is not being vandalized as heavily, and based on tree value has surpassed the $1\frac{1}{2}$-in. bare-root stock for the same species. Nursery B, although cheaper initially, is not a good source of planting stock for species 1. Overall, species 2 is a better investment for the community, both in terms of establishment cost and net value added to the tree population. Balled-and-burlapped stock has an overall higher survival rate, but that is offset by the lower initial cost of bare-root stock.

Additional analyses that may be made on the planting program are to vary management activities to determine the impact on the cost per surviving tree and net value added to the tree population. These analyses could include initial and supplemental fertilization, staking, corrective pruning, pest control programs, or other planting and maintenance activities. Site analysis may also be included as a factor influencing tree survival and establishment costs. Species analysis can lead to the elimination of too many species as acceptable for planting, and this must be weighed against the need for diversity in the overall population.

Contract planting. Many communities elect to contract tree planting. Tate (1984) reports 63 percent of cities surveyed in New Jersey contract tree planting, with Robson (1984) reporting similar results in a survey of eight urban forestry programs in the Chicago suburbs. Although it appears that contract planting constitutes the preferred method of establishing street trees, survival of these trees will vary with the contractor. Foster and Blaine (1978) report that four contract tree planters in Boston had survival rates varying from 97 percent to 62 percent after one year. They attribute the high loss rate of the worst contractor to careless handling of planting stock.

Contract planting should be scrutinized as intensely as city planting, using cost per surviving tree after the establishment period as the measure of success. The city forester must have the right to reject any bid from a contractor with a history of poor performance. Contractors should guarantee survival for one year by replacing all trees that die due to causes other than external problems such as vandalism, auto impact, and so on. Contract bids with survival guarantees may include watering by the contractor for one growing season to ensure survival.

TABLE 11–7 COST ANALYSIS OF TREE PLANTING AT END OF ESTABLISHMENT PERIOD

	Species 1						Species 2					
	Nursery A			Nursery B			Nursery A			Nursery B		
Costs	1-in. BR[a]	1½-in. BR	2-in. BB[b]	1-in. BR	1½-in. BR	2-in. BB	1-in. BR	1½-in. BR	2-in. BB	1-in. BR	1½-in. BR	2-in. BB
Procurement	$ 19.00	$ 39.00	$ 76.00	$ 16.00	$ 34.00	$ 70.00	$ 18.00	$ 27.00	$ 78.00	$ 15.00	$ 21.00	$ 74.00
Planting	22.00	26.00	35.00	22.00	26.00	35.00	22.00	26.00	35.00	22.00	26.00	35.00
Fertilizing	1.00	1.00	1.50	1.00	1.00	1.50	1.00	1.00	1.50	1.00	1.00	1.50
Staking	6.00	6.00	8.00	6.00	6.00	8.00	6.00	6.00	8.00	6.00	6.00	8.00
Mulching	3.00	3.00	3.00	3.00	3.00	3.00	3.00	3.00	3.00	3.00	3.00	3.00
Pruning	4.00	4.00	5.00	4.00	4.00	5.00	4.00	4.00	5.00	4.00	4.00	5.00
Protection	—	—	—	—	—	—	3.00	3.00	3.50	3.00	3. 00	3.50
Added fertilizer	1.00	1.00	1.00	1.00	1.00	1.00	1.00	1.00	1.00	1.00	1.00	1.00
Stake removal	4.00	4.00	6.00	4.00	4.00	6.00	4.00	4.00	6.00	4.00	4.00	6.00
Total cost	$ 60.00	$ 84.00	$135.50	$ 57.00	$ 79.00	$129.50	$ 62.00	$ 75.00	$141.00	$ 59.00	$ 69.00	$137.00
Survival (%)	50	76	84	60	85	90	80	74	82	83	69	92
Cost/surviving tree	$120.00	$110.52	$161.31	$ 95.00	$ 92.94	$143.89	$ 77.50	$101.35	$171.95	$ 71.08	$100.00	$148.91
Average appraised value	$200.00	$248.00	$272.00	$165.00	$221.00	$228.00	$195.00	$262.00	$169.00	$215.00	$275.00	$256.00
Net value/tree	$ 80.00	$137.48	$110.69	$ 70.00	$128.06	$ 84.11	$117.50	$160.65	$160.35	$143.92	$175.00	$107.09

[a]BR, bare-root stock.
[b]BB, balled-and-burlapped stock.

Two types of bids for tree planting contracts are generally used: lump sum or per unit. In a lump-sum bid a price is given to complete the entire contract, with no itemized individual components of the bid. If the contractor cannot fulfill a portion of the contract or is asked to add to it, determining the fair cost of these changes becomes difficult. A per unit bid lists costs of individual components of the contract, simplifying changes made once the contract is in effect (Foster, 1978).

Permits. Tree planting in the public right-of-way by adjacent property owners is generally permitted in communities, and in some cases represents the only source of street trees. In most communities the street tree population has its source in both the public and private sector. Public control of private tree planting in the public right-of-way is essential for an effective management plan and is best controlled by a permit system administered by the municipal forestry department.

The first consideration of a permit system is public safety. Planting location is controlled by spacing and setback requirements as specified in the municipal tree ordinance. Species that have been identified as creating public safety problems or nuisance conditions are prohibited from being planted on city streets. If the community does not have a scheduled public tree maintenance program, the adjacent property owner may be responsible for pruning in the interest of public safety. This includes street and sidewalk clearance, traffic signal and sign obstruction, and blocking of street lighting.

A permit system usually has a list of approved trees for the property owner to select from and a list of species prohibited by city ordinance. Communities with a Master Street Tree Plan can be even more restrictive in granting a permit, offering a very limited selection of species permitted in accordance with the plan. Future maintenance responsibilities for the tree should be on the permit, and either place all responsibility for maintenance on the property owner, or state that the city assumes ownership and all maintenance responsibilities for the tree (Figure 11–9).

Easements. Streets with narrow tree lawns, low overhead wires, or with no space for public trees can be planted behind the sidewalk if the tree ordinance permits an easement for planting on private property (Figure 11–10). Easements are typically voluntary, with the city planting and performing routine maintenance on the trees. The city retains ownership, and the property owner cannot remove, prune, or otherwise alter the tree without permission from the city forester.

Containers. The use of containers for street tree planting (Figure 11–11) is becoming more common, especially in central business districts of cities where limited suitable sites are available (Williams, 1978). However, the use of containers for trees and other landscape plants is expensive, and if improperly designed will detract from aesthetics and damage plants. Trees in containers are subject to limited soil, excessive drying and wetting, winter injury to roots, nutrient deficiencies, scorch, and stress-related problems. These problems, although severe, can be overcome to some degree through proper container design.

PARK RECREATION AND FORESTRY DEPARTMENT
PERMIT FOR PLANTING A STREET TREE

Permit No. ______________

Name and Address of Permittee

Approximate Planting Date

Location (Address)

Overhead Wires Present? ____________________

Species: __________________________________

Conditions: The tree planted will become the property of the city, and the city assumes full responsibility for management. This includes pruning, insect and disease control, other necessary management, and removal when it becomes necessary.

APPROVED: ______________________ City Forester ________________ Date

Figure 11–9 Tree planting permit for planting in the public right-of-way.

Cervelli (1984) recommends that containers be no less than 8 to 10 ft. (2.4 to 3 m) in diameter, with a minimum depth of 3 ft. (0.9 m) for small tree species. In addition, he recommends that containers be fixed in place, aesthetically arranged, and designed to blend with surrounding structures. Williams (1978) suggests amending container soils to improve drainage, below-grade drains, and regular fertilization and watering. In cold climates roots of normally cold hardy species are damaged by the extreme cold in above ground containers, and insulation is recommended. Standing water will freeze in cold climates and damage the containers if not properly drained. Larger specimens in containers may need permanent guying to prevent uprooting or leaning.

Containers are popular components of urban beautification efforts, but must be designed carefully and are expensive to maintain. They require scheduled fertilization

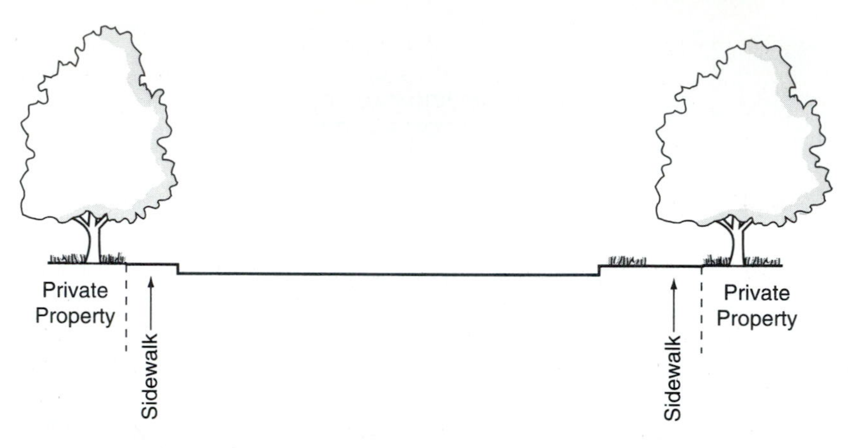

Figure 11–10 Trees planted in an easement on private property where the tree lawn is too narrow or is nonexistent.

Figure 11–11 Street trees planted in above ground containers. (Richard Rideout, Milwaukee Bureau of Forestry.)

and watering, winter draining in cold climates, and intensive maintenance of trees and other plants due to the stresses associated with a limited growing medium. Before using containers, adequate budgets must be available for maintenance, or what was intended as a community asset will quickly become a liability.

Planting Equipment

Equipment used in planting operations can vary from the simple to the highly complex, depending on the magnitude of the operation and local budgets. Hand planting uses the common tools of spades, picks, rakes, and a truck to haul the planting stock. To prevent drying, the haul truck should be covered if the trees have begun to leaf out. General supplies for planting in all cases include mulch, and stakes and bindings if staking is to be done. A source of water such as a watering truck or water tank on the haul truck is essential for transplant survival.

Larger-scale operations utilize augers to dig the planting hole, but caution should be exercised in heavy soils, as augers will glaze the sides of the hole, making root penetration difficult. Some communities use their stump grinders to dig the planting hole, thus avoiding glazing problems and getting extra use from the equipment. In all cases extreme caution should be exercised in digging the planting hole, as buried utility lines can represent a real hazard to crews. At the very least a severed line is expensive to repair and results in interrupted service. Most utilities now have a "diggers' hotline" service that will mark utility lines at no charge to municipalities or residents.

Hydraulic tree spades have become a popular method for transplanting in recent years (Figure 11–12). However, tree spade usefulness can be limited by tree lawn width, soil type, and wet weather in nurseries. A municipal forestry department should have its own nursery or other source of trees, and should anticipate a lot of use before investing in a tree spade. When used to move one tree at a time, a tree spade can be inefficient, as much equipment and crew time is spent transporting trees. This is especially true when the distance between the source and the planting

Figure 11–12 Hydraulic tree planting spades can be used to plant street trees where no underground utilities are present. (Courtesy of Burkeen Manufacturing Company.)

site is great. However, tree spades are very efficient when used to lift and put nursery stock in wire baskets for truck transport to the planting site.

A holding area for planting stock is necessary when the trees are not planted immediately, as in the case when a large shipment from a commercial nursery is delivered. Balled-and-burlapped stock must have the balls covered with mulch to prevent drying, and be watered on a scheduled basis. Bare-root stock should be placed under a shelter out of the wind and have their roots kept moist at all times. Container-grown stock needs to be watered and to have the outside row of containers screened from direct sunlight.

LITERATURE CITED

AMERICAN NATIONAL STANDARDS INSTITUTE (ANSI Z60.1). 1986. *American Standard for Nursery Stock.* Washington, D.C.: American National Standards Institute.

BARKER, P. A. 1975. "Ordinance Control of Street Trees." *J. Arbor.* 1(11):212–216.

BLACK, M. E. 1978. "Tree Vandalism: Some Solutions." *J. Arbor.* 4(5):114–116.

BLONIARZ, D. V., and H. D. P. RYAN III. 1993. "Designing Alternatives to Avoid Street Tree Conflicts." *J. Arbor.* 19(3):152–156.

CERVELLI, J. A. 1984. "Container Tree Plantings in the City." *J. Arbor.* 10(3):83–86.

CLARK, J. R., and R. K. KJELGREN. 1989. "Environmental Factors Affecting Urban Tree Growth." *Proc. Fourth Natl. Urban For. Conf.* Washington, D.C.: Am. For. Assoc., pp. 88–92.

COUNCIL OF TREE AND LANDSCAPE APPRAISERS. 1992. *Guide for Plant Appraisal.* Savoy, Ill.: International Society of Arboriculture.

DAWSON, J. O., and M. A. KHAWAJA. "Change in Street Tree Composition of Two Urbana, Illinois Neighborhoods After Fifty Years: 1932–1982." *J. Arbor.* 11(11):344–348.

FOSTER, R. 1978. "City Tree Planting." *Am. Nurseryman* 158(7):13, 117–120.

FOSTER, R. S., and J. BLAINE. 1978. "Urban Tree Survival: Trees in the Sidewalk." *J. Arbor.* 4(1):14–17.

GERHOLD, H. D., and K. C. STEINER. 1976. "Selection Practices of Municipal Arborists." *Better Trees for Metropolitan Landscapes,* USDA-For. Serv. Gen. Tech. Rep. NE-22, pp. 159–166.

GETZ, D. A., A. KAROW, and J. J. KIELBASO. 1982. "Inner City Preferences for Trees and Urban Forestry Programs." *J. Arbor.* 8(10):258–263.

GURIES, R. P., and E. B. SMALLEY. 1985. "Elms for Urban Forests." *Wis. Arbor.* 4(2):1–2.

GRIFFITH, D. M., and ASSOCIATES. 1993. *Comprehensive Review of the Services, Operations, Organization, Financing, Management, and Staffing of the Bureau of Forestry.* Milwaukee.

HARRIS, R. W. 1983. *Arboriculture: Care of Trees, Shrubs, and Vines in the Landscape.* Englewood Cliffs, N.J.: Prentice Hall.

HARRIS, R. W. 1992. *Arboriculture: Integrated Management of Landscape Trees, Shrubs, and Vines.* Englewood Cliffs, N.J.: Prentice Hall.

KALMBACH, K. L., and J. J. KIELBASO. 1979. "Resident Attitudes toward Selected Characteristics of Street Tree Plantings." *J. Arbor.* 5(6):124–129.

KIELBASO, J.J. 1988. *Trends in Urban Forestry Management.* Baseline Data Report, vol. 20, no. 1. Washington, D.C.: International City Management Association.

LITZOW, M., and H. PELLETT. 1982. "Establishment Rates for Different Bareroot Grades of Trees." *J. Arbor.* 8(10):264–266.

MILLER, R. H. 1989. "Frequency of Use and Survival of Selected Street Tree Taxa in Three Wisconsin Communities." M.S. thesis, Univ. Wisc., Stevens Point.

MILLER, R. H., and R. W. MILLER. 1991. "Planting Survival of Selected Street Tree Taxa." *J. Arbor.* 17(7):185–191.

MILLER, R. W., and T. R. BATE. 1978. "National Implication of an Urban Forestry Survey in Wisconsin." *J. Arbor.* 4(6):125–127.

MILLER, R. W., and M. S. MARANO. 1986. "Urban Forest: A Street Tree Management Simulation." *Proc. For. Microcomputer Software Symp.,* Morgantown, W. Va., June 30–July 2.

RICHARDS, N. A. (1982–83). "Diversity and Stability in a Street Tree Population." *Urban Ecol.* 7:159–171.

RICHARDS, N. A. 1993. "Optimum Stocking of Urban Trees." *J. Arbor.* 18(2):64–68.

ROBSON, H. L. 1984. "Urban Forestry in the Chicago Suburbs." *J. Arbor.* 10(4):113–116.

SCHROEDER, H. W., and W. N. CANNON. 1987. "Visual Quality of Residential Streets: Both Street and Yard Trees Make a Difference." *J. Arbor.* 13(10):236–239.

SHIGO, A. L. 1991. *Modern Arboriculture.* Durham, N.H.: Shigo and Trees Associates.

TALARCHEK, G. M. 1987. "Indicators of Urban Forest Condition in New Orleans." *J. Arbor.* 13(9):217–224.

TATE, R. L. 1977. "The Worth of Municipally-Owned Tree Nurseries." *J. Arbor.* 3(9):169–171.

TATE, R. L. 1984. "Status and Operating Costs of Selected, Municipally-Owned Tree Nurseries in the Northeast United States." *J. Arbor.* 10(10):286–288.

TSCHANTZ, B. A., and P. L. SACAMANO. 1994. *Municipal Tree Management in the United States.* Kent, Ohio: Davey Tree Expert Co.

WAGAR, J. A., and P. A. BARKER. 1983. "Tree Root Damage to Sidewalks and Curbs." *J. Arbor.* 9(7):177–181.

WARE, G. H. 1994. "Ecological Basis for Selecting Urban Trees." *J. Arbor.* 20(2):98–103.

WILLIAMS, D. J. 1978. "Handling Plants in Landscape Containers." *J. Arbor.* 4(8):184–186.

12

Management of Street Trees: Maintenance

Forest-grown trees are self-maintaining unless a specific forest management objective requires otherwise. Street trees, on the other hand, are forest trees and cultivars of forest trees transplanted to an alien environment, thus demanding intensive management (arboriculture). Street trees grow in poor soils, must compete with sod for nutrients and water, are subject to various pollutants, must develop roots under impervious surface covers, be resistant to pest problems, and must withstand physical abuse from autos, lawn mowers, and people. Street trees must also add to the aesthetics and value of the community. Some of these factors can be partially ameliorated through tree breeding and species selection, but the balance is attained through cultural practices. These practices include pruning, fertilization, protection, cabling and bracing, and wound and cavity treatment.

PRUNING

The city is responsible for public safety in the public right-of-way, and this includes street trees. Some communities issue planting permits and assign maintenance responsibilities for street trees to adjacent property owners, but in the event of injury or property damage the community or its employees may still be held libel (Chapter 5). Forest-grown trees self-prune as the result of close spacing and shading of lower branches, but city trees are open-grown and live branches low on the crown will not self-prune. As a result street trees obstruct traffic, signs, and lighting, and present other hazards due to rapid crown expansion. These problems must be reduced or eliminated by frequent pruning using proper pruning techniques. (Improper pruning, on the other hand, may make trees even more hazardous in the long run.)

Street trees receive two general types of pruning: training and maintenance. Training is done primarily on young trees to develop a branching habit that ensures structural strength and low maintenance as the tree matures. This consists of developing a single leader trunk, selecting permanent scaffold branches with strong attachment to the trunk, keeping temporary branches clear of the walk and street, and starting the process of lifting or raising the crown for permanent street and walk clearance. Maintenance pruning removes dead, dying, diseased, crowded, weakly attached, or low-vigor branches and epicormic sprouts from the tree. It also includes continued raising the crown for permanent street and walk clearance, and crown thinning (International Society of Arboriculture, 1995).

All branch removal is made by cutting just outside the branch bark ridge and branch collar, as depicted in Figure 12–1. Lifting cuts remove lower branches that obstruct traffic in accordance with right-of-way clearance standards, and are made primarily during the training period. Scaffold cuts are training cuts made to develop a structurally sound crown through the retention of branches well spaced along the main leader. Thinning cuts are made in training and maintenance pruning to reduce the number of branches in the crown, which allows remaining branches to become more vigorous.

Hazard pruning removes those portions of the crown that present a public hazard, such as impending structural failure. Deadwood cuts remove partially or completely dead portions of the crown for safety and wound closure. Deadwooding is done mostly on more mature specimens, but all dead or dying branches should also be removed in the training period. Poor species selection or a change in site characteristics sometimes necessitates a reduction in the size of a crown. This is done through a technique known as drop crotching, or pruning a main leader back to a subordinate lateral that is at least one-third the diameter of the leader. Avoid topping or dehorning, as this technique disfigures the tree and results in permanent structural damage (Figure 12–2).

Pruning for vigor is usually done after some damage to the tree, particularly to the root system. Disturbance of the soil around the tree often results in root damage, and this can be reflected in crown dieback. Removing dead and declining branches will reduce the potential for extensive decay as the tree recovers from the disturbance. Damage to branches that results in hazards or potential decline of that branch should be corrected through removal of the damaged portion. Pest problems may be controlled through removal of infested portions of the crown and by maintaining a vigorous crown utilizing the various pruning cuts described above. The reader is referred to the texts *Arboriculture: Integrated Management of Landscape Trees, Shrubs, and Vines* (Harris, 1992) and *Modern Arboriculture* (Shigo, 1991).

Municipal trees that interfere with energized utility lines must be pruned to

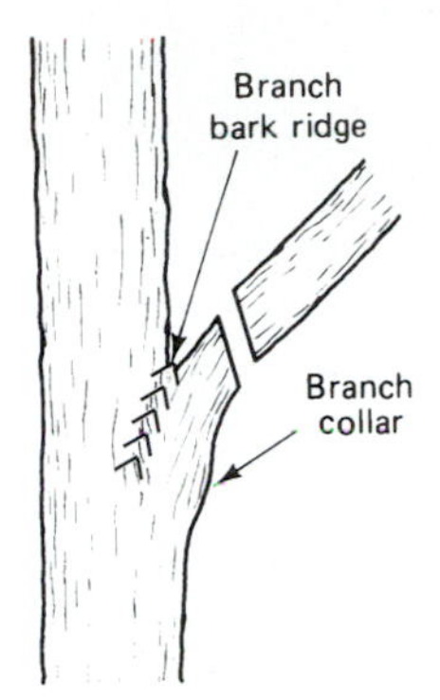

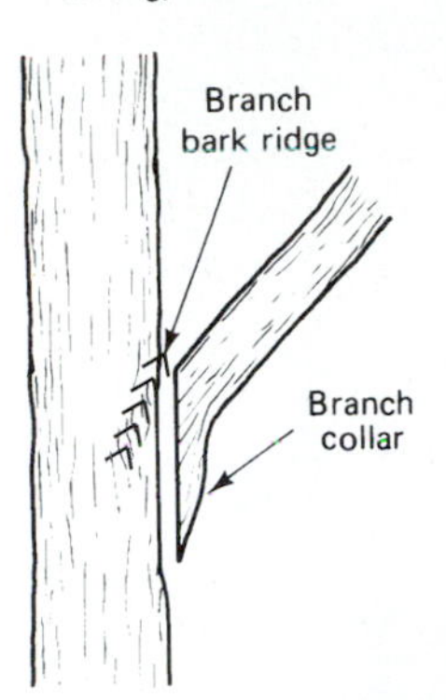

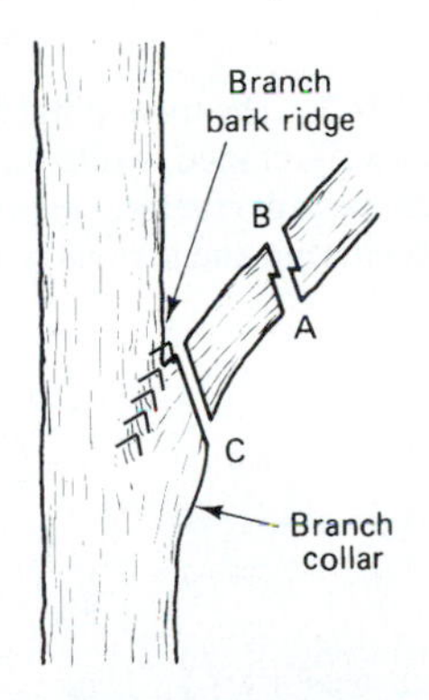

Figure 12–1 Proper pruning involves cuts A, B, and C made outside the branch bark ridge and branch collar.

within a specified distance of those lines. However, municipal pruning crews or contractors do not prune near energized lines for reasons of safety and liability. Line clearance and vegetation management in a utility right-of-way is the responsibility of utility arborists and foresters and is the subject of Chapter 15.

Municipal Pruning

Of all municipal tree management activities pruning is the most essential for long-term tree safety and survival. Yet in 1986 only 39 percent of municipalities surveyed reported a systematic approach to tree management versus crisis management, down from 50 percent in 1980 (Kielbaso, 1989). In 1994, 44 percent of respondents to a question concerning pruning in a survey of municipal forestry programs reported having an annual or seasonal pruning schedule; however, neither "annual" or "seasonal" were defined (Tschantz and Sacamano, 1994).

Figure 12–2 The tree on the left has been properly pruned, while the tree below has been topped, resulting in undesirable heavy suckering and serious wounds.

Municipal pruning operations by city employees are scheduled utilizing one or some combination of five approaches: request pruning, crisis pruning, task pruning, species pruning, and programmed maintenance (Yamamoto, 1985). Request pruning is scheduling based on citizen requests for pruning of trees adjacent to their property. Crisis pruning is removal of a clear hazard that must be attended to immediately. Task pruning is for a particular reason, such as removing branches infested with a pest, or clearing the right-of-way of a street. Species pruning involves the grouping of all trees within a species for similar treatment.

Programmed maintenance (also referred to as area, grid, or scheduled maintenance) is the servicing of all trees in a given area of the community on a rotational basis. This includes making all necessary pruning cuts on each tree in the area, and may include other maintenance activities, such as cabling and bracing. Obviously, service requests, hazards, tasks, and species needs cannot be ignored, but programmed pruning scheduled with reasonable frequency will greatly reduce the need to schedule pruning for these reasons.

The city of Modesto, California, switched from scheduling pruning based on request, crises, tasks, and species needs to programmed maintenance on a seven-year cycle. The first step in this process involved a strong public education effort to inform residents of the advantages of programmed pruning versus the popular request system. City crews also had to be informed of the benefits of programmed pruning so that they could explain the system to residents. The effort was successful, with 70 percent of the tree maintenance being done on a programmed basis. Of greater significance, the productivity of crews doubled as measured by the number of trees pruned. This increase in productivity is due primarily to reduced transportation costs and more efficient task scheduling (Yamamoto, 1985). The city is now on a three-year cycle, with young trees pruned as often as every two years. Gilstrap (1990) reports that among other benefits the three-year cycle allows for more frequent tree inspections, smaller pruning cuts, better public relations, a dramatic reduction in tree service requests, fewer tree failures, and reduced task pruning for clearance and control of mistletoe infestations.

Hudson (1990) compared programmed or scheduled pruning time to request pruning time in the city of Santa Maria, California, over a period of eleven years. Scheduled pruning took an average of 1.03 hours per tree versus 2.38 hours, the difference being attributed to unproductive crew travel and setup time.

Pruning Cycle

The pruning cycle is the number of years it takes to prune all street trees in the community using programmed maintenance. A city on a six-year pruning cycle will prune one-sixth of the street trees each year, with each tree receiving maintenance once every six years. The length of a pruning cycle will depend primarily on the number of trees in the community and funds available for maintenance. The optimum pruning cycle, on the other hand, will vary based on the condition, species, and age of the tree population, and the climatic characteristics of the region.

The condition, species mix, and age of the tree population influence the optimum pruning cycle for that population. A high condition rating as described by the Council of Tree and Landscape Appraisers necessitates the tree having a well-developed crown, good branch arrangement, and small pruning wounds. Frequent pruning, especially when the tree is young, will allow a tree to develop a high condition rating, and in turn that tree will demand less corrective pruning at maturity. Species influences the frequency of pruning, as there is great variation in growth rate, branch arrangement, wood strength, and rates of decay among species. Age of the tree also relates to the frequency of pruning. Generally, young trees need more frequent pruning because of rapid growth rates, the need for training, and low branches in the right-of-way. Trees growing in warm climates with abundant moisture will grow faster and need more pruning than their counterparts in less favorable climates.

A method to determine the optimum pruning cycle for a community is to compare the marginal cost of pruning to its marginal return. A portion of Milwaukee, Wisconsin, was inventoried using a computerized system that recorded tree condition and calculated tree value. Since condition class influences tree value, the date of last pruning and average condition class for each work unit inventoried was subjected to regression analysis to determine relationship between pruning and condition class (Figure 12–3). Marginal costs were calculated using condition classes generated by the regression equation to compute the loss in tree value for each one-year extension of the pruning cycle. Marginal returns are the savings in pruning costs for each one-year extension of the pruning cycle (Table 12–1). The relationship between marginal cost and return (Figure 12–4) indicates that the optimum pruning cycle for the city of

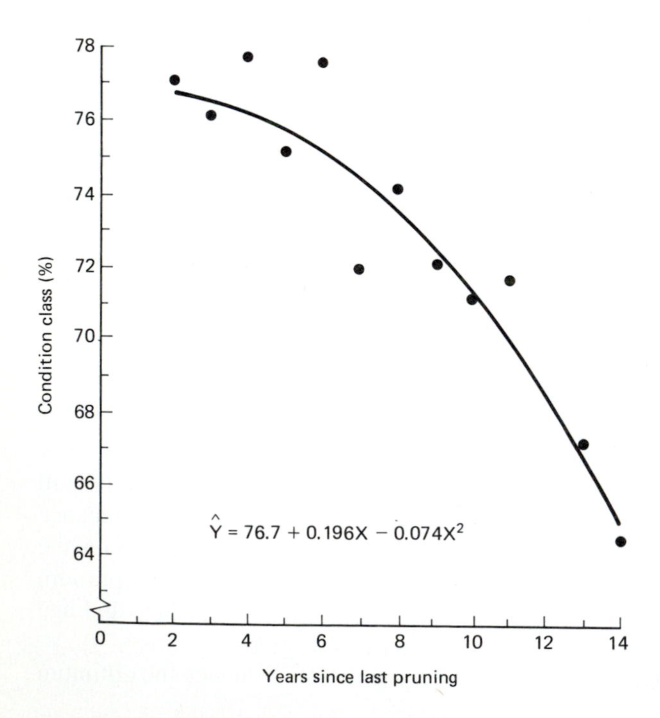

Figure 12–3 Relationship between average tree condition class and number of years since last pruning (sig. [.005]). (From Miller and Sylvester, 1981.)

TABLE 12–1 TREE VALUE AND PRUNING COSTS FOR VARIOUS PRUNING CYCLES, BASED ON 40,808 STREET TREES IN MILWAUKEE, WISCONSIN (From Miller and Sylvester, 1981.)

Pruning Cycle (yr)	Average Condition Class for Specified Pruning Cycle (%)	Tree Value for Specified Pruning Cycle	Marginal Cost	Annual Pruning Cost for Specified Pruning cycle[a]	Marginal Return
2	76.8	$20,381,000	—	$337,000	—
3	76.7	20,358,000	$ 23,000	224,000	$113,000
4	76.6	20,321,000	37,000	168,000	56,000
5	76.4	20,272,000	49,000	135,000	33,000
6	76.2	20,210,000	62,000	112,000	23,000
7	75.9	20,134,000	76,000	96,000	16,000
8	75.5	20,046,000	88,000	84,000	12,000
9	75.2	19,944,000	102,000	75,000	9,000
10	74.7	19,829,000	115,000	67,000	8,000
11	74.2	19,702,000	127,000	61,000	6,000
12	73.7	19,561,000	141,000	56,000	5,000
13	73.1	19,407,000	154,000	52,000	4,000
14	72.5	19,239,000	168,000	48,000	4,000

[a]Assume average pruning cost of $16.50 per tree.

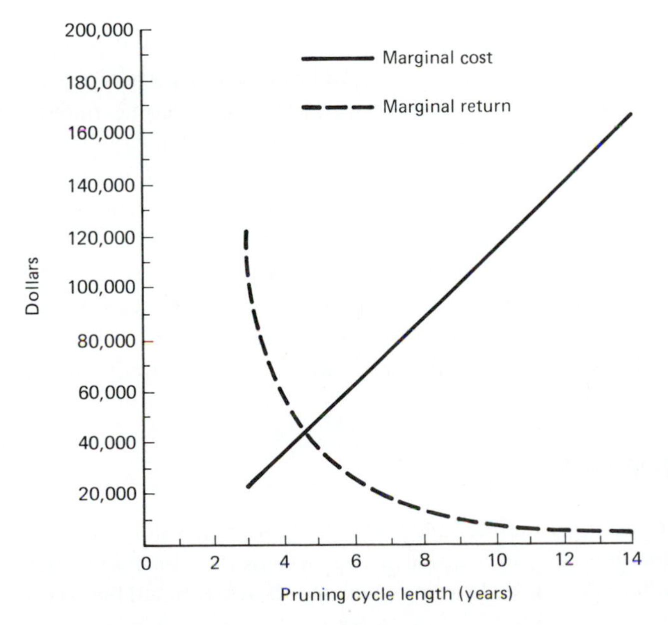

Figure 12–4 Comparison of loss in tree value versus savings in pruning costs for various pruning cycles in Milwaukee, Wisconsin. (From Miller and Sylvester, 1981.)

Milwaukee is five years, assuming that the management goal is to provide the highest-value tree population for dollars expended (Miller and Sylvester, 1981).

Based on the calculations above, a pruning cycle of five years will produce optimum tree values, at least in the city of Milwaukee. However, a cycle in the vicinity of five years is probably a reasonable goal for most communities in north temperate climates, especially if there is not a preponderance of young trees. Young trees often need more frequent pruning than do mature specimens, for training and to keep streets and sidewalks clear of obstruction. Some communities use two pruning cycles: for example, three years for young trees, and six years for mature trees. Pruning-cycle length is ultimately a local decision based on the specific tree population, management priorities, and budgetary constraints.

Sisinni et al. (1995) reported the effects of an ice storm on different species in Rochester, New York. Based on their observations they recommend a five- to ten-year pruning cycle on mature trees and a three-year pruning cycle on young trees to minimize the impact of storms.

Once a general pruning cycle has been decided upon, the spatial arrangement of the street tree population by size and species influences the assignment of pruning areas and crew configuration. Crew size and equipment needs are a function of the size and species of trees to be pruned on a work assignment basis. A Master Street Tree Plan that specifies the development of age classes and species themes by blocks or street segments simplifies making crew and equipment assignments. A street tree inventory is useful when initiating programmed maintenance by providing information descriptive of size and species mix by block or work unit. These data are used to subdivide the city into working areas of equal work load, and to prioritize these areas for pruning sequence. A system of programmed maintenance should produce equivalent work loads for crews and equipment on an annual basis.

If two pruning cycles are used for different size classes, the process is the same. All mature tree areas are divided by their pruning cycle and tree training areas are divided by their cycle (Figure 12–5). The point where training trees become mature trees with a longer pruning cycle will be influenced by species and local growth rates. A guide to assist in making that decision is to change the cycle once the major scaffold branches have been developed and pruning needs become a matter of primarily removing deadwood, epicormic branches, and minor corrective cuts.

Once a pruning cycle is well established in a community, some trees may not need pruning or other maintenance when a given management unit is serviced. However, a pruning cycle will allow crews to inspect all trees on a regular basis to determine if hazards or other problems exist, an important consideration in maintaining public safety.

Pruning Budgets and Equipment

A well-planned system of programmed maintenance will generate equivalent budgets each year. Annual pruning budgets are determined using inventory information descriptive of tree frequency by diameter class. Pruning costs are then estimated based on

UNIT 1	UNIT 2	UNIT 3
1997 D,M 2000 D 2003 D,M 2006 D	1998 D,M 2001 D 2004 D,M 2007 D	1999 D,M 2002 D 2005 D,M 2008 D
1997 D 2000 D,M 2003 D 2006 D,M	1998 D 2001 D,M 2004 D 2007 D,M	1999 D 2002 D,M 2005 D 2008 D,M
UNIT 4	UNIT 5	UNIT 6

Figure 12–5 Three-year pruning cycle on developing (D) trees and six-year cycle on mature (M) trees; six work-unit community.

time and equipment needs for each diameter class. Tree frequency by diameter class is multiplied by the average pruning cost for that class, these values are summed, and divided by the pruning cycle (Table 12–2). If the pruning cycle is dictated by the budget, this method can also be used to determine the length of the pruning cycle. Annual pruning cost estimates for the entire community can be subdivided by management work unit to schedule pruning over the cycle. Table 12–3 depicts a community with twenty work units, and an annual pruning budget of $20,000 on a five-year pruning cycle.

Although it is desirable to do most pruning on a scheduled basis, there will always be a need for some request, task, and crisis pruning. Once the cost to prune on a specific cycle is determined, some additional funds should be requested to service out-of-cycle pruning needs. As discussed earlier, the amount needed for this kind of pruning should decrease over time as a city completes its first time through a pruning cycle, or as the city moves through a shorter pruning cycle. Out-of-cycle pruning budget requests can be based on previous years' experiences, and can be modified over time as pruning cycles change.

Equipment needs for crews performing programmed maintenance will vary based on the characteristics of the street tree population. A typical work crew assigned maintenance of trees will need hand and chain saws, fuel, oil, hard hats, safety goggles, pole pruners, ladders, and pruning shears. A truck will be needed to transport crews and equipment, haul debris from the work site, and will need to be equipped with traffic barriers, first-aid kits, and spare tools. Spare tools as backup equipment are essential for efficient operation, as a lot of money is wasted when crews sit idle because of equipment breakdown. If climbing is the technique to be used, ropes and saddles, safety lines, and bull ropes for lowering large limbs will be needed (Figure 12–6).

Trucks with aerial towers are often used to prune street trees because of reduced labor costs (Figure 12–7). However, it is still good practice to have climbing equipment on hand for emergencies or to prune portions of the crown inaccessible by hydraulic boom. Most aerial towers are now equipped with hydraulic power sources in the bucket, and these may be used to power chain saws and shears. Chippers often accompany crews, as they reduce the volume and facilitate loading of pruning debris.

TABLE 12–2 CALCULATION OF ANNUAL PRUNING COST

Dbh Class	Number of Trees	Pruning Cost per Tree	Pruning Cost by Diameter Class[a]
1	92	$10	$ 920
2	2973	10	29,730
4	2147	11	23,617
6	1119	11	12,309
8	435	14	6,090
10	307	16	4,912
12	254	19	4,826
14	224	22	4,928
16	206	25	5,150
18	232	28	6,496
20	229	31	7,099
22	220	33	7,260
24	185	34	6,290
26	150	35	5,250
28	90	35	3,150
30	52	36	1,872
32	19	36	684
34	15	37	555
36	10	38	380
38	2	40	80
		Total	$131,598

Cost to prune all trees	$131,598
5-year pruning cycle	÷ 5
Annual pruning expense	$ 26,320

[a]Number of trees × cost per tree.

Chips are also in demand as mulch and ground covers, and even if given away will save the cost of landfilling. Pruning crews are configured by the size of trees being pruned and the equipment used. Usually one or two people working on the ground stacking and/or chipping brush and controlling traffic can service several climbers.

Small trees may be pruned from the ground using hand saws, shears, loppers, and pole pruners. Hydraulic power units can be mounted on the back of pickup trucks and used to power hydraulic pole saws and shears (Figure 12–8). Two people per crew is adequate when pruning from the ground—one to do the actual pruning, and one to handle brush and control traffic. These crews also have the standard safety and backup equipment as assigned to crews pruning large trees.

A well-equipped maintenance shop is essential to a tree maintenance operation.

TABLE 12–3 MAINTENANCE CYCLE BUDGET FOR A FIVE-YEAR PRUNING CYCLE

Work Unit	Cost Per Unit	Annual Cost
1	$5,000	
2	4,000	
3	6,000	$20,000
4	5,000	
5	5,000	
6	8,000	$20,000
7	7,000	
8	4,000	
9	3,000	
10	4,000	$20,000
11	3,000	
12	6,000	
13	10,000	
14	2,000	
15	3,000	$20,000
16	5,000	
	3,000	
17	5,000	
18	2,000	
19	5,000	$20,000
20	5,000	
Total Pruning Cost	$100,000	

Equipment should be serviced daily and be sent to the field in good working order. A supply of common spare parts should be on hand to reduce downtime. Equipment in poor working condition is no excuse for idle crews. Poorly maintained equipment and idle crews create a very negative public image for the forestry operation, and in the long run could affect funding.

Contract Maintenance

Many communities contract pruning and other programmed maintenance activities. Tate (1984) found that 53 percent of ninety-one New Jersey cities surveyed contracted pruning, while Robson (1984) reports that five of eight suburban Chicago

Figure 12–6 Pruning street trees using ropes and saddles.

Figure 12–7 Pruning street trees using an aerial tower.

Figure 12–8 Pruning street trees using a hydraulic pole saw and a pickup-mounted power source. (Richard Rideout, Milwaukee Bureau of Forestry.)

communities contract all or some pruning. The prebid notification describes the area to be pruned and contains a summary of diameter class and species frequency. The notification and subsequent contract should specify in detail the pruning standards to be followed. Pruning standards that can be used in contracts are available from the American National Standards Institute (ANSI A-300) (1995). These standards, developed in conjunction with the National Arborist Association and the International Society of Arboriculture describe appropriate tools, equipment, and pruning cuts, and differentiate between young tree pruning, mature tree pruning, and utility pruning. The following is a brief summary of the three types of pruning standards, and the reader is urged to obtain a complete copy prior to advertising for a bid or entering into a contract.

Young Tree Pruning

1. At planting: Remove all damaged and dead branches, and competing leaders.
2. First three years: Begin to select scaffold branches and remove branches that cross or that have narrow angles of attachment and/or included bark.
3. Years four to six: Continue to select scaffold branches and begin to raise the crown.

Mature Tree Pruning

1. Do not remove more than one-quarter of the foliage in one season.
2. One-half of the foliage should be retained in the lower two-thirds of the crown.
3. Specify the minimum diameter or branches to be removed.
4. Specify objectives for pruning.
   a. Hazard reduction
   b. Crown cleaning to remove dead, dying, diseased, weak, and epicormic branches
   c. Crown thinning to remove limbs for air and light penetration and weight reduction
   d. Crown raising for clearance
   e. Crown reduction to reduce the size of the crown
   f. Vista pruning, including thinning, for a view from a specified point
   g. Crown restoration to repair from past damage

Utility Pruning

1. Minimize the number of cuts to achieve the goal.
2. Trees under wires: remove the tree or prune entire branches from the tree.
3. Trees not under wires: remove entire branches or prune to a lateral that will direct growth away from wires.

The ANSI A-300 standards also specify specialty pruning for palms, espalier, and pollarding.

Contract pruning is a common and effective method of maintaining street trees, particularly in smaller communities where equipment costs cannot be justified by the amount of use each year. A pruning cycle is used with the appropriate area of the community advertised for bids each year. However, it is essential to have a good contract that describes what is expected by each party, sets work standards, and allows for on-site inspections.

Future Pruning Costs

Developing a pruning budget for the present is a matter of determining the cost to prune all trees in the city and dividing by the desired pruning cycle as is shown in Table 12–2. However, as trees grow pruning costs will increase over time, necessitating the need for increased funding or a change in management priorities. Predicting future pruning costs can be made from present pruning costs, growth rates, and the inventory. Churack et al. (1994) analyzed pruning time, waste wood yield, and growth rates in the city of Milwaukee to estimate future pruning budget needs. Pruning crews were timed across seven diameter classes pruning four species of trees. The diameter classes were in 2-in. (5 cm) increments (4-, 6-, 8- . . . 16-in. classes), and the species were honeylocust (*Gleditsia triacanthos*), green ash (*Fraxinus pennsylvanica*), littleleaf linden (*Tilia cordata*), and Norway maple (*Acer platanoides*). Pruning time varied slightly by species, but when combined pruning times were found to increase at a rate of six minutes per diameter inch (2.5 cm) (Figure 12–9).

An increment borer was used to determine the growth rates of sample trees. For combined species the average annual diameter growth rate is 0.5 in. (1.27 cm). The growth rate of individual trees differs from the growth rate of a population of trees because of tree mortality and replacement (i.e., large trees are removed and replaced with small trees). The street tree simulation described in Appendix B was used to determine the population growth rate, using an inventory file of 9,000 street trees, an annual diameter growth rate of 0.5 in., annual mortality rates for Milwaukee, and planting to replace removals. The tree population mean diameter at the start of the simulation was 7.4 in. (18.5 cm). The population diameter peaked at 14.2 in. (35.5 cm) in sixty-five years and stabilized at 12.9 in. (32 cm) by year eighty-five when attrition in the larger diameter classes was balanced by planting of small replacement trees (Figure 12–10). The peak in the curve is due to the preponderance of small-diameter trees at the start of the

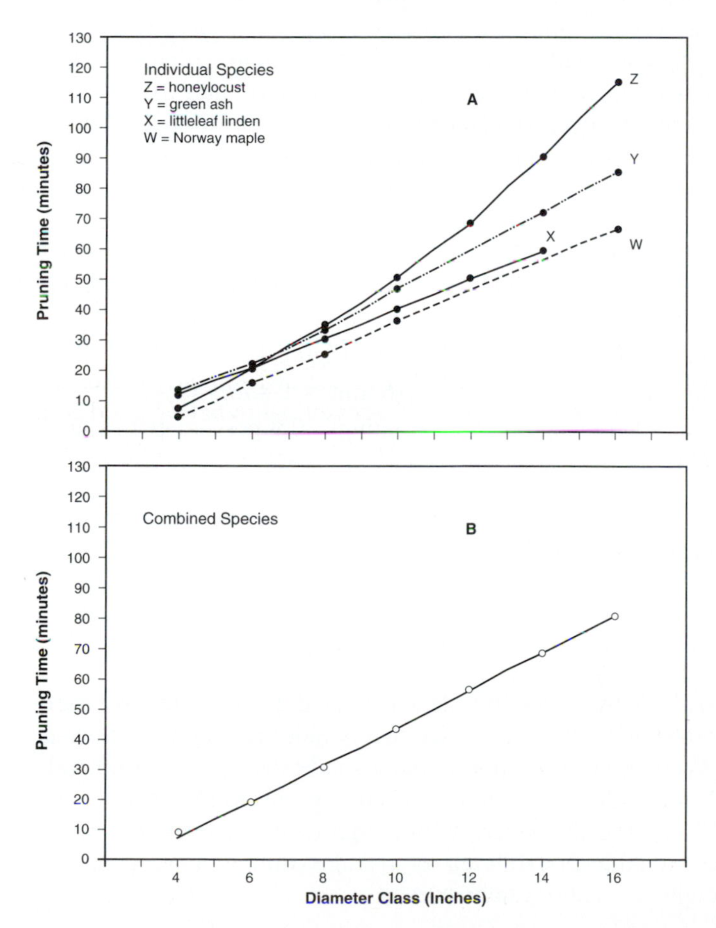

Figure 12–9 Relationship between pruning time and diameter for individual species (A) of honeylocust (z); green ash (y); littleleaf linden (x); and Norway maple (w); and all species combined (B).

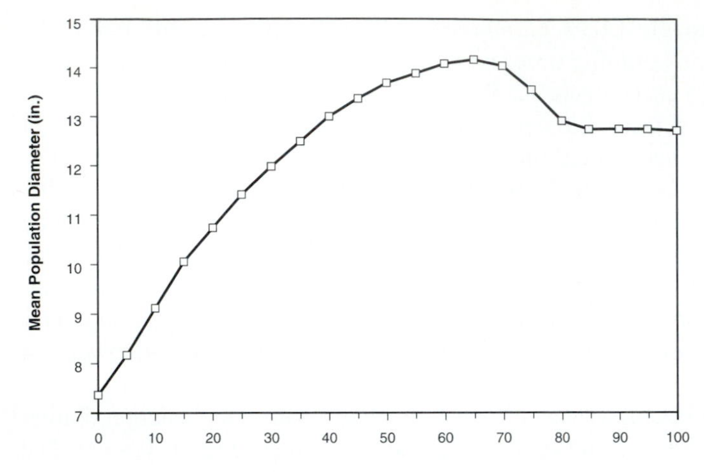

Figure 12–10 Simulated tree population mean diameter changes over time, removals = replacements.

simulation. It is not likely that a tree population will precisely follow such a curve because of perturbations by storms and epidemics. However, a population growth curve can be used to project pruning costs within the long-term planning horizon of one median life span, as described in Chapter 10. For example, if you are pruning 10,000 trees per year (50,000 trees on a five-year cycle) and the present mean population diameter is 8 inches, what will your pruning cost be in person years in ten years?

Pruning time per tree = 0.5 person hours (Figure 12–9)
Number of hours worked per person per year = 2,000
Person years for pruning = 2.5 (10,000 trees × 0.5 = 5,000 hours pruning/2,000 hours)
Increase in mean diameter in 10 years = 1.8 inches (Figure 12–10)
Increase in pruning time per tree = 0.18 hours [(6 min. × 1.8 in.)/60]
Additional personnel needed in 10 years for pruning = 0.9 persons [(0.18 hr. × 10,000 hr. per year)/2,000]

Note: The above calculations do not include the time of a ground person. Usually one ground person can service a number of climbers.

FERTILIZATION

Fertilization of established shade trees can provide accelerated growth rates. A faster-growing tree is more vigorous and will reach a size that is more resistant to injury in a shorter period. Van de Werken (1981) reports young shade trees growing on soils "with medium levels of P and K (30 and 20 ppm respectively), respond to surface application of nitrogen at 120 to 150 lbs N/acre with a significant increase in growth rate." Trees receiving these treatments exhibited an average canopy area increase of 32 percent over controls eight years after establishment.

Figure 12–11 Fertilization of street trees using a needle and water-soluble fertilizer. (Richard Rideout, Milwaukee Bureau of Forestry.)

Surface applications of fertilizer have limitations when applied to street trees. The abutting property owner may complain about the growth response of grass in the tree lawn to a fertilizer high in nitrogen. Surface applications also run the risk of burning the sod if too much is used. Injection of liquid fertilizer below the sod, although more expensive, will prevent excess stimulation of the sod and reduce the risk of sod damage (Figure 12–11).

PROTECTION

Control of insect and disease problems on street trees is best accomplished through a program of integrated pest management (IPM). The process begins with selecting species and cultivars resistant to insect and disease problems. Although the Master Street Tree Plan may call for street segments having the same species on them, it should direct the development of an overall population diverse in species to prevent catastrophic losses of the tree population. A schedule of programmed maintenance will aid in maintaining a healthier tree population through pruning weakened or pest-infested branches and by developing sound vigorous crowns. Prompt removal of declining and diseased trees along with enforcement of community ordinances dealing with pest problems provides additional cultural controls.

Monitoring the tree population for potential problems is advised, as is monitoring for existing insect and disease problems, to determine if and how much control is needed. Another step in IPM is the establishment of damage thresholds for each pest in which control action is warranted. Once action is to be implemented, a public information and input program should precede actual selection of the control measure. If spraying is necessary, select the control measure that is the most cost-effective and most acceptable to the public. IPM is being utilized successfully to control a number of serious pest problems, including gypsy moth (*Lymantria dispar* L.) (Shaw, 1982) and Dutch elm disease.

The International Society of Arboriculture has developed a system of pest management termed Plant Health Care (PHC) that is an outgrowth of IPM (Smith et al., 1995). PHC differs from IPM in that the primary focus is on individual plants rather than on groups of plants such as is found in agriculture. Although primarily designed for use by commercial arborists, PHC is applicable to street tree management through the selection of appropriate species for site conditions, and cultural practices that emphasize plant health.

An important consideration of any pest management program is the cost of control in relation to benefits derived. In North America the pest that has caused the greatest problem for municipal foresters is Dutch elm disease. Despite losses incurred from this disease, much has been learned about the economics of its control, and these lessons can be applied to other pest management problems. Cannon et al. (1982) suggest that any Dutch elm disease control technique that costs less than elm removal and reduces the number of trees removed saves money for the community. They further report that a method of sanitation, which includes early detection, girdling of diseased trees, and prompt removal, saves money and elms.

A computer simulation was used in three Wisconsin communities to determine future net values of street tree populations under four Dutch elm disease control scenarios. The four control levels were: no control (18 percent annual mortality), fair control (5 percent annual mortality), good control (3.5 percent annual mortality), and best control (1 percent annual mortality). Simulations were run for twenty years, trees pruned on a five-year cycle, and trees replaced as removed. Tree values (Council of Tree and Landscape Appraisers [CTLA], 1983, formula) and management costs for planting, pruning, and removal were generated by the computer, with disease control costs for each scenario added to management costs. In all three communities net tree values (tree value – management costs) were highest under the best control scenario throughout twenty years of simulation, including District 2 in the city of Milwaukee (Figure 12–12) (Miller and Schuman, 1981).

Benefit-cost analysis compares annual discounted benefits with annual discounted costs of programs to determine if the costs are justified. This analysis technique has been used to assess the effectiveness of pest management programs. Sherwood and Betters (1981) conducted benefit-cost analyses on different Dutch elm disease control strategies in selected Colorado municipalities. They found that when comparing discounted control costs to the discounted benefits (CTLA formula) of more surviving elms, benefits exceeded costs. In addition, they conclude that the more intense the control program, the greater the ratio of benefits to costs.

Benefit-cost and other economic analysis techniques can be useful tools for de-

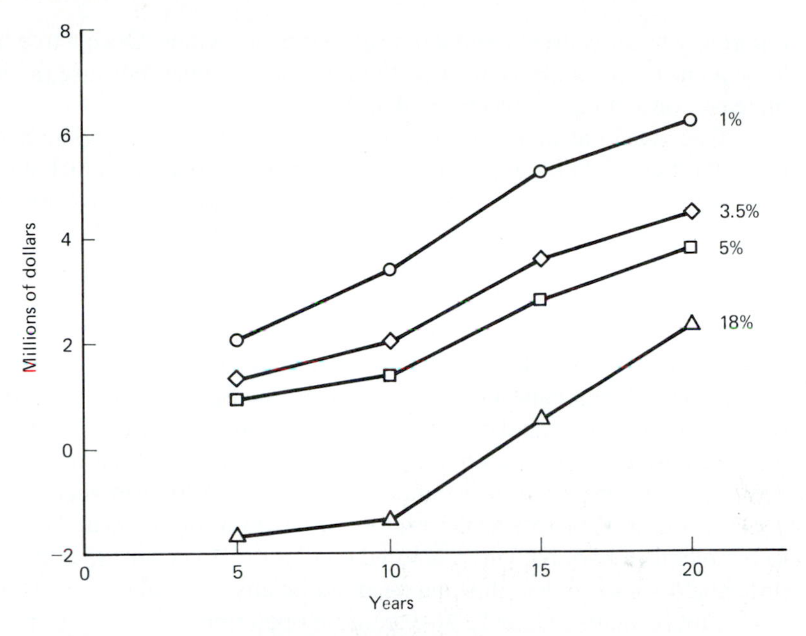

Figure 12–12 Predicted net forest values over time; replace removals scenario for various American elm annual loss rates (Milwaukee, District 2). Net forest value = total forest value − (management costs + D.E.D. control costs). Based on 32,229 trees, of which 3178 are American elm. (From Miller and Schuman, 1981.)

cision making, but should be tempered with good judgment and an understanding of community values. Some communities place higher value on their vegetation than appraisals indicate and are willing to pay for pest control measures in excess of damages incurred. Other communities place low or negative value on species that cause high pest control costs and would as soon be rid of them.

OTHER STREET TREE MAINTENANCE

Other street tree maintenance includes hazard inspection, cabling and bracing, wound treatment, cavity treatment, and root pruning. Hazard inspection should be done a minimum of once per year by a person qualified to evaluate tree safety. A hazard tree must meet two criteria: the presence of a structural problem that may cause the tree to fail, and a target (Matheny and Clark, 1994). Each street tree certainly has many targets, as do park trees in areas where people are frequently present. Inspection for hazards can lead to the removal of the tree, or recommendations for corrective action such as pruning or cabling to alleviate the problem. Phone calls from the public expressing concern over a perceived hazard should be dealt with quickly by an on-site inspection. Likewise, all forestry employees and contractors should be instructed to report any hazards they observe during routine activities. Once a tree has been judged to be a hazard it should be dealt with in a timely manner because the level of liability increases once a hazard has been identified.

One outcome of a hazard inspection is the recommendation to cable and/or brace the tree. Cabling and bracing is an expensive operation that should be weighed against pruning or tree replacement to correct the problem. A vigorous tree that will have its useful life span extended many years by cabling and bracing is an excellent candidate for this work. However, if there is any question as to the safety of the tree, even if cabled and braced, it should be removed immediately. It is important to remember that it is far cheaper to prune offending branches when the tree is being trained than to correct the problem later through cabling and bracing.

Wound treatments such as painting or tracing following normal pruning operations are unnecessary and can be detrimental to the tree (Shigo, 1983). Wound dressings are cosmetic and if used should be applied in a thin coat only. The exception to this is if a pruning or other wound is attractive to a disease vector and painting renders the site unattractive to the vector. Other trunk injuries can be treated by tracing to remove torn bark back to uninjured cambium. However, tracing should remove as little healthy bark as possible and need not be any particular shape (Harris, 1992).

Cavity treatment can be an expensive operation, especially if it includes cleaning and filling or covering. High labor costs and recent findings (Shigo, 1983) that cavity treatment does little to prevent decay or extend tree longevity has reduced the amount of this work being done. In street tree maintenance cavity treatment primarily involves removal of some decayed wood (being cautious not to break compartment walls) and installation of drain pipes in response to complaints. It should be noted, however, that drain pipes do penetrate compartment walls and lead to additional decay (Shigo, 1983).

As a general rule the city is responsible for curb maintenance and the abutting property owner is responsible for sidewalk maintenance, including repair and/or replacement when tree roots lift or break pavements. Curb and sidewalk damage by tree roots is usually a function of poor species selection or a reduction in the width of the tree lawn due to sidewalk or street widening once trees are established. In most municipalities city engineering departments make periodic inspections to identify sections of curbs and sidewalks in need of repair. Although standards may vary, once sidewalk displacement reaches the point where a public hazard exists, replacement is ordered, work is performed by a city crew or contractor, and the homeowner is billed for the work.

Root pruning, before sidewalk replacement is needed, will save the homeowner money and avoid negative feelings toward the tree or the forestry department. However, root-cutting machines (Figure 12–13) can sever roots to a depth of 36 in. (0.9 m), causing excessive physiologic damage to the tree. Deep root cutting can result in tree instability if many support roots are cut. The depth of root pruning should not exceed minimum ordinance standards, that is, 6 in. (15 cm) below the sidewalk base. If there is concern about tree stability following root cutting, removal is warranted. Root pruning is frequently done by city engineering department crews and/or

Figure 12–13 Root cutting of a street tree to prevent sidewalk damage. (Richard Rideout, Milwaukee Bureau of Forestry.)

contractors and it is essential that the city forester maintain close liaison with the city engineer concerning this procedure.

MANAGEMENT NEEDS OVER TIME

As trees increase in size they need different management or care. It can be useful to divide trees into three size categories and prescribe different levels and types of maintenance for each size class. The following is an example of a size-related management system:

Small Trees (less than 6 in. [15 cm] in diameter)
- Watering as needed
- Fertilizing
- Training pruning
- Pest management

Developing Trees (6 to 12 in. [15 to 30 cm] in diameter)
- Training and maintenance pruning
- Pest management
- Root pruning

Mature Trees (larger than 12 in. [30 cm] in diameter)
- Maintenance pruning
- Pest management
- Cabling and bracing

The above size classes can be altered by region and growth rates, as can the number of size classes. However, it is important to recognize that trees need different services as they mature, and that a program of early and frequent maintenance will develop a tree that is structurally sound, adapted to its site, and demands less maintenance.

MANAGEMENT BY ZONE

Street trees are found in a variety of land-use categories in a community, and it may be useful to prescribe different management efforts by zone. Different land uses in a community can be placed in three general zones: residential, commercial, and industrial. The type of tree selected and level of maintenance prescribed for each zone can vary according to the goals and priorities established for each zone. Table 12–4 is an example of a community zoned by general land use for tree management activities. In this example a community has placed the highest planting priority on establishing full stocking in residential neighborhoods, and will allocate most of the planting budget for residential planting. Species selected for residential areas will be different than species selected for commercial or industrial areas, and their growth rates and habits demand different levels of maintenance. Residential trees will be selected for ease of training and to match available growing space. Species planted in commercial areas will be selected for their ornamental value, strong wood to resist people pressure, and slow growth so as not to quickly outgrow available above and below ground space. The commercial zone trees will likely need annual maintenance, and are on a short rotation because of their relatively short median life span. Species planted in industrial areas are selected for fast growth with the anticipation of short rotations based on their median life span, and pruning specifications are selected primarily to ensure safety.

Priorities and management activities will vary by community, and over time

TABLE 12–4 A COMMUNITY ZONED BY LAND USE FOR DIFFERENT MANAGEMENT ACTIVITIES.

Zone	Planting Priority	Species Selection	Maintenance Cycle	Median Life Span	Service Requests[a]
Residential	High	Moderate growth rate, single leader	3 and 6 years, training and maintenance	30 years	High Priority
Commercial	Medium	Slow growth, ornamental & strong wood	Annual, training and maintenance	8 years	High Priority
Industrial	Low	Fast growth, pollution resistant	6 years, maintenance	15 years	Low Priority

[a]Nonemergency

within a community as management goals evolve and the tree population changes. Levels of service and budgets can be more effectively and efficiently allocated by recognizing that different areas of a community need different levels of service. Master Street Plans can be subdivided by zone, species matched to site characteristics within zones, and management prescriptions developed for each zone. Within each zone, specific street segments can be managed based on the general size category and species characteristics of the tree population.

REMOVAL AND REPLACEMENT

All street trees planted will ultimately be removed and probably be replaced. Tree removal includes the removal of stumps, as they present obvious pedestrian hazards and can interfere with future planting. Stump removal is done by grinding to below grade and backfilling with soil and grindings to slightly above grade level to allow for settling (Figure 12–14). Typical causes for removal include death, decline, public hazard, damage, nuisance, street widening, and urban redevelopment projects. Obviously, trees that are dead or nearly so must be removed for the public safety. Trees that are in early stages of decline or are not obviously hazardous present a more difficult situation.

Figure 12–14 Stump grinding following removal of a street tree. (Richard Rideout, Milwaukee Bureau of Forestry.)

Removal may be necessary, but public opinion may induce pressure to save the tree. However, if in the professional judgment of the city forester the tree is a hazard, the reasons for removal must be made known to public, and the tree removed.

Nuisance trees are another difficult problem, as public opinion can be divided on whether a tree must be removed. If the tree is in violation of the city ordinance and the ordinance has historically been enforced, removal will probably be necessary. If the tree is not in violation of the ordinance and majority public opinion favors retention, the tree should probably be retained. Do not selectively enforce an ordinance, as the political consequences of doing so can be severe.

Street widening and other construction activities take a heavy toll on street trees. Foster and Blaine (1978) report that continuing construction in Boston kills more trees than any other cause by damaging the tree or by altering the growing environment. Trees can be protected during construction through fencing and other protection measures and have a good chance for survival. However, if it is obvious that extensive damage is imminent and unavoidable, and that the trees will probably decline and die within a few years after construction ceases, removal of existing trees and purchase of replacements will be part of the construction budget. However, removal of dying trees and replacement after the fact will usually come from forestry funds.

Ideally, annual removal takes place at a relative constant rate and cost over time. Under normal circumstances this will be the rule, as tree attrition combined with replacement will develop a broad spectrum of age classes. The exception is in communities where large-scale planting (especially of the same species) produces an even-aged population that becomes senescent at about the same time.

Replacement is guided by objectives and overall management plans as described in the Master Street Tree Plan. Street tree spacing standards as described by the tree ordinance will determine if a tree is to be replaced following removal. The actual date of planting is then determined from land-use and site priorities.

Equipment utilized in removals is generally the same as that used in pruning. Additional equipment such as log or front end loaders and flatbed trucks may be needed for large specimens. Large trees that overtop wires and structures sometimes require the use of cranes for safe removal. Utility companies will drop and deenergize wires for removals rather than risk injury or damage.

Storms and Epidemics

Storms and lethal pest epidemics present abnormal circumstances resulting in widespread tree losses. Contingency plans must exist for storm damage to provide rapid restoration of utility services, street clearing, and hazard removal or repair. Crews need to be available for rapid mobilization, and contingency funds available for overtime pay. In the event that damage is more than can be handled by city crews, a list of cooperating private contractors should be available to assist in storm cleanup. Emergency funding is often available from higher levels of government to assist communities with catastrophic damage.

Storm disaster planning and mitigation consists of the following three chronological phases (Burban and Andresen, 1994):

1. Preparation—disaster planning and warning activities. Examples of activities include the identification of an early warning system for severe weather, development of a disaster response plan, identification of roles of various individuals and municipal departments during disasters, and identification of groups or communities to contact when necessary.

2. Response—immediate activity during and after the disaster. Examples of activities include: tree damage clean up, clearance, identification of methods of communication from the field to the office, determination of debris disposal options, and use of efficient record-keeping methods.

3. Recovery—activities after the disaster that attempt to restore conditions prior to the disaster. Examples of activities include: public and private tree planting and care, training, tree planting awareness events and celebrations, and recognition activities for volunteers, citizens, municipal workers, and others involved.

Tree species selection and maintenance is an important part of the planning process for storms. City foresters, arborists, and scientists routinely report that trees that have been properly trained through pruning are less susceptible to storm damage than trees improperly pruned or not pruned at all (Sisinni, 1995). Trees that have been previously topped with much of the crown consisting of large suckers attached to decayed trunks are especially vulnerable to storm damage. Appendix H contains the disaster relief plan for Oak Park, Illinois.

Pest epidemic losses are of a longer-term nature than storms but can be equally destructive. The most destructive epidemic in most North American cities was and is Dutch elm disease (Figure 12–15). Loss rates from this disease vary based on the frequency, spacing, and distribution of elms in the community, and control efforts. An elm population can be decimated in as little as twelve years with no control, or can be made to last many decades through control efforts (Cannon and Worley, 1979). When such an epidemic occurs, management goals, priorities, and plans are temporarily or permanently altered to deal with the problem. In communities with Dutch elm disease epidemics, funds are shifted from planting, pruning, and other management activities to disease control and tree removal until losses reach manageable levels.

Utilization

The value of wood residue from tree management activities is discussed in Chapter 5. Utilization of wood products from street tree removal is more a means to dispose of residue economically than a profitable venture, as processing costs often exceed wood value. However, it may be cheaper to sell wood at a loss or give it away than to haul it to a landfill and pay tipping fees (Figure 12–16). The city forestry department in Cincinnati, Ohio, removed thirty-six overmature, hazardous red oak (*Quercus rubra*) street trees and sold the residue as firewood, chips, and saw logs for $2550 rather than landfill the residue for about the same cost (Gulick, 1985). The cost of removal and processing obviously exceeded the return from utilization, but landfill savings plus product sales resulted in a savings of $5000 for the city.

State and national governments are increasingly enacting laws that prohibit placing

Figure 12–15 Dutch elm disease (*Ceratocystis ulmi*) killed American elms. Much has been learned about diversity and spacing of street trees following the disastrous losses to this disease.

wood and other landscape residue in landfills. More and more municipal forestry departments will need to dispose of residue through sales or giving away of products ranging from logs to composted wood chips and other landscape residue. Churack et al. (1994) reported that each minute spent pruning street trees in Milwaukee, Wisconsin,

Figure 12–16 Firewood sales provide utilization of pruning and removal work wood, and may help to offset management costs and landfilling fees. (Courtesy of Milwaukee Bureau of Forestry.)

yielded 3 pounds (1.4 kg) of wood residue. The same study reported an increase of six minutes per diameter inch (2.5 cm) or pruning time. As street tree populations grow in size and numbers, the volume of wood residue will increase proportionately.

Removal Contracts

As in the case of other tree management activities, tree and stump removal can be carried out by city employees or by contract. Tate (1984) reports that 69 percent of ninety-one communities surveyed in New Jersey contracted tree removal, and 43 percent contracted stump removal. Robson (1984) found that six of eight suburban Chicago communities contracted all or some tree and stump removals.

Tree removal contracts should follow specifications as described in Chapter 10. Exactly what is to be removed and when should be described in the contract, as should specification for removal. In the case of insect or disease control, the contractor must perform removals in the period specified by the ordinance and/or contract. Disposal of infested or diseased residues should be described specifically in the contract.

TASK SCHEDULING

Task scheduling on an annual basis is designed to meet short-, medium-, and long-term objectives as described in the overall Master Street Tree Plan. Task scheduling must also meet daily objectives of handling emergencies, carrying out work schedules, and efficient allocation of crews and equipment. From an arboricultural perspective there are optimum times to carry out tree management activities, and these times, plus efficient work load allocation, are used to plan seasonal activities. However, it is not always possible to perform tree management activities at the most optimum time or to schedule work with complete efficiency. These two factors influence each other, and compromise is necessary to get the job done.

In temperate climates, deciduous trees are best transplanted in fall or spring, pruned when dormant, and removed anytime (Harris, 1983). However, there are notable exceptions by species to these guides. Some species transplant better in fall or spring, or may not be available when desired. The time of pruning can affect the spread of insects or diseases and the growth rate of trees. Removal for sanitation purposes must be done when the problem is diagnosed and be followed by prompt residue disposal.

Priorities

Priorities for street tree management activities must be established and used as an overall guide to management planning, and to schedule daily work activities. The three major activities in street tree management are planting, pruning, and removal. Removal must be the first priority in a street tree management plan. Dead, dying, and damaged trees present a public hazard for which the city or its employees are libel in the event of damage or injury. Infested or diseased trees must be removed quickly when the method of control involves sanitation.

Pruning and other maintenance activities should be the second priority when planning for budgeting. It is imperative that existing trees be maintained at an acceptable safety and aesthetic level before adding additional trees to the population. Within an overall maintenance program, hazards followed by service requests take higher priority than programmed maintenance; however, programmed maintenance will reduce hazards and service requests. Training young trees is more important than general pruning of mature trees, the reason for a two-level pruning program. A tree trained well when young will demand far less pruning when mature.

The lowest priority in task scheduling and budget allocation should be planting. If there are not enough funds to maintain an existing tree population at professional standards, planting monies should be diverted to maintenance. Adding trees to a population already undermaintained will only compound management problems in the future. Unfortunately, from a political perspective planting will often have a top-priority rating among elected officials. The impact of new trees on a street is easy to see and justify. On the other hand, the effect of pruning and other maintenance activities are more difficult to see, and the negative impact of not pruning or long pruning cycles takes many years to become apparent. The item in a municipal budget most likely to be cut is pruning, and the item least likely to be cut is planting.

Under ideal circumstances, planting, pruning, and other maintenance activities, and removal, will be adequately funded, with elected officials supporting management priorities. Education of public officials and the community is necessary to obtain adequate funding along priority lines. When budgets must be cut, they should be cut following management priorities.

Annual Tasks

Once priorities are established and funding obtained, annual work schedules may be developed. On a daily basis work assignments should attend to priorities as established for public safety, but must also be guided by overall long-term management goals and arboricultural principles. The following is a generalized task schedule for a community in a north temperature climate:

January–February: equipment maintenance and repair, some programmed maintenance on milder days

March–April: programmed maintenance, fertilization, tree planting, lifting in municipal nursery, and early pest management activities

May–June: finish planting, programmed maintenance, begin summer pest surveys

July–August: programmed maintenance, pest control, removals

September–October: programmed maintenance, removals, pest control assessment, fall planting

November–December: programmed maintenance, removals, finish planting, fertilization, equipment maintenance and repair

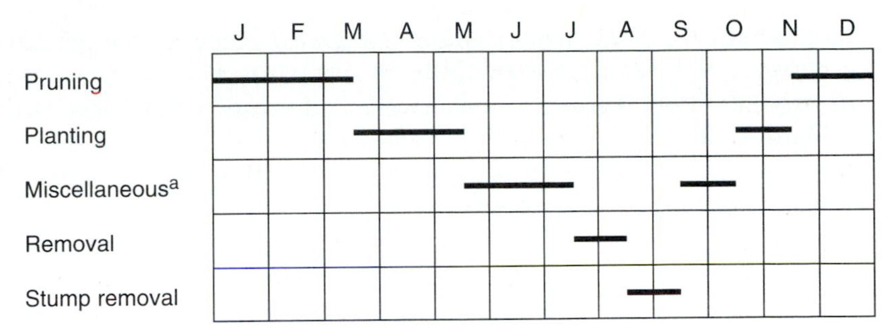

Figure 12–17 Annual task scheduling for the city of Milwaukee Forestry Bureau. (From Ottman, 1986.)

An example of annual task scheduling for the city of Milwaukee Forestry Bureau is presented in Figure 12–17.

LITERATURE CITED

AMERICAN NATIONAL STANDARDS INSTITUTE (ANSI A-300). 1995. *For Tree Care Operations: Tree, Shrub and Other Woody Plant Maintenance Standard Practices.* Washington, D.C.: American National Standards Institute.

BURBAN, L. L., and J. W. ANDRESEN. 1994, *Storms Over the Urban Forest,* 2nd ed. Saint Paul, Minn.: USDA Forest Service, Northeastern Area.

CANNON, W. N., and D. P. WORLEY. 1979. "Dutch Elm Disease Control: Performance and Costs, Update to 1979." USDA-For. Serv. Res. Paper NE-457.

CANNON, W. N., J. H. BARGER, and D. D. WORLEY. 1982. "Dutch Elm Disease Control: Economics of Girdling Diseased Elms to Improve Sanitation Performance." *J. Arbor.,* 8(5):129–135.

CHURACK, P. L., R. W. MILLER, K. OTTMAN, and C. KOVAL. 1994. "Relationship Between Street Tree Diameter Growth and Projected Pruning and Waste Wood Management Costs." *J. Arbor.* 20(4):231–236.

COUNCIL OF TREES AND LANDSCAPE APPRAISERS. 1983. *Guide for Establishing Values of Trees and Other Landscape Plants.* Urbana, Ill.: Int. Soc. Arbor.

FOSTER, R. S., and J. BLAINE. 1978. "Urban Tree Survival: Trees in the Sidewalk." *J. Arbor.* 4(1):14–17.

GILSTRAP, C. 1990. Personal communication. Parks and Recreation Department, Modesto, Calif.

GULICK, J. 1985. "The Elephants of Savannah Ave." *Am. For.* 91(5):34–36.

HARRIS, R. W. 1992. *Arboriculture: Integrated Management of Landscape Trees, Shrubs, and Vines.* Englewood Cliffs, N.J.: Prentice Hall.

HUDSON, B. 1990. Personal communication. Recreation and Parks Department, Santa Maria, Calif.

INTERNATIONAL SOCIETY OF ARBORICULTURE. 1995. *Tree-Pruning Guidelines.* Savoy, Ill.: International Society of Arboriculture.

KIELBASO, J. J. 1989. "City Tree Care Programs: A Status Report." In G. Moll and S. Ebenreck, eds., *Shading Our Cities.* Washington, D.C.: Island Press.

MATHENY, N. P., and J. R. CLARK. 1994. *A Photographic Guide to the Evaluation of Hazard Trees in Urban Areas,* 2nd ed. Savoy, Ill.: International Society of Arboriculture.

MILLER, R. W., and S. P. SCHUMAN. 1981. "Economic Impact of Dutch Elm Disease Control as Determined by Computer Simulation." *Proc. Dutch Elm Dis. Symp. Workshop,* October 5–9, 1981. Winnipeg: Manitoba Dept. Nat. Res., pp. 325–344.

MILLER, R. W., and W. A. SYLVESTER. 1981. "An Economic Evaluation of the Pruning Cycle." *J. Arbor.* 7(4):109–111.

OTTMAN, K. 1986. Personal communication. Bureau of Forestry, Milwaukee, Wisc.

ROBSON, H. L. 1984. "Urban Forestry in the Chicago Suburbs." *J. Arbor.* 10(4):113–116.

SHAW, D. C. 1982. "The Gypsy Moth Problem as a County Arborist Sees It." *Proc. Sec. Natl. Urban For. Conf.* Washington, D.C.: Am. For. Assoc., pp. 282–286.

SHERWOOD, S. C., and D. R. BETTERS. 1981. "Benefit-Cost Analysis of Municipal Dutch Elm Disease Control Programs in Colorado." *J. Arbor.* 7(11):291–298.

SHIGO, A. L. 1983. "Targets for Proper Tree Care." *J. Arbor.* 9(11):285–294.

SHIGO, A. L. 1991. *Modern Arboriculture.* Durham, N.H.: Shigo and Trees Associates.

SISINNI, S. M., W. C. ZIPPERER, and A. G. PLENINGER. 1995. "Impacts from a Major Ice Storm: Street Tree Damage in Rochester, New York." *J. Arbor.* 21(3):156–167.

SMITH, M. A. L., ET AL. 1995. *A Guide to the Plant Health Care Management System.* Savoy, Ill.: International Society of Arboriculture.

TATE, R. L. 1984. "Municipal Tree Management in New Jersey." *J. Arbor.* 10(8):229–233.

TSCHANTZ, B. A., and P. L. SACAMANO. 1994. *Municipal Tree Management in the United States.* Kent, Ohio: Davey Tree Expert Co.

VAN DE WERKEN, H. 1981. "Fertilization and Other Factors Enhancing the Growth Rate of Young Shade Trees." *J. Arbor.* 7(2):33–37.

YAMAMOTO, S. T. 1985. "Programmed Tree Pruning and Public Liability." *J. Arbor.* 11(1):15–17.

13

Program Administration and Analysis

Previous chapters dealt with the planning, establishment, and management of public trees through community forestry programs. This chapter addresses the administration and management of urban forestry programs by focusing on funding, organization, program analysis, personnel administration, and public relations. Although the focus is primarily on municipal forestry, much of what is covered also applies to administration and management of all organizations dealing with urban forests including commercial arboriculture and utility forestry. In all three areas management should stress cost effectiveness, efficiency, and public or customer satisfaction.

FUNDING

Funds for street tree management programs are available from a variety of sources; however, funds for tree planting are generally easier to obtain than funds for maintenance. There are many sources of funding for tree planting because planting trees has an immediate visual impact, gains recognition for the agency, group, or person doing it, and is valued by urban residents. Sources of funds for other street tree management activities are much more limited because of minimal visual impact, low recognition for maintenance, and the fact that justifications are poorly understood by urban residents (VanderWeit and Miller, 1985). Street tree program funds are obtained from local, state, and federal agencies and from private sources. Most of the following sources provide money for tree planting but not for other management activities.

Local Funding

Most funding for the overall management of street trees comes from local government sources. These funds are available because of community support for a street tree program, the need to keep streets clear of obstructions, and concerns for general community welfare and safety. As discussed in Chapter 5, local government exists by permission of the state, and this permission includes the right to levy local taxes. The funding sources described below are generally permitted by most state governments, with the sources requiring special enabling legislation so noted.

General revenues. The sources of general revenue funds is a local property tax. The city forester or tree board submits an annual budget request directly to the city council or through a parent agency, such as a park or public works department. General funds are used for the entire program, including planting, maintenance, and removal. Under normal conditions general revenues are the surest sources of funding for activities beyond tree planting. However, in recent years general funds have declined relative to other sources of funding for municipal forestry programs. Kielbaso (1988) reported that 94 percent of communities received municipal general funds to operate programs in 1986, while Tschantz and Sacamano (1994) report that by 1994 the percentage had declined to 66.6 percent.

Local bonds. Local bonds can be sold by a community to raise capital for a particular need, provided that a majority of voters approve. This source of funding is not normally used to fund tree programs except in the event of a catastrophe that decimates the street tree population, such as a wind or ice storm. Revenues from bonds are used primarily to plant trees.

Capital improvement funds. These funds come from general revenues and are project specific. Capital improvement funds are used to landscape public buildings and parks, or to plant trees as part of a redevelopment project. The city of Milwaukee, Wisconsin, regards trees as an integral part of the urban infrastructure, and includes trees in its road building budget (Figure 13–1). This approach makes the inclusion of trees a minor part of the construction budget (2.2 percent of the entire cost of the project) and shifts the burden of cost for trees onto the Engineering Department (Skiera, 1988).

Permits. The abutting property owner pays for a tree permit and the city plants and maintains a street tree in front of their property. This can be a voluntary or a mandatory program. Permits cover a portion or the entire cost of the tree. Maintenance is provided at no additional special assessment.

Frontage tax. A frontage tax is levied on the street frontage of private property owners, and must have permissive legislation from state government before being implemented. Funds are collected on an annual basis and used for the entire

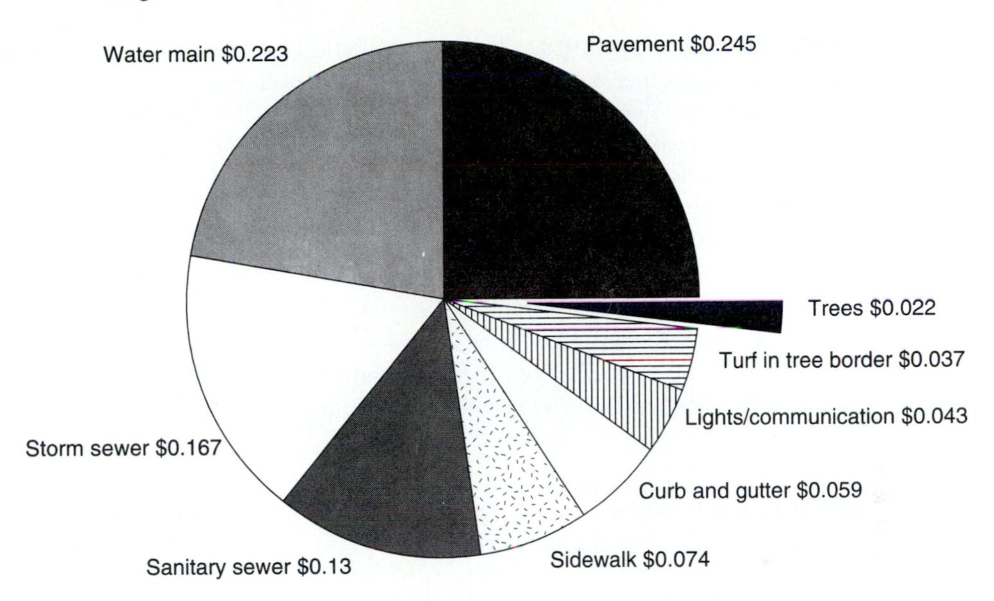

Figure 13–1 Integration of tree planting into the road construction budget in Milwaukee, Wisconsin, on the basis of each dollar spent.

street tree management program. If permitted by state government, this tax may be levied by vote of the city council.

Benefit district assessment. Residents of a particular block or neighborhood may establish a "benefit district" by majority vote of the property owners and tax themselves to pay for a street tree program (Nighswonger, 1982). This tax is usually assessed by frontage and must have permissive legislation. A benefit district tax is used to pay for all street tree management activities. A shortcoming of this method is that neighborhoods with a high proportion of the housing owned by absentee landlords are unlikely to have enough votes to impose the tax.

Tax incremental district. Some states permit cities to designate a portion of the community as a Tax Incremental District for urban redevelopment. Redevelopment bonds are issued and property taxes collected within the district are used to pay off the bonds rather than go into general revenue funds. Part of the redevelopment process may include landscape design and tree planting.

Building permit assessment. A fee is added to each building permit and applied to the purchase and planting of a tree in front of the new building. If planting is not possible, the funds are used to plant trees in another part of the community (Nighswonger, 1982).

Compensatory payments. Utility installation or relocation, street widening, construction, and other damage to public trees can serve as the source of funds for forestry programs. The city of Cincinnati, Ohio, uses the Council of Tree and Landscape Appraisers formula (Chapter 5) to assess the dollar value of public trees damaged or lost to these activities, and receives approximately $15,000 per year in compensatory payments (Sandfort and Runck, 1986). Other communities use a per diameter inch rate as a penalty for damage or destruction of public trees regardless of species, location, or condition.

Subdivision ordinances. Subdivision or landscape ordinances may require developers to plant street trees in new subdivisions (Chapter 9). The builder must meet city specifications for planting and guarantee tree survival for a given period.

Wood residue sales. Revenues generated from chips, firewood, or sawlog sales are used to fund tree planting operations.

State Funding

State funding for urban forestry programs varies from little support to extensive assistance programs. In most states, the funds that are available are administered through the state forestry program, with lesser amounts available from other sources, such as state highway funds.

State forest services. Many state forest services are quite involved with community forestry. State forestry agencies support urban forestry programs by providing technical assistance, research, nursery stock, grants, training, and distributing federal monies specified for community forestry assistance. Some states now designate a portion of their operating budgets specifically for community forestry assistance.

State highway funds. Highway funds are generated by gasoline taxes and must be invested in transportation. A portion of these funds is available for highway beautification, including landscaping in transportation corridors. Municipalities have access to state highway funds for maintenance of urban streets, boulevards, and parkways, and in some states this includes landscaping during construction and reconstruction.

Emergency and special funds. Wind and ice storms and other natural disasters can cause extensive damage to community tree populations. State emergency funds are available to assist with cleanup operations and sometimes are available for replanting lost trees. Special funding has been made available to communities in some states with severe Dutch elm disease. From 1975 through 1982 the state of

Minnesota provided matching grants in excess of $60 million to assist communities in controlling the disease and replanting city streets (Willeke, 1982).

Federal Funding

There are a variety of funds available for urban forestry from federal agencies, especially for tree planting. The greatest source of these funds is from agencies involved in community redevelopment and economic assistance. A complete list of federal funds available to communities is the Catalog of Federal Domestic Assistance, published annually by the Office of Management and Budget, available from the Superintendent of Documents (Tate, 1982). The following is a brief description of some federal funding sources available for urban vegetation.

Cooperative forestry. The U.S. Forest Service distributes cooperative forest management funds to state forestry agencies through the state and private forestry program. A portion of these funds is designated for distribution to communities by state forestry agencies through the community forestry assistance program. Most of these funds are given to communities under a matching grant program on an annual basis. Occasionally, other funds are made available through cooperative forestry. In the late 1970s funds were made available through state and private forestry for the Dutch elm disease demonstration project in cooperation with several state forestry agencies and communities.

Community development block grants. The Community Development Agency within Housing and Urban Development administers the community development block grant program—long-term grants designed to assist communities with urban renewal within deprived portions of the city. Communities may spend a portion of these grants for urban tree planting in areas designated as eligible for funding.

Economic development grants. The Economic Development Agency of the Department of Commerce administers a grant program designated for redevelopment of city housing. These grants are made on an annual basis, and portions of these funds may be used for landscape design and tree planting as part of development efforts.

Federal highway funds. Federal highway funds are collected via a federal gasoline tax. These funds are available to states to maintain state and national highways. These monies can also be used for landscaping and beautification of all transportation corridors including bikeways and walking paths.

Emergency and other federal funds. As with state emergency funds, communities may be designated a disaster area by the federal government. If so designated, low-interest loans are available to the community to assist in cleanup efforts and for reconstruction. Loan monies may be used for replanting areas where trees have been lost due to the disaster. Other sources of federal funding for urban trees include resource conservation and development funds.

Private Sources of Funds

Funds for street trees are available from a variety of private sources. Civic and service organizations look for projects that benefit the community. Tree planting has high visibility and can be accomplished in a short period of time, making it an ideal service project. Local businesses and organizations will often contribute to tree planting funds for community beautification. Merchants' associations in many communities are willing to fund tree planting in commercial districts to make them more attractive to shoppers (Figure 13–2). Neighborhood associations have accomplished the same on residential streets. Tree trusts can be established in communities and interest earned on private donations used to manage street trees.

Funding Strategies

Funding strategies for a particular community will be based on local characteristics and funding sources. However, in all cases management priorities must guide the street tree management program. It is not good management to plant more trees than can properly be maintained just because short-term money is available. Most sources described above provide money on a one-time basis, primarily for tree planting and other landscaping activities. Funds that are available on a recurring annual basis are the backbone of an urban forestry program. These funds come from general revenues, special taxes designated for street tree management, permits, and tree trusts. Other sources are useful and sometimes necessary to establish street trees, but each new tree will increase management costs over time, and budget increases will be necessary to provide proper maintenance.

City foresters should be involved in the planning and design of urban landscapes that include vegetation to be maintained by forestry personnel. A poorly designed landscape lacking sufficient maintenance funds will quickly become a

Figure 13–2 These trees in Stuttgart, Germany, were planted by the local business association to enhance the quality of shopping on this commercial street.

community liability rather than an asset. Such areas evoke negative public opinion and could affect future planting programs. Landscapes that demand high maintenance are fine but must have a future budgetary commitment to pay for that maintenance. If budgetary commitments are not made, the landscape should be redesigned to fit the current maintenance budget. Similarly, poor species selection can greatly increase maintenance costs even though the basic landscape design can be efficiently maintained. To keep future maintenance costs within reasonable limits, city forestry personnel must be involved in species selection as well as design.

ORGANIZATION

Most public agencies and private businesses are based on the classical Weberian vertical or line and staff organization (Figure 13–3). Hudson (1983) describes the following as six general functions of vertical organizations:

1. Provides for formal lines of communication both upward and downward
2. Facilitates the control of resources and activities
3. Allows for the collection of unitized operational data

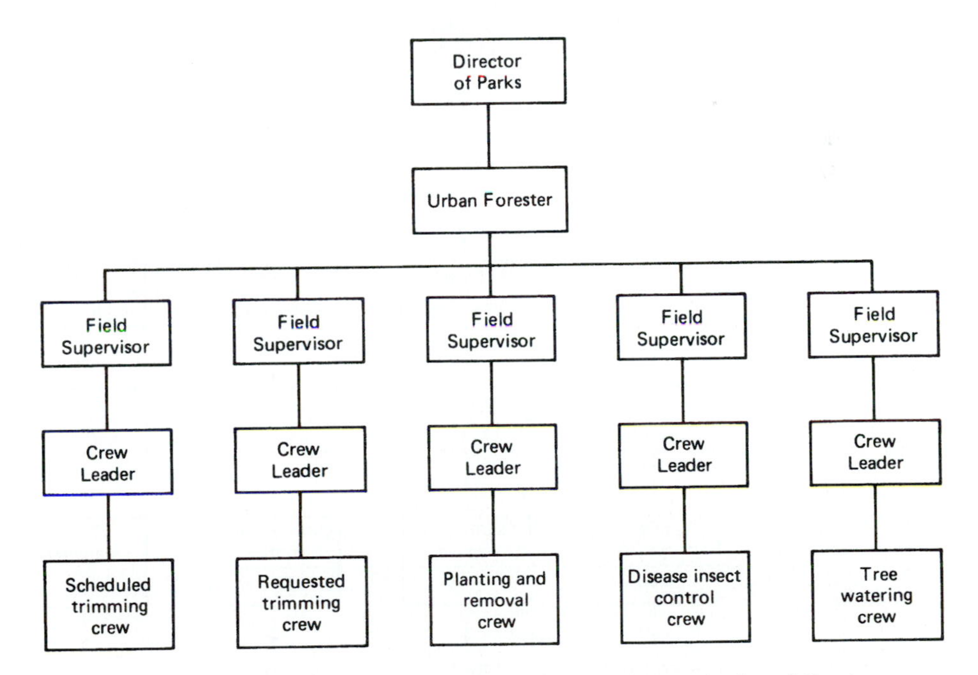

Figure 13–3 Typical municipal forestry department organization, following the classical vertical organization structure. This structure provides for interval analysis only and appears to be self-serving. (From Hudson, 1983.)

4. Controls the behavior of employees
5. Facilitates the division of labor
6. Encourages internal organizational analysis

Vertical organizations serve these goals very well, but Hudson suggests that they also create tunnel vision in municipal forestry programs because they:

1. Do not encourage planning and analysis outside the organization to determine service user needs
2. Do not identify where labor and other resources should be tapped across organizational lines for major in-house projects
3. Do not reflect public benefits

Hudson further suggests that municipal forestry departments complement their structure with the establishment of an internal horizontal structure designed to relate external program concerns to public needs and benefits (Hudson, 1983).

Horizontal structure within the classical vertical organization is important to the private as well as the public sector (Figure 13–4). An understanding of the goals and

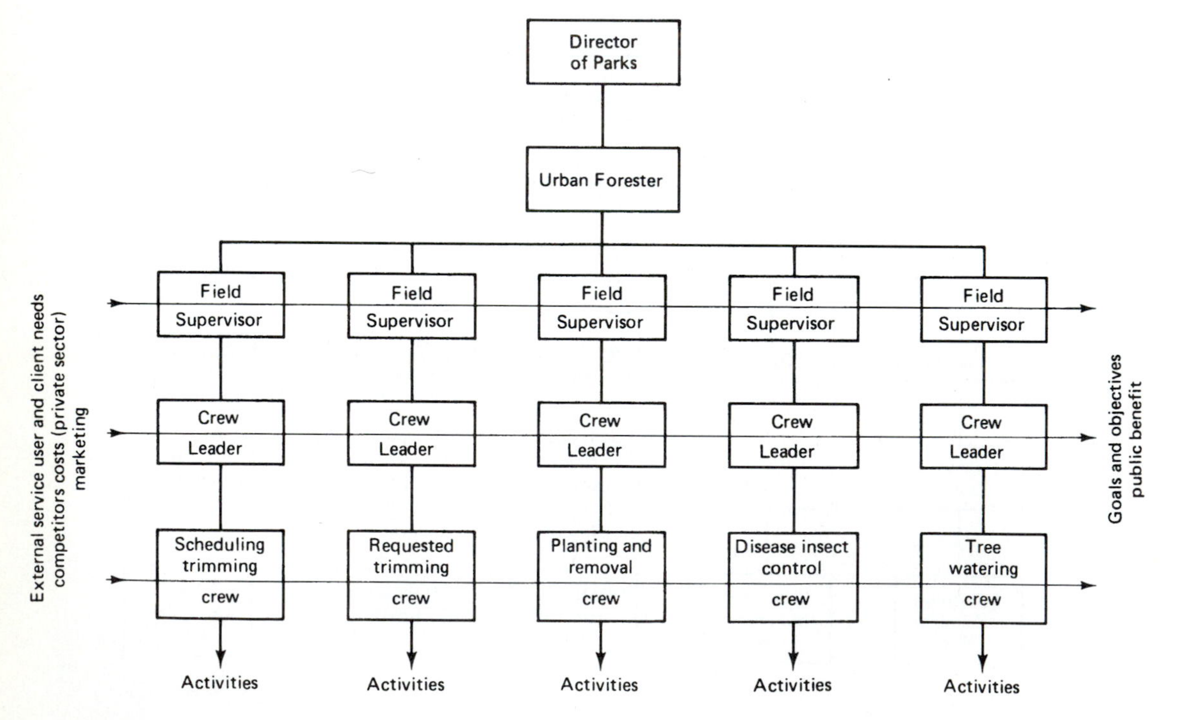

Figure 13–4 Same as Figure 13–3, but provides for both internal and external environment analysis. (From Hudson, 1983).

objectives of the entire organization and sensitivity to public (customer) needs and concerns by all employees will contribute to a more efficient and, in the private sector, a more profitable organization. Established lines of internal communication will assist in resolving employee differences and will allow for more efficient employee and equipment allocation.

Organizational charts can be used to describe the general functions of personnel and establish lines of authority and responsibility, but when rigidly applied can also be counterproductive. Job descriptions in organizational charts are often written for the person occupying that slot at the time of development and may not be relevant when personnel changes take place. The organization can be described, but it must be flexible to allow people to develop their particular skills and talents, which in the long run will better serve the company or agency (Peters and Waterman, 1982).

Municipal Forestry Program Organization

Municipalities are governed by two predominant forces, the legislative branch (city council) and the executive branch (mayor or city manager). The executive branch in most communities is either an elected mayor who is independent of the city council or a city manager who is appointed by the city council. The legislative branch (city council) consists of elected representatives who serve as the primary policymaking body, including setting tax rates, passing ordinances, and approving budgets (Florida Division of Forestry, 1971).

Community forestry programs can fit into municipal government organizations in a number of ways. In many cities an independent tree or park board exists to serve in an advisory or policymaking capacity. Park boards serve forestry programs which are independent entities or are located in a different municipal department. Advisory boards are appointed by elected officials to serve in an advisory capacity concerning matters of policy and budgets. Elected boards or commissions function independent of the city commission and have the power to set policy, budgets, and levy taxes (Minnesota Department of Agriculture, 1982).

The organizational location of city forestry programs varies greatly. Typical locations will be within a parks department, public works department, streets and sanitation department, or an independent forestry department. Johnson (1982) conducted an analysis of twelve urban forestry programs representing six geographic regions in the United States. He found a variety of organizational locations for these programs. Location in the city parks and recreation department was found to be most compatible with the goals of the forestry programs surveyed (Figure 13–5), but forestry agencies located within public works or a similar service department had more resources and were more viable (Figure 13–6). This greater viability is related to access to more personnel and equipment when needed. Johnson (1982) further reports that it did not appear bureaucratically feasible to establish autonomous forestry agencies, although he did find some successful autonomous operations.

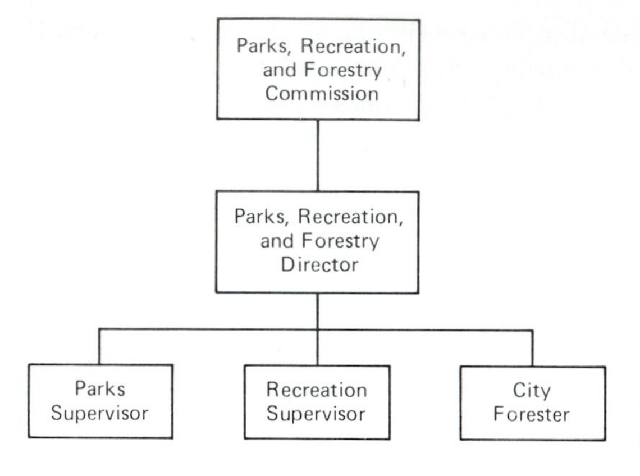

Figure 13–5 Municipal forestry department within a parks department that reports to a park commission.

Beyond the management of street trees, city forestry programs can have a wide variety of management responsibilities. A 1994 survey of municipal forestry programs in the United States found nineteen other areas of responsibilities in the survey cities (Table 13–1).

Beyond the Organization

Wellman and Tipple (1989) interviewed administrators of ten highly regarded and successful urban forestry programs and describe a number of characteristics common to all ten programs. First and foremost, they found that technical competence is a necessary, but not the only, component of successful programs. Programs that are highly regarded by their communities link with larger social values such as environmental concerns, quality of community life, and community image. Likewise these programs work across administrative boundaries to include other departments in decision making, and their administrators have the ability to adjust to changes when elections brought in different mayors and city councils with different ideologies.

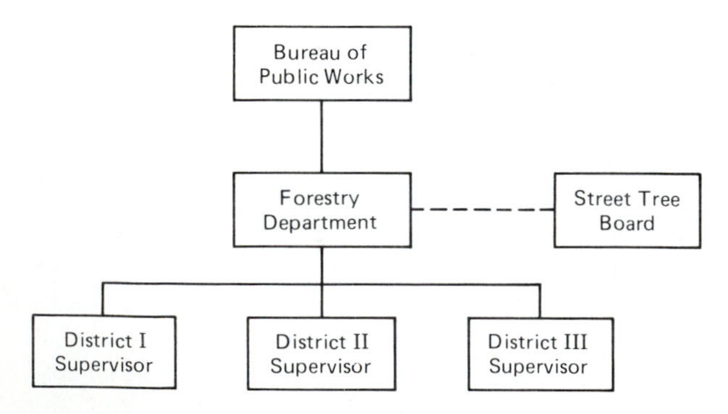

Figure 13–6 Forestry department within a public works bureau and served by an advisory board.

TABLE 13–1 ADDITIONAL RESPONSIBILITIES OF DEPARTMENTS RESPONSIBLE FOR TREE MANAGEMENT. (From Tschantz and Sacamano, 1994.)

Responsibility	Percentage of Respondents
Parks	51.7
Recreation Area Maintenance	28.2
Other	25.1
Landscaping	21.5
Streets	21.2
Water Control	16.8
Building Maintenance	15.4
Turf Maintenance	14.8
Median/Roadway/Right-of-Way Maintenance	13.7
General Maintenance	10.3
Cemeteries	10.1
Administration	9.5
Sewers	7.0
Special Programs	7.0
New Construction	5.9
Sanitation	5.9
Holiday Decorating	3.6
Snow Removal	3.1
Line Clearance	0.8

Skiera (1994) recommends city forestry personnel understand and do the following relative to the politics of urban forestry:

1. The political planning horizon is only as long as the time between elections.
2. The effectiveness of the forestry program is related to budget approval.
3. Keep policy makers informed—no surprises.
4. Spend time with newly elected officials to teach them about urban forestry.
5. Spend time in the field with political management.
6. Stay ahead of actions with news releases.
7. Use graphics when making budget requests.
8. Work with citizens groups.
9. Sell the forestry program at every opportunity, such as Arbor Day.
10. Don't make bureaucratic or political enemies.
11. Take advantage of significant events: storms, diseases, and so on.
12. Don't be a bad loser—live to fight another day.
13. Be prompt and professional at all times.
14. Develop a reputation of trust and follow-through.

PROGRAM ANALYSIS

There are a variety of methods available to assist in decision making and to analyze program effectiveness. This section deals with cost analysis, performance standards, benefit/cost analysis, and internal rate of return as tools to aid in decision making. It also explores budget forecasting and allocations.

Cost Analysis

Municipal forestry differs from commercial arboriculture by not being profit oriented, but is similar in that it must satisfy the client (taxpayers). Hudson (1983) suggests that while municipal forestry is not profit oriented, it can benefit from the application of business analogies to increase efficiency. The first step is to analyze internal operating expenses to determine precisely what each activity costs and to identify potential savings. These costs fall into two categories, overhead and fringe benefits.

1. Overhead costs
   a. Management
   b. Rent and utilities
   c. Motorized equipment
   d. Hand tools
   e. Material
   f. Training
   g. Uniforms
   h. Wages
   i. Marketing and advertisement (information and education)
2. Fringe benefits
   a. Health insurance
   b. Life insurance
   c. Retirement
   d. Dental insurance
   e. Long-term disability insurance
   f. Vacation hours
   g. Paid holidays
   h. Sick leave

Daily operation records describing labor and equipment hours on each job will then provide an assessment of where budgets are actually being spent and will identify where inefficiencies exist (Hudson, 1983).

An efficiency study of the forestry division in the city of Grand Rapids, Michigan, followed a similar procedure (Overbeek, 1979). Crews were asked to keep daily records of work location, work performed, labor and equipment records, and units of work completed. Five variables were manipulated to determine their effect on crew efficiency: crew size, crew makeup or personality, equipment allocations, job routing, and type of work performed. All costs, including overhead and fringe benefits and variables, were subjected to analysis by a computerized cost accounting system to determine the most efficient allocation of these variables. At the end of a two-year study the following results were obtained.

1. Tree-trimming operations increased by 11 percent and costs decreased by 9.5 percent.
2. Tree removals increased by 21.7 percent, with a cost reduction of 4 percent.
3. Tree planting operations increased by 52.2 percent at a cost per unit reduction of 4.7 percent.
4. The number of trees maintained increased by nearly 100 percent with the addition of only one person and an aerial tower. The cost per work unit dropped 11.4 percent.
5. Root cutting increased by 19.6 percent at a cost reduction of 66.3 percent.

It can be useful to determine the cost of tree damage from construction or other activities when arguing for a greater degree of tree protection. A study of construction damage to street trees in Milwaukee revealed that replacement of curbs and sidewalks, as well as street widening, caused significantly higher mortality and lowered tree condition ratings when compared to trees on streets where there was no construction (Hauer et al., 1994). The city has a population of 200,000 street trees with an appraised value $220 million, calculated using the Council of Tree and Landscape Appraisers formula. Approximately 6,000 trees valued at $6.6 million are damaged per year. Tree damage results in a loss of 6.1 percent in the condition rating of these trees for a value loss of $521,500 per year. Mortality rates for damaged trees exceeds normal mortality by 4.1 percent for an additional loss of $270,600. Assuming damage to 6,000 trees per year, the annual loss to street and sidewalk reconstruction in Milwaukee is $792,100 ($521,500 + 270,600).

Cost analysis can also be applied to individual trees or species to determine if expected costs justify their continued use, or if it is best to replace them. Tulip trees (*Liriodendron tulipifera*) have not proved adaptable to climatic and site conditions in the city of Berkeley, California, yielding high maintenance costs. The decision to replace 400 tulip trees with 400 London plane trees (*Platanus acerifolia*) was based on an anticipated twenty-year cost for keeping the tulip trees of $354,000 compared to a twenty-year cost of $223,000 to replace them (Dreistadt and Dahlsten, 1986).

Performance Standards

Performance standards can be developed by keeping records of crew activities relative to jobs performed (Figure 13–7). The number of person hours needed to plant different size or types of tree transplants, or the number of hours needed to prune dif-

Daily Work Report
Division of Parks and Forestry

Page 1 of ___

Date: ___

Crew Members	Time Reg	Time O.T.
Crewleader:		
Crew:		

Equipment Trucks:	Saws:

Break Time	Start	Finish
A.M.		
Lunch		
Mileage		
P.M.		

CREWLEADER SIGNATURE

SUPERVISOR SIGNATURE

Ass. #	Serial Number Assignment Description	Work Code*	Assignment Time Arrive	Assignment Time Depart	Clock Hours	Crew Hours	Mileage Arrive	Travel Time Totals
1								
2								
3								
4								
Page 1 Total								

*Work Code Explanation:

1 Street Trim
2 Park Trim
3 Alley Trim
4 Street Removal
5 Park Removal
6 Alley Removal
7 Stump Removal
8 Pick-Ups
9 Hangers
10 Woodlot
11 Storm Dmg. Reg.
12 Storm Dmg. OT
13 Snow Reg.
14 Snow OT
15 Tree Planting
16 Nursery Maintenance
17 Elmdale Time (Cleanup Inventory)

Special Projects

18 Banners
19 Wood Delivery
20 Brush Removal
21 Paintings/Street Light Removal
22 Equipment Downtime (Breakdowns)
23 Downtime
24 Equipment Pickup
25 Meetings
26 Training
27 Benches
28 Newplant Maint.
29 Planting Preparation

Figure 13–7 Work card used by the Toledo, Ohio, Division of Parks and Forestry. (From O'Brien and Joehlin, 1992.)

ferent species and/or sizes of trees can serve as standards to evaluate crew and individual performance. Performance standards can also be used to evaluate seasonal differences in work activities.

O'Brien et al. (1992) used a database of 6,272 work records from the Forestry Division of Toledo, Ohio, to develop performance standards for tree care operations. They found that diameter class significantly influenced time spent both pruning and removing trees and stumps, but that species of tree had no influence. They also found significant seasonal differences in tree and stump removal. Tree removal hours were higher in the spring and summer, especially for trees larger than 12 in. (30 cm) in diameter. Stump removal is more expensive in the winter due to frozen ground and a higher rate of equipment failure.

Once performance standards have been developed, then the level of service can be measured by expected outcomes. If the goal is to increase stocking of street trees by a given number of trees each year, or to shift to a shorter pruning cycle, we can prepare budgets to meet those goals. Budgets that describe a range of outcomes and citizen satisfaction based on different funding levels can be presented to policy makers to enable them to choose the level of service they desire for the public (Cole, 1993).

Benefit/Cost Analysis

Summarizing the expected costs associated with maintaining trees over time is useful, but ignores the effect of time on money. Money is generally worth more at the present than in the future, and that is why people are compensated with interest payments for deferring to use money in the present. Likewise money in the future is worth less, and to determine its present value it must be discounted at the prevailing interest rate back to the present time. If we wish to compare money spent or received at different points in time, we must discount the money to a common point in time to make a valid comparison. The formula used to calculate the present value (PV) of a future value (FV) of money is:

$$PV = \frac{FV}{(1+r)^n}$$

where n = the number of years in the future
r = discount rate

If we wish to compare the present cost of something with expected future benefits, we can discount the future benefits back to the present to see if the expenditure is justified. If the present value of the future benefit exceeds the present cost, then the expense can be considered justifiable. When we divide the present benefit by the present cost we have what economists describe as the benefit/cost (B/C) ratio. If the ratio exceeds one, then it is considered a wise expenditure, but if it is less than one, then the cost cannot be justified by the benefit. For example, is it a wise investment to install a root barrier to protect a sidewalk from lifting by tree roots for a cost of $50 when we plant the tree, if it saves us $50 for root pruning in year 15 and $200 for sidewalk repair in year 25? Using a 5 percent discount rate, the present value of $50 savings in year 15 is $24, and the present value of the $200 sidewalk repair in year 25 is $60. The total benefit of $84 divided by the cost of $50 yields a B/C ratio of 1.68, a favorable ratio. Benefit/cost analysis can be further used to compare alternative management strategies so the most economically rational approach can be selected (Schwarz and Wagar, 1987).

Sherwood and Betters (1981) used benefit/cost analysis to select the most cost-effective Dutch elm disease (*Ceratocystis ulmi*) control alternative for several communities in the state of Colorado. They used the Council of Tree and Landscape Appraisal (CTLA) method to determine the present value (benefit) of American elms (*Ulmus americana*) saved, and discounted future control and removal costs. They found the most favorable B/C ratio with the most intensive control alternative (Table 13–2).

TABLE 13–2 BENEFIT/COST ANALYSIS OF THREE DUTCH ELM DISEASE (*CERATOCYSTIS ULMI*) CONTROL SCENARIOS. (From Sherwood and Betters, 1981.)

Alternative	Total discounted cost	Trees saved	Total discounted benefits	B/C ratio
B	$ 15,328	8	$ 18,970	1.24
C	60,416	54	95,880	1.62
D	128,276	152	309,577	3.41

A more extensive benefit/cost analysis was conducted as part of the USDA Forest Service's Chicago Urban Forest Climate Project (McPherson, 1994). A computer model was developed to compare a thirty-year stream of benefits and costs associated with planting 95,000 trees on streets, highways, parks, residential yards, and public housing sites. Costs included planting, pruning, removal, waste disposal, inspection, infrastructure repair, and program administration (Table 13–3). Benefits included energy savings, improved air quality, avoided and sequestered carbon dioxide, reduced water runoff, water saved in energy generation, and other benefits (Table 13–4). The other benefits value was calculated using the CTLA formula to calculate the annual incremental tree value minus a value equivalent to all the benefits from energy savings, sequestered and avoided carbon dioxide, and so on. Positive benefit/cost ratios were obtained for discount rates of 4, 7, and 10 percent. Projected benefit/cost ratios using a 7 percent discount rate were largest for residential and public housing sites (3.5) and least for parks (2.1) and highways (2.3). Discounted payback periods for tree planting and maintenance ranged from nine to fifteen years, depending on the planting site.

Internal Rate of Return

Selecting the appropriate discount or interest rate for benefit-cost analysis can be difficult as there is no way to predict what the actual rates will be in the future. An alternative method of analysis is to use the internal rate of return. This method calculates the interest or discount rate that equalizes the present value of both costs and benefits in the following manner:

$$\frac{\text{Present Value of Benefits}}{\text{Present Value of Costs}} = 1$$

To determine if a present expenditure will result in annual savings (n) years into the future the formula will be

$$\frac{\frac{\text{annual savings }((1+r)^n-1)}{r(1+r)^n}}{\text{initial investment}}=1$$

where (r) is the discount rate. Multiple calculations are made using a variety of discount rates until the numerator is equal to the denominator, at which point that rate is the internal rate of return. Different formulas must be used if future annual benefits are to be discounted or if future benefits and costs are periodic rather than annual. Internal rate of return is time consuming for hand calculation, but there are a number of

TABLE 13–3 PROJECTED PRESENT VALUE OF COSTS FOR TREE PLANTINGS IN CHICAGO (30-year analysis, 7 percent discount rate, in thousands of dollars). (From McPherson, 1994.)

	Tree Location					
Cost Category	Park	Yard	Street	Highway	Housing	Total
Planting[a]	4,258	5,484	7,107	1,097	272	18,218
Pruning[b]	346	192	585	75	57	1,255
Removal[c]						
Tree	221	105	547	18	36	927
Stump	27	15	90	9	4	145
Subtotal	248	120	637	27	40	1,072
Tree waste disposal[d]	31	0	0	0	0	31
Inspection[e]	3	0	13	0	1	17
Infrastructure repair[f]						
Sewer/water	3	14	8	0	1	26
Sidewalk/curb	5	7	27	1	1	41
Subtotal	8	21	35	1	2	67
Liability/litigation[g]	0	6	11	1	0	18
Program administration[h]	15	0	0	13	87	115
Total	4,909	5,823	8,388	1,214	459	20,793

[a]Reported cost of trees, site preparation, planting, and initial watering.
[b]Reported cost of standard Class II pruning. Pruning frequency varied by location.
[c]Reported cost of tree and stump removal. Frequency of removals varied by location.
[d]Tree waste disposal fee $40/ton. Value of wood waste recycled as compost and mulch assumed to offset recycling costs where no net cost shown.
[e]Reported labor and material costs for systematic tree inspection.
[f]Cost of infrastructure repair due to damage from tree roots assumed to vary by location.
[g]Cost of litigation/liability as reported or based on data from other cities when unavailable.
[h]Salaries of administrative personnel and other program administration expenditures. Administrative costs were incorporated in other reported costs for residential street trees.

TABLE 13–4 PROJECTED PRESENT VALUE OF BENEFITS FOR TREE PLANTINGS IN CHICAGO (30-year analysis, 7 percent discount rate, in thousands of dollars). (From McPherson, 1994.)

	Tree Location					
Benefit Category	Park	Yard	Street	Highway	Housing	Total
Energy[a]						
Shade	233	984	1,184	91	75	2,567
ET cooling	340	1,296	1,676	135	105	3,552
Wind reduction	1,479	5,648	7,302	586	457	15,472
Subtotal	2,052	7,928	10,162	812	637	21,591
Air quality[b]						
PM10	8	11	11	2	1	33
Ozone	1	2	1	0	0	4
Nitrogen dioxide	8	19	18	2	2	49
Sulfur dioxide	8	23	21	2	2	56
Carbon monoxide	1	1	1	0	0	3
Subtotal	26	56	52	6	5	145
Carbon dioxide[c]						
Sequestered	37	65	82	12	5	201
Avoided	92	359	465	37	27	980
Subtotal	129	424	547	49	32	1,181
Hydrologic[d]						
Runoff avoided	46	170	494	24	15	749
Saved at power plant	6	26	32	3	2	69
Subtotal	52	196	526	27	17	818
Other benefits[e]	8,242	11,854	12,262	1,926	923	35,207
Total	10,501	20,458	23,549	2,820	1,614	58,942

[a]Net heating and cooling savings estimated using Chicago weather data and utility prices of $0.12 per kWh and $5 per MBtu. Heating costs due to winter shade from trees are included in this analysis.
[b]Implied values calculated using traditional costs of pollution control.
[c]Implied values calculated using traditional costs of control ($0.11/lb) and carbon emission rates of 0.11 lb/kWh and 29.9 lb per MBtu.
[d]Implied values calculated using typical retention/detention basin costs for stormwater runoff control ($0.02/gal) and potable water cost of ($0.00175/gal) for avoided power plant water consumption.
[e]Based on tree replacement costs (CTLA value—other benefits).

computer programs available that allow for instant calculation (Schwarz and Wagar, 1987).

Internal rate of return allows the manager to decide if the return rate is high enough to warrant the investment. This can be done by comparing the discount rate with current interest rates or with alternative management plans. For example, you are a utility forester and are considering a program in which your company will remove large trees interfering with electric lines and replace them with trees that have a small mature height. This will cost your company $350 per tree, but will save $85 every four years in pruning costs per tree. The replacement tree is expected to last thirty-two years. Is this a good investment when compared to an alternative rate of return of 3.5 percent? The internal rate of return for the program is 4.1 percent, making it an attractive investment.

A volunteer organization, Trees for Tucson/Global ReLeaf, established a goal in 1989 of planting 500,000 desert-adapted trees in Tucson, Arizona. McPherson (1991) developed a case study to compare annual costs of establishing and maintaining these trees over forty years to expected annual benefits from energy conservation, dust interception, and runoff reduction. He found that costs exceeded benefits for the first five years, but that benefits exceeded costs in the next twenty-five years. During the final ten years costs begin to catch up to benefits as more large trees die and are removed (Figure 13–8). The internal rate of return for the entire project was calculated at 7.1 percent, with the return for residential trees being 14 percent, park trees at 5 percent, and street trees at 2 percent.

Forecasting Work Loads

Chapter 12 discussed forecasting future pruning costs based on the average growth rate of an entire street tree population. A computer program, CityTrees: A Street Tree Management Simulator (Appendix B), can further predict future management costs by accepting an existing inventory and management costs for planting, pruning, and removal of street trees, and simulating management of that population over time.

Wagar (1991) suggests using computer records of workloads gathered by crews performing maintenance and inventory updating to develop performance profiles of species by diameter class. Records of time spent to prune a species by diameter class can be summarized to determine the hours per size class, and this information, coupled with anticipated growth rate, can be used to predict size distributions by species in the future for budget planning purposes.

The city of Milwaukee Bureau of Forestry has made a substantial change in its pruning cycle as the result of changes in the tree population; shifting from a five-year cycle to a two-stage cycle of three/six years (Griffith and Associates, 1993). From 1980 to 1992 the mean diameter of the tree population increased by 2 in. (5 cm) and the number of street trees increased by 26,000 (Table 13–5). For more than a decade the city was on a five-year pruning cycle, but recent years has seen the cycle gradually increase to six years, partially the result of more trees, tree growth, and reductions in the number of arborists. The twelve-year period has also seen an increase in the number of out-of-cycle request prunes. An analysis of requests for

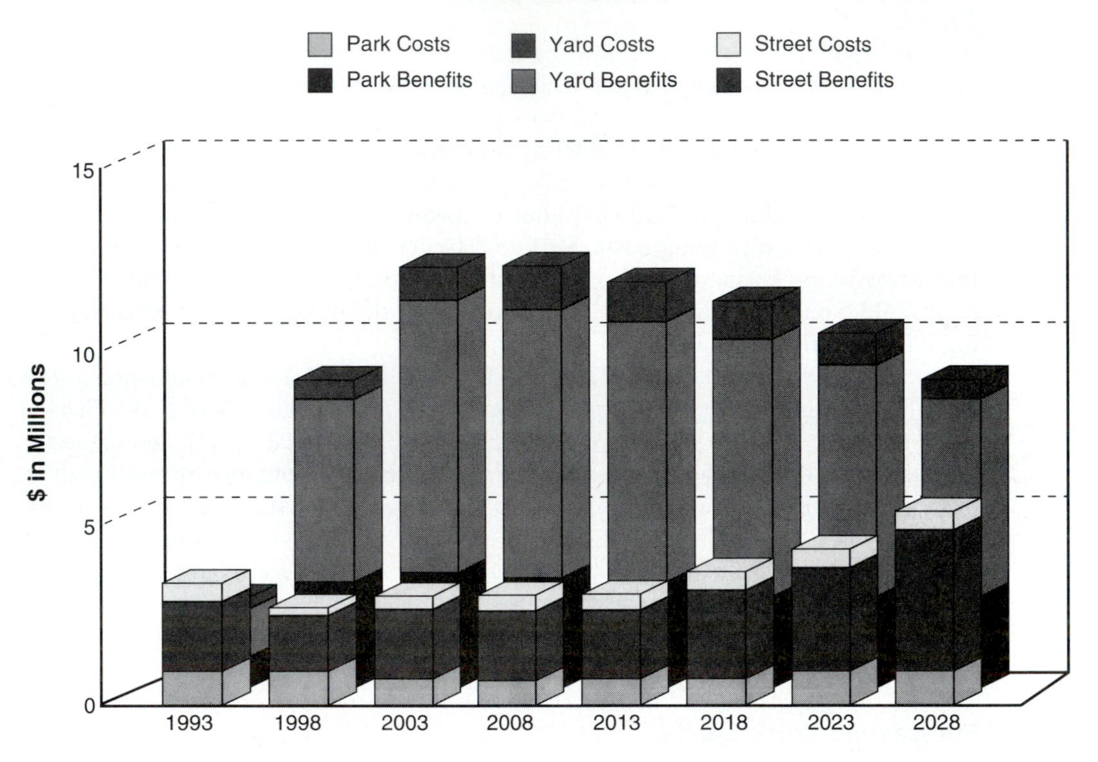

Figure 13–8 Projected average annual benefits and costs for trees planted in park, yard, and residential street locations. (From McPherson, 1991.)

TABLE 13–5 CHANGES IN THE MILWAUKEE TREE POPULATION AND MANAGEMENT, 1980–1992. (From Griffith and Assoc., 1993.)

	1980	1992	Change
Number of Trees	174,000	200,000	+26,000
Value	$202 million	$280 million	+$78 million
Pruning Cycle	5 years	6 years	+1 year
% Request Pruning	22%	26%	+4%
Mean Diameter	8 in. (20 cm)	10 in. (24 cm)	+2 inc. (5 cm)
Planting Survival (18 mo.)	70%	83%	+13%
Planting		3,600 trees	
Removals		3,490 trees	

pruning indicated most requests were in the 10 to 14 in. (25 to 35 cm) diameter classes, indicating that rapid crown expansion in these classes were demanding more frequent pruning than the five-year cycle (Figure 13–9). A profile of the population by diameter class revealed many trees grew into these classes over the twelve-year period, and the potential for a much larger number of trees in these diameter classes in the next decade (Figure 13–10).

Several pruning scenarios were examined to determine if smaller trees could receive more frequent pruning for training and if problem trees could be pruned more frequently to reduce the number of request prunes. Using a timed pruning study in Milwaukee by Churack et al. (1994), pruning hours per year were calculated for five- and six-year cycles, and for three/six-year and four/eight-year cycles. The two staged cycles involve pruning all trees 12 in. (30 cm) and smaller on the short cycle and all trees on the long cycle. Table 13–6 summarizes the estimated hours spent annually per cycle. Based on this analysis it appears that the six-year cycle is most cost effective. However, since request prunes are more than twice as expensive as cycle prunes (see Chapter 12), a different cycle that reduces service requests might be more cost effective. For example, in 1992, 26 percent of all trees pruned were request prunes.

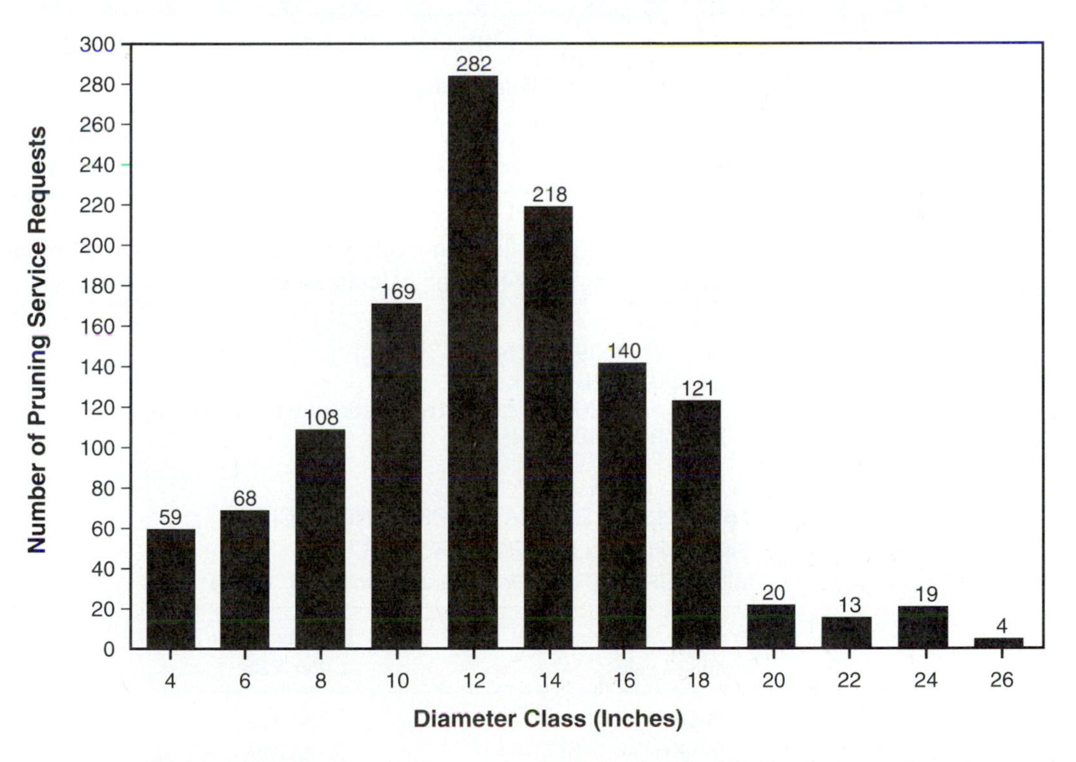

Figure 13–9 Pruning service requests in Milwaukee for 1992 by diameter class. (From Griffith and Associates, 1993.)

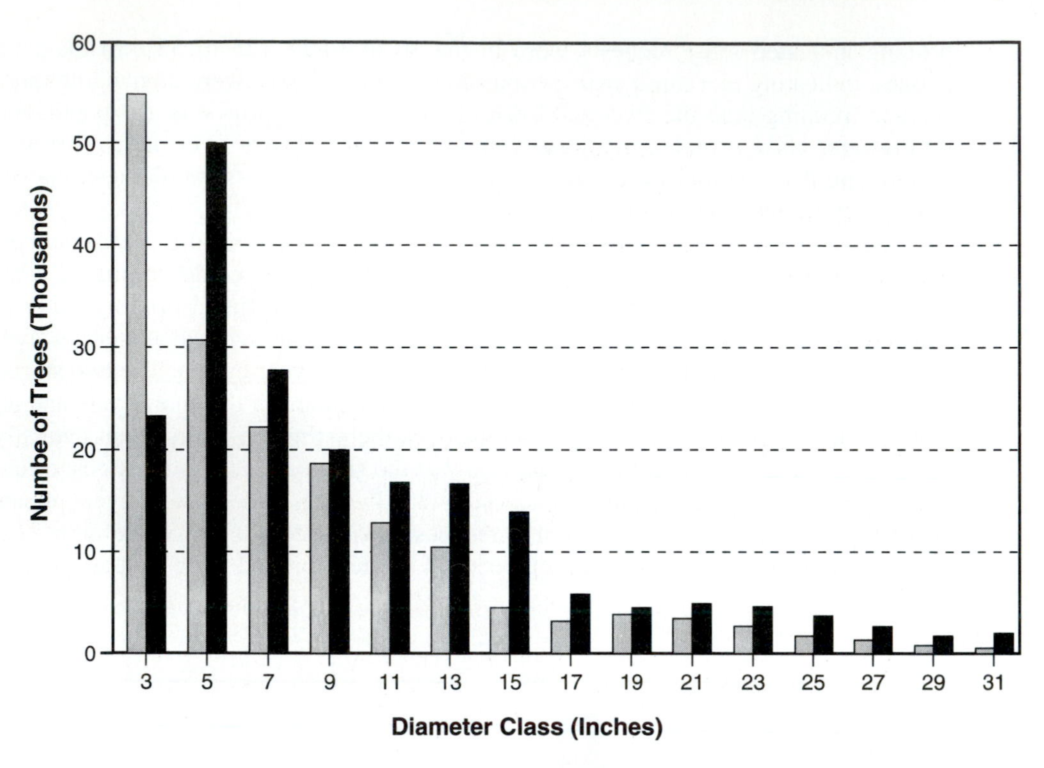

Figure 13–10 Diameter distribution of street trees in Milwaukee, 1980 and 1992. (From Griffith and Associates, 1993.)

Applying 74 percent (100 – 26) to the six-year cycle yields a total time for all pruning of 49,100 hours per year using the following calculations:

$$
\begin{aligned}
&28{,}690/.74 &&= 39{,}040 \text{ hours}\\
&39{,}040 - 28{,}890 &&= 10{,}150 \text{ hours}\\
&10{,}150 \times 2 &&= 20{,}300 \text{ hours (requests take twice as long)}\\
&28{,}890 + 20{,}300 &&= 49{,}190 \text{ hours}
\end{aligned}
$$

TABLE 13–6 ESTIMATED PRUNING CYCLE SCENARIOS IN HOURS PER YEAR FOR MILWAUKEE. (From Griffith and Associates, 1993.)

Pruning Cycle	Pruning Hours Per Year
5-year	34,428
6-year	28,890
3- and 6-year	43,130
4- and 8-year	32,356

The total of 49,190 is very close to the actual number of hours spent pruning as reported by the tree division in 1992. If we conservatively assume that a two-stage cycle reduces pruning requests by 50 percent, then the total time spent pruning for the four pruning scenarios is presented in Table 13–7. The three/six year cycle is close in cost to the six-year cycle, will do a better job of training young trees, and will reduce the number of request prunes. It is likely that the cost will be even less than anticipated because a request prune often costs more than two times the cost of a cycle prune, and that there will likely be more than a 50 percent reduction in request prunes.

Budget Allocations

Maintaining a population of street trees involves following a Master Street Tree Plan and planting, maintaining, and removing/replacing trees. One way to assess the effectiveness of a program is to determine the percentage of funds allocated to these various activities. Kielbaso (1988) in a 1986 survey of municipal forestry programs in the United States compares the actual allocation of forestry funds to what might be an ideal situation with adequate funding:

	Ideal	Actual
Pruning/other maintenance	40%	28%
Removals	14%	22%
Planting	10%	14%
Other	36%	36%

The "ideal" allocation is based on those communities surveyed known to have highly regarded programs, while "actual" is based on national averages. This suggests that for every dollar spent on tree planting there should be four dollars in the budget for maintenance. Underfunded programs spend proportionately less on maintenance.

A survey by the author of urban forestry programs in six large cities in the midwestern United States revealed a wide range of funding and management (Table 13–8). Cities A, B, and C are at or close to full stocking, plant more trees than they remove, and are on scheduled pruning cycles of eight years or less. Table 13–9 compares these same cities based on expenditure per tree and number of trees per employee. From these data

TABLE 13–7 ESTIMATED TOTAL PRUNING HOURS PER YEAR INCLUDING PRUNING REQUESTS FOR MILWAUKEE. (From Griffith and Associates, 1993.)

Pruning Cycle	Cycle Prune	Request Prune	Total Hours
6-year	28,890	20,300	49,190
3- and 6-year	43,130	10,150*	53,280
4- and 8-year	32,356	10,150*	42,506

*50% reduction in request pruning assumed with two-stage pruning cycle.

TABLE 13–8 URBAN FORESTRY PROGRAM COMPARISON OF SIX LARGE MIDWESTERN CITIES IN THE UNITED STATES. (From Griffith and Associates, 1993.)

City	Miles of Streets	Street Tree Population	Vacant Spaces	Planting (1992)	Removals (1992)	Pruning (1992)	Pruning Cycle	1992 Annual Budget	Full-Time Employees	Percent Stocking	Trees Per Mile
A	1,000	150,000	12,000	4,700	2,200	33,000	4.5	5.9 million	91.5	92.6%	150.0
B	3,800	470,000	64,000[a]	13,000	8,000	60,000	7.8	12 million	318	88.0%[a]	123.7
C	1,400	200,000	4,000	3,621	3,490	40,000	6	5.0 million	120	98.0%	142.9
D	1,200	71,000	31,000	540	1,300	5,400	13.1	1.5 million	32	69.6%	59.2
E	1,000	62,000	30,000	5,100	700	5,500	11.3	1.06 million	9[b]	67.4%	62.0
F	3,400	120,000	240,000	1,000	1,500	500	None	1.3 million	14	33.3%	35.3

[a]Estimated based on 140 trees per street mile
[b]All contract maintenance

TABLE 13–9 PER TREE EXPENDITURES AND TREES PER EMPLOYEE IN SIX LARGE MIDWESTERN CITIES IN THE UNITED STATES. (From Griffith and Associates, 1993.)

City	1992 Per Tree Expenditure	1992 Trees Per Employee
A	$39.33[a]	1,639
B	25.53	1,478
C	25.00	1,923
D	21.13	2,219
E	17.10	6,889
F	10.83	8,571

[a]Includes other expenditures that could not be readily separated.

it appears that a program funded annually at $25 per tree (1992 dollars) with 1,500 to 2,000 trees per full-time employee can be regarded as receiving enough support to maintain full stocking of well-maintained street trees. Kielbaso (1988) reports that in 1986 the average number of trees per employee was 3,798, but this number does not separate cities that maintain their trees by contract service.

PERSONNEL

Effective management of a business or agency requires three basic skills: technical, human, and conceptual (Figure 13–11). In urban forestry and arboriculture, technical skills include the knowledge and ability to perform assigned vegetation management tasks, an essential attribute of professionals entering the field. Human skills are equally important and include public relations, personnel management, conflict resolution, communications, teaching, and motivating. Conceptual skills are not that important initially, but become increasingly important as a manager develops and advances in an organization. A crew leader needs a good mix of both technical and human skills to supervise actual field work, with a conceptual understanding of organizational policies and goals. Personnel at higher levels of management devote less time to day-to-day technical and human matters, with a correspondingly greater emphasis of planning, coordination, and policy setting (Lynch, 1985).

Time Management

Time is a major cost item for commercial arborists and municipal foresters—estimated to constitute about 50 percent of controllable expenses. Efficient utilization of time will result in substantial savings in operating budgets. To manage time more efficiently, three things must be known (Long and Love, 1983):

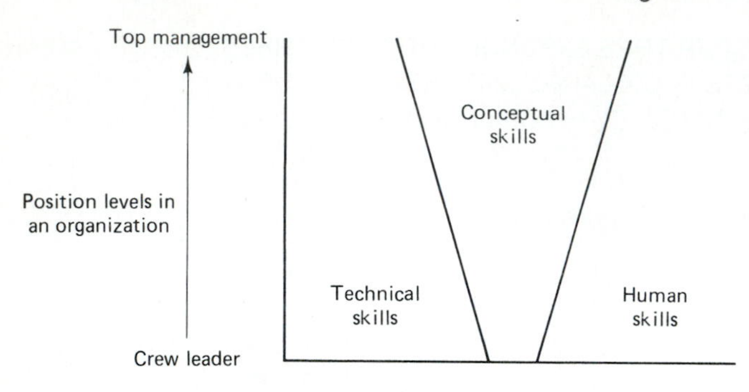

Figure 13–11 Mix of administrative skills needed within an organization. (From Lynch, 1985.)

1. How time should be used
2. How it is actually being used
3. Techniques to improve time management

Time use. Time use on the job may be divided into three general categories: nonproductive, productive, and management. Nonproductive time is time spent doing things on the job that contribute nothing to the operation. Productive time is time used to contribute to the objectives and/or profits of the agency or firm. Management is that time spent in routine or supervisory work, and in planning and decision making. Routine or supervisory activities are of a routine or recurring nature, such as scheduling work crews or ordering supplies. Planning and decision making is time used to develop long-range goals for the firm or agency, and strategies to achieve these goals. Figure 13–12 describes how time is allocated within an organization based on supervisory level (Long and Love, 1983).

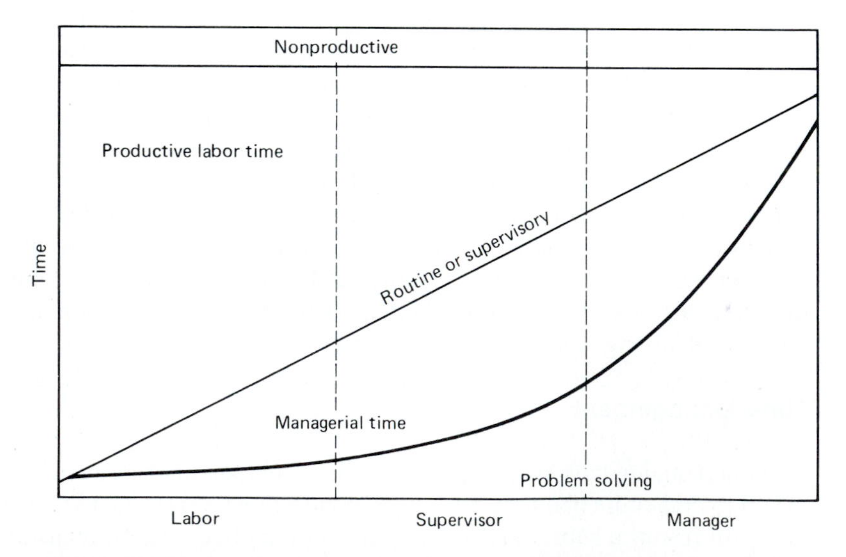

Figure 13–12 How people use working time. (From Long and Love, 1983.)

A problem arises in time management when people are spending time doing things that are productive, but a level below the level of work they are being paid to do. An example is when a supervisor is doing field work when he or she should be supervising employees or preparing reports. The work needs to be done, but time (and money) is wasted when a high-salaried employee is doing work that should be done by a lower-paid crew member. How time should be spent is determined by management. How time is actually spent can be determined by a time-use diary. Time spent on different tasks, task descriptions, and type of time use each task constitutes is recorded in the diary and summarized after several days (Long and Love, 1983).

Time-use improvement. If time is not being used in an optimum manner for an employee's level of responsibility, three techniques can produce significant improvement in time usage (Long and Love, 1983):

1. *Increase delegation.* Supervisors and managers can and should delegate tasks to subordinates. The following criteria may be used to evaluate those tasks considered for delegation:
   When to delegate:
   - when someone else can do it better than you,
   - when you might do poorly because of limited time,
   - when a subordinate can do the job well enough for the time and cost involved,
   - when cost time and risk are small, and it will develop a subordinate, or
   - when the job is repetitive.

   When not to delegate:
   - when no else can do the job satisfactorily,
   - when your prestige is needed,
   - when you wish to set an example,
   - when it is confidential or beyond the authority and responsibility of a subordinate, or
   - when you need to become familiar with the job.
2. *Employee training and development.* Employees who are well trained and understand what they are supposed to be doing and why they are doing it are efficient employees. Time and money spent on employee training and development usually pays off in the long run if the training is in line with long-term personnel needs of the organization.
3. *Organized records.* Problems in many organizations are re-solved time and again because no one bothered to keep records of past solutions, or if records were kept, they cannot be found. An organized management information system is essential and easy to develop and maintain with today's computer technology. Computer hardware is inexpensive to purchase and easy to use. User-friendly packages and packages adaptable to a variety or organizations are now available for arboricultural firms.

When delegating authority, a supervisor does not delegate responsibility, so delegation must be made with care. If an employee of a tree care firm is given the authority to spray trees for pest control and inadvertently uses an herbicide, the owner or manager of the firm is ultimately responsible for the employee's action. Similarly, if a municipal forestry crew leader has the authority to remove street trees but causes property damage in the process, the city forester must bear the responsibility.

Management by Objectives

Management by objectives (MBO) was defined by Mali (1972) as "a strategy of planning and getting results in the direction that management wishes and needs to take while meeting the goals and satisfaction of its participants." All too often, employees of a firm or agency work in directions counter to one another, resulting in conflict and loss of productivity. MBO seeks to blend individual plans and needs of managers toward a common goal within a specific time frame, and to simplify management processes within the organization. MBO consists of four basic concepts: the setting of objectives, development of a time strategy, total management, and individual motivation (Mali, 1972).

1. *Objectives.* Objectives are those goals or accomplishments that are set forth in the overall management plan of the organization. An arboricultural firm might set a goal of increasing sales by 15 percent, while a municipal forestry department might wish to shift from a system of service requests to programmed maintenance for street trees.

2. *Time strategy.* Time strategy is the establishment of a timetable for coordinating the efforts of managers to meet the objectives of the management plan. The objective of increased sales might be within a one-year time frame, with new customer contacts made during inclement weather or the off-season. Programmed maintenance of street trees could have a timetable equal to the length of the maintenance cycle (i.e., a pruning cycle of five years will take five years to implement fully). Time strategy also includes having the necessary work completed when the components are needed. If the vegetation cover type beneath a utility transmission line is to be changed, the existing vegetation must be eliminated before new planting can begin.

3. *Total management.* Total management is coordinating the efforts of individual managers to reach the stated objectives. Crew leaders that are particularly talented at obtaining new customers should be given more time to do so, while crew leaders more comfortable with field supervision should have lower new-contract objectives and increased supervisory responsibility. Municipal tree maintenance crew leaders and field operations supervisors should plan the most efficient scheduling of programmed maintenance, with a decrease in the need for service requests being offset by more intensive programmed maintenance.

4. *Individual motivation.* The basic strategy of MBO is to motivate employees to work toward common goals by developing a desire and willingness to achieve. This is realized through the involvement of all personnel in overall planning, objective setting, and management process. The objective of increasing the customer list by a given percent must have the input and support of field supervisors, and must be feasible in the time frame established. Similarly, a programmed maintenance schedule for street trees will need the input and support of field supervisors prior to implementation. Involvement in management planning will aid in setting realistic objectives and time strategies, and will yield greater motivation in achieving those objectives.

The process of MBO involves five basic steps: (1) identifying objectives, (2) setting objectives, (3) validating objectives, (4) implementing objectives, and (5) feedback (Mali, 1972). Objectives are identified and set through the planning process as described in Chapter 8 (i.e., what we have versus what we want). These objectives are validated at all levels of the organization by participation in setting them, by determining the feasibility of the time strategy, and through commitment of personnel to achieve the objectives. Implementation is using the management plan to work toward the objectives, and feedback monitors progress toward meeting objectives and is sensitive to changes in value states which can change objectives.

Job analysis. Job analysis is a useful management tool for implementing MBO. Once the objectives have been established and validated, the job analysis process should begin with the top manager and proceed down the organizational chart, ultimately including all employees. MBO will be fully successful only if all employees are motivated to achieve the objectives. Employee dissatisfaction is often tied to not knowing what the organization's objectives are and what their supervisors expect of them. The job analysis technique involves four basic parts: duties, authorities, controls, and resources (Love and Long, 1983).

1. *Duties.* Describe the duties of each employee in terms of goals and time frames. All employees must know what they are expected to do and minimum acceptable performance levels in both quantity and quality of work.
2. *Authority.* Supervisors and crew members must know how much authority they have to make decisions affecting their jobs. Once duties are assigned, personnel must have enough authority to carry out their duties. This involves the delegation of authority to subordinates, a difficult task for most people.
3. *Controls.* Controls are used to evaluate employee performance in meeting management objectives (feedback). In the case of supervisors, controls are used to evaluate how well delegated authority has been used to meet management goals. Formal reports, spot checks, planned observations, and informal reports are all methods used to evaluate performance.
4. *Resources needed.* Resources needed includes what supervisors and crew members need in the way of training, equipment, and supplies to perform tasks assigned. It may also be used to evaluate what is needed if controls indicate that work performance is falling short of task assignments.

Job analysis provides a means of communications with employees by clearly stating what the organization's objectives are, what is expected of them, and what authority they have to do their jobs. It is an evaluation and motivational tool that can be used to measure employee knowledge and performance, and to provide incentives and rewards. Job analysis is, finally, a training tool that ensures that each employee has the knowledge, skills, and tools to perform his or her job. Figure 13–13 provides an example of how job analysis can be used to describe the duties, authorities, controls, and resources needed by a crew leader of a tree care firm to implement objectives as determined by the MBO process (Love and Long, 1983).

Training

Employee motivation and performance can be enhanced through in-service training programs. However, an effective training program needs to go beyond merely providing technical information; it should be a part of the overall employee management program. Employees should not only know how to perform a task, but should know why it is being done and what will happen if it is done differently or not at all. Tate (1981) suggests the development of an in-service training program be divided into three elements: what to teach, preparation, and delivery. A fourth factor can be added: evaluation of the program.

What to teach. The subject matter of an in-service training program should be relevant to job needs and should be made at the start of the specific season where learned skills will be implemented immediately, thus reinforcing learning. Training programs should be prioritized by needed skills, deficiencies, and where the greatest efficiencies can be obtained on the job. General subject matter of in-service training courses for urban forestry and arboricultural professionals include:

1. Transplanting
2. Pruning—training young trees
3. Pruning—mature trees by rope and saddle
4. Pruning—mature trees by aerial tower
5. Pest management—diseases
6. Pest management—insects
7. Pesticide labeling, mixing, application, rate determination, and safety procedures
8. Cabling, bracing, and wound treatment
9. Traffic safety
10. Chain saw use, service, and safety
11. Vehicle use, service, and safety
12. Chipper use, service, and safety

Job analysis

Position: Crew Leader/Part-time Salesperson ______

Person: ______ Jay Bay ______

Duties	Authorities	Controls	Resources and Training Needed
1. Produce $15,000 revenue/week	To repair or have equipment repaired	Job orders and sales charges	Labor supervisory skills; equipment; trained workers
2. $60 revenue per man hour	To supervise; dismiss employee with manager's permission	Monthly operating statements	Knowledge of equipment and work procedures
3. Service 400 customers this year	Contact each customer (5 min. call) and log details	Spot checks and customers' sales list	Ability meet people and explain work
4. Full time crew employed throughout year	Plan and list rainy day jobs	Amount of unproductive down time (% of total time)	List duties and job needs; set priorities
5. Cost estimate overruns less than 10% of total	Check job orders and complete to specs	Work diaries	Job estimation and labor supervision
6. Labor costs less than 50% of total labor costs	To schedule and supervise labor assigned to department	Labor reports and monthly operating statements	Adequate equipment; supervisory and job planning skills
7. 100 new customers contacted in addition to all old customers of the past two years (crew chief will contact 10% of these)	To contact 40 new prospects in March and April	Completion of prospects' list	Sales and planning skills
8. Daily work schedule and plan	To meet with management and schedule work	Spot checks	Knowledge of area and job requirements to minimize travel and job time

Figure 13–13 Example of the job analysis technique. (Love and Long, 1983.)

13. First aid
14. Occupational Safety and Health Act (OSHA) rules
15. Environmental Protection Agency (EPA) rules
16. Organization and objectives of the firm/agency
17. Regularly scheduled safety programs
18. Public relations

Most of the subjects listed above can be divided into two distinct areas of training, theoretical and practical. For example, sessions on pruning might include the concept of decay compartmentalization from a theoretical standpoint, and be followed by how compartmentalization relates to pruning and wound treatment. There are other subjects, to be sure, that will need to be covered by in-service training sessions.

Preparation. Tate (1981) suggests that once the subject matter has been selected, in-service training programs be prepared following five key elements:

1. *Objectives.* These should be explicit and realistic for the time allotted, and can include employee input. The objectives are also used at the end of the program to evaluate its effectiveness.
2. *Course content.* This will be based on the subject matter, available time, new technologies, and employee needs. As in setting objectives, selecting course content should have employee input.
3. *Course outline.* Material to be included should be prioritized to be sure the most important information is covered. Least difficult subjects should be taught first, and theory covered before practical applications.
4. *Instructional materials and training aids.* Information on the above subjects is available from a variety of sources including books, professional journals, government publications, and universities. This information could be handed out as is or summarized in a course outline for each participant. Training aids include chalkboards, overhead transparencies, slides, and films. Canned programs are available from professional societies and audiovisual supply firms. Training aids assist in retaining student interest and allow for more rapid coverage of the subject material.
5. *The lesson plan.* A lesson plan contains a title, objectives, introduction, presentation, and summary. When well designed, lesson plans organize subject matter in a logical fashion and give confidence to the instructor.

Presentation. The ability to teach will vary greatly among individuals. Some people seem to have a knack for getting up in front of a group, while others experience severe anxiety at the very thought of making a short presentation. For most people, teaching is something that improves with experience, which in turn builds self-confidence. A well-thought-out training session supported by a solid lesson plan and instructional materials will go a long way in providing the necessary confidence, even for a person making his or her first public presentation.

Evaluation. Once a training session is completed, it should be evaluated to determine if the objectives have been met and what improvements can be made before it is taught again. One way to evaluate a training session is an exam for participants following course completion. Exams can give rise to employee anxiety, but if used to evaluate the course not the participants, this anxiety can be kept to a minimum. The simplest way to do this is to have employees not put their names on the exams. Other methods of evaluation include a course questionnaire for employees following the session, or evaluation of field work before and after training.

On-the-job training. Employee training should not be limited to formal classroom sessions. New employees should receive continuous supervision and training from their immediate supervisors. How a task is to be done should be carefully explained, as should why it is done. Employees who are trained as to how and why tasks are performed, given the opportunity to make decisions, and provided with advancement and recognition will develop into long-term assets to an organization. These employees will also represent the firm/agency well to the customers and general public.

COMMUNICATIONS AND PUBLIC RELATIONS

Communications and public relations involve two levels of communication: communications with other professionals and communications with the public. Johnson (1982), in a survey of political and administrative factors in urban forestry programs, reports that a need exists for increased contact between urban foresters and improved skills in public outreach and involvement in urban forestry programs. Both professional contact and public involvement take communications skills, and successful urban foresters must not only be professionally competent but must be good communicators.

Professional Communication

Professional communication is an ongoing process that occurs in both formal and informal circumstances. Formal communications generally take place through professional organizations such as those described in Chapter 3. These organizations sponsor such functions as professional journals, newsletters, conferences, workshops, and symposia. Membership and participation in these organizations is essential for urban foresters and arborists to remain current in their profession. A formal education that prepares a person to enter urban forestry or arboriculture only provides a reference frame for professional development. Growth and development of competent managers takes place on the job through learned skills from co-workers and supervisors, and through communications with other professionals.

Benefits from professional communication are numerous. Managers face common problems regardless of location, but a great deal of time is wasted solving these problems on an individual basis. Much can be learned from others who deal with the same problems from a different perspective. New technologies evolve with increasing frequency, and managers must keep current to apply them when they result in better performance and/or cost savings. It is through professional communications channels that developing managers learn of new and more challenging opportunities, and existing managers use these channels to recruit new personnel.

Informal communications often take place through channels established by formal communications. Informal groups of municipal foresters and commercial and utility arborists are common on a local basis. These groups get together informally to discuss common interests. Other groups form to address a particular problem and dissolve once the problem has been resolved. Communications technologies allow networks of professionals to span large geographic areas. These networks serve to keep participants current on issues that affect them. Sometimes existing professional organizations form short- or long-term coalitions with other professional groups to address common issues and/or problems.

Regardless of the formality or informality of the group, organization, coalition, or network, this is where the information exchange occurs. Professional managers can choose to operate in isolation and hope enough information comes their way to do their jobs, or they can choose to be active and involved in professional communications, thereby guaranteeing access to information.

Public Relations

Public relations is something all organizations that interact with the public have regardless of whether it is intentional or not. Municipal forestry departments, commercial arborists, landscape contractors, utility forestry departments, and so on, all have a public image. This image may be real or imagined, but it exists just the same.

The first aspect of a good image is to do a good job. No matter how hard a public relations program works to convince the public that an agency or firm is doing a good job, it will be only a matter of time for the image to wear thin if shoddy work is the rule. On the other hand, doing an excellent job will not necessarily guarantee a good public image unless the public is aware of what is being done. No image is as much an image as a good or a bad image.

The first step in thinking about public relations is for managers to ask themselves what their business is. Does the city forester take care of trees or provide amenities to the public? Does the commercial arborist take care of trees or manage people's landscapes? Is the utility forester responsible for keeping transmission lines clear or for managing vegetation in utility corridors? The answer in all three cases is the latter, but some managers behave as if they believe it is the former. Most individuals enter the urban forestry or arboriculture professions because of an interest in vegetation, but like it or not, they must also take an interest in people, specifically their clients.

Public relations, or information education as it is sometimes called in public agencies, is something that must be a part of the job description of all personnel who have contact with clients or the public. This includes managers, office staff, field supervisors, and crew members. The best tree service job possible will not compensate for a rude employee. Well-trained personnel will be able to answer, in a pleasant and intelligent manner, questions concerning what they are doing and why they are doing it. If they do not know the answer to a question, they should know where to find the answer or to whom to refer the person.

Public relations is a part of the everyday work experience of personnel, and they are creating an image even when they are not in direct contact with the public. In spite of the often dirty nature of tree work, crews should be well groomed and present a positive image to the public. Company or agency uniforms or coveralls, with a uniform allowance as part of employee compensation, can help to create a favorable impression. Equipment should be kept clean and serviced on a scheduled basis, and vehicles identified with a logo describing the agency or firm. Power equipment can be noisy and disruptive, therefore should be well muffled. Decibel levels should be specified when requesting bids for purchase. A city or commercial arborist crew arriving on the job in a filthy truck only to sit around for an hour because the chain saw won't start will not inspire the confidence of people paying the bill—the taxpayer or customer.

Programmed maintenance or tree service should never surprise citizens or customers. Residents of work units scheduled for programmed maintenance should be notified, in the local newspaper or by direct mail, of work to be performed, along with an explanation of why it is being done, prior to city or contract crew arrival. It is especially important to notify abutting property owners of pending removals and to explain carefully why removal is necessary. Commercial work on private property may be scheduled months in advance, but as a courtesy clients should be notified several days before work is to begin so as to avoid conflicts. In all cases work should be scheduled for minimum disruption. Tree crews blocking traffic during rush hours or using chain saws at 6:00 A.M. will not be greeted with much enthusiasm from the public.

Public Relations and Municipal Forestry

No public program, regardless of technical compentancy, will last long without public understanding, support, and financial commitment. Funding may be easy to obtain during good economic times, but when a municipal government needs to cut costs the programs that suffer most are those perceived by the public and elected officials as having a low priority. A community aware of the public tree program and knowledgeable as to the benefits provided will place that program in a higher priority when allocating funds. Public employees serve a series of publics interested in the services provided, and those publics become active when they perceive these services to be below an acceptable level (Tate, 1976). Publics served by city forestry programs include people who love and are constantly aware of trees and parks, nature organizations such as Audubon or the Sierra Club, garden clubs, people alienated by our actions (tree cutting, programmed maintenance, etc.), and the general public who desires services provided. These are the people and/or groups who are to be reached in a municipal forestry public relations program, and this program should include not only what the department provides, but should also involve the public in setting management goals and priorities (Chapter 8). A good public relations program is a two-way street used to disseminate information and to gather public input.

The dissemination of information to the public through a public relations program consists of enlightening the public as to the benefits, problems, and costs of an urban forest. This information includes two major parts (Tate, 1976):

1. Specific tree care information including planting, pest management, and maintenance practices. This information not only enlightens the public as to work needed on public trees, but also provides them with information on how to manage their own vegetation.
2. Exposure of the municipal forestry program to the public so they can observe the benefits of tax dollars spent on tree care and in providing public assistance.

The most effective public relations occur through direct face-to-face contact with the public. However, due to time constraints the amount of direct public contact will be limited, so a need exists to use the local media for indirect contact. Public relations, then, should occur on two levels: direct and indirect.

Direct public relations. As stated previously, all employees who have contact with the public should be trained to interact in a professional and courteous manner. Phone service requests should be handled promptly by forestry personnel and be followed up by reporting back to the concerned citizen. Requests that are not within the jurisdiction of the forestry department should be referred to the appropriate agency.

City crews attract a lot of attention when working on street trees, and will receive numerous queries as to what they are doing and why they are doing it, as well as informational questions concerning private vegetation. These questions must be dealt with in a courteous manner, but crews cannot interrupt assigned work for extended periods to engage in prolonged conversations. Brief, courteous answers are usually sufficient for most people. More difficult problems or irate citizens should be referred to the forestry office for further explanation or discussion. Handouts describing common tree maintenance activities serve as good public relations and educational tools, and should be carried by all crews. Subjects such as proper pruning techniques, how to plant trees and shrubs, recommended species, pest management, and Dutch elm disease are of common interest to the public and will be well received (Figure 13–14).

Public contact can be difficult due to the nature of some tree management activities. Informing citizens that the street trees on their block must come down due to disease can be a traumatic experience. Worse yet, the condemnation of a tree on private property can lead to outright confrontation. This requires the utmost patience,

Figure 13–14 Handouts describing proper post-planting care help tree survival and promote urban forestry. (Courtesy of Minneapolis Park and Recreation Board.)

tact, and listening skill. Nobody will benefit if a city employee gives an unreasonable citizen a "piece of his or her mind." An employee might feel better afterward, but bad feelings often persist for a long time. Sometimes city employees bear the brunt of anguish over the pending loss of a private elm, but that is a part of the job.

Another level of direct contact is through organizations such as neighborhood groups, service clubs, garden and/or nature clubs, or schools. Neighborhood organizations should be contacted prior to special projects such as tree planting, pest management, or redevelopment, which includes landscaping. These organizations serve as excellent sources of public input in developing plans and setting management goals. Often, neighborhood groups will assist in actual tree planting and followup maintenance if they are made a part of the process. Involved people will identify with the new landscape and will be diligent in protecting new plantings.

Service clubs are always looking for speakers, and in many cases membership interest in trees is high. Contact officers of these groups and volunteer to speak at future meetings on a variety of topics relevant to municipal forestry. It is surprising how much interest in the community forestry program can be generated through a good presentation to a service organization. Service clubs are often looking for community improvement projects, and tree planting is a highly visible activity that will generate community support and recognition for those participating. Community leaders from both the public and private sectors frequently are members of service clubs, and this provides an additional means to sell forestry.

Garden and/or nature clubs by definition will have an interest in the work of a community forestry department. Beyond simply making presentations to these groups, they can be called upon to assist with major vegetation improvement projects, and will often provide long-term follow-up maintenance. These groups will also support forestry from a political perspective provided that the appropriate relationship has been developed. It should be cautioned that some of these groups have strong feelings about the use of pesticides, and will take an adversarial role against their use if they feel that it is not justified or that the forestry department has not considered appropriate alternatives. In all cases these groups should be considered and involved when making pest management decisions.

Neighborhood groups, service clubs, and garden and/or nature clubs respond well to clinics dealing with tree care techniques. Presentations and clinics should be well prepared and, if possible, include high-quality visual aids to assist in getting the message across. A poorly planned, prepared, and executed presentation will not win much support for tree management activities.

Schools provide excellent opportunities to initiate positive attitudes toward urban vegetation. Programs designed to develop an appreciation and understanding of urban trees can reduce vandalism and in the long run result in adults who support community forestry. Schoolchildren also provide a conduit to get information into homes. Kids love to tell parents what they learned in school and are eager to carry high-quality information home with them. However, when addressing schoolchildren it is important to remember three things: Keep it short, consider their age and potential understanding, and make it entertaining with visual aids.

Volunteers in urban forestry. Volunteers have played an increasing role in community forestry programs in recent years. Although it is not likely there will be substantial cost savings by using volunteers, volunteers do increase the level of support for community forestry. Stiegler (1989) conducted a survey of residents in a suburban community of Minneapolis to determine their perceptions of urban forestry. He reports that the concept of the urban forest is poorly recognized and understood by his study group, and he concludes that there is a great need for the urban forestry professionals to educate the public. Volunteer programs can increase the level of awareness of urban forestry in a community.

Volunteers can plant trees, provide early care for them, and even provide some fundamental level of pruning for small trees. TreePeople, a citizens volunteer organization, has planted and cared for millions of trees in the Los Angeles area since 1973

(Lipkus and Lipkus, 1990). Sommer et al. (1994) compared the level of satisfaction with street trees of Fresno, California, residents who planted their own trees with residents on streets where the city did the planting. They found residents involved in planting were more satisfied with the outcome than those who were not, and that residents who paid for their trees were more satisfied.

However, volunteers need to be educated and supervised if they are to make a contribution to a community forestry program, and this takes time and resources. The Minnesota Extension Service and Minnesota Division of Forestry sponsored an urban forestry training program for 52 volunteers that included 30 classroom hours of training and a pledge of 50 hours of volunteer community service by participants (Johnson, 1995). Since completing the course the volunteers have provided more than 2,000 hours of urban forestry service and program assistance to their communities. The payback of investing in volunteers is a core of educated citizens who will provide substantial political support for urban forestry, and a substantial contribution to urban forestry in their communities.

Indirect public relations. Since it is next to impossible to establish personal contact with all residents or groups within a community, contact can be made indirectly through the media and other means. Indirect public relations through the media include working with local newspapers and television and radio stations. Other means of public contact can be made through exhibits, newsletters, special mailings, literature distribution, flyers, posters, billboards, public transit advertisements, street signs, heritage or big tree programs, user surveys, annual reports, and Arbor Day (Hudson, 1983; Minnesota Department of Agriculture, 1982; and Tate, 1976).

Public media. The first step in reaching the public through the media is to develop a good relationship with local newspaper, television, and radio personnel. This involves not only an introduction, but includes providing interesting and high-quality information. The public is interested in vegetation management, and media personnel know this, but poorly prepared news releases, insignificant news items, and excessive contact will give cause for avoidance. A number of methods of public contact are available through the public media.

News. News is information that is of public interest in a short time frame. Examples include tree planting, programmed maintenance for a specific area, tree dedications, Arbor Day activities, tree damage, and so on. Two approaches are open to reach the public through the news: the press release, and news tips. Press releases are written by forestry personnel and distributed to local newspapers and radio and television stations. A news release should be written in a simple nontechnical manner and answer the following six questions: who, what, when, where, why, and how. Addressing these questions will ensure that the necessary information is provided in the news release, as shown in the example in Figure 13–15.

A news tip is notifying news personnel of upcoming events that may have news merit. Be sure that the subject does have news interest and do not overkill with ex-

For Immediate Release

Tree planting throughout the city will begin early next week, according to city forester Carol Jones. Four hundred trees will be placed on tree lawns as part of a continuing program to replace elms lost in recent years, and to provide street trees in new subdivisions.

The city incurred serious tree losses in the past two decades from Dutch elm disease. Approximately 900 elms remain, but these are dying at a rate of 5 percent per year. The Master Street Tree Plan calls for planting 400 trees per year for the next 8 years to replace elms, plant new subdivisions, and replace other trees lost to a variety of problems. After 8 years planting will drop to about 200 trees per year, a normal replacement rate.

This year's planting list includes greenspire linden, corktree, green ash, ginkgo, Norway maple, and flowering crab. City forester Jones said that the city is planting a variety of species to avoid problems in the future that can result from overuse of a single species. The American elm was overplanted in the past, resulting in present problems with Dutch elm disease. The current planting plan restricts the use of any variety to under 5 percent of the total street tree population.

According to Jones, fall is a good time to plant trees, as tree roots continue to grow until the soil freezes. This provides new trees with a head start to begin growth in the spring.

Figure 13–15. Sample municipal forestry news release.

cessive tips on items of interest to forestry personnel only. Do not be pushy, and let news personnel decide if the item has merit. A lot will depend on how busy they are on a particular day. Slow news days will yield a lot of interest in events that do not seem too exciting, while a major fire will eclipse most other news items for the day. As in a news release, media personnel will be interested in who, what, when, where, why, and how.

Features. Feature stories are frequently carried in Sunday papers and in local radio or television magazines. Features are not really news items, but rather deal with items or situations of general public interest. Examples of potential feature stories concerning municipal forestry include management of a greenbelt forest, an overview of municipal forestry, public involvement in tree planting, Dutch elm disease, pest management, and historic trees.

Columns. Some papers have regular columns that deal with gardening and other facts of landscape establishment and maintenance. Municipal forestry personnel often make regular contributions to these columns with suggestions on proper planting, pruning, pest management, and other vegetation management items.

Talk shows. Many radio and television stations feature local talk shows where individuals are interviewed by the program host and then respond to questions from listeners and viewers. These shows provide an excellent format to inform the public about the urban forestry program in general, specific projects such as tree planting or Arbor Day, and proper tree maintenance in general. Public interest in landscaping subjects is seasonal and guest appearances should coincide with this interest.

Public service announcements. Television and radio stations in the United States are required by federal regulation to air public service announcements. These announcements range from fifteen to thirty seconds, and if carefully prepared can provide the public with a lot of information on tree management. Public service announcements can be prepared locally or be obtained from public resource agencies and private organizations such as the National Arbor Day Foundation.

Exhibits. Local fairs, conferences, expositions, garden shows, Arbor Day festivities, and arboretums will often provide space for educational exhibits at no charge. An exhibit featuring municipal forestry activities will meet the educational criteria for free space and will get a lot of exposure to the public. The exhibit should be easy to transport and to set up, attractive, and carry a simple informational message. The public will soon tire of the same message appearing time and time again, so the exhibit should be constructed in a manner that allows the message to be changed to suit the location, theme, and season. The message should include not only information about the forestry program, but also information to assist homeowners with managing their own landscapes. Before developing an exhibit it is best to consult literature dealing with the subject, or, if funds permit, hire a consultant to assist in development. There are firms that will design and construct exhibits on contract.

Literature. Articles dealing with urban forestry and arboricultural subjects can be published in local newsletters put out by garden clubs, nature organizations, and arboretums. Literature in the form of flyers or handouts is available from resource and university extension agencies, and from private foundations, at a nominal or no cost (Figure 13–16). These can be given to citizens who desire detailed information on some aspect of tree maintenance.

Municipal forestry departments can also develop their own handouts dealing with subjects not available from other sources or to meet specific local problems or conditions. Tree planting programs can be preceded with mailings to abutting property owners explaining what is being done and how to care for the tree once planted. Pest management activities preceded by mailed announcements describing what the problem is and how it is being handled will help to alleviate public concern over the problem and techniques used for management.

Urban Forestry: Bringing the Forest to Our Communities

Robert W. Miller

At first, the term urban forestry may be confusing, even contradictory. The city and the forest seem incompatible and, in a practical sense, they are. The urban environment is an inhospitable place for trees and other vegetation to grow.

Yet we desire the beauty, shade and shelter of the forest near where we live. Over 80 percent of Americans now live in and around cities. Rising energy costs will keep us closer to home in the future, restricting our contact with nature outside the city limits. Thus, the trees and plants that grow on our streets, boulevards, parks, lawns, empty lots, golf courses, cemeteries, waterways and business districts – known collectively as the urban forest – bring nature into the city for our benefit.

THE VALUE OF THE URBAN FOREST

City trees, in addition to their beauty, shade our homes and reduce air conditioning costs in summer. And, in winter, they shelter us from cold winds and reduce heating expenses. Trees help filter dust and pollution from the air and reduce soil erosion. Vegetation softens the harsh lines and surfaces of the urban environment. The urban forest provides habitat for birds and wildlife, and sparks imaginary kingdoms in the minds of children.

Researchers have found that shade trees on a developed lot will add as much as 20 percent to the selling price of a home. In short, trees make our communities more livable places.

THREATS TO THE URBAN FOREST

The American elm was a rugged street tree. It was easy to establish and it tolerated such urban stresses as soil compaction, drought, salt and pollution. It survived root and bark injury. The American elm tree did not survive Dutch elm disease, however, and efforts to save it usually proved fruitless.

Communities that had been nestled under the graceful, sweeping elm branches were left bare because no other tree species had been planted. Therefore, a major goal of urban forestry is to diversify the number of tree species planted, so that when a disease does strike, it claims only a fraction of a community's trees, not all.

The trees planted in the wake of Dutch elm disease will require more care because the replacement trees are not as well adapted to the city. Plus, the stresses placed on trees by the unnatural, urban environment are greater because of increased pollution, road salt, traffic and human impact. Diseases are threatening many of the trees planted to replace the elms. Road salt, used to de-ice our streets in winter, damages the roots of many shade trees. Street widening and construction take their toll, either by removing trees or damaging root systems so that the tree is fatally injured. Trees are damaged by people who excavate utilities near them, cover their roots with asphalt, hit them with cars and lawn mowers. People expect trees to grow in limited rooting space, and pay little attention to their need for watering, fertilizing and pruning.

Without regular attention and care, urban trees have little chance to survive more than a few years. Removing and replanting these trees is expensive – much more expensive than good management.

MANAGING THE URBAN FOREST

Urban forestry is the scientific management of vegetation under urban conditions. It differs from traditional forestry in that urban forestry focuses primarily on caring for the individual trees and developing the urban forest for the enjoyment and benefit of the community, rather than on forest products and conservation of a mass of trees.

A number of different professionals fit under the umbrella of urban forestry. City foresters are usually responsible for the establishment and maintenance of city-owned street

Figure 13–16 Sample University Extension urban forestry handout.

trees and vegetation on other public land such as parks. Commercial arborists provide tree care for property owners and are sometimes hired to maintain trees on public property. Nursery operators provide the plants for private property owners, and, in many instances, trees for public land. Landscape architects develop landscaping plans for both the public and private sectors. Horticulturists, entomologists and plant pathologists are involved in cultivating landscape plants and developing techniques for their care and protection. These experts prepared the extension publications listed at the end of this publication to help people with no experience to care for their trees.

YOUR ROLE IN URBAN FORESTRY

The success of an urban forestry program depends on the concern and interest of the people of the community. If your community has an urban forestry program, find out who your urban forester is. If your community doesn't have an urban forestry program, find out who is responsible for the street trees. Many communities cannot support a full-time urban forester, so they may share with a neighboring community or occasionally hire a consultant. Many communities maintain healthy shade tree populations through the actions of service and youth clubs and neighborhood organizations. Arbor Day, celebrated on the last Friday in April in Wisconsin and many other states, provides the ideal opportunity for a community tree planting program. A planting program must be planned with the foresight and funds for maintaining those trees as they grow.

Trees add beauty, increase economic value and improve the livability of our communities. The close relationship between people and trees can be traced throughout recorded history from the hanging gardens of Babylon to the tree that grew in Brooklyn. The bleakest suburban development on an abandoned farm field soon becomes dotted with young trees planted by the new residents. Within a few years, the new community is nestled within a vigorous young urban forest. For this forest and the urban forest of the entire community to remain vigorous it must be maintained through pruning, control of insect and disease problems, removal of dead and diseased trees, and replanting trees as they are lost. Urban trees, when properly cared for, are an economic asset to the entire community. Without care, the trees are an economic liability and the community never fully realizes the benefits of the urban forest.

For more information on tree selection and care, consult the following University of Wisconsin–Extension publications:

A3070	*How Trees Benefit Communities*
A3066	*Landscaping for Energy Conservation*
A3071	*Dutch Elm Disease—A Lesson in Urban Forestry Planning*
A3067	*Selecting, Planting and Caring for Your Shade Trees*
A1817	*Caring for Your Established Shade Trees*
A3072	*Preserving Trees During House Construction*
A3073	*Identifying Shade Tree Problems*
A2079	*Recognizing Common Shade Tree Insects*
A7791106	*How to Develop a Community Forestry Program*
A2014	*Street Trees for Wisconsin*
A2865	*A Guide to Selecting Landscape Plants for Wisconsin*
A2392	*Dutch Elm Disease in Wisconsin*
A2842	*Homeowner's Guide to Controlling Dutch Elm Disease with Systemic Fungicides*
A1771	*Deciduous Shrubs: Pruning and Care*
A1730	*Evergreens—Planting and Care*
G1609	*Landscape Plants that Attract Birds*
A2970	*Salt Injury to Landscape Plants*
A2308	*Fertilization of Trees and Shrubs*

Please contact your county Extension agent for films, slide sets, posters and other publications about trees.

This publication is part of the urban forestry series, developed to help in the reforestation of our towns and cities struck by Dutch elm disease.

Robert W. Miller is associate professor of urban forestry at the College of Natural Resources, University of Wisconsin-Stevens Point.

UNIVERSITY OF WISCONSIN–EXTENSION

University of Wisconsin-Extension, Gale L. VandeBerg, director, in cooperation with the United States Department of Agriculture and Wisconsin counties, publishes this information to further the purpose of the May 8 and June 30, 1914 Acts of Congress; and provides equal opportunities in employment and programming including Title IX requirements. This publication is available to **Wisconsin residents** from county Extension agents. It's available to **out-of-state purchasers** from Agricultural Bulletin Building, 1535 Observatory Drive, Madison, Wisconsin 53706. **Editors**, before publicizing, should contact the Agricultural Bulletin Building to determine its availability. **Order by serial number and title**; payment should include price plus postage.

MARCH 1980

5¢

A7791025 **URBAN FORESTRY: Bringing the Forest to Our Communities**

Figure 13–16 *(continued)*

Public signs. Posters dealing with specific tree management topics are available from resource agencies and private foundations, or may be locally produced. These posters can be placed in schools, public buildings, parks, and public transit carriers (Figure 13–17). Billboard space can be rented, or is sometimes donated, to display important public services message to motorists (Figure 13–18).

Another opportunity exists when performing tree management activities. Signs explaining what is being done and why will not only answer questions, but will inform the public as to what proper tree care is all about. These signs should be simple, self-explanatory, and should be placed where they do not create traffic hazards.

Heritage or big tree program. Heritage trees are trees of historic or extraordinary aesthetic significance to a community. These trees can be identified to the public through brochures and informational signs describing significant attributes such as historic events in their presence, age, or prominence in the landscape. Dedi-

Figure 13–17. Poster promoting urban tree planting. (Courtesy of Minnesota Department of Agriculture.)

Figure 13–18. Billboard in Minneapolis promoting urban tree care. (Courtesy of Minneapolis Park and Recreation Board.)

cation of such trees will usually attract media interest and can be used to promote community tree management.

The American Forestry Association has a National Register of Big Trees that lists the name, size, and location of the largest known specimen of each species. Other big tree registers exist in some individual states. These registers can be used to sponsor a local search for big trees in the community for placement on state and national registers, or a community can sponsor its own big tree program. Once identified, dedication ceremonies and placement of an informational plaque provide excellent opportunities for involvement of community groups and the media.

Tree walks. A number of communities have developed a series of tree-walk guides for distribution to the public. These guides are self-instructional, containing a map of the walk, tree location, species identification, and a discussion of the special attributes of each tree. Tree walk guides are used by school groups, garden clubs, and individuals interested in tree identification and community forestry. Appendix H contains an example of tree walks developed for the city of Madison, Wisconsin.

Arbor Day. Arbor Day had its beginning in Nebraska when J. Sterling Morton introduced a resolution at a meeting of the State Board of Agriculture on January 4, 1872, calling for a day to be "especially set apart and consecrated to tree planting in the State of Nebraska and the State Board of Agriculture hereby name it Arbor Day" (Everard, 1926). The resolution passed, and over a million trees were planted that first Arbor Day in Nebraska. Arbor Day is now observed by all 50 states in the United States, with some states designating Arbor Week or even Arbor Month.

The function of Arbor Day is to call attention to the need to plant trees in communities, on farms, and in forests. In communities Arbor Day provides an excellent opportunity to involve the public in tree planting and to inform the public of the need to plant and maintain trees on both public and private property. Well-planned and well-executed Arbor Day festivities will attract the attention of the news media and can lead to feature stories on the community forestry program. Community leaders, schools, politicians, service or business organizations, natural resource agencies, senior citizen groups, youth organizations, garden clubs, and the news media can all be invited to participate in tree planting programs. Some typical Arbor Day activities include:

1. School tree planting and seedling give-away programs
2. Tree planting ceremonies in parks or other public places
3. Public seminars or workshops on proper arboricultural techniques
4. Street or park tree planting projects
5. Planting memorial trees
6. Arbor Day races to a tree planting site

Information on these and other Arbor Day projects and programs is available from the National Arbor Day Foundation in Lincoln, Nebraska, and state and national forestry agencies.

Public Relations and the Private Sector

Public relations in the private sector is as important as it is in the public sector, if not more important. The impact of public relations in the private sector is immediate through the loss of customers and the establishment of a poor reputation. Opportunities for public relations activities are more limited than for the public sector, but opportunities do exist. Commercial firms can participate in any of the informational activities carried out by the municipal forestry agency, and sponsor their own programs. Advertising that includes tree management information will serve as a good public relations tool for arboricultural and other firms dealing with urban vegetation establishment and management.

Tree management demonstrations, free and low-cost trees, flyers and brochures, posters, and public signs describing work being performed all assist in getting the message out to the public, and also advertise services. Customer news letters describing tree management techniques and containing articles about other facts of landscape management will encourage more interest in management services. Appendix I contains an example of a newsletter published by a commercial arboricultural firm.

The most important public relations tool for a commercial firm consists of high-quality work at a fair price carried out by competent personnel. This includes the use of clean, well-maintained equipment by courteous and informed work crews. It is often said that the best advertisement is a satisfied customer, and nowhere is this more true than in the landscape establishment and maintenance business.

LITERATURE CITED

CHURACK, P. L., R. W. MILLER, K. OTTMAN, and C. KOVAL. 1994. "Relationship Between Street Tree Diameter Growth and Projected Pruning and Waste Wood Management Costs." *J Arbor.* 20(4):231–236.

COLE, P.D. 1993. "Reinventing Municipal Urban Forestry Budgets." *Proc. Sixth Urban For. Conf.* Washington, D.C.: Am. For. Assoc., pp. 26–29.

DREISTADT, S. H., and D. L. DAHLSTEN. 1986. "Replacing a Problem Prone Street Tree Saves Money: A Case Study of the Tuliptree in Berkeley, California." *J. Arbor.* 12(6):146–149.

EVERARD, L. C. 1926. *Arbor Day: Its Purpose and Observance.* U.S. Dept. Agric. Farmers Bull. No. 1492.

FLORIDA DIVISION OF FORESTRY. 1971. *Urban Forestry Handbook.* Tallahassee: Fla. Div. For.

GRIFFITH, D. M., and ASSOCIATES. 1993. *Comprehensive Review of the Services, Operations, Organization, Financing, Management, and Staffing of the Bureau of Forestry.* Milwaukee, Wisc.

HAUER, R. J., R. W. MILLER, and D. M. OUIMET. 1994. "Street Tree Decline and Construction Damage." *J Arbor.* 29(2):94–97.

HUDSON, B. 1983. "Private Sector Business Analogies Applied in Urban Forestry." *J. Arbor.* 9(10):253–258.

JOHNSON, C. 1982. "Political and Administrative Factors in Urban-Forestry Programs." *J. Arbor.* 8(6):160–163.

JOHNSON, G. R. 1995. "Tree Care Advisor: A Voluntary Stewardship Program." *J. Arbor.* 21(1):25–32.

KIELBASO, J. J. 1988. *Trends in Urban Forestry Management.* Baseline Data Report 20(1). Washington, D.C.: International City Management Association.

LIPKUS, A., and K. LIPKUS. 1990. *The Simple Act of Planting a Tree.* Los Angeles: Jeremy P. Tarcher.

LONG, D. L., and H. G. LOVE. 1983. "Time Management." *J. Arbor.* 9(2):54–56.

LOVE, H. G., and D. L. LONG. 1983. "Job Analysis: A Management Tool for Instituting Management by Objectives." *J. Arbor.* 9(2):44–47.

LYNCH, D. L. 1985. "Developing Administrative Skills." *J. Arbor.* 11(2):50–53.

MALI, P. 1972. *Managing by Objectives.* New York: John Wiley & Sons.

MCGREGOR, D. 1960. *The Human Side of Enterprise.* New York: McGraw-Hill Book Company.

MCPHERSON, E. G. 1991. "Environmental Benefits and Costs of the Urban Forest." *Proc. Sixth Natl. Urban For. Conf.*, Washington, D.C.: Am. For. Assoc., pp 52–54.

MCPHERSON, E. G. 1994. *Benefits and Cost of Tree Planting and Care in Chicago.* In E. G. McPherson, D. J. Nowak, R. A. Rowntree, eds. *Chicago's Urban Forest Ecosystem: Results of the Chicago Urban Forest Climate Project.* Gen. Tech. Rep. NE-186. Radnor, Pa: USDA Forest Service, NEFES.

MINNESOTA DEPARTMENT OF AGRICULTURE, SHADE TREE PROGRAM. 1982. *A Guide: Community Forestry,* MN7080/Misc.

NIGHSWONGER, J. 1982. "Urban-Community Forestry: Any Which Way You Can." *Proc. Sec. Natl. Urban For. Conf.* Washington, D.C.: Am. For. Assoc., pp. 317–321.

O'BRIEN, P. R., and K. A. JOEHLIN. 1992. "Use of Municipal Tree Standards." *J. Arbor.* 18(5):273–277.

O'BRIEN, P. R., K. A. JOEHLIN, and D. J. O'BRIEN. 1992. "Performance Standards for Municipal Tree Maintenance." *J. Arbor.* 18(6):307–315.

OVERBEEK, J. A. 1979. "Increased Efficiency in Urban Forestry." *J. Arbor.* 5(11):262–264.

PETERS, T. J., and R. H. WATERMAN. 1982. *In Search of Excellence.* New York: Warner Books.

SANDFORT, S., and R. C. RUNCK III. 1986. "Trees Need Respect, Too!" *J. Arbor.* 12 (6):141–145.

SCHWARZ, C. F., and J. A. WAGAR. 1987. "Street Tree Maintenance: How Much Should You Spend Now to Save Later?" *J Arbor.* 13(11):257–261.

SHERWOOD, S. C., and D. R. BETTERS, 1981. "Benefit-Cost Analysis of Municipal Dutch Elm Disease Control Programs in Colorado." *J. Arbor.* 7(11):291–298.

SKIERA, R. W. 1988. Personal communication. Bureau Chief (retired), Milwaukee Bureau of Forestry.

SKIERA, R. W. 1994. Personal communication. Consulting Urban Forester, Milwaukee.

SOMMER, R., F. LEAREY, J. SUMMIT, and M. TIRRELL. 1994. "The Social Benefits of Resident Involvement in Tree Planting." *J. Arbor.* 20(3):170–175.

STIEGLER, J. H. 1989. "Public Preception of the Urban Forest." *Proc. Fourth Natl. Urban For. Conf.* Washington, D. C.: Am. For. Assoc., pp. 40–45.

TATE, R. L. 1976. "Public Relations in Urban Forestry." *J. Arbor.* 2(8):170–172.

TATE, R. L. 1981. "Guidelines for In-Service Training for Urban Tree Managers." *J. Arbor.* 7(7):188–190.

TATE, R. L. 1982. "Applying for Federal Funding Grants for Urban Tree Management Activities." *J. Arbor.* 8(4):107–109.

TSCHANTZ, B. A., and P. L. SACAMANO. 1994. *Municipal Tree Management in the United States.* Kent, Ohio: Davey Tree Expert Company.

VANDERWEIT, W. J., and R. W. MILLER. 1986. "The Wooded Lot: Homeowner and Builder Knowledge and Perception." *J. Arbor.* 12(5):129–134.

WAGAR, J. A. 1991. "Computerized Management of Urban Trees." *Proc. Fifth Natl. Urban For. Conf.* Washington, D.C.: Am. For. Assoc., pp. 151–155.

WELLMAN, J. D., and T. J. TIPPLE. 1989. "A Model for Successful Management of Urban Forests." *Proc. Fourth Urban For. Conf.* Washington, D.C.: Am. For. Assoc., pp. 145–151.

WILLEKE, D. 1982. "The Springtime of Urban Forestry." *Proc. Sec. Natl. Urban For. Conf.* Washington, D.C.: Am. For. Assoc., pp. 207–214.

14

Management of Park and Open Space Vegetation

Urban forest vegetation is described in Chapter 3 as existing on a land-use continuum. This continuum ranges from individual trees through landscaped parks to true forests and associated ecosystems. The management of individual trees is the essence of arboriculture, while the management of forest ecosystems is silviculture. Chapters 10 to 13 dealt with the management of street trees collectively, using arboricultural principles. Following a discussion of public needs this chapter revisits land use planning from an ecosystem perspective and uses this perspective to guide land use allocations. Park and open space vegetation management are then presented using long-term ecosystem management concepts.

The urban forest continuum exists on both public and private lands. Management of forest vegetation on public land is primarily by public agencies, and management of private land is by the owner or consultant hired by the owner. The discussion of forest vegetation management in this chapter is primarily from the perspective of public land management, but the silvicultural principles apply to private lands as well. In both cases management planning must focus on the needs of the owners, prescribe management plans compatible with existing vegetation, and be economically feasible.

PUBLIC NEEDS

Managing parks and open-space vegetation in urban areas involves the development of comprehensive management plans. The planning model described in Chapter 8 serves as the basis of management planning as described in this chapter. To review, the model consists of the following:

1. What do we have?
2. What do we want?
3. How do we get what we want?
4. Feedback.

"What do we have?" includes a vegetation inventory as described in Chapter 7, plus an assessment of public needs. "What do we want?" includes goals developed from the inventories, and "How do we get what we want?" is the management plan that employs arboricultural and silvicultural techniques. Feedback provides for monitoring and updating of management plans as needed.

Public needs are determined through an assessment of all public needs in a community that can be met through the management of the urban forest. The magnitude and priority assigned to each public need will vary in accordance with local conditions, but will fall into general categories that can often be met through multiple use of urban forests. Multiple use does not imply that all lands are used for all functions; rather, it means that where possible, multiple functions will be assigned. Some uses are incompatible, and other uses demand total dedication of a particular area.

National legislation in Germany mandates multiple use of forests throughout the country, and these mandates apply to urban forests as well. Multiple-use management in Germany identifies three basic functions of forests in meeting human needs: recreational, environmental, and product (Schabel, 1980). This management model provides an excellent basis for developing urban forest management plans and is used in the ensuing discussion.

Recreation

The use of urban park and forest land for recreation activities is frequently a top priority for managers. These activities range from intensive recreation in specially prepared high-maintenance sites to extensive recreation in areas receiving low levels of maintenance. There are a broad array of recreational opportunities that can be provided by managers, depending on characteristics of the local population.

Talbot and Kaplan (1984) report that inner-city residents in Detroit, Michigan, valued opportunities to enjoy the outdoors, but felt more comfortable in areas that were well maintained and contained built-in features such as picnic shelters and ball fields. Additionally, respondents felt that densely wooded areas were disorderly and unsafe, describing them as having too many trees and shrubs. These findings are reinforced by Schroeder (1982), who reports that long-term urban residents in Chicago prefer more developed sites than do their suburban or rural counterparts. He found further that trees enhance site quality for users, but that poor maintenance made sites undesirable to visit. Dwyer (1983) feels that urban residents prefer a recreational landscape that contains a mix of forests, scattered trees, meadows, and water.

Whatever individual preferences are, the overall need appears to be for a mix of various cover types for different recreational functions. Driver and Rosenthal (1978) recommend that a variety of green spaces be made available to support different uses, and that they be near neighborhoods where needed. They further state: "Any one area should not be managed to 'be all things to all people.' Instead, each area should be designed and managed to provide those benefits (or experience opportunities) that are most appropriate for it."

Environmental Protection

The use of urban forests for environmental protection functions is described in Chapter 4. In developing forest management plans, German foresters recognize soil, water, emission, visual, climatic, and landscape protection as legitimate concerns in forest management, especially in urban areas (Miller, 1983). Forest belts around Stuttgart, Germany, provide all these functions, especially climatic protection. Rising hot air in the urban heat island draws fresh air into the city through green corridors that penetrate to the city core (Figure 14–1).

Forest Products

Products available from urban forests include wood fiber, wildlife (consumptive and nonconsumptive), and water. Other products sometimes available from urban forest lands are minerals, earth materials, and forage. Urban forests in North America are not often managed for specific forest products such as wood fiber, but rather, serve more recreational and environmental functions. However, harvesting forest products from urban forests is a common activity in other parts of the world, where wood is scarce. Oslo, Norway, is surrounded by a forest of 180,000 ha (450,000 acres), 70 percent of which is in private ownership. Management of this forest is based on legislated objectives to:

1. Provide optimum conditions to give as many people as possible the opportunity pursue the widest range of outdoor activities in rich and varied natural surroundings.
2. Maintain the forests in a biological, technical, and economically sound manner tempered by constructive nature conservation.
3. Adjust the management of forests to met the wishes and requirements of water and health authorities, who are charged to protect drinking water supplies.

Figure 14–1 Stuttgart, Germany, showing urban areas, peripheral forest belt, and green spokes radiating from the city center.

4. Provide ecologically precise conditions for wildlife conservation and sport fishing maintenance.

Constrained by these mandates the forest produces 380,000 cubic meters of pulp and timber per year (Mjaaland and Andresen, 1986).

Timber management is common to nearly all forests in Germany, a country that imports 50 percent of its wood fiber. In peripheral urban forests timber management is strongly influenced by recreational and environmental functions, but is practiced on most lands (Figure 14–2). Silvicultural systems are selected that are compatible with other uses, with revenues from timber sales supporting all other management activities, including recreation (Miller, 1983).

Wildlife as a product of urban forest management has a high value to urban residents. Most use of urban forest wildlife is nonconsumptive, including activities such as bird and other animal watching, feeding, and identification. Consumptive use (hunting and collecting) is not done in most North American urban forests, but is in Europe. Like all hunting in Europe, hunting in urban forests is carefully controlled. Hunting licenses are expensive and involve passing a stringent examination, animals

Figure 14–2 Timber harvesting is carried out in urban greenbelt forests throughout Germany.

are often preselected for harvest, and fees are charged per animal if the hunter chooses to keep the meat. Hunting is deemed necessary to control species where natural predators have been eliminated, with fees helping to support management.

Regulations against hunting can create serious management problems for urban forest managers. Deer in the Cook County Forest Preserve District (Chicago) are found in herds of sixty to seventy individuals, and annual road kills number 400 to 500 animals. Some woodlands are completely browsed up to a height of 4 feet. Additionally, beavers have flooded roadways to the point where some roadbeds had to be raised to accommodate higher water levels (Buck, 1982).

Some urban forests were originally acquired as municipal watershed lands, but have become important for other uses beyond water supply. On lands where water quality is a top priority, other management must accommodate this use. In the German management model these lands are assigned water protection status under the environmental protection function. Beyond protection, vegetation management can affect water quality in a positive manner. Smith (1969) suggests that conifers be established on stream and reservoir banks to keep hardwood leaves out of the water. Hardwood leaf decomposition products react with chlorine to impart a disagreeable taste to drinking water. Timber harvesting can increase water yields for a short period by temporarily reducing transpiration, a critical factor where demand for water is exceeding supplies (Smith, 1969).

Within a few years more than 50 percent of the world population will reside in urban areas. Urban forests can play a crucial role in meeting the needs of residents in cities and towns in developing counties. Kuchelmeister (1993) suggests that although developed countries regard urban forests primarily from an amenity perspective, urban forests in developing countries can and must be used to help alleviate poverty and improve the living environment of residents. He divides the benefits of urban forests into those provided form conventional forestry and social forestry (Table 14–1), and suggests that all activities must be acceptable to the local population.

Land-Use Planning

Planning for open space and urban forests is a function of land-use planning, as discussed in Chapter 8. French and Sharpe (1976) state: "Urban forestry embodies the concept of multiple-land use and promotes the sustained-yield value of forestry. It is also a part of the holistic approach necessary for managing our total ecosystem, and must not be viewed in urban isolation." The multiple-use law in Germany applies to all forest land, regardless of ownership. Landscape protection provisions call for designation of landscape protection zones in which land use cannot be changed unless permitted by planning authorities. This includes major changes in vegetative cover (i.e., forest to agriculture or meadow to forest).

TABLE 14–1 MAJOR BENEFITS OF THE URBAN FOREST IN CONVENTIONAL AND SOCIAL FORESTRY. (From Kuchelmeister, 1993.)

Multipurpose Urban Development Forestry	
Conventional Forestry	Social Forestry
• reduces noise	• provides food
• reduces air pollution	• provides fuel
• reduces climatic extremes	• provides fodder
• cools cities and planet	• provides timber
• conserves energy	• provides fencing material
• provides beauty and shade	• provides medicine, oil
• improves water quality	• provides raw material, fiber, etc.
• controls water runoff	• increases cash/subsistence income
• provides habitat for wildlife	• provides employment
• increases recreation value	• improves gardening conditions
• increases health/well being	• plus all functions of conventional forestry

Overall forest management in Germany begins with a mapping system that identifies the primary function of all forest land within each management unit. Timber management is ubiquitous, but not necessarily the prime function unless specifically identified. Primary functions identified for each management unit receive top priority within that unit, and any other uses must conform to the management objectives for that function (Figure 14–3).

Open-space planning in North America primarily involves public acquisition of the land or development rights to that land. The city of Seattle, Washington, has de facto forested greenbelts in private ownership within the city due to construction difficulties on steep slopes. However, new construction technologies have made these lands developable, so the city is now purchasing some of these lands as permanent greenbelts (Black, 1982). The Cook County Forest Preserve District ownership goal is 75,000 acres, or 12 percent of the county. As of 1982 the district had achieved 88 percent of that goal (Buck, 1982).

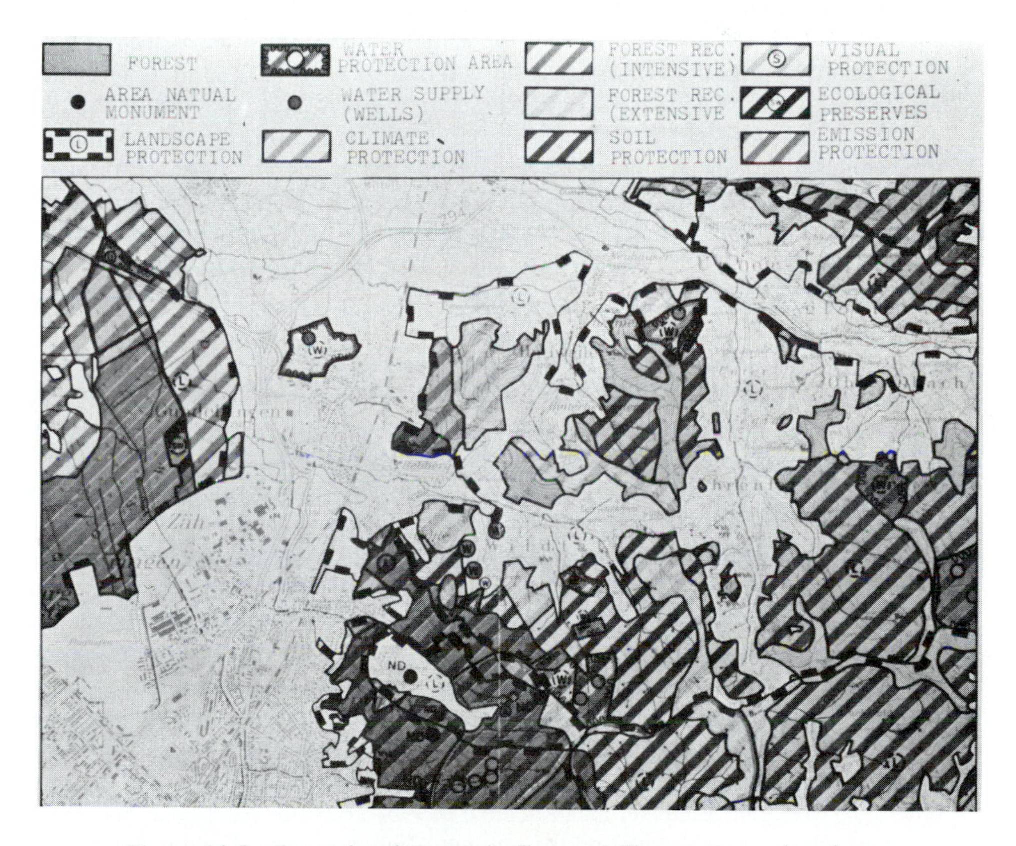

Figure 14–3 Forest function map in Germany. These maps are based on a complete resource inventory and identification of "people" needs. Management plans emphasize the primary function in each area.

McHarg (1969) recommends that land-use planning decisions be made on map overlays that depict development limitations by degree of shading for each resource identified as critical. For example, areas to be developed for housing may consider topography, soils, water table, drainage, and transportation access and other necessary infrastructures as critical resources. When all overlays are placed on the base map, areas shaded most deeply should be protected, and areas shaded least, designated for development. This method of allocating land use can also be used to allocate uses within a management area. A publicly owned urban forest will have heavy and conflicting use demands. Specific activities may be designated to the most appropriate sites through this technique, and sites capable of supporting multiple nonconflicting uses identified.

Planning and diversity. Conservation biology is broadly defined as the science of preserving biological diversity and includes compositional, structural, and functional diversity. Compositional diversity includes species richness, genetic richness within a species, and diversity of ecosystems in the landscape. Structural diversity is how the diversity of species, habitats, and communities are arranged in time and space across the landscape, and within a specific ecosystem. Functional diversity includes the broad ecological processes in the landscape such as disturbance and interactions between ecosystem (Anderson, 1993). Historically, ecosystems have been subjected to repeated disturbance regimes by fire, flood, wind, ice, climatic change, insects, disease, and human influences. Species in these systems have likewise evolved either to tolerate or to take advantage of disturbances. Modern land use, management, and development works against conservation of species by fragmenting habitats, introducing exotic species, and interrupting cycles of natural landscape disturbance or introducing different cycles.

Urban development will cause the loss of species such as large predators that are not compatible with humans, or species not adaptable to fragmentation or changes in disturbance regimes. Anderson (1993) observes that ecosystems become simpler in function and composition as human influences increase in the landscape, reaching the greatest simplicity in the urban core. However, careful planning and management will protect, restore, and enhance some level of urban biodiversity.

Land-use planning can identify and protect critical habitats during development, and consider the number, size, and location of greenspaces in the overall land-use plan. Diversity is better preserved by fewer large reserves rather than many smaller areas. Likewise connecting greenspaces with corridors can increase habitat size and diversity, especially if they follow riparian corridors. Planners should try to represent all native ecosystems in an area, and managers manipulate vegetation to provide all seral stages. Likewise, structural diversity and control of exotics within ecosystems should be a management goal (Anderson, 1993).

Milligan Raedeke, and Raedeke (1995) consider patch size and shape important relative to diversity and the amount of edge present. Edge is defined as the area where two different ecosystems abut one another, and are normally considered species-rich habitants. This can be beneficial if the species present enhance diversity and are considered desirable, or can be detrimental if species attracted work against existing rare species. A large patch that is round will have much less edge than a similar size patch that has an irregular shape. They likewise recommend managing smaller patches of habitat for greater structural diversity to increase biological diversity.

Dunster and Dunster (1992) recommend that planners identify environmentally sensitive areas (ESAs) as they develop land use plans. They define an ESA as:

> Any parcel of land, large of small, under public or private control, that already has, or with remedial action could achieve, desirable environmental attributes. These attributes contribute to the retention and or creation of wildlife habitat, soils stability, water retention or recharge, vegetative cover, and similar vital ecological functions. Environmentally sensitive areas range in size from small patches to extensive landscape features. They can include rare or common habitats, plants, and animals. Taken together, a well-defined and protected network of environmentally sensitive areas performs necessary ecological functions within urban and rural landscapes. This network makes a very important contribution to the overall quality of life for all species living in and around the area, and plays a particularly important role in maintaining the health and livability of city and urban landscapes.

Following this definition, they recommend ESAs and greenspaces be linked into a network to maintain larger continuous open spaces, preserve ecological continuity, and attempt to achieve no net increase in water runoff.

Protecting biological diversity can extend to the individual development plan by the inclusion of greenspace patterns that connect to other greenspaces. Homeowners and other property owners can be encouraged to contribute by their personal landscape plans. Nassauer (1993) observes that weed-free lawns with few trees and shrubs contribute little to wildlife habitat, but that less conventional landscaping that includes native plant communities and beds of diverse species will enhance diversity, especially if contiguous with neighboring property of similar design. Education that leads to a better understanding of ecological processes by property owners can lead to their adopting alternative landscaping provided that the landscape meets community aesthetic expectations such as the appearance of intent and care, and provided that some portion of the property retains a more conventional landscape.

Planning in the watershed. Urbanization can dramatically alter the quantity and quality of water in watersheds. Urban development generally increases water runoff and pollution, and decreases evapotranspiration and infiltration. This results in surges of water during precipitation and less water between storms to maintain stream flow and aquatic ecosystems. High runoff can erode stream channels, adding to pollution, and carry additional pollutants from paved surfaces. As urban areas grow, careful planning can reduce these problems by protecting riparian areas and wetlands, maintaining and establishing as much plant cover as possible, and by using porous hard surfaces and retention ponds to increase infiltration.

PARK VEGETATION MANAGEMENT

Park vegetation is defined for this discussion as a landscape consisting of scattered shade trees, open lawns, and shrub and flower beds. Management focuses on scheduling activities, maintenance rotations, and development of stable systems. Specific horticultural techniques are described in texts dealing with turf management, arboriculture, floriculture, and ornamental horticulture.

Management of park vegetation must be carefully weighed against uses, costs, and user preferences. All too often the vegetation management scheme is based on historic management, not present uses and needs; in other words, "We've always done it this way." As a general rule it is cheaper to maintain a natural ecosystem than an ecosystem not naturally found in the region. A forest in a humid temperate climate is a self-maintaining ecosystem, but a savanna (park) must have succession held back through mowing and/or other management to prevent a forest from developing. Similarly, lawns, parks, or forests in arid regions require irrigation, whereas plant communities adapted to local conditions will be less expensive to maintain.

When developing vegetation management plans for parks, the use of each area should be considered before selecting the appropriate vegetative cover. Mowing vast expanses of lawn is justified if use requires turf, or if the community prefers it that way and is willing to bear the cost. However, other ground covers such as shrub beds may be cheaper to maintain, and allowing some areas to revert to natural conditions will produce substantial savings. Flower beds are expensive to establish and maintain, but if they produce high user satisfaction the cost is justified. Many park vegetation schemes were developed when labor costs were much lower. Subsequent budget squeezes have produced substandard management by attempting to hang onto the old system. In the face of present labor costs and limited finances, creative vegetation management can produce high-quality park landscapes at a reasonable cost.

Park Vegetation and Crime

Statistics from the United States Department of Justice indicate that nearly 5 percent of violent crime occurs in public parks or similar recreation areas. This has led to extensive removal of vegetation in many urban parks to reduce cover for criminals and hopefully improve safety. It has also led to the loss of aesthetics and wildlife habitat in these parks. A study of crime and vegetation in parks by Hull and Michael (1995) indicates that vegetation can contribute to crime in parks, and they suggest the following management approach to parks where criminal activity is a problem.

1. The wholesale removal of vegetation is unnecessary. Vegetation is rarely the sole cause of crime and often causes no problem.
2. Law enforcement is the primary deterrent to crime.
3. Park personnel should focus vegetation maintenance on specific areas and create safe places people can use with confidence.
4. Be sure to target problem areas. Some areas of parks become "hot spots" of re-

peated criminal activity. Police or park records can identify these locations. Vegetation management might be one part of a joint effort with enforcement, transportation access, and scheduling.

5. Open and maintain sight corridors. Views from corridor roads or buildings into strategic areas might enhance safety by enabling surveillance.
6. Minimize vegetation near circulation patterns that offers opportunity for concealment. Keep vegetation low around parking lots, water fountains, pathways and other places where park users congregate.
7. Minimize escape routes by converting thick vegetated patches into several thin strips. Thick patches offer a maze of cover for fleeing criminals, strips of vegetation can still provide the visual massing of natural elements that make parks attractive.

Park Tree Management

Management of individual park trees is similar to street tree management. Trees must be planted, trained, protected, pruned, repaired, and ultimately replaced. Environmental conditions in parks are not normally as harsh as on city streets, so park trees usually require less overall maintenance, except in areas of very high public use.

Planting. Planting trees in parks should be based on the desired overall tree cover for each area. Not all areas should be planted, as open spaces are often used for activities such as nonorganized sports. A newly planted tree in the middle of a popular site for these activities will probably have an extremely short life span.

Prior to tree planting, a park vegetation inventory should be taken. A park tree inventory should identify the number of trees by species in each size class to determine the present age structure. Long-term management plans for park trees should be based on maintenance of optimum stocking, and development of diversity in species and age classes.

Green (1984) describes an inventory system designed to assess the health and longevity of park trees (see Chapter 7). A condition rating assigned to each tree describes its size and health, estimates longevity, recommends immediate removal if necessary, and identifies planting spaces. When tested in a city park, the system estimated that over 53 percent of the existing trees will be gone in twenty years. The tree management program recommended for the park included:

1. Planting in excess of losses to allow for transplanting mortality and to build up the tree stock
2. Planting areas where highest losses are anticipated
3. Stressing diversity in replanting (i.e., no more than 10 to 15 percent of any one species)
4. Emphasizing native species adapted to local soils and climate
5. Corrective maintenance of declining trees to reduce the loss rate

Green further suggests that in parks where high losses are anticipated, the planting rate should exceed replacement without changing the character of the landscape.

It may be impracticable to plant trees as individual replacements in a grove of trees. Even tree species that are tolerant of shade will grow slowly when shaded by the canopies of surrounding trees. In landscapes characterized by groves and open areas it may be better to plant new groves in open areas as replacements for groves that are beginning to decline. In time the declining grove can be completely removed and converted to open area.

Maintenance. A newly planted part tree demands the same training period as is necessary for a street tree. Pruning for sidewalk and street clearance is not necessary, but the crown must be lifted as the tree develops so as not to interfere with pedestrian traffic and mowing crews. A programmed maintenance cycle should be established, but frequency may not be as important as for street trees, due to fewer traffic restrictions caused by park trees. However, safety considerations are just as important, and a poorly maintained tree causing damage or injury can be cause for a lawsuit.

A young park tree needs frequent pruning to develop a sound structure, while mature specimen maintenance needs are met by less frequent deadwooding, light pruning, and minor repairs. Turf fertilization will normally provide needed nutrients, and nonscheduled repair work should be assigned on an as-needed basis. Programmed maintenance budgets may be estimated from inventory data and previous maintenance budgets.

Removal/replacement. Parks attract a great number of visitors, and the municipal parks department has the responsibility to maintain trees in a reasonably safe manner. Park trees can present significant hazards when two conditions are met: There must be a structural defect that is likely to fail, and there must be a target. Dead and dying trees, split crotches, extensive cavities, trunk splits, evidence to root decay, large dead limbs, and other defects are indications of potential failure. The presence of a target is also necessary for a tree to be deemed a hazard. A defective tree in a forested grove that is not in or near an area where people frequent may not be a hazard at all. However, if that tree is leaning over a trail, or it is near a building, picnic area, or other developed site, then it is a hazard. The key is to consider where it will land if it fails, and what it will hit. This does not mean that every tree with a crack, cavity, or other defect is likely to fail, but it does mean the tree should be monitored to determine if the condition is becoming more serious. Annual inspections of park trees in risk areas should be made by a knowledgeable employee or consultant to assess hazards and to make recommendations regarding repair of removal. It is advisable to make additional inspections following storms that may have caused tree damage. If a tree is determined to be a hazard and it cannot be repaired, it should be removed as soon as possible, and replaced if appropriate. Sometimes it is easier and cheaper to move a picnic table then to remove a tree. Comprehensive guides to hazard tree evaluations are available from the International Society of Arboriculture,

P.O. Box GG, Savoy, IL 61874, and from forestry and parks departments. Annual removal budgets may be estimated from previous removals and from inspections.

Turf Maintenance

Turf management activities include fertilization, dethatching, aeration, mowing, and pest management. All turf management costs can be kept lower by proper planning of those areas to be maintained in grass cover. Most maintenance activities are now done with specialized equipment, and design that allows for efficient use of this equipment results in lower maintenance costs. Some sites may not be suitable for sod due to shading or soil conditions. These areas can be placed in a mulch or vegetative ground cover.

Turf fertilizers are generally high in nitrogen, with lesser amounts of other elements. Fall application with a slow-release fertilizer is recommended to allow for cool weather growth and fast start in the spring. Additional applications may be made in spring and early summer with faster-release fertilizers. Remember, frequent fertilization combined the irrigation will produce very rapid growth, and this will mean more frequent mowing. Fertilization and irrigation should be in balance with mowing budgets.

Rapid turf growth and frequent mowing result in a buildup of partially decomposed glass cuttings known as thatch. If thatch becomes too thick, it may cause disease problems and reduce water penetration. Thatch can be mechanically removed with a dethatcher, usually in the spring. However, if turf can be maintained without thatch buildup by using less fertilizer and irrigation, these funds can be diverted to other activities.

Heavy soils subject to intensive use are prone to compaction. Sod growing on these sites will decline in vigor due to lack of oxygen in the soil and poor water penetration. Mechanical aeration alleviates this problem by punching holes in the upper soil horizons. However, aerators should be kept away from trees as root damage occurs when used close to the trunk. General use of mechanical aerators is often not necessary on lighter soils or on areas subject to light use. Observation of turf conditions and use of soil survey maps of field checks will determine those areas in need of treatment. Aeration is usually done once a year in spring, but may be done any time if conditions warrant.

Mowing should be scheduled throughout the growing season, but frequency can be controlled somewhat be fertilization and/or irrigation rates. Areas subject to more intensive use will receive more intensive management to maintain a healthy turf, including more frequent mowing. As a general rule, turf can be cut shorter in cool weather without harm to the plant. However, hot weather mowing should leave longer stalks to keep the soil moister. Maintenance of each area of turf should be governed by the level of use, aesthetic goals, and budgetary constraints.

Pest management in turf involves the control of unwanted insects, disease, and plants. Traditional turf management prescribes broadcast applications of fertilizers, insecticides, fungicides, and herbicides in combination on a preventive maintenance basis. Applications are made several times during the growing season, the frequency depending on the local climate. Recent controversies over this practice warrant a shift to integrated pest management (IPM) to control turf pest problems, especially on public lands.

Shrub and Ground Cover Maintenance

Shrubs, while attractive individually, are easier to maintain and can be more attractive when massed in beds. Mowing costs will be lower if mowers pass on the outside of a designated shrub bed rather than mowing between individual plants. Shrubs will grow faster and be more vigorous if they do not have to compete with sod for elements and water. Shrub bed maintenance costs can be kept down by maintenance of a mulch layer within the bed. This improves soil moisture conditions and reduces invasion of the bed by weeds.

As in tree management, shrubs must be planted, pruned, and replaced as necessary. Pruning is needed more frequently than on trees, often once per year to remove deadwood, thin, control size, and reinvigorate the plant. Timing of pruning flowering shrubs is critical. Shrubs that set flower buds during the previous growing season should not be pruned until after flowering is finished, while shrubs that flower from buds set during the current growing season should be pruned when dormant. A shrub that needs to be pruned several times during the growing season may be the wrong species for the growing space, unless a shaped hedge is the desired landscape feature. Broadcast fertilization on the mulch is an efficient and effective means of getting nutrients to the plants. Fertilizer timing for shrubs is the same as for trees, fall and/or spring.

Ground covers provide an attractive alternative to turf in the proper location. These areas must be protected from pedestrian traffic, as they are not nearly as resilient as turf. During establishment frequent weeding is necessary, but once well established, ground covers are inexpensive to maintain, needing only occasional weeding and fertilization. Establishment costs can be reduced through the use of an appropriate herbicide prior to planting, and mulching between plants after planting.

Flower Bed Maintenance

Flowers beds are the most expensive vegetation to maintain a per unit area basis, but elicit the most favorable response from the public. Flower beds for annuals must be tilled, planted, fertilized, cultivated, weeded, and pests controlled when necessary, all of which is labor intensive. Perennial flowers and bulbs do not need replanting each year, but the other costs are the same. However, some bulbs must be redug each year for separation and/or winter storage.

Flower beds should receive weekly maintenance to remove weeds, prune, and monitor for pests. Maintenance costs can be reduced by selecting species that naturalize and become self-maintaining. Naturalized flower beds are informal and should be in borders and under trees and shrubs rather than in planters or mixed with maintained turf. Flower beds are best placed in prominent locations for viewing by the maximum number of people (i.e., park entrance, in front of municipal buildings, or in boulevard medians).

URBAN SILVICULTURE

Urban silviculture is defined as the art of reproducing and managing forests continuously to obtain sustained yields of forest benefits in urban regions through the application of ecologic principles. Traditional silviculture places emphasis on wood production, while urban silviculture has as primary functions recreations and environmental protection, but does not preclude wood fiber production. The transition in management concepts from arboriculture to silviculture becomes somewhat arbitrary in urban forest management. Care of individual trees is arboriculture and management of tree communities is silviculture, but in urban forestry a forest community may be manipulated as a whole, while a tree in that community receives individual attention. From a spatial perspective it is also difficult to distinguish where management of park vegetation becomes silviculture, especially the transition from informal tree, lawn, and shrub cover to a woods or forest community. There will be areas where both silviculture and horticultural principles apply, and these should be so designated in the management plan. With an understanding of this overlap, the following discussion of urban silviculture deals primarily with the manipulation of forest communities.

Silviculture, Ecosystem Management, and the Urban Forest

Forest ecology forms the basis for the practice of silviculture, which ranges from an understanding of the ecological needs of individual species to the functioning of tree communities and the interactions of forest communities across broad landscapes. Traditionally, the practice of forestry involves focusing on each individual stand of trees and seeking to optimize the desired values we hope to obtain from that stand. While this approach does well in providing high yields of forest products, it sometimes works against the maintenance of biological diversity in the landscape. Protecting species is best approached from the perspective of protecting their habitats or ecosystems on a broad landscape scale, rather than attempting to rescue each individual species when it has been determined it is becoming scarce (Franklin, 1993). In forest planning this involves an assessment of existing conditions, attempting to understand historic conditions and disturbances that shaped systems before "management," and designing plans that meet the spatial and temporal needs of all species in the landscape. This concept is termed ecosystem management, and it now serves as the basis for forest management on most public lands, and increasingly on private lands.

The level of disturbance and development found in urban regions may not meet all the criteria necessary for ecosystem management as conceptualized in rural landscapes, but as discussed earlier in this chapter under the section of planning and diversity, aspects of conservation biology can be applied to the planning and management of urban forest ecosystems. Though often unnoticed, urban areas contain a great deal of plant and animal diversity in parks, cemeteries, wetlands, riparian corridors, industrial sites, residential neighborhoods, and undeveloped areas. The location, size, and shape of greenspaces, their connections to each other and to the intentional use of vegetation on public and private land all affect biodiversity in urban regions. The manipulation of vegetation within this matrix through silviculture can also enhance diversity and can also be used to educate the urban public in the science of ecology and ecosystem management.

Forest management plans for urban forests first involve setting desired goals and objectives for the forest and then subdividing the forest into stands for specific silvicultural prescriptions. The stand is the smallest management unit is silviculture, and is a portion of the landscape with similar site conditions containing a plant community of similar species and age classes. The most important ecological concept is to understand is the dynamic nature of forest communities. Change is the rule rather than the exception in forest stands, as plant succession proceeds or is periodically interrupted by natural and human-induced events.

Unfortunately, the dynamic nature of forest communities is a concept poorly understood by the general public. The public tends to view a stand of trees as a permanent feature of the landscape rather than a transitory community that will change into something else even if untouched. However, this perception by the public is also the opportunity for the urban forester to establish a dialogue with the public and interest groups when developing management plans for the forest. Through the planning process the forester can gain knowledge of what the public wants from the forest and how they desire the forest to be in the long term. Likewise, the public will gain a better understanding of the dynamic nature of ecosystems, the role disturbance plays in ecosystems, and the need to manage ecosystems. Complete exclusion of disturbance in urban forest stands will, over time, result in the loss of many vegetation types that are dependent on disturbance, and with them will go wildlife species that can survive only in that type. Silvicultural applications can be used to set back succession, accelerate succession, change vegetation types, accelerate the process of developing old growth conditions, and regenerate stands that have begun to decline. Silviculture will replicate natural disturbance events to a degree, although removal of wood is different than leaving it behind as would happen following wind or fire. Many urban residents may not at first consider manipulation of forest stands by cutting appropriate management, but many consider managed rural forests aesthetically more pleasing than unmanaged forests.

Brush (1981) conducted a forest aesthetics survey of people in Massachusetts who own forestland for recreational purposes. When asked to score twenty photographs of forest stands for their aesthetic quality, respondents rated stands that had been under management for several decades as the most attractive. These stands contained large trees, were more open, and had trunks clear of low branches, conditions readily obtained through thinning, improvement cutting, and pruning. Public acceptance of urban silviculture is a matter of forest ecology education, as aesthetics can be enhanced rather than damaged through cutting. The silviculture system prescribed for an urban forest must be carefully selected with aesthetics a prime consideration. No amount of education will yield public acceptance of a large clearcut in an urban forest, even though it may be ecologically sound.

Odum (1971), in discussing the ecological principles and urban forests, suggests that diversity is the most important consideration in management followed by species selection. He contrasts urban greenbelt forests with commercial forests in terms of ecological characteristics and management goals (Table 14–2), recommending later succession stage and careful site/species matching. However, it is important to recognize that the concept of diversity applies not only to stands, but to the entire forest as well. An urban forest is more diverse and perhaps most stable when it consists of a mosaic of stands representing many different stages of succession, including nonforest. Some individual stands may approach monocultures, a situation not atypical with early successional species. It is not only important to consider diversity within stands and among stands, it is also important for biological diversity to plan for the amount and location of different forest types across the entire landscape. These patterns can shift over time with succession and disturbance as long as enough critical habitat for rare species exists at any one point in time for their survival.

Wildlife considerations. Wildlife is an important part of urban forests, and a highly diverse forest consisting of many successional stages will provide the greatest diversity in habitat, and a diverse wildlife population (Leedy et al., 1978). Edges between diverse communities are important, as are highly stratified communities. When developing urban forest management plans, the following steps should be taken to ensure a diverse population of wildlife for enjoyment by urban residents (Leedy et al., 1978):

TABLE 14–2 ECOLOGICAL CHARACTERISTICS OF COMMERCIAL FORESTS AS CONTRASTED WITH PROTECTIVE GREENBELT VEGETATION. (From Odum, 1971.)

Features	Commercial Forest	Greenbelt Vegetation
Species diversity	Low (usually monoculture)	High (mixed species)
Age structure	Even-aged	Multiaged
Annual growth increment	High	Low
Stratification	One-layered (mostly canopy trees)	Multilayered (understory, and ground cover well developed)
Mineral cycles	More open (losses from leaching and runoff)	More closed (retention and recycling within stand)
Selection pressure	For rapidly growing, sunadapted species (often softwoods)	For slower-growing, shade-tolerant species (more hardwoods)
Maintenance costs (replanting, fertilization, pest control, thinning, etc.)	High (requires “management”)	Low (self-maintaining)
Stability (resistance to outside perturbations such as storm, pest outbreaks, etc.)	Low	High
Overall function	Production of marketable products	Protection of the quality of man’s environment

1. Identify habitats and determine their relative value for wildlife.
2. Identify habitats of threatened and endangered species.
3. Identify groups of plant species of value to wildlife.
4. Analyze adjacent land uses.
5. Develop a continuous open space/wildlife corridor systems.

Site plans. Prior to initiating vegetation management plans, a complete resource inventory of the entire area is necessary to provide a database for development of management plans. In many regions habitat classification systems have been developed that not only identify the present community, but also provide successional tendencies and management recommendations (see Chapter 7). Each stand is identified and inventoried to determine species present, successional stage, soil characteristics, wildlife habitat, water resources, aesthetic characteristics, and any other unique features that influence the management plan. Stands may then be assigned primary and or multiple-use categories and the appropriate vegetation management system developed using silviculture techniques described below. There may be unique ecosystems or areas where no manipulation is desirable, and this should be described in the management plan.

The management plan must recognize the dynamic nature of each ecosystem, and silviculture prescriptions used to shape ecosystems in accordance with management goals. It may be desirable to accelerate succession, set back succession, or maintain the status quo of each community. Noyes (1971) suggests that when making land-use allocations the basic approach include:

1. A determination of the location and amount of land available or that could be set aside for each use
2. Decisions as to what major use or combination of uses such lands will be devoted
3. The potential for manipulating or adapting vegetation to these purposes

In most cases it will be less expensive to manage communities that are self-maintaining ecosystems rather than work against ecologic tendencies such as succession, competition, diversity, and stratification.

Long-term impacts. Ecosystems by their very nature change over time. Climax communities that are described for broad climatic regions develop, in theory, as plant communities interact with xeric (dry) and hydric (wet) sites to produce mesic (intermediate) conditions. In reality the time frame in which this occurs will often be far beyond the range of management planning. Exceptions occur when external events accelerate succession, such as when excess nutrients cause rapid eutrophication of aquatic ecosystems. However, under normal circumstances and planning horizons, succession occurs up to the point where a self-maintaining community reaches stasis with a specific site. When normal succession is disturbed the community will respond in a variety of ways. Alteration of vegetation will have four general long-range impacts on forest communities:

1. Canopy reduction will accelerate growth on remaining trees, but will also encourage the development of additional strata and may accelerate succession.
2. Introduction of a new species into the ecosystem will result in a reduction in frequency of one or more existing species.
3. Each change made in the present stand will probably cause a permanent alteration in the future stand.
4. Wildlife will be tied to all silvicultural manipulations.

Special considerations. As stated earlier in this chapter, carefully selected silvicultural systems will produce stands that are aesthetically pleasing to urban residents. However, a word of caution is necessary before making final silvicultural prescriptions for urban forest stands. Attitudes and perceptions of professionally trained foresters may differ with those of the general public when viewing the visual impact of silvicultural activities. Willhite and Sise (1974) compared responses of forestry and nonforestry students to photographs of different silvicultural treatments. They found that forestry students were likely to analyze each photograph from an ecological perspective prior to making a management judgment while nonforestry students rated management primarily on aesthetics. They conclude that the public judges the competency of foresters by the appearance of the landscape, and that silviculture must consider scenic qualities to be acceptable to the public. Nowhere is this more critical than when manipulating vegetation in urban forests.

Timing and operational methods are important social considerations in urban silviculture. Chain saws and crawler tractors do not mix well with people seeking solitude or recreation in urban forests, not to mention obvious safety problems. Tree cutting should be scheduled during periods of low use, and skidding done when there will be minimal site scarring, such as when the ground is frozen or during dry weather. Power equipment that is well muffled will attract much less attention, and hand tools, although more labor intensive, may be a useful alternative.

Management plans for private holdings in urban areas or for rural land owned by urban residents will probably be more acceptable if foresters understand the motives for owning land. Timber production does not provide enough monetary return to justify owning land for that purpose in urban regions, as purchase price and other costs exceed income from timber sales when subjected to soil rent analysis. This land is held for other reasons, such as recreation, aesthetics, or speculation. The forester must be aware of these reasons before making management recommendations. In many cases aesthetics will be a prime consideration when writing management plants (Miller et al., 1978).

Even-Aged versus Uneven-Aged Stands

Once the forest has been subdivided into stands and land uses defined, the manager must decide if each vegetative community can and should be maintained as it is, or if it should be changed to something different. Knowing the age structure of each stand is essential to understanding how it will respond to use or change, and this can be determined through the stand-analysis technique, in which diameter classes are plotted by frequencies (Figure 14–4). Stand age structure will then form the base for the development of specific management plans by describing successional tendencies.

Two basic systems of stand management are available to the manager: even-age management or uneven-age management. Even-age stands contain trees that are within one-fifth of the rotation age of each other; rotation being defined as how long the trees will be allowed to grow. In some cases an even-aged stand will contain two distinct age classes, a mature overstory and an understory. Uneven-age stands contain three or more distinct age classes, and may be all-aged with all age classes represented, from seedlings to mature specimens. Even-aged stands result from disturbance, are usually composed of shade-intolerant species, and, through succession, eventually become uneven-aged. As a management concept, intolerant to intermediate-tolerant species are managed in even-aged stands, while intermediate to tolerant species are managed in uneven-aged stands. Table 14–3 is a summary of understory tolerance for common North American tree species.

A number of silvicultural treatments are available to the urban forest manager. These range from establishment techniques, through intermediate cuttings, to harvesting and regeneration systems; all of which will maintain or regenerate existing communities, or will change the community to a different ecosystem. Other silvicultural techniques discussed include sanitation/salvage cutting, fire management, and protection of unique ecosystems.

Establishment

Nonforest plant communities exist as meadows or former pastures, wetlands, prairies, and abandoned fields that have been invaded by forbs, grasses, and shrubs. In the past, urban forest managers quickly planted trees on suitable sites following procurement in order to establish forest communities. However, in recent years many managers have been attempting to keep these areas free of forest vegetation so as to provide landscape and habitat diversity. In the American Midwest many communities now have active prairie restoration projects designed to preserve small samples of what was the dominant plant cover of the region prior to agriculture (Figure 14–5). Three management approaches can be taken with nonforest plant communities: the existing community

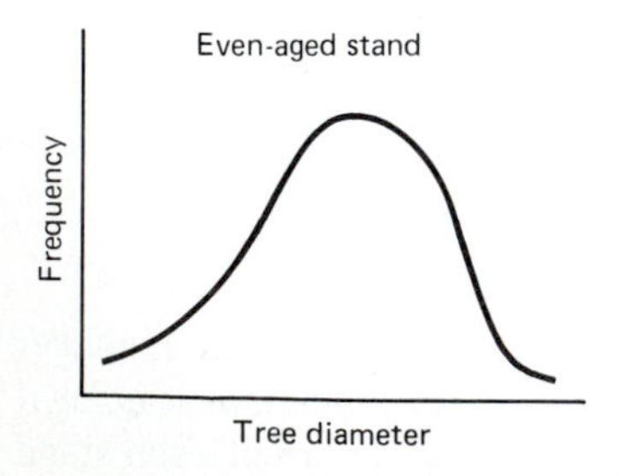

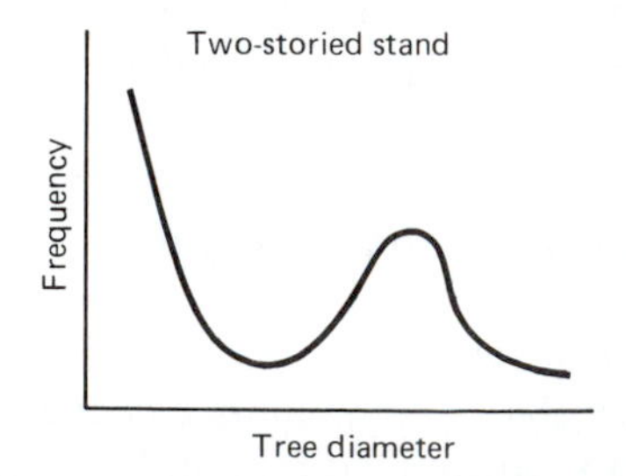

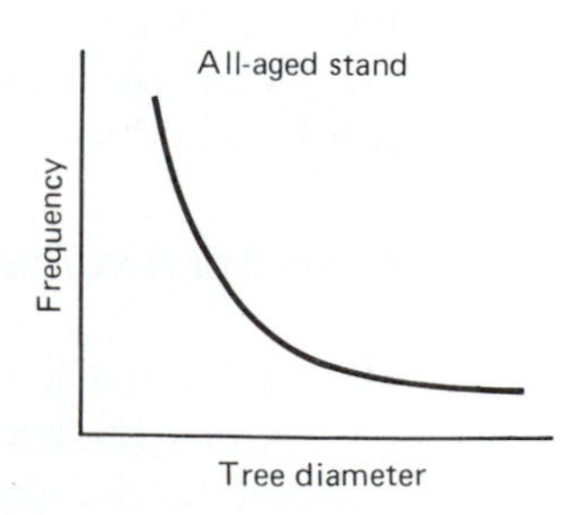

Figure 14–4 Stand profiles of even-aged, two-storied, and all-age stands.

TABLE 14–3 RELATIVE UNDERSTORY TOLERANCE OF SELECTED NORTH AMERICAN FOREST TREES.[a] (From Spurr and Barnes, 1980.)

Eastern North America		
Gymnosperms	Angiosperms	Western North America
	Very Tolerant	
Eastern hemlock	Flowering dogwood	Western hemlock
Balsam fir	Hophornbeam	Pacific silver fir
Red spruce	Sugar maple	Pacific yew
	American beech	
	Tolerant	
White spruce	Basswood	Spruces
Black spruce	Red maple	Western red cedar
		White fir
		Grand fir
		Alpine fir
		Redwood
	Intermediate	
Eastern white pine	Yellow birch	Western white pine
Slash pine	Silver maple	Sugar pine
	Most oaks	Douglas fir
	Hickories	Noble fir
	White ash	
	Elms	
	Intolerant	
Red pine	Black cherry	Ponderosa pine
Shortleaf pine	Yellow poplar	Junipers
Loblolly pine	Sweet gum	Red alder
Eastern red cedar	Sycamore	Madrone
	Black walnut	
	Scarlet oak	
	Sassafras	
	Very Intolerant	
Jack pine	Paper birch	Lodgepole pine
Longleaf pine	Aspens	Whitebark pine
Virginia pine	Black locust	Digger pine
Tamarack	Eastern cottonwood	Western larch
	Pin cherry	Cottonwoods

[a]Based on representative site conditions for the respective species. Survival in the understory is related to light irradiance, moisture stress, and other factors. As a general guide to the light irradiance component, we estimate the range of minimum percentage of full sunlight for a species to survive in the understory at each of the five arbitrary levels of tolerance: *Very tolerant* species may occur when light irradiance is as low as 1 to 3 percent of full sunlight; *tolerant* species typically require 3 to 10 percent of full sunlight; *intermediate* species 10 to 30 percent; *intolerant* species, 30 to 60 percent; *very intolerant* species, at least 60 percent. For example, an intolerant species competing in the understory is unlikely to survive with less than about 30 percent of full sunlight (unless other compensating factors are favorable).

maintained as it is, the community allowed to change through natural succession, or the community changed through direct intervention by planting.

Nonforest maintenance. In earlier times people and nature maintained nonforest communities through periodic wildfires or grazing in regions where forests are the natural cover. These tools are as effective as ever, but may be difficult to use or to gain acceptance of urban areas. Controlled burning is widely practiced in rural forestry in many parts of the world today, but public acceptance and air pollution standards make this practice difficult in urban situations. However, controlled burning to maintain prairie or other plant communities is now being done in some cities. The city of Madison, Wisconsin, permits controlled burning of prairies in the University of Wisconsin Arboretum. There is obvious risk involved with controlled burning, and fire should be used only with extreme caution. Standby fire suppression crews must be on site, and experienced fire management personnel should plan and implement the burn.

Grazing by domestic or wild animals will also maintain meadow conditions. In Germany grazing permits are issued to maintain forest openings in both rural and urban areas. Grazing by domestic livestock provides a surer system of maintaining meadows than grazing by wildlife, which may also bring nuisance complaints if the population becomes too large for the range. However, selective or overgrazing by domestic livestock can degrade prairies and other plant communities.

Periodic mowing is another method of maintaining nonforest communities. This need not be done too frequently, once every year or two being sufficient. Spring mowing is desirable, as it will cut invading woody species but will not damage

Figure 14–5 Prairie restoration is becoming common in American midwestern urban parks and greenbelts. Agriculture has eliminated much of the native prairies, and projects such as the one above help to preserve these unique ecosystems.

desired grasses and forbs. However, the best mowing time to control woody species is in midsummer, when food reserves in the plants are at their lowest. Landscape protection laws in Germany require private landowners in landscape protection zones to mow fields annually to keep them from reverting to forest.

Herbicides can be used to control woody vegetation directly through direct spraying of foliage, basal spraying, frilling, injection, or stump treatment after cutting. Direct foliar spraying of woody plants is inexpensive and effective, but entails the risk of drift to nontarget organisms and is the most controversial means of application. Basal spraying, frilling (a ring of overlapping axe cuts treated with an herbicide), and injection alleviate drift problems, but flagging and dying trees are unsightly in the landscape. Cutting and stump treatment puts the herbicide directly into the plant, preventing resprouting, drift, and flagging, making it a highly desirable method in urban forests. Selective systemic herbicides are also applied in granular or tablet form. These herbicides are broadcast around the base of unwanted woody vegetation, resulting in selective mortality. However, there is a strong public sentiment in some communities against the use of herbicides on urban forest and parklands, with many of these communities having ordinances banning such uses.

Natural succession. A second approach to managing nonforest vegetation is benign neglect. These areas will gradually be invaded by woody shrubs and trees, ultimately resulting in a forest stand at some point in the future (Figure 14–6). The composition of the future stand will be determined by the species available to colonize the site and adaptability of these species to site conditions once established. Species composition can be shaped through intervention in the colonization process through planting and intermediate treatments.

Planting. In most cases planting is done by use of seedlings, but sometimes direct or broadcast seeding is an effective method of establishing forest stands. Planting is used in forestry to establish trees where none exist, to change species

Figure 14–6 Invasion of an old field by white pine (*Pinus strobis*) will allow the gradual conversion to a forest stand in this urban greenbelt.

composition, to supplement natural regeneration, and to shorten the regeneration period. Urban forest managers have additional objectives when planting trees on urban forest lands. Noyes (1971) describes four of these objectives:

1. To change an open landscape into a semiopen or closed landscape
2. To increase the density of a landscape composed of single trees and shrubs or groups of trees and shrubs
3. To improve species composition on clearcut areas or in openings by planting more valuable or ornamental species
4. To enhance vistas along roads, water bodies, and fields

Planting in urban forests should follow a predetermined landscape scheme that stresses diversity, naturalness, and aesthetics (Figure 14–7). Tree planting in the Cook County Forest Preserve District, Illinois, is designed to produce deciduous forests indigenous to the region (Buck, 1982). Hardwood seedlings are raised in the district nursery, and machine-planted on preserve land. Machine planting of straight rows in geometric clearings has no place in urban forests and is questionable from an aesthetic perspective in rural forestry. Openings need not be made in geometric shapes, and tree planters pulled in curvilinear patterns help to eliminate the appearance of trees in rows.

Intermediate Treatment

Once trees have been established, additional intervention can dramatically influence the future composition and character of the stand. This intervention takes the form of intermediate treatments. Intermediate treatments include cutting, fire, herbicide, and other activities used to manipulate forest ecosystems between the time of establishment

Figure 14–7. A recent tree planting project in this urban greenbelt will convert an old farm field to a white pine (*Pinus strobus*) stand.

and final harvest. General objectives of intermediate treatments in forest management are to improve forest composition and quality, and to accelerate the growth of crop trees by reducing competition. In urban forests additional objectives include:

1. Favoring individual trees and species with special aesthetic qualities
2. Favoring species resistant to local pollutants
3. Favoring species with few urban-stress-induced pest problems
4. Favoring species with high wildlife value
5. Improving diversity
6. Reducing public safety hazards
7. Identifying specimen trees and reducing neighboring competition. Specimen trees may be individual trees defined as wolf trees, or large trees that hinder the growth of surrounding trees but have little commercial value. Specimen trees, however, can be of a high aesthetic and wildlife value.

Intermediate treatments include cleaning, pruning, and improvement cuttings and thinning.

Cleaning and release. Cleaning, or weeding, is prescribed for sapling stands of the same age and height to improve species composition and to remove undesirable stems (i.e., crooked or otherwise deformed stems). In urban forests undesirable stems will be those that will develop safety problems, while a crooked or otherwise deformed stem may be sound and possess aesthetic interest. Release cutting removes trees overtopping regeneration, such as residuals from a recent harvest cut or fast-growing undesirable specimens. Cleaning and release cutting provides an opportunity to favor species and trees with special aesthetic attributes such as crown form, bark, foliage, fall coloration, twig coloration, and flowers (Noyes, 1971). Cleaning and release-cutting have a profound impact on the future stand, as species composition can be greatly altered through these methods. Two general methods of cleaning are commonly used, mechanical and chemical.

Mechanical cleaning is a labor-intensive operation that requires crews trained to identify species and stems to be favored, and to understand spacing requirements. Hand tools normally used include axes, brush axes, hand saws, and machetes. Trees to be favored have surrounding competing trees lopped back to a subordinate position (Figure 14–8). It is not necessary to cut the entire tree, as lopping the upper portion will give the favored specimen the competitive advantage necessary to survive. Power tools are usually not necessary in mechanical cleaning, but may be useful in felling residual trees in release-cutting. Very dense stands of regeneration will sometimes stagnate, resulting in very slow growth of individual trees. If the objective is to alleviate stagnation rather than improve composition, a crawler tractor or other mechanical device can be used to cut or knock down swaths of regeneration, thus allowing adjacent trees to gain a competitive advantage.

Chemical treatment with herbicides is an inexpensive and effective method of

Figure 14–8 A cleaning in this newly planted forest stand will ensure the ultimate survival and dominance of the eastern larch (*Larix laricina*) seedings.

cleaning and release. If conifers are to be favored, selective herbicides such as the phenoxyes will kill hardwoods, with little damage to conifers if applied at the proper rate and time of year. These herbicides can be broadcast-sprayed over the regeneration area, applied as a basal spray, or injected into individual trees. Although inexpensive and effective, this practice is the subject of controversy in rural forestry, and could become a highly volatile issue if used on urban forests.

Pruning. The objective of pruning in forestry is to produce high-quality knot-free lumber by removing lower branches from the trunk when the tree is young. Pruning is also done to remove low branches as a fire-hazard-reduction technique in stands subject to wildfire damage. The timber management approach to pruning has aesthetic merit, as pruned stands were given higher aesthetic scores by respondents to a survey by Brush (1981). However, removal of lower branches can produce both positive and negative effects in urban forest stands.

Forest pruning will open stands to more public use by making them more attractive and by removing the physical barrier of low branches (Figure 14–9). Low branches are hazards along trails and in heavily used areas, and should be removed for public safety. Scenic vistas can be enhanced within the stand and by allowing the visitor to see beyond the stand. From a negative standpoint, pruning will encourage more intensive use of areas, where it may not be desirable. Fragile soil can be damaged, soil erosion increased, and ground cover destroyed. Pruning the edge of a stand next to a nonforested area will reduce fire hazards, but desirable screening will also be lost. Finally, pruning live lower branches along forest edges removes valuable wildlife cover.

The concept of pruning forest stands should be expanded in urban forests to include elements of arboricultural pruning. Areas subject to intensive recreational use—such as picnic sites, campgrounds, and public trails—must be kept free of hazards. Individual trees should be inspected annually for dead or damaged branches

Figure 14–9 Trees along this urban greenbelt trail will need pruning for safety reasons and to allow vistas of the lake.

need of removal. Historic, specimen, or landscape feature trees should be identified and receive programmed maintenance.

Improvement cutting and thinning. Improvement cutting is the removal of dominant trees of undesirable species or form while thinnings are designed to improve the growth rate of remaining trees. In practice, an improvement cut is often the first cut made on a previously unmanaged stand to improve both stand quality and growth rate. Thinnings are scheduled throughout the rotation of a managed stand and trees competing with preselected crop trees are removed. Along with improving composition, growth rate, and vigor, improvement cutting and thinning can provide the following additional benefits in urban forest stands (Noyes, 1971; Rudolf, 1967):

1. Improve stand diversity
2. Remove hazardous trees
3. Open up vistas
4. Increase stratification to improve wildlife habitat, screening, and to initiate development of a forest regeneration layer
5. Improve aesthetics by encouraging shrubs and wildflowers
6. Stimulate specimen trees
7. Remove unattractive trees (a value judgment)

8. Develop wind-firm trees where future recreational use will require open stands
9. Accelerate succession by releasing a desirable understory and creating a multi-aged stand

Stand age structure is an important consideration when making improvement and thinning cuts. When even-aged stands are thinned, a regeneration layer forms that is usually composed of species more tolerant than the overstory. With regular thinnings this layer will persist and form the next stand when the existing overstory is finally harvested. If this is desirable, nothing need be done. However, if the existing stand is to be reproduced, lighter, less frequent thinnings are recommended to hinder formation of this regeneration. Other controls may be used, such as controlled burns to suppress the next successional stage, or mechanical or chemical control following harvest operations.

If management plans call for maintenance of a specific stand as long as possible, light or no thinning might be desirable to slow maturation. However, very intolerant stands may be lost prematurely if low vigor associated with overcrowding results in insect or disease outbreaks.

Estimates of basal area [area of wood in square feet (meters) per acre (hectare) at breast height (4.5 ft. or 1.4 m)] provide a guide to thinning intensity for different species. State or local forestry offices will have thinning guides based on basal area for commercially valuable species. These guides are based on optimum timber production and may need to be altered for urban forest management objectives. Thinning guides may not be available for stands composed of species with little or no commercial value. However, guides for species with similar ecologic (silvical) characteristics and site requirements can prove useful for managing noncommercial stands.

Uneven-age or all-age stands also benefit from thinning and improvement cutting. Managed uneven-aged stands contain a variety of age classes resulting from periodic harvest operations. It is usually most economical to thin groups of immature trees during harvesting when the necessary equipment is on site. Thinning and management guides for tolerant species are also available from local or state forestry offices.

The overall maintenance of established forest stands through improvement cutting and thinning will be tied to land-use objectives. High stand densities may be desirable to screen unsightly vistas, while low stand densities are necessary to encourage an understory resistant to heavy recreational use (Richards, 1974). Rudolf (1967) recommends a mixture of open and dense stands to provide more attractive settings for recreation, and to provide better wildlife habitat.

The dead tree. We readily recognized the importance of living trees in a forest for their potential as forest products, their aesthetic value, and wildlife habitat value. However, it is easy to regard a dead tree as having little or no value and to remove it. In reality a standing dead tree will contribute to the forest ecosystem for decades by providing habitat for a host of plants and animals that ultimately recycle the tree back into the living community. After falling, the tree will continue to provide important habitat as it slowly decomposes into the forest floor. If a dead tree is located where it poses little or no hazard to forest users, there is little need to cut it down, and good reason to leave it standing. Likewise, when living trees are eliminated in a silvicultural operation some can be girdled and left standing for the overall health and diversity of the ecosystem.

Regeneration Systems

Forests managed for wood fiber production are harvested on a regular basis, the allowable annual cut being governed by annual growth and economics. As discussed previously, urban forests are also managed for timber production in many countries, although this is often subordinate to recreational use and environmental protection. In the United States most urban forests are not managed for wood production, but rather serve primarily as recreational areas, greenbelts, and watershed protection. This is a function of a present wood surplus and a public familiar with the overexploitation of American forests a century ago. It is also an indication of the relatively short period in which American cities have owned and managed greenbelt forests. As these forests mature and existing stands decline, there will be increased interest in manipulation of these ecosystems through silviculture.

There are other compelling reasons to consider timber harvest operations in urban forests, projected global wood shortages within a few decades being high among them. From the present standpoint of how best to manage forest ecosystems in urban areas, the following are important considerations:

1. Stand sanitation and salvage. Controlling the spread of insect and disease epidemics is often enhanced by prompt removal of infested or infected trees. These trees must be disposed of, and can frequently be sold if there is a local market.
2. Overmature trees deteriorate over time and must eventually be removed. Removal while the wood is still marketable will provide needed income for forest management, or at least to the responsible unit of government.
3. Declining and unsafe trees must be removed in the interest of public safety.
4. If additional clearings are needed as part of the overall management plan, timber harvests will utilize wood from the clearings, and income generated can be used to maintain the clearing free of trees.
5. Without cutting, succession will drive forest ecosystems to late-successional communities, resulting in a reduction in cove types, aesthetics, and wildlife diversity.
6. Overmature stands will decline and die, becoming a hazard to recreational users.
7. Desired forest stands can be reproduced by the proper silvicultural system.
8. Cover types can be changed.
9. Urban forests can be used to demonstrate good forest management for an increasingly urban public.

All timber management activities in urban forests must be coupled with an intensive information and education program designed to make urban residents aware of the renewable nature of forests. They must also be aware of the desirability of diverse cover types, and the ultimate decline of stands without intervention.

Five general types of silvicultural systems are recognized: single tree selection, group selection, shelterwood, seed tree, and clearcutting. Selection of the appropriate system is related to the shade or understory tolerance or successional level of the community being managed, with the most tolerant species managed by single tree selection, and the most intolerant species managed by the clearcut or seed tree method. Species of intermediate tolerance are managed by group selection or shelterwood cutting. Both types of selection cutting produce uneven-aged stands, while shelterwood, seed tree, and clearcutting yield even-aged stands.

Aesthetics is the prime consideration in both selection and application of a silvicultural system to urban forest stands. Minimal visual impact from regeneration cutting can be obtained by regulating the size and location of cutting units in relation to public-use areas. Management plans should emphasize long rotations and the development of attractive stands composed of large trees. The following discussion of silvicultural systems is in order of increasing visual impact. However, this does not imply that the latter methods must never be used, but rather, they must be used with the utmost consideration of their visual impact.

Single tree selection. This method of harvest and regeneration is most adaptable to tolerant and very tolerant species. Individual trees are marked for removal, with the resulting opening in the canopy allowing enough light for regeneration or release of only the most tolerant species (Figure 14–10). Allowable cut is based on the annual growth of the stand times the number of years between harvest cuts (cutting cycle). Single tree selection will have minimal visual impacts, but widely applied will gradually convert the entire forest to late successional communities.

Group selection. Trees are harvested in small groups not exceeding one-fifth acre (0.08 ha) in group selection, yielding uneven-age stands (Figure 14–11). The microclimate of these openings is modified by the surrounding stand, but the clearing receives enough light to seed in with tolerant and intermediate-tolerant species. Cutting is visually more evident than with single tree selection, but the overall impact is minimal due to the small size of the openings. As with single tree selection, allowable cut is based on accumulated growth over a specified cutting cycle.

Shelterwood. Shelterwood harvesting regenerates even-aged stands by removing the overstory in two or more cuts made within one-fifth or less of the rotation age (Figure 14–12). These harvest cuts progressively open the forest floor to seeding from the overstory, yielding advanced regeneration by the time the final harvest cut is made. Intermediate-to-intolerant species respond well to this system. The visual impact is stronger than the two selection methods, but the presence of advanced regeneration softens this impact. Advanced regeneration also reassures the public that good management is being prescribed.

Figure 14–10 Single tree selection silviculture in this stand is producing multiple age classes of tolerant species. (James Johnson.)

Figure 14–11 Group selection silviculture in this stand is creating openings in the canopy that fill with tolerant to intermediate-tolerant species. (James Johnson.)

Figure 14–12 Shelterwood management of this paper birch (*Betula papyrifera*) stand is successfully reproducing paper birch. (James Johnson.)

Seed tree and clearcutting. Both methods remove all or most of the overstory in a single harvest cut. In the seed tree method five to ten trees are left per acre as a source of seed. Seed trees are then removed once regeneration has been established. Clearcuts seed in from surrounding stands, are planted, or regenerate from sprouts or residual seed in the forest floor (Figure 14–13). Very intolerant species are readily

Figure 14–13 Clearcutting quaking aspen (*Populus tremuloides*) has resulted in abundant regeneration. (James Johnson.)

reproduced by these methods, but the visual impact is profound. These methods are necessary to regenerate early successional species if deemed necessary by the management plan. Seed tree and clearcutting systems can be visually acceptable by keeping cutting units small, nongeometric in shape, and away from areas receiving heavy use.

A problem frequently encountered when regenerating very intolerant to intermediate tolerant species is the presence of advanced regeneration consisting of the next successional stage. As stated previously, controlled burning will suppress this regeneration if the overstory is composed of thick-barked fire-resistant species [e.g., southern pine species (*Pinus* spp.) in the United States]. When dealing with species readily damaged by fire or in situations where the use of fire is impractical, other methods can be used. Mechanical scarification during or following logging and subsequent cleaning will assist the establishment of desired species. Herbicides can also be useful in this regard, but as previously mentioned, public reaction may be unfavorable.

It is sometimes necessary to protect areas subject to heavy use during the regeneration period. In the case of human use, posting or closing access points will reduce disturbance. If the area cannot be protected in this manner, temporary fencing may be necessary. High populations of browsing wildlife will virtually eliminate regeneration, making control necessary. Hunting is used in many European urban forests to keep deer populations in check (Figure 14–14), as are fences and repellants. General public access to hunting, and antihunting sentiment, make hunting impractical in North American urban forests as a means to control deer and other browsers. When populations of these animals get too high for regeneration to take place in these forests, repellents, fencing, and trapping and relocation are necessary.

Slash disposal may be necessary following logging in urban forest stands. Slash from cutting in selection and shelterwood system can often be disposed of by lopping and allowing it to decompose naturally. Limbs and small-diameter logs can be sold as firewood, thus reducing the overall volume of debris. Heavy accumulations of slash will often create fire hazard and aesthetic problems. Broadcast burning or piling and burning, where permitted and acceptable by the public, is an efficient method of slash disposal. However, if large accumulations cannot be dealt with and are not acceptable, a silvicultural system should be selected that will not create this problem.

Remnant Forests

Following land development there are often remnant forests behind as small parks and greenbelts in developments. These remnants are often left over from larger closed forest systems and begin to change once their edges are exposed. Exotics invade the understory and exposure of forest trees to the wind and other climatic changes can result in windthrow and crown dieback. Likewise more sunlight on the edge will stimulate understory development and growth. To retain the natural character of remnant forests Clark and Matheny (1991) recommend control of exotic species and regeneration of overstory/dominant species.

Forests near cities frequently comprise young, fast-growing, even-aged stands subject to windthrow on exposed edges. Agee (1995) recommends whorl thinning

Figure 14–14. This high stand in an urban forest in Germany is used to hunt deer. (H. Schabel.)

(removing every other whorl of branches) of crown on conifers to reduce wind resistance, and thinning the edge of the stand to reduce the "wall" effect. He further suggests planting vegetation on the edge to lift the wind over the top of the stand. Interior portions of the stands can be thinned to allow better growth and crown development of residual trees, and to stimulate regeneration of replacement trees. Dunster (1995) reports that many remnant forests in developments near Vancouver, British Columbia, are too small and residual trees too spindly to last long after the canopy is opened. He recommends that all trees within striking distance of buildings be removed before construction, but that the land be retained with minimal disturbance and new trees established that will adapt to the changed realities of the environment.

Watershed Management

Municipal watersheds differ from greenbelt or recreational forests, as their primary objective is the production of water for urban use. In many cases, however, these lands are subjected to heavy recreational demand, particularly if they are close to the community. Management of these lands will have the protection and production of high-quality water as the primary goal, with timber management and recreation as secondary objectives.

Baltimore's watershed. The city of Baltimore owns 17,580 acres (7032 ha) of watershed land surrounding three municipal reservoirs (Hartley and Spencer, 1978). Protection of water quality is the primary management goal, but multiple-use forest management has been practiced on these lands since 1948. Recreational use of these lands places aesthetics as a primary concern in timber management activities. Six techniques are used to reduce the visual and water quality impact of silvicultural operations (Hartley and Spencer, 1978):

1. *Cutting systems.* Single tree and group selection are used on all hardwood stands to lessen the visual impact and protect water quality. Conifer plantations are initially row-thinned (every third row), with subsequent thinning being selective. Clearcutting is used on a limited basis for intolerant conifer stands. Clearcuts are kept small and away from public roads and reservoirs.
2. *Screening and buffer strips.* Buffer strips are maintained along all public thoroughfares, streams, and reservoirs. These strips vary from 50 to 200 ft. (15 to 60 m) in width, depending on the cutting location and terrain.
3. *Postlogging operations.* After logging has been completed, the following activities are implemented:
   - **a.** Severely damaged trees are cut.
   - **b.** Slash is lopped to lie close to the ground.
   - **c.** Tops are placed in skid paths to reduce erosion.
   - **d.** Log deck debris is removed.
   - **e.** Access roads are regraded and water bars replaced.
   - **f.** Log decks and logging roads are seeded.
4. *Visual improvements.* Scattered large trees are left on sites cleared for planting.
5. *Public information.* Signs are erected in cutting areas to inform the public as to the intent of the silvicultural operation.
6. *Compromise.* Management is flexible, allowing for public input and compromise. Stands that possess unique beauty or are heavily used by the public are set aside, with only dead trees logged on a salvage basis.

Fire Management

Fire management consists of fire prevention, fire suppression, and the use of fire as a management tool. Cities located in regions where periodic weather conditions are coupled with flammable vegetation types suffer devastating wildfires, resulting in the loss of property and life (Figure 14–15). In 1991 a wildfire burned through densely populated Oakland–Berkeley Hills, California, killing twenty-five people and destroying 3,011 homes (Svihra, 1992). The historic approach to fire management was immediate suppression and prevention through public education programs. Unfortunately, this approach does not recognize that many ecosystems are strongly shaped by fire, and in fact are regenerated by the very fires that destroy them. Under natural circumstances these ecosystems burn at periodic intervals, thus eliminating fuel buildup and reducing the risk of intense fires. Long-term suppression and prevention will allow fuels to accumulate to the point where under extreme weather conditions catastrophic fires occur that are impossible to control.

In the case of the Berkeley–Oakland Hills fire, presettlement vegetation consisted mostly of grasses. The vegetation that burned was either planted or had escaped cultivation and consisted mostly of introduced fire-prone species (Svihra, 1992). Clark (1995) recommends the following for a fire-safe environment near homes and other structures (adapted from Baptiste, 1992):

1. Separating vegetation from structures
2. Reducing the overall amount of fire fuels on the property
3. Preventing vertical spread by eliminating fire ladders (low branches)
4. Reducing the continuity of fire through fuel breaks

Figure 14–15. Heavy fuel buildup and home construction in flammable ecosystems often results in catastrophic loss of life and property. (Courtesy of Wisconsin Department of Natural Resources.)

5. Integrating fire management across both landscape and structure
6. Being familiar with fire potential, including topography

Fire prevention includes outdoor burning bans, woods closings, and continuous public education of wildfire hazards and fire prevention. Public education should also include the need for periodic controlled burning and reasons for nonsuppression of beneficial wildfire. Fire suppression is the ability to detect and control wildfire when it is unwanted. Fire suppression is also knowing when not to suppress wildfire. A wildfire that is not threatening but is reducing fuel buildup and altering ecosystems in a manner consistent with the overall management plan should be allowed to burn.

Fire as a management tool can be used to manipulate ecosystems in regions where these ecosystems have evolved with wildfire as an environmental factor. To be sure, wildfire is dangerous and can result in tremendous losses when out of control. However, where fire is an integral part of the ecosystem it is necessary for the survival of the system. Fire management includes the controlled burning of ecosystems with a frequency that prevents the dangerous buildup of fuels and is beneficial to the plant community. It is also nonsuppression of wildfire for reasons stated above. Other aspects of fire management include the maintenance of fuel breaks for controlled burning and fire suppression, maintenance of management access roads, and the establishment and management of nonflammable ecosystems in high-risk areas.

Palo Alto fire management plan. The city of Palo Alto, California, developed an Open Space Fire Management Program for 8000 acres (3200 ha) of city-owned property: land covered with mountain forests, brush, and grass. The program consists of the following (Deitch, 1978):

1. Maintaining fuel breaks covered with low-growing fire-resistant species
2. Replacing highly flammable introduced annual grasses with more fire-resistant native perennial grasses
3. Prescribed burning to reduce fuel buildup
4. Ordinances to enforce fire management practices on private property
5. Public education as to the necessity of prescribed burning
6. Coordinating efforts with neighboring fire management districts to develop a regional fire management plan

East Everglades fire management plan. The Everglades region of southern Florida is a vast wetland composed of a variety of fire-influenced vegetative communities. This area is subject to periodic droughts and subsequent intense wildfires that threaten nearby cities and damage wildlife habitats. The Florida Division of Forestry developed a fire management plan for eastern Everglades, a 154,880-acre (61,952-ha) area of public and private land between Everglades National Park and the Miami metropolitan area (Newburne and Duty, 1981). The management goals of this plan are to:

1. Maintain the status quo of the various natural communities
2. Obtain hazard reduction, thereby reducing damage caused by untimely wildfire
3. Improve wildlife habitats in the East Everglades area

The plan identifies five major plant associations and describes timing and weather conditions under which each community will be burned. The burning cycle is three to six years, depending on the vegetation type. Wildfires will be allowed to burn or will be suppressed, depending on their location and conditions surrounding the fire.

Protecting Urban Forests

Beyond fire management other protection measures are needed in urban forests. Insects, diseases, and overuse are common problems for urban managers. Insect and disease outbreaks in urban forests are similar to those experienced in rural forests, although pollution and overuse-induced stresses may make these problems more pronounced. A variety of texts are available describing the control of forest and shade tree pests, as is assistance from local and state forestry offices.

Problems associated with overuse of urban forests must also be dealt with. Soil compaction, sheet erosion, destruction of ground cover, and ultimate decline of forest communities are all related to heavy use, fertilization, mulching, and rotation of use through several locations. Changing the ground cover to species more resistant to use is advised, as is canopy density reduction to stimulate forest floor vegetation.

Special problems exist for unique or endangered ecosystems. These areas are often attractive to the public, resulting in overuse and damage to the system. Protection measures include limiting use, maintenance of physical barriers to keep users on access trails, mulching exposed and/or compacted soil, and periods of nonuse to allow areas to recover. In some cases it may be necessary to close these areas to the general public.

LITERATURE CITED

AGEE, J. K. 1995. "Management of Greenbelts and Forest Remnants in Urban Forest Landscapes." In G. A. Bradley, ed., *Urban Forest Lanscapes: Integrating Multidisciplinary Perspectives.* Seattle: Univ. Wash. Press, pp. 139–149.

ANDERSON, E. M. "Conservation Biology and the Urban Forest." *Proc. of the 6th Natl. Urban For. Conf.* Washington, D.C.: Am. For. Assoc. pp. 234–238.

BAPTISTE, L. 1992. *Firescape: Landscaping to Reduce Fire Hazard.* Oakland, Calif.: East Bay Municipal Utility District.

BLACK, M. E. 1982. "The Seattle Urban Forestry Experience." *Proc. Sec. Natl. Urban For. Conf.* Washington, D.C.: Am. For. Assoc. pp. 38–41.

BRUSH, R. O. 1982. "Forest Esthetics: As the Owners See It." *Am. For.* May, pp. 15–19.

BUCK, R. L. 1982. "The Forest Preserve District of Cook County: The Success Story of a Major Forest Recreation System." *Proc. Sec. Natl. Urban For. Conf.* Washington, D.C.: Am For. Assoc. pp. 157–161.

CLARK, J. R. 1995. "Fire-Safe Landscapes." In G. A. Bradley, ed., *Urban Forest Landscapes: Integrating Multidisciplinary Perspectives.* Seattle: Univ. Wash. pp. 164–172.

CLARK, J. R., and N. P. MATHENY. 1991. "Management Concepts for 'Natural' Urban Forests." *Proc. Fifth Natl. Urban For. Conf.* Washington, D. C.: For. Assoc., pp. 22–29.

DEITCH, J. 1978. "Fire Management." *J. Arbor.* 4(11):259–260.

DRIVER, B. L., and D. ROSENTHAL. 1978. "Social Benefits of Urban Forests and Related Green Spaces in Cities." *Proc. Natl. Urban For. Conf.,* ESF Pub. 80-003. Syracuse SUNY, pp. 98–111.

DUNSTER, J. 1995. "Effective Tree Retention in New Developments: An Undisturbed Landbase Is the Key to Success." *Proc. Trees and Buildings Conf.* Savoy, Ill.: International Society of Arboriculture.

DUNSTER, K., and J. DUNSTER. 1992. *The Nature of Burnaby: An Environmentally Sensitive Areas Strategy.* Bowen Island, B.C., Canada: Dunster and Associates.

DWYER, J. F. 1983. "Management Technologies for Outlying Forests: A Summary and Synthesis." Proceedings of the seminar *Management of Outlying Forests for Metropolitan Populations.* Man and the Biosphere, Washington, D.C., pp. 27–31.

FRANKLIN, J. F. 1993. "Preserving Biodiversity: Species, Ecosystems, or Landscapes?" *Ecological Applications* 3(2):202–205.

FRENCH, J. R. J., and R. SHARPE. 1976. "Urban Forests for Australian Cities." *Trees and Forests for Human Settlements.* Toronto: Univ. Toronto, pp. 123–135.

GREEN, T. L. 1984. "Maintaining and Preserving Wooded Parks." *J. Arbor.* 10(7):193–197.

HARTLEY, B. A., and W. G. SPENCER. 1978. "Management Problems and Techniques on a City Watershed." *Proc. Natl. Urban For. Conf.,* ESF Pub. 80-003. Syracuse: SUNY, pp. 410–420.

HULL, R. B., IV, and S. E. MICHAEL. 1995. *Paving Over Paradise: How to Decrease Crime in Public Areas.* News release, 2/1/95, International Society of Arboriculture, Savoy, Ill.

KUCHELMEISTER, G. 1993. "Trees, Settlements and People in Developing Countries." *Arboric. J.* 17:399–411.

LEEDY, D. L., R. M. MAESTRO, and T. M. FRANKLIN. 1978. *Planning for Wildlife in Cities and Suburbs.* USDI, Fish Wildl. Serv.

MCHARG, I. L. 1969. *Design with Nature.* Garden City, N.Y.: Doubleday & Company.

MILLER, R. W. 1983. "Multiple Use Urban Forest Management in the Federal Republic of Germany." Proceedings of the seminar *Management of Outlying Forests for Metropolitan Populations.* Man and the Biosphere, Washington, D. C., pp. 21–23.

MILLER, R. W., R. S. BOND, and B. R. PAYNE. 1978. "Land and Timber Values in an Urban Region." *J. For.* 76(3):165–166.

MILLIGAN RAEDEKE, D. A., and K. J. RAEDEKE. 1995. "Wildlife Habitat Design in Urban

Forest Landscapes." In G. A. Bradley, ed., *Urban Forest Landscapes: Integrating Multidisciplinary Perspectives.* Seattle: Univ. Wash. Press, pp. 139–149.

MJAALAND, B., and J. W. ANDRESEN. 1986. "Amenity and Service Functions of Oslo's Municipal Forest." *Arboric. J.* 10:101–112.

NASSAUER, J. I. 1992. "Ecological Function and the Perception of Suburban Residential Landscapes." In P. H. Gobster, ed. *Managing Urban and High-Use Recreation Settings.* St. Paul, Minn.: USDA Forest Sevice Gen. Tech. Rep. NC-163, pp. 55–60.

NEWBURNE, R., and R. DUTY. 1981. *East Everglades Management Area: A Prescribed Burning Plan.* Tallahassee; FL. Div. For.

NOYES, J. H. 1971. "Managing Trees and Woodlands to Improve the Aesthetics of Communities." *Trees and Forests in an Urban Environment,* Plann. Res. Dev. Ser. No. 18. Amherst: Univ. Mass., pp. 115–120.

ODUM, E. P. 1971. "Ecological Principles and the Urban Forest." *Proc. Symp. Role Trees South's Urban Environ.* Athens: Univ. Ga., pp. 78–80.

RICHARDS, N. A. 1974. "Greenspace Silviculture." *Proc. 1974 Natl. Conv.* Washington, D.C.: Soc. Am. For., pp 80–84.

RUDOLF, P. O. 1967. "Silviculture for Recreation Area Management." *J. For.* 65(6):385–390.

SCHABEL, H. G. 1980. "Urban Forestry in the Federal Republic of Germany." *J. Arbor.* 6(11):281–286.

SCHROEDER, H. W. 1982. "Preferred Features of Urban Parks and Forests." *J. Arbor.* 8(12):317–322.

SMITH, D. M. 1969. "Adapting Forestry to Megalopolitan Southern New England." *The Massachusetts Heritage* 7(2), Amherst, Univ. Mass.

SPURR, S. H. and B. U. BARNES. 1980. *Forest Ecology,* 3rd ed. New York: John Wiley & Sons.

SVIHRA, P. 1992. "The Oakland–Berkeley Hills Fire: Lessons for the Arborist." *J. Arbor.* 18(5):257–261.

TALBOT, J. F., and R. KAPLAN. 1984. "Needs and Fears: The Response to Trees and Nature in the Inner City." *J. Arbor.* 10(8):222–228.

WILLHITE, R. G., and W. R. SISE. 1974. "Measurement of Reaction to Forest Practices." *J. For.* 71(9):567–571.

15

Commercial and Utility Arboriculture

The principles and techniques of urban forest management as described throughout this text apply to both the public and private sectors. At times the emphasis is on public urban forestry and arboriculture, but most methodologies have application to commercial and utility arboriculture. For example, inventories, programmed maintenance, silvicultural techniques, and many other activities are performed by firms offering arboricultural series.

The purpose of this chapter is to focus on the private sector by describing its current status, trends, and innovations, and basic management principles unique to the industry. The distinction between commercial and utility arboriculture is somewhat vague, with many arboricultural firms providing line clearance work by contract for utility companies. However, for discussion purposes this chapter treats the subject matter in two distinct categories: commercial arboriculture, and utility arboriculture.

COMMERCIAL ARBORICULTURE

If commercial arboriculture is defined as performing services on trees for a fee, then the profession has been around as long as trees have been in communities. Anyone hired to do anything to a tree, such as cut it down or saw off a branch, was, and is, a commercial arborist by these standards. However, if commercial arboriculture is defined as providing professional tree management services, then commercial arboriculture is a more recent profession. Felix (1978) describes commercial arboriculture as being about 100 years old in the United States. The key term in this definition is "professional tree management," indicating that the work performed has a solid foundation in basic tree science, is performed by qualified personnel, and is usually a part of an overall landscape management plan.

Arboricultural Services

From a historical pe.spective, professional arboricultural services were initially performed on large estates of the well-to-do and primarily consisted of cavity work on large, old, decayed trees. Some early arborists, recognizing the limitations of this market, specialized in providing line clearance services for the newly evolving electric and telephone industries. Following World War II, suburban development created a demand for landscape management services, creating an expanding market for commercial arborists. This, coupled with increasing environmental awareness in the 1960s and a better understanding of the role that trees play in the human environment, has led to a high demand for arboricultural services (Abbott and Joy, 1978; Felix, 1978).

Today, most tree work consists of pruning, spraying, and removal of trees on $\frac{1}{2}$- to 1-acre suburban house lots. The National Arborist Association (NAA), a trade association representing the industry, conducted a survey of the industry in 1975 and reported service statistics by dollar volume percentages (Table 15–1). The value of services provided in 1995 by more than 13,000 arboricultural firms to residential, commercial, and public-owned landscapes in the United States was estimated at $2 billion annually, with services to utilities valued at more than $1.5 billion (Felix, 1995).

Other arboricultural services include tree protection, landscape management planning, landscaping, and community forestry consultation. A number of communities now have tree protection ordinances designed to protect existing trees and other vegetation during land development (Chapter 9). Some arboricultural firms contract with land developers to protect trees from damage during construction and to assist in tree recovery work after disturbance.

TABLE 15–1 PERCENTAGE OF SERVICES PERFORMED BY COMMERCIAL ARBORISTS IN THE UNITED STATES BASED ON DOLLAR VOLUME IN SALES (From Abbott and Joy, 1978.)

Pruning	37%	Landscaping	7%	Planting	2%
Spraying	18%	Fertilizing	6%	Cavity work	1%
Removal	16%	Tree moving	2%	Diagnosis	1%
Utility trimming	8%	Bracing	2%		

Large-scale developers often dedicate a substantial portion of the area developed to public open use, especially when constructing planned unit developments or cluster housing (Chapter 8). Commercial arboricultural or urban forestry firms are sometimes hired by these developers to select the best areas and vegetation types for preservation and recreational use. Once areas have been designated for open space and recreational use, vegetation management plans are necessary to preserve existing vegetation, change cover types if necessary, or establish new vegetation. These plans may also include management recommendations for newly established landscapes near residences, on boulevards and public spaces, and on streets. However, most actual landscaping in the United States is carried out by firms specializing in that activity, with arboricultural firms providing a small portion of landscape establishment service (Abbott and Joy, 1978).

Small communities are often in a position where they wish to develop street tree management plans but cannot justify hiring a city forester on a full-time basis. Street tree management services can be purchased by these communities from arboricultural firms specializing in consulting urban forestry. These services include street tree inventories, street tree management plans, planting and maintenance specifications, and development of ordinances.

Marketing. Services provided by arboricultural firms can be divided into three general areas: removal, pruning, and health care. Most companies start as removal services and, if they stay in business, usually evolve into full-service firms. Little investment is required to start a removal service, but as higher levels of service are offered investment in equipment, training, and personnel increase considerably. With the exception of hazardous removals, competing in the removal market is difficult for high-quality, established firms because the consumer is interested primarily in the lowest bid to get rid of an unwanted tree. The lowest bid will likely come from a company with low overhead costs (Ball, 1992).

Pruning offered as topping or cutting off an offending branch is viewed by the consumer as work not requiring much knowledge on the part of the arborist. As in removals, the full-service professional arborist will have difficulty competing with low-overhead companies that offer substandard service unless they convince the consumer they offer a better, more professional service for a higher cost. Customers who value their trees can be sold quality pruning services by salespeople with a thorough knowledge of trees and their care (Ball, 1992).

Tree health care includes fertilizing, aerating, mulching, pest management, and other services designed to maintain healthy, high-quality trees for consumers. Price is less of an issue than quality tree care because these customers are truly interested in their landscape and are willing to pay for quality services. Successful arboricultural firms have a high percentage of customers who desire quality services and become long-term clients. Ball (1992) suggests that sending a company newsletter to existing customers is the best advertising device to maintain an existing customer base (Appendix I). Advertising in local newspapers can be effective, especially if used to remind potential customers of periodic tree needs. Arborists who willingly become public speakers for service organizations and who volunteer for interviews on talk shows become known as tree experts in a community. They also get calls from people who hear them and are interested in quality landscape management (Ball, 1992).

Estimating tree care jobs. Salespeople not only have to sell the job, they must estimate a price that is both competitive and profitable for the company. Abbott and Miller (1987) recommend accurate record keeping of crew and equipment time relative to tree size and service performed to develop a basis for estimating jobs. They suggest each job be broken in to fixed and variable factors in writing bids (Table 15–2). Fixed factors are dependent on tree size and service provided, while variable factors are based on job location, and tree and/or site characteristics. The following is a list of some variable factors (Abbott and Miller, 1987):

1. Electric wires at outside edge of tree canopy; increase person-hours by 10 percent.
2. Electric wires about halfway between edge of canopy and tree trunk; increase person-hours by 20 percent.

TABLE 15–2 FIXED AND VARIABLE FACTORS USED TO ESTIMATE TREE CARE JOBS.

Task	Fixed factor	Variable factor
1. Job assignment, pick up tools, gas up equipment	X	
2. Travel time to job site		X
3. Work site organization, tools, traffic signs	X	
4. Actual work operation climbing tree, cutting branches for a particular size tree and pruning class	X	
5. Presence or absence of electric wires, traffic, buildings		X
6. Clean up of work site, chipping brush, cutting up wood for a particular pruning class or activity	X	
7. Disposal of debris at dump site		X
8. Travel time back to crew headquarters		X

3. Electric wires near trunk of tree; increase person-hours by 40 percent.
4. Tree within 25 feet of building; increase person-hours by 20 percent.
5. Tree within 5 feet of building; increase person-hours by 40 percent.
6. Traffic (medium volume) flagperson necessary, part-time; increase person-hours by 25 percent.
7. Traffic (heavy volume) flagperson necessary, full time; increase person-hours by 50 percent.
8. Brush and wood must be dragged from backyard for chipping; increase person-hours by 100 percent.
9. Major decay affecting 30 to 50 percent of trunk area; increase person-hours by 20 percent.
10. Extensive decay affecting 60 to 90 percent of trunk area; increase person-hours by 50 percent.
11. If the tree canopy is exceptionally sparse or thick; increase person-hours by 20 percent.
12. Client desires firewood cut to length and left; increase person-hours by 20 percent and an additional 10 percent if must locate and stack.
13. Deadwood in 25 to 50 percent of tree canopy; increase person-hours by 25 percent.
14. Deadwood in more than 50 percent of tree canopy; increase person-hours by 60 percent.

When making an estimate the salesperson should determine from the property owner exactly what is to be done. The trees can then be measured, variable factors applied, and an estimate developed on the person-hours needed to perform the task. The customer should then sign the order. When the service is provided, records should be kept to see how long it actually takes the crew to do the work, which provides information on both crew performance and bid accuracy (Abbott and Miller, 1987).

Current Status of Commercial Arboriculture

As a business enterprise, commercial arboriculture has experienced generally steady growth in the United States since its inception. Based on current interests of the public in their personal landscapes and communitywide efforts to improve the quality of urban life, the industry should experience continued growth. Abbott and Joy (1978) characterize the industry in the following manner:

1. The industry consists of many small firms employing two to five people on a seasonal basis. The smaller companies report that tree removals represent their largest sales volume.
2. Most firms complement tree care with landscaping, nursery, or garden center operations.

3. Most services are bid on a competitive basis.
4. The industry is not united to deal with professional problems, with only a small portion belonging to the National Arborist Association. However, those that do belong to the NAA represent 50 percent of industry sales volume.

Trends

There are a number of significant trends that are or will affect the arboriculture industry in the near and more distant future. These trends are important, and are summarized as follows:

1. The highest concentration of homeowners with enough income to afford professional tree care services now live in homes that are relatively new (Black, 1981). The landscapes around these homes are beginning to mature, and many trees and shrubs are in desperate need of pruning and pest management (Figure 15–1). However, the more traditional practices of cabling and bracing, cavity work, and removal will not be in high demand on these properties in the near future, due to the youth of the landscape.

2. There is an increased interest in all aspects of nature, including landscaping. This has manifested itself in a great deal of pride and care of individual landscapes by urban and suburban residents (Black, 1981). The impact on the tree care industry

Figure 15–1 The maturing landscape around this twenty-year-old home will demand increasing arboricultural services.

will be an increasing demand for traditional services as well as a demand for specialized, unique, or "natural" landscaping.

3. A number of household tasks residents formerly did themselves are now being turned over to professional service companies, including landscape maintenance (Black, 1981). This trend will continue to encourage the growth of the landscape maintenance industry.

4. Controversy over the use of pesticides will persist with increasing awareness and caution on the part of homeowners (Black, 1981). Commercial arborists are sensitive to these concerns with many adopting the principles of Plant Health Care (PHC) as their primary approach to dealing with pest problems. PHC was developed by the International Society of Arboriculture and consists of selecting, growing, and maintaining healthy, pest-resistant plants in the landscape. When pests reach a predetermined threshold (Figure 15–2) where they are considered a problem, a variety of methods are available for control, including biological, cultural, and chemical. Firms that have adopted PHC have experienced a dramatic reduction in the volume of chemical pesticides used (Felix, 1995). Information on PHC can be obtained from the International Society of Arboriculture, P.O. Box GG, Savoy, IL 61874.

5. Labor costs will continue to increase, so there is an increasing need for mechanization to keep labor costs at a reasonable level (Abbott and Joy, 1978). The development of aerial towers, chippers, hydraulic saws and shears, stump grinders,

Figure 15–2 Plant Health Care (PHC) monitoring on a residential property.

tree spades, and so on, have all served to reduce labor costs. Continued innovations in equipment are needed to keep services affordable by customers.

6. Computers are now affordable by most firms. Software is available for most business tasks and is so user-friendly that minimal effort is needed to learn how to use it. Many computer companies furnish self-instructional programs to train buyers in how to use the hardware and software. There are now programs available that have been specifically written for the tree care and landscape maintenance industry. A number of general business application programs are described below.

a. Database management systems. These programs serve the function of storing information in an electronic filing system, retrieving information, searching files, and providing file summaries.

b. General ledger accounting, accounts payable, and accounts receivable programs.

c. Inventory control programs to keep track of parts and supplies.

d. Payroll programs designed to compute payrolls and make necessary deductions.

e. Word processing programs to take the place of the typewriter. These programs will also store customer mailing lists and letters (Massey, 1983).

f. Inventory programs with graphic capabilities are available for commercial arborists. These programs allow the user to quickly draw a map of the property, locate and number trees, record information describing the tree and management needs, and write work orders. These programs can also be used to keep work records and generate reports with graphics for customer mailings.

The arboricultural industry provides professional tree and landscape services. There is increasing interest in these services by the public, and this will exert a profound impact on the industry during the next several decades. There is, and will continue to be, rapid growth in the industry, providing opportunities for entrepreneurs and well-trained individuals. Firms, both large and small, that offer good-quality service, hire courteous and qualified employees, charge a fair price, and stand behind their work will be successful. These firms will adopt and develop new technologies that result in savings to the customer, while providing increased quality and efficiency on the job.

UTILITY ARBORICULTURE

Utility arboriculture or forestry consists primarily of managing vegetation in electric transmission and distribution line networks. Transmission lines carry electricity from power plants to consumers in corridors characterized by high-voltage lines over land the utility either owns outright or has an easement to cross and manage. Distribution networks connect to transmission lines and distribute the electricity directly to customers through underground or overhead wires. Vegetation in transmission corridors is managed to exclude trees that might either grow up into or fall across wires. Vegetation in above-ground distribution networks is generally owned by individuals or a municipality, and is managed by encouraging use of small mature-sized trees under or near lines, or by pruning larger trees to reduce the risk of power outages or injuries. The primary goals of vegetation management by utilities are to provide reliable service and insure public safety. Regardless of who owns the vegetation under or near utility lines, utility companies have the legal authority to access their lines and to prune trees that will interfere with service.

Tree Trimming

Frequently the largest budget item for utilities in right-of-way vegetation management is trimming or pruning trees that interfere or threaten to interfere with service. Perry (1977) states that utility companies are obligated to control costs, and this means "building, maintaining, and operating lines as economically as possible with careful consideration to safety and continuity of service." Maintenance costs are a deduction from income; therefore, tree-trimming budgets must be balanced against the reliability of service and trimming acceptability by customers. Less frequent but more severe pruning will save money, but the visual impact on trees may be unacceptable. Perry (1977) points out there are three factors to consider when deciding on the tree trimming budget:

1. What is the impact of this cost on the total operating budget?
2. What is the relationship between the cost of trimming and the frequency of service interruptions?
3. What is the acceptable level of reliable service?

Ulrich (1983) addressed the issue of the impact of tree trimming budgets on the level of service for the Metropolitan Edison Company in Pennsylvania. Obviously, a point of diminishing returns is reached when attempting to reduce the number of tree-caused outages to zero. However, there is a point where a certain cost of trimming is justified to keep outages to an acceptable level. For example, the Cleveland Electric Illuminating Company reports that it costs them $2 million annually to keep tree-caused outages at 8 percent of their total outages (Perry, 1977). Ulrich (1983) reports that by cutting tree-trimming budgets during a period of austerity, Metropolitan Edison experienced a dramatic increase in power outages caused by trees (Figure 15–3). Smith (1984) reports a similar relationship between trimming expenditures and outages for the Wisconsin Power and Light Company. In this study increasing expenditures resulted in lower customer outages (Figure 15–4).

Tree-trimming specifications are essential to keeping maintenance costs at an acceptable level, whether by contract or by in-house crews. The historic approach to line clearing involved either heading back all interfering branches on a tree to within a specified distance from wires, or rounding over to keep trees under wires from growing up in to lines (Figures 15–5 and 15–6). In both cases the results are heavy

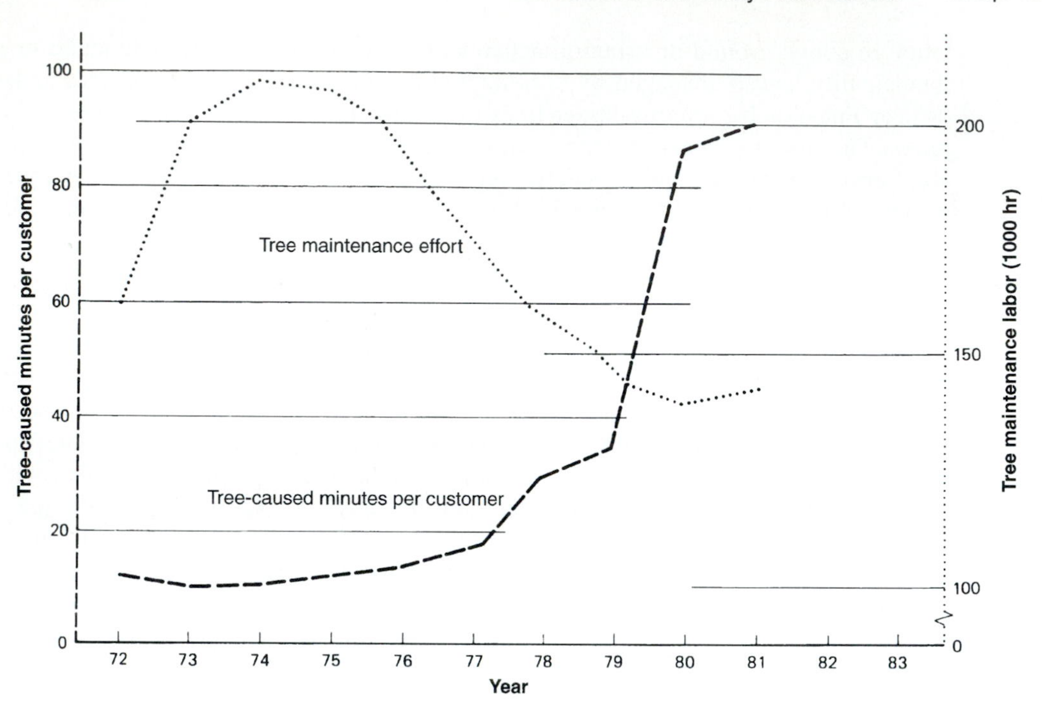

Figure 15–3 Tree-caused interrupted minutes per customer compared to tree maintenance effort using five-year rolling averages for the Metropolitan Edison Distribution System. (From Ulrich, 1983.)

suckering, serious damage to the tree, the need to retrim within a short period of time, and high long-term costs. Directional pruning involves removing branches to laterals (drop-crotching) that will direct future growth away from utility lines, resulting in less disfigurement and damage to trees, less suckering, and less future pruning need. Four general types of trimming are done utilizing directional pruning: top trimming, side trimming, under trimming, and through trimming (Figure 15–7) (Johnstone, 1983; Holewinski et al., 1983). Shigo (1990) suggests that it may be better to remove entire offending limbs rather than use directional pruning, especially if directional pruning will result in rapid regrowth back into lines, or if so much living foliage is removed that the pruned branch will die.

Directional pruning has been found to be more cost effective than rounding over. Goodfellow et al. (1987) measured less regrowth following directional pruning versus roundover pruning. Delmarva Power Company, an electric utility serving parts of Maryland, Virginia, and Delaware, instituted a program of directional pruning, more efficient crew scheduling, and growth regulators to control pruning costs. Growth regulators are a class of chemicals that when applied by trunk injection or

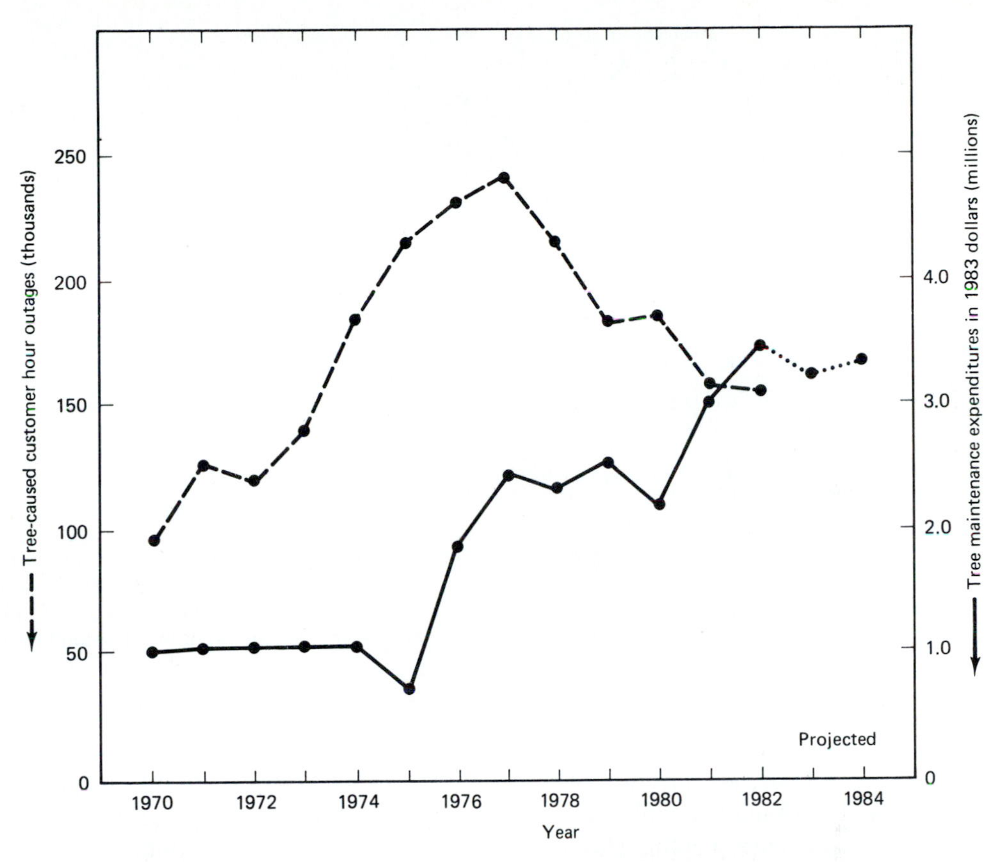

Figure 15–4 The twelve-year customer-hour outages versus tree maintenance expenditures at Wisconsin Power and Light Company. (From Smith, 1984.)

soil application reduce the growth rates of trees. The program reduced the number of trees pruned per year by 18 percent, reduced pruning time per tree by 30 percent, and reduced service interruptions by 56 percent (Johnstone, 1988). Goodfellow (1995) suggests that the amount of pruning per tree can be reduced by using utility pole systems that hang wires all on one side on an extended arm, or by using insulated wires that can be hung more closely together. The extended arm will place wires further from a tree while insulated wires spaced close together occupy less space, requiring less pruning.

Planting trees that grow slowly and have a small mature height under or near wires is an alternative to continued pruning. Fast-growing tree species under multiple utility lines demand trimming nearly on an annual basis. Repeated trimming often results in an unsightly tree that can become a safety hazard (Figure 15–8). Penelec, a utility company that serves the community of Erie, Pennsylvania, planted 3,000 compatible tree species under their wires in 1960 following extensive consultation with

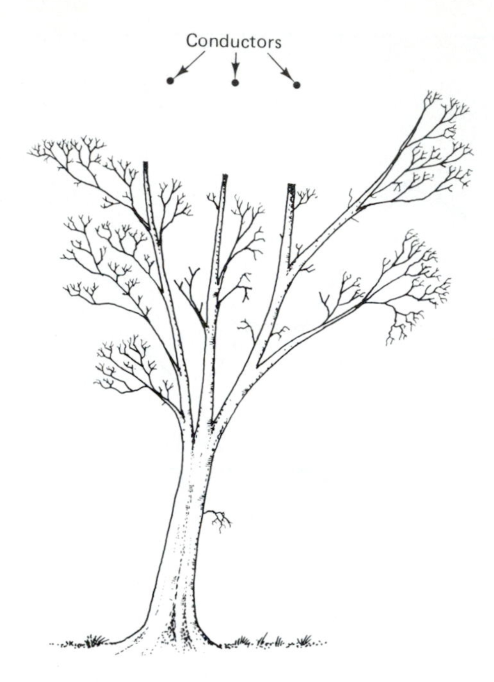

Figure 15–5 An undesirable means of line clearance is heading or topping of interfering trees. This will result in serious decay and heavy suckering, leading to future structural hazards.

residents, local government, and the Erie Shade Tree Commission. Most trees were planted between the sidewalk and curb. A survey twenty-five years later found 39 percent of the trees still present and in good condition. None of the trees had required pruning since planting (Rossman and Harrington, 1986). A number of utilities offer tree replacement programs for property owners and municipalities willing to have existing large trees removed and replaced with utility-compatible species. The utility pays for the removal and replacement, but saves money in the long term by reducing power outages and repair, and by having lower pruning costs (Barnes and Greenlee, 1991; Bauer et al., 1990).

Utility companies are responsible for line clearing on city streets, and this includes trimming city-owned street trees. Utility arborists should work with municipal foresters and other city officials by providing input into the species selection process for street tree planting under utility lines. In communities where most street tree planting is done by the city forestry department, a cooperative effort can be used to develop an approved species list. The community Master Street Tree Plan should recommend utility-compatible species for those streets where trees and wires occupy the same space. In communities where street trees are planted by property owners the local tree ordinance should require a permit for tree planting, and specify a list of acceptable species be developed by the municipality. Utility pruning is necessary, but can upset property owners who find their street tree severely pruned. Morell (1988) recommends notification of property owners in advance of utility pruning, with adequate explanation of why it is necessary.

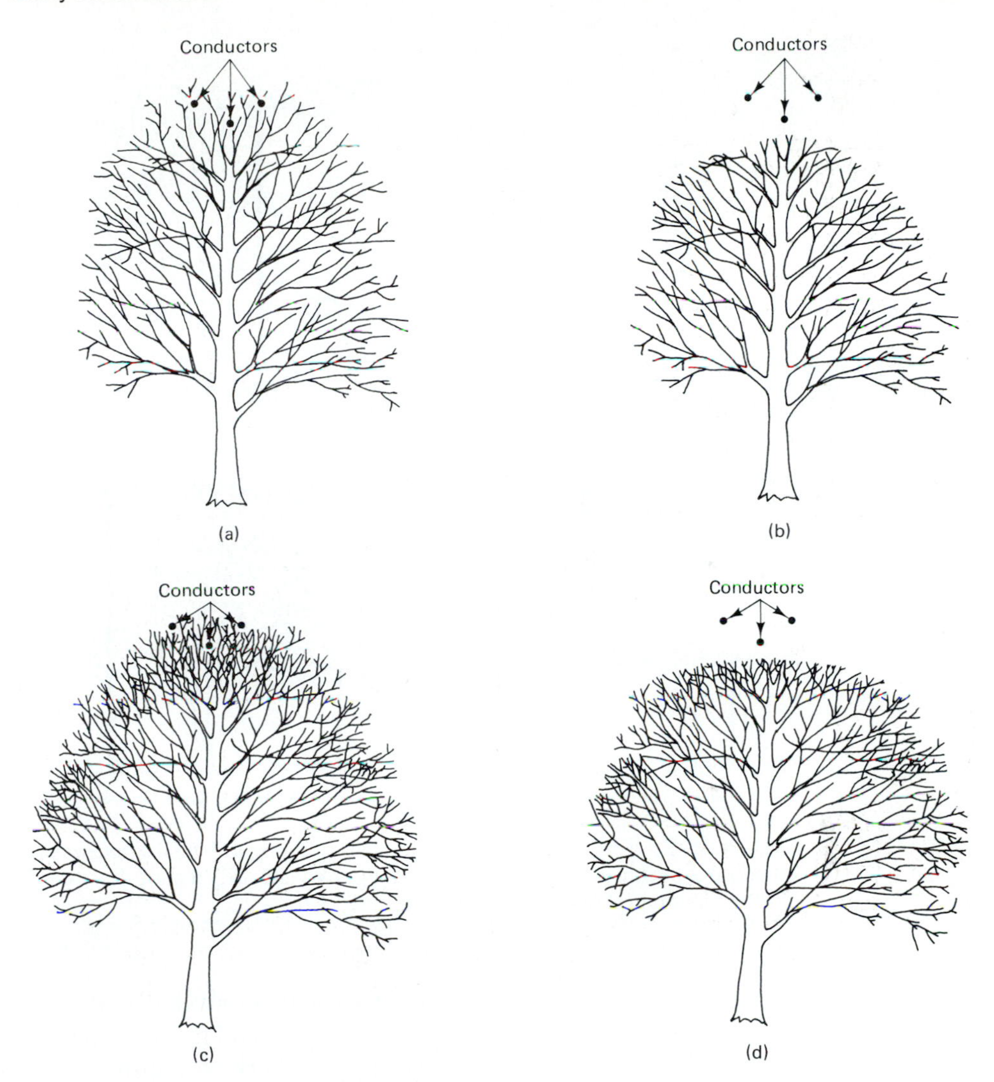

Figure 15–6 Rounding over a tree (a) produces a tree (b) that does not interfere with conductors. However, within a few years suckering (c) will cause interference with wires, resulting in a repeat of rounding over (d). Rounding over is an undesirable means of keeping branches away from conductors, for both the tree and utility service reliability. (From Holewinski et al., 1983.)

Right-of-Way Management

Right-of-way management includes tree trimming and other vegetation management activities, such as maintenance of desirable plant communities under transmission lines, vegetation-type conversion, and herbicide use (Figure 15–9). A system of

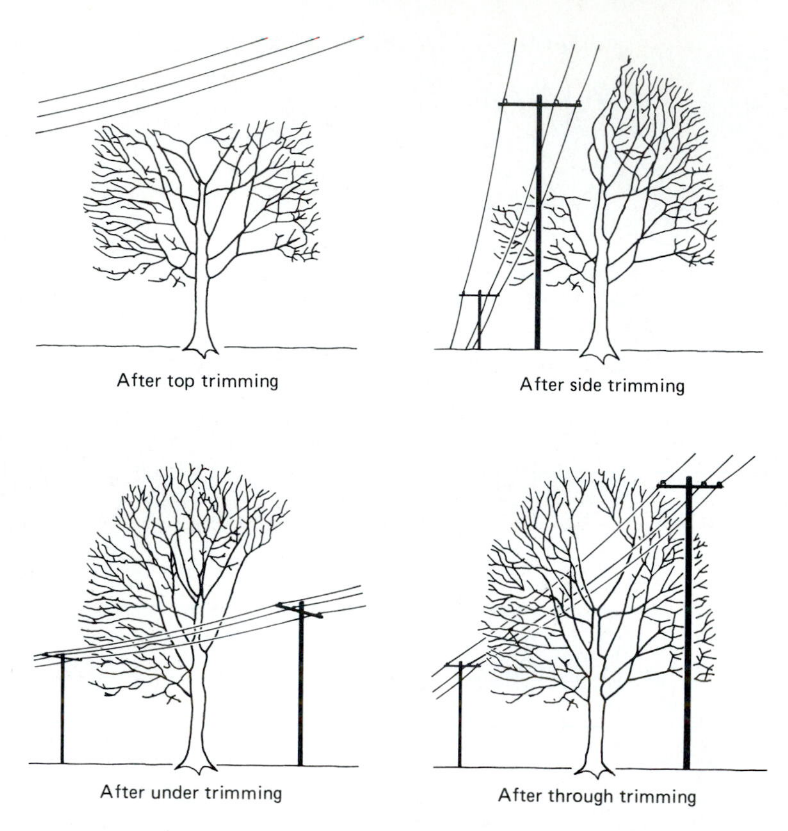

Figure 15–7 Top, side, under, and through trimming to lateral branches will produce a healthier, more natural-appearing tree, and will require less drastic future trimming. (From Holewinski et al., 1983.)

vegetation management in utility corridors starts with an inventory of existing vegetation within the utility right-of-way. This inventory should include trees or groups of trees near distribution lines and general cover types beneath transmission lines. The inventory should also identify those areas in need of immediate attention, estimate future maintenance needs based on regrowth, and identify areas where cover-type conversion will be cost-effective.

Ontario Hydro in Canada devised a distribution line vegetation management system based on inventories of nine management unit categories.

1. *Agriculture.* Intensive farming with few trees present to interfere with lines
2. *Rangeland.* Marginal farming with more potential brush and trees present
3. *Urban residential.* High population density with many mature trees present
4. *Rural estate.* Lower population density, but higher tree density than urban residential

Figure 15–8 These trees have been excessively and poorly pruned in the past to clear utility lines. It may be desirable to replace these trees with a species that attains a small mature size.

Figure 15–9 The plant community beneath this transmission line is attractive, yet little maintenance is required to keep it out of the wires, due to its composition of shrubs and small trees species. (John Goodfellow.)

5. *Mixed-hardwood (on-road).* Nonagricultural land with a high density of trees and brush in roadside right-of-ways
6. *Mixed-hardwood (off-road).* Dense hardwood and brush in cross-country rights-of-way
7. *Mixed-conifer (on-road).*
8. *Mixed-conifer (off-road).*
9. *Inaccessible.* Areas where there is no access by mechanical equipment

Each of these cover types is treated as a separate management unit with different prescriptions and crew and equipment assignments (Griffiths, 1984).

Programmed maintenance. Programmed maintenance is the establishment of a specific time frame or cycle for maintenance of vegetation. In utility arboriculture this consists of tree trimming, brush cutting, and herbicide spraying cycles. These cycles should be established according to vegetation cover types, potential for outages, and estimates of regrowth following treatments. Frequency of treatment should also be influenced by the number of customers affected should a tree-caused outage occur.

The frequency of maintenance will vary according to local rates of vegetation growth and acceptable levels of service. Van Bossuyt (1984) recommends tree-trimming cycles of three years based on experience in the wet and mild climate of Portland, Oregon; while Ulrich (1983) suggests that four- to five-year cycles are sufficient in colder Pennsylvania. Trimming cycles of two and one-half years in the city and three years in adjacent rural areas are used in Memphis, Tennessee (Wallace and Heuberger, 1984).

Right-of-way cover conversion and maintenance. In forested regions transmission-line corridors are cleared of trees prior to installation of utility lines. However, vegetation successional tendencies will cause continuous reinvasion of the right-of-way by tree species, creating the necessity for continuous control measures. The development and maintenance in these corridors of plant communities that are resistant to tree invasion is an important management program in right-of-way management. Bramble and Byrnes (1983), reporting on the results of thirty years of right-of-way vegetation management in Pennsylvania, found that the selective use of herbicides resulted in stable communities of herbs, shrubs, and grasses that resist invasion by forest trees (Figure 15–10). Selective spraying is spraying of individual trees, usually applied to the trunk or stump. This technique kills the entire tree including the root system, thus prevents sprouting. Mowing is an alternative to spraying, but costs more and has fewer benefits. Johnstone (1990) suggests selective use of herbicides on individual trees is superior to mowing for the following reasons:

1. Herbicides are safer for workers than using hand tools, chainsaws, and brushaxes.
2. Spraying eliminates suckering of trees and allows aesthetically pleasing communities of wildflowers, ferns, and shrubs to become established.

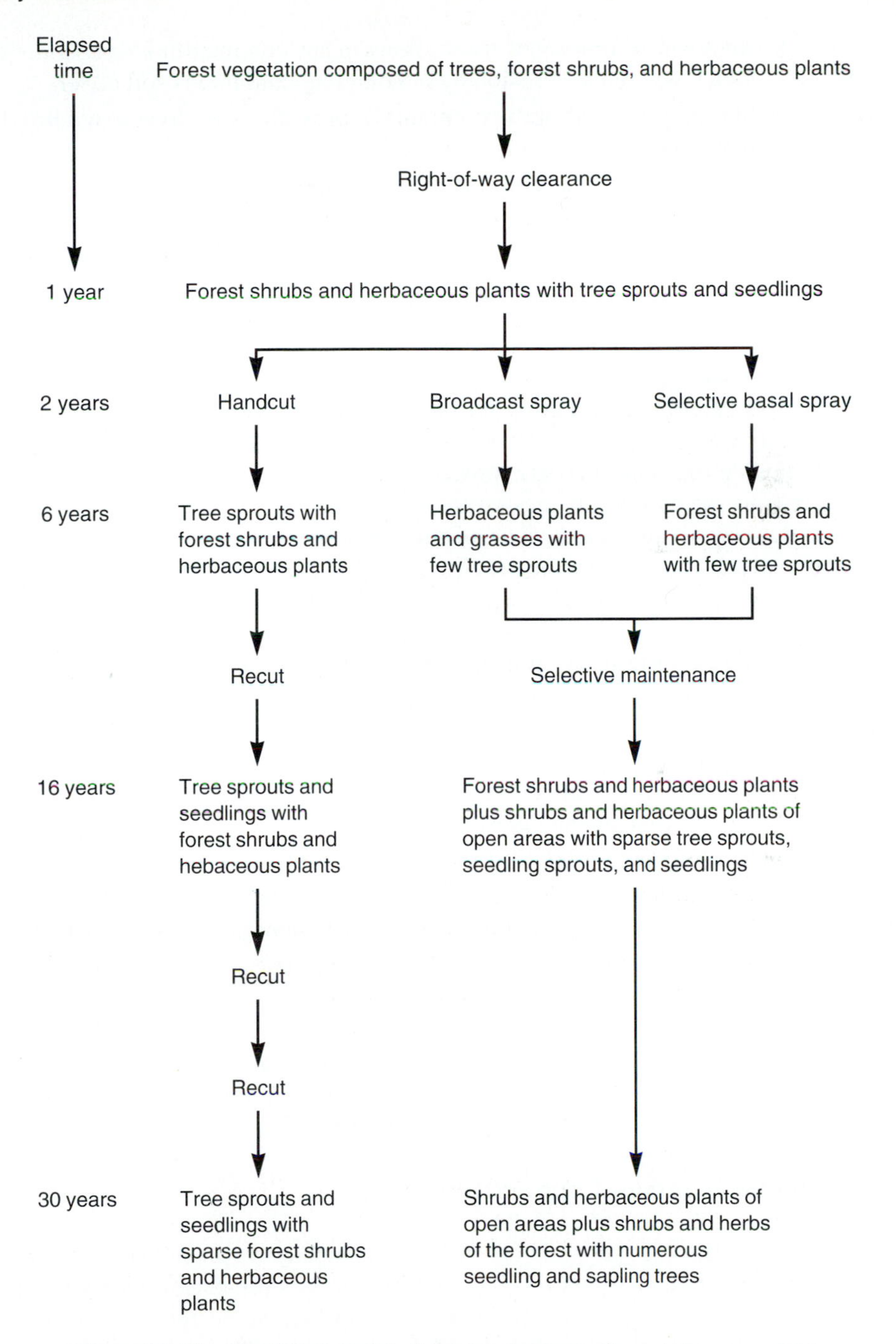

Figure 15–10 Simplified model of the development of vegetation on a right-of-way, showing dominant life forms. (Bramble and Burns, 1983.)

3. Removal of unwanted trees allows plant communities to become established, which will reduce erosion by maintaining continuous soil cover.
4. Shrub, grass, and herb communities provide more diverse wildlife habitat than tree suckers.
5. Herbicides are cheaper than cutting or mowing.

However, public concern over the safety of herbicides has led some utilities to reduce spraying in favor of mowing. Nowak et al. (1993) caution that important wildlife habitat and aesthetic values will be lost if mechanical control becomes the primary method of vegetation management.

Nonforest vegetation could also be maintained in utility rights-of-way by fire. Arner (1979) reports that prescribed burning of utility corridor vegetation to control woody vegetation is cost-effective in Mississippi and Alabama compared to mowing or herbicide use. Fire lines are plowed along the edges of the corridors with fire breaks crossing the corridor in one-fourth- to one-half-mile (0.4 to 0.8 km) intervals. Burning is done on a three-year cycle, and fire lanes on steep slopes are seeded with grasses to prevent erosion. As in the case described above, this technique not only controlled woody vegetation, but also improved wildlife habitat within the right-of-way.

Management planning. Many utility companies are developing long-term management plans for all company-owned lands. Ontario Hydro, Canada, developed the following objectives for its land management plan:

1. To collect and maintain an inventory of pertinent information necessary for sound management decisions
2. To compile and analyze inventoried data within the context of current social values and corporate objectives
3. To recommend optimal resource management practices on a site-specific basis
4. To develop guidelines to assist regional and area personnel to implement the management plan recommendations

Once implemented, land management plans are monitored to determine if management objectives are being met, and all plans are reviewed at five-year intervals and updated when necessary (O'Connor, 1983).

Company versus Contract Crews

Utility companies rely on both in-house company crews and contract crews to maintain vegetation on rights-of-way. The percent of work performed by company or contract crews varies greatly from utility to utility. Some rely heavily on contractors, while others perform essentially all maintenance with company crews, the final mix being a management decision guided by company policies and local needs.

In general, contract crews are less expensive than company crews per unit serviced, due primarily to competitive bidding. Higgins (1981) reports that contract crews clear lines at a savings of 30 percent over company crews working for Central Hudson Gas and Electric Corporation in New York. In spite of these savings, the company maintains a number of their own crews because they can respond more quickly in emergencies, are familiar with the local area, are more sensitive to customer needs, and represent a stable work force. In addition, linemen can be assigned to trimming crews during periods of reduced line work.

Three basic forms of contractual agreements are available to bidders on utility line clearing contracts (Goodfellow, 1985):

1. *Time and materials contract.* Hourly rates for labor and equipment, and unit prices for materials, are used by bidders to obtain the contract. These costs are charged to the company on a scheduled basis.
2. *Unit price contract.* Fixed prices are bid per unit of work performed such as linear distance cleared, trees removed by diameter class, acres, and so on.
3. *Fixed-price contract.* An entire job is placed up for bid and a single price charged for the entire contract.

The traditional method of contracting utility clearance work is the use of time and materials contracts; however, Goodfellow (1985) reports fixed-price bidding to be a cost-effective alternative to this method. However, he cautions that the project specification must be carefully written and that the scope of each project be of a size to encourage the maximum number of qualified bidders. Small projects are not worth bidding on, while very large contracts will discourage many potential bidders.

In all cases, whether by company or contract crew, utility arborists and foresters must balance reliability of service against costs and professional tree and other vegetation management standards. Company crews should be well trained not only in what constitutes a safe conductor, but also in the professional management of trees and other vegetation in the right-of-way. Utility customers are not concerned about electricity until it is shut off, but a poor trimming job on their favorite tree will remind them of their dissatisfaction with the utility company for a long time. Similarly, the cheapest bidder may save money for the company in the short run through substandard work, but the savings will probably not match the loss in goodwill from dissatisfied customers.

LITERATURE CITED

ABBOTT, R. E. and J. W. JOY. 1978. "Services of Commercial and Utility Arborists." *Proc. Natl. Urban For. Conf.*, ESF Pub. 80-003. Syracuse: SUNY, pp. 448–456.

ABBOTT, R. E., and K. C. MILLER. 1987. "Estimating and Pricing Tree Care Jobs." *J. Arbor.* 13(4):118–120.

ARNER, D. H. 1979. "The Use of Fire in Right-of-Way Maintenance." *J. Arbor.* 5(4):93–96.

BALL, J. 1992. "Marketing Arboricultural Services." *J. Arbor.* 18(4):205–208.

BARNES, B., and J. GREENLEE. 1991. "Seattle City Light Urban Tree Replacement." *J. Arbor.* 17(4):98–102.

BAUER, I. O., JR., T. D. MAYER, and W. T. REES, JR. 1990. "Tree Replacement Program at Baltimore Gas and Electric Company." *J. Arbor.* 16(2):42–44.

BLACK, W. M. 1981. "Innovations in the Tree Care Industry." *J. Arbor.* 7(12):326–328.

BRAMBLE, W. C., and W. R. BYRNES. 1983. "Thirty Years of Research on Development of Plant Cover on an Electric Transmission Right-of-Way." *J. Arbor.* 9(3):67–74.

FELIX, R. 1978. "The Arborists Role." *Proc. Natl. Urban For. Conf.,* ESF Pub. 80-003. Syracuse: SUNY, pp. 607–614.

FELIX, R. 1995. National Arborist Association, Executive Director. Personal communication.

GOODFELLOW, J. W. 1985. "Fixed Price Bidding of Distribution Line Clearance Work: Another Look." *J. Arbor.* 11(4):116–120.

GOODFELLOW, J. W. 1995. "Engineering and Construction Alternatives to Line Clearance Tree Work." *J. Arbor.* 21(1):41–49.

GOODFELLOW, J. W., B. BLUMREICH, and G. NOWACKI. 1987. "Tree Growth Response to Line Clearance Pruning." *J. Arbor.* 13(8):196–200.

GRIFFITHS, S. T. 1984. "A Managed System for Distribution Forestry." *J. Arbor.* 10(6):184–187.

HIGGINS, A. L. 1981. "Contract vs. Company Tree Crews." *J. Arbor.* 7(4):96–98.

HOLEWINSKI, D. E., J. W. ORR, and J. P. GILLON. 1983. "Development of Improved Tree-Trimming Equipment and Techniques." *J. Arbor.* 9(5):137–140.

JOHNSTONE, R. A. 1983. "Management Techniques for Utility Tree Maintenance." *J. Arbor.* 9(1):17–20.

JOHNSTONE, R. A. 1988. "Economics of Utility Lateral Pruning." *J. Arbor.* 14(3):74–77.

JOHNSTONE, R. A. 1990. "Vegetation Management: Mowing to Spraying." *J. Arbor.* 16(7):186–189.

MASSEY, J. 1983. "The Use of Small Computers in the Tree Care Business." *J. Arbor.* 9(2):39–41.

MORELL, J. D. 1988. "Utility and Municipal Communications Relating to the Urban Forest." *J. Arbor.* 14(11):273–275.

NOWAK, C. A., L. P. ABRAHAMSON, and D. J. RAYNAL. 1993. "Powerline Corridor Vegetation Management Trends in New York State: Has a Post-Herbicide Era Begun?" *J. Arbor.* 19(1):20–25.

O'CONNER, W. R. 1983. "Transmission Right-of-Way Management Plans." *J. Arbor.* 9(2):48–50.

PERRY, P. B. 1977. "Management's View of the Tree Trimming Budget." *J. Arbor.* 3(8):157–160.

ROSSMAN, W. R., and C. J. HARRINGTON. 1986. "An Attractive Alternative to Tree Trimming for Line Clearance." *J. Arbor.* 12(1):20–23.

SHIGO, A. L. 1990. *Pruning Trees Near Electric Utility Lines.* Durham, N.H.: Shigo and Trees Associates.

SMITH, T. K. 1984. "Determining Line Clearance Needs at Wisconsin Power and Light Company." *J. Arbor.* 10(7):205–208.

ULRICH, E. 1983. "Correlating Tree Disturbances, Tree Work, and Tree Budgets." *J. Arbor.* 9(3):79–84.

VAN BOSSUYT, D. P. 1984. "Restructuring the PGE Line Clearance Program." *J. Arbor.* 10(7):198–201.

WALLACE, J., and M. HEUBERGER. 1984. "Municipal Line Clearance." *J. Arbor.* 10(3):95–96.

Appendix A

*Computer Software for Urban Forest Management: A Buyer's Guide**

Sample Data Collection Instructions and Data Sheet

Sample Summary Tables

As microcomputer prices have fallen and become affordable for mid-sized cities, many urban forestry departments have computerized their tree information systems. When a department purchases a computer, it must decide which software to use. Different software functions serve different forestry needs.

Urban forestry software systems designed for municipalities have six major functions: The first three are data files and the last three are ways the files can be manipulated.

DATA FILES

The first data file manages tree inventory data. The information on individual trees must be easily updatable as the condition of the tree or planting site changes. There must be provision for adding newly planted trees and deleting trees that are removed.

*E. Thomas Shirley. From G. Moll and S. Ebenreck (Eds.). 1989. *Shading Our Cities.* Washington, D.C.: Island Press.

In order to keep the inventory current, data must be added regularly. Prior to purchasing a system, assess the ease of changing the individual tree information.

The second capability of most systems is the ability to record work conducted on trees. This work history file should keep records of what work has been completed, when, by whom, what equipment was used, how many hours were required, and what materials were used. These factors can be summarized and used to help with budget preparation, species selection, comparison of crew/contractor efficiency, and determination of equipment requirements. It can also be valuable in tree-related legal disputes.

The third data file is for service requests. Most municipal forestry departments are deluged with requests for tree work from citizens who notice problems. Hand recording and keeping track of inspection cards can be a monumental task during the spring and summer months. Computerizing service requests greatly simplifies matters and allows the inspection orders, tree inventory data, and work histories to be linked together.

MANIPULATING FILES

The fourth system capability is that of numerically summarizing tree inventory data, work history data, and service request data. For example, if you need to know how many dead trees there are in the inventory database with a DBH (diameter at breast height) greater than twelve inches, the computer should be able to display that number. This type of numerical information can then be used to prepare annual and/or monthly reports on the condition of municipal trees, how much work has been completed, and how many service requests have been received.

The fifth capability of most systems is the ability to generate listings of trees. This allows foresters to pinpoint the location of problem trees or to produce a work order. For day-to-day operations, it's best to have listings based on maintenance requirements. If you need to schedule the removal of all dead trees greater than twelve inches DBH, the computer can list the address or location of each tree in this category. Listings of trees on which work has been completed and listings of service requests are also available in many systems.

Computer mapping of tree locations can be an integral part of a computer system. Computer-generated maps are a valuable aid in locating park and boulevard trees. If the database includes underground and overhead utilities as well as building setback distances, computer maps can be a valuable aid in selecting the right species for a given location.

Prior to purchasing a system, look at these six capabilities and decide which you require. Systems should also be assessed for "friendliness," or general ease of use. If you have special needs, make sure they can be added to the system.

The majority of these systems were developed for IBM personal computers or compatible systems (see Table A–1). One has been developed for the Apple IIe and three for minicomputers.

System capabilities vary among the programs (see Tables A–2 and A–3). Most appear to store and manipulate tree inventory data in similar ways and can generate

TABLE A–1 SOFTWARE PACKAGES: WHO PRODUCES THEM, WHAT COMPUTERS DO THEY WORK ON, AND HOW MANY TREES CAN THEY HANDLE?

Package Code	System Name	Developer/ Organization	Hardware/Software Required	Capacity Floppy	Capacity 10mb Hard Disk
CFIP	Community Forestry Inventory Program	Helburg, Hoefer/CO State University	IBM PC/Word Proc. Apple IIe/Word Proc.	?	?
CTI	Central Tree Inventory	Reidel/Nat'l Park Serv.	Unify/Unix	100K	
CT	Compu-Tree (TM)	?/Systemics	IBM PC/MSDOS	1,800	50K
dT	dTree	Jones, Dossin/dTree	IBM PC/MSDOS	1,500	30K
GCTM	Golden Coast Tree Mgmt. Software System	Giedraitis, York/Golden Coast Env. Services	IBM PC/MSDOS Unix/Rel DB Mgr.	?	30K
OUF	Oakland U.F. Data Mgmt. System	?/USFS, Oakland, CA	IBM PC/dBase III	500+	50K
SMUF	Santa Maria U.F. Data Mgmt. System	?/USFS, Santa Maria, CA	HP 3000/IMAGE	?	?
TB	Trebase	Miller, Andrews/Univ. of Wisconsin	IBM PC/dBase III	4,000	?
TI	Tree Inventory	?/Michigan State Univ.	IBM PC/MSDOS	2,500	75K
TIMS	Tree Inventory and Mgmt. System	Maggio/Texas A&M	IBM PC/DOS, Lotus	10,000	?
TIS	Tree Inventory System	McCarter, Baker/Utah State University	IBM PC/MSDOS	6,800	200K
TM	Tree Manager (TM)	Joehlin/ACRT, Inc.	IBM PC/dBase III	1,300	50K
UTIP	Urban Tree Inventory Prog.	?/OK Forestry Division	IBM PC/MSDOS	?	?

tree listings, work orders, and numerical summaries. Work histories can be recorded in nine of the systems, and maps can be generated by one system. The UTIP (Urban Tree Inventory Program) system is intended for use in a "windshield"-type partial inventory. UTIP provides information to make decisions on the needs of a city forestry department, but it is not intended for daily operations.

The costs of systems vary greatly (see Table A–4). Off-the-shelf software systems tend to be less expensive. Due to the high cost of customization, systems that

TABLE A–2 SYSTEM CAPABILITIES: WHAT DO THEY DO?

Package	*Tree Inventory Data*					*Work Order (WO)*		*Work History (WH)*				
	Updated		Location System			Generated by		Record by		*Numerical Summaries*		
Code	Daily	Batch	Address	Grid	Other	Block	Tree	Block	Tree	Inventory	WO	WH
CFIP	X	X	X	X	X	X				X	X	
CTI	X	X		X		X	X	X	X	X	X	X
CT	X	X	X	X	X	X	X	X	X	X	X	X
dT	X	X	X	X		X	X			X	X	
GCTM	X	X	X		X	X	X	X	X	X	X	X
OUF	X	X	X			X	X	X	X	X	X	X
SMUF	X	X	X			X	X	X	X	X	X	X
TB	X		X		X			X	X	X		X
TI	X		X			X	X			X		
TIMS	X	X	X		X	X	X	X		X		
TIS	X		X	X		X	X	X	X	X		
TM	X	X	X	X	X	X	X		X	X	X	X
UTIP		X								X		

TABLE A–3 SOFTWARE PACKAGES: WHAT DO THEY DO?

Package	*Tree Listings by*			
Code	Location	Maintenance Requirements	Other	Maps
CFIP	X	X		
CTI	X	X	X	
CT	X	X	X	
dT	X	X	X	
GCTM	X	X	X	
OUF	X		X	
SMUF	X	X	X	
TB	X	X	X	
TI	X	X	X	
TIMS	X	X	X	X
TIS	X	X	X	
TM	X	X	X	
UTIP				

TABLE A–4 SOFTWARE PACKAGES: WHO IS USING THEM AND HOW MUCH DO THEY COST?

Package Code	No. of Cities Using System	*Approximate Costs*		
		Software	Installation	Support
CFIP	3	?	?	?
CTI	0	?	?	?
CT	5+		$10,500	
dT	0	$200	—	—
GCTM	4	$3,500	$600	$600[a]
OUF	1	?	?	?
SMUF	1	?	?	?
TB	11	$175	?	NC[b]
TI	1	$300	$300+	$30/call
TIMS	1	$9,000	$2,500	$500[a]
TIS	1	$100	?	NC[b]
TM	5+	$3,500	$600	$600[a]
UTIP	28	NC	NC	NC

[a]Installation costs do not include travel or hardware.
[b]No charge.

are adapted to existing management needs are more expensive. Some systems include installation, which consists of an expert visiting the municipality to make sure the software works on its computers. Training of personnel who will use the system is also included in the installation expense. Support costs are for the right to call the software provider and have them solve problems that arise during the use of the software. This is helpful for departments without staff knowledgeable about computers.

To find out more about computer systems, contact the organizations that provide software or software developers listed in Table A–5.

TABLE A–5 PROVIDERS OF TREE MANAGEMENT COMPUTER SOFTWARE

Organization	Contact Person	Address	Phone
Commercial			
ACRT, Inc.	Beth Buchanan	P.O. Box 219 Kent, OH 44240	(216) 673–8272
dTree	Dan Dossin	8601 Roberts Dr., Suite 4–1 Dunwoody, GA 30338	(404) 993–0831
Golden Coast Env. Service	Thomas Pehrson	2736 W. Orangethorpe Ave., Suite 5 Fullerton, CA 92633	(714) 441–1308

(*continued*)

TABLE A–5 CONTINUED

Organization	Contact Person	Address	Phone
Commercial (cont'd)			
Michigan State University	J. James Kielbaso	Department of Forestry, MSU E. Lansing, MI 48824	(517) 355–7533
Systemics	Arthur Costonis	43 Green St. Foxborough, MA 02035	(617) 543–4557
Texas A&M	Robert Maggio	Department of Forest Science College Station, TX 777843	(409) 845–5069
Utah State University	Fred Baker	Department of Forest Resources Logan, UT 84322	(801) 750–2550
University of Wisconsin	Robert Miller	College of Natural Resources Stevens Point, WI 54481	(715) 346–4189
Noncommercial			
Colorado State University	Larry Helburg	Fort Collins, CO 80523	(303) 491–6303
Oklahoma Forestry Division	Rob Doye	2800 N. Lincoln Blvd. Oklahoma City, OK	(405) 521–3864
U.S. Forest Service	Philip Barker	P.O. Box 245 Berkeley, CA 94701	(415) 486–3927

SAMPLE DATA COLLECTION INSTRUCTIONS AND SAMPLE INVENTORY SHEET

1. *Work unit:* 3 characters (numeric)
2. *Street:* 20 characters (alpha and numeric)

   Note: Coding may be used in the field (i.e., 2 for Baker Street). However, if the name is used, the code must be used. The twenty characters available will be used by the machine for a street code file created by the user.

   Example:

Code	Street
1	Algoma Street
2	Baker Street
3	16th Street
—	etc.

3. *Record data:* 6 characters (numeric, always 2 digits)

   Note:

Year	Month	Day
84	06	14

4. *Action:* 20 characters (alpha or numeric)
Note: These data may be tailored by the user to maintain a history of service to the tree by action and date. The date recorded will be the date in item 3 above and will automatically be assigned to the action in the file. The user will develop an action file which may contain some, all, or more than those listed below.

Numeric Code	Action
1	Inventory
2	Plant
3	Prune
4	Fertilize
5	Remove
6	Service Call
7	Damage Insect
8	Damage Disease
9	Damage Vandalism
	etc.

For the initial inventory, INVENTORY has been inserted to describe the action. If other *Action* data are to be taken during the initial inventory, they should be placed on a separate sheet.

5. *Crew:* 20 characters (alpha or numeric)
Note: Use the name of the individual collection inventory data or crew leader of the crew performing maintenance.

6. *Block or address:* 6 characters (numeric)
Note: If an address is not available, use the block number and number of trees sequentially from the center of the city outward (see item 7).

7. *Tree number:* 2 characters (numeric)
Note: Use if more than one tree at an address or if using Block in item 6.

8. *Lawn width:* 2 characters (numeric, 1–99)
Width of the tree lawn in feet or meters.

9. *Wire height:* 2 characters (numeric, 1–99)
Height of overhead wires, if present.

10. *Tree present:* 1 character (numeric)
Note: 1 if present, 0 if vacant.

11. *Planting priority:* 1 character (numeric)

12. *Species:* 20 characters (alpha or numeric)
Note: Coding should be used in the field (i.e., abbreviations or numbers). The 20 characters available will be used by the computer for a tree species code file. The attached data sheet allows 6 characters for code.

13. *Diameter class:* 2 characters (numeric)
1, 2, 4, 6, 8, 10, . . . , 38

14. *Condition class:* 3 characters (numeric)
100, 80, 60, 40, 20, 0

Work Unit ____________ Street ______________________________ ____________
Name Code

Crew ______________________________ Date ____ /____ /____
yr mo day

Address or block						Tree Number		Lawn Width		Wire Height		Tree Pres.	Pl. Prior	Species						Dia. Class		Cond. Class			Action Code		Corner Street Code		Corner Address					
1	2	3	4	5	6	7	8	9	10	11	12	13	14	15	16	17	18	19	20	21	22	23	24	25	26	27	28	29	30	31	32	33	34	35
1	2	3	4	5	6	7	8	9	10	11	12	13	14	15	16	17	18	19	20	21	22	23	24	25	26	27	28	29	30	31	32	33	34	35

SAMPLE SUMMARY TABLES

02/11/85

SUMMARY BY SPECIES

SPECIES	AVG DIAM	FREQUENCY	AVG COND	VALUE
NORWAY MAPLE	10.8	849	59.4	824823.280
BUCKEYE	4.7	446	70.7	72800.562
RED MAPLE	15.1	395	55.9	773057.215
GINKGO	19.1	388	58.9	1140659.045
WHITE ASH	2.7	241	82.0	12865.476
SILVER MAPLE	9.8	163	55.0	49071.910
LITTLELEAF LINDEN	2.9	100	79.3	7974.387
RED OAK	6.7	98	67.8	50536.855
AMERICAN ELM	1.7	91	71.5	1522.488
HACKBERRY	13.5	87	55.3	109502.403
ENGLISH OAK	4.7	86	51.9	28948.198
HORSECHESTNUT	15.7	81	61.8	103917.285
GREEN ASH	6.7	78	65.7	34143.446
TREE OF HEAVEN	4.0	73	67.5	8295.955
BASSWOOD	5.6	45	72.9	12833.440
SUGAR MAPLE	16.1	43	62.2	110098.391
CALLERY PEAR	12.2	40	47.3	40208.981
HONEYLOCUST	3.3	39	70.3	7699.363
BUR OAK	4.3	32	79.8	8672.165
CATALPA	13.4	30	57.8	17155.862
POPLAR SP	1.4	28	74.8	141.683
BOXELDER	5.9	13	58.4	848.164
MAPLE, NORWAY COLUMN	13.8	11	65.4	18181.856
SIBERIAN ELM	2.0	9	66.0	57.422
CRABAPPLE	7.5	8	54.7	2896.645
PIN OAK	7.4	7	68.5	3452.618
SYCAMORE	2.0	5	84.0	162.476
MISC HARDWOOD	2.0	1	100.0	19.342
** TOTAL **				
		3487		3440546.913

Sample species summary table for a work unit. Also available for a city summary.

02/11/85

SUMMARY BY DIAMETER CLASS

DIAM	FREQUENCY	AVG COND.	VALUE
1	176	63.6	1023.212
2	735	74.6	17352.429
4	404	72.0	40071.166
6	206	59.3	46117.600
8	232	61.6	94851.865
10	336	59.0	204572.475
12	275	56.7	239776.248
14	247	58.0	303516.226
16	223	57.5	366298.513
18	194	55.5	394723.015
20	135	57.4	347457.202
22	121	60.0	387504.915
24	74	60.0	298431.471
26	52	53.5	221220.479
28	29	60.9	161880.414
30	18	53.2	102490.542
32	10	60.0	72294.154
34	9	63.8	80355.583
36	6	53.3	50135.500
38	1	60.0	10473.909
** TOTAL **			
	3483		3440546.922

Sample diameter summary table for a work unit. Also available for a city summary.

Appendix B

*CityTrees! Lite©**

INTRODUCTION

What Is CityTrees! Lite?

CityTrees! Lite v1.0 is a street tree management simulation designed to serve two purposes. It can be used as a management simulation for city foresters to project the long term implications of management decisions regarding street trees, and it can be as a training tool for students and employees. **CityTrees!** is an upgraded version of URBAN FOREST (1985) that is easier to use, readily accepts inventories and management information from communities, and runs in a Windows environment.

*Robert W. Miller and Robert Rogers, College of Natural Resources, University of Wisconsin—Stevens Point, Stevens Point, WI 54481.

CityTrees! operates on an annual cycle growing trees and killing trees, and it allows the user to remove dead trees, to plant trees, and to prune trees. At the end of each run changes to the population are summarized, as are the costs of management activities and tree values.

What Can It Be Used For?

City foresters who have a street tree inventory and desire to use **CityTrees!** as a management simulation can enter their inventory summary directly into a matrix, or can enter it from a spreadsheet. Inventory parameters include species, diameter classes (1, 2, 4, 6 ... 38+ inches), and CTLA condition classes by species/diameter class (Council of Tree and Landscape Appraisers). Inventory systems that are in larger diameter classes can be arrayed across multiple classes, and inventory systems with mean condition classes by species or for the entire population can also be used.

As a training tool, **CityTrees!** has a game function that introduces six random events into the simulation. Each year the simulation is run there is a possibility of a random event occurring that will alter the population mortality rates or will alter the management budget.

GETTING STARTED

Equipment Needed

Hardware

- IBM or IBM compatible computer with a 80836-based, 80486-based, or Pentium processor.
- **CityTrees!** Lite must be installed on a hard disk. The space required for the program and supplementary files is 1240K.
- **CityTrees! Lite** can be run with the minimum RAM supported by MS Windows.
- A color or gray-scale VGA monitor or better.
- A mouse and printer are optional, but highly recommended.

Software

- Microsoft Windows Version 3.1 or higher.

Installing *CityTrees! Lite* on Your Computer

Before you can use **CityTrees! Lite** on your computer you must install it on your hard disk. However, it is a good idea to make copies of your original disks and use the copies to install **CityTrees!** on your computer.

To start installing **CityTrees! Lite** on your computer, follow these steps:

1. Insert **CityTrees! Lite** Installation Disk 1 in drive A
2. If it is not already running, start Windows 3.1 or later.
3. Choose Run from the File Menu in Program Manager.
4. In the Command Line entry field type `a:\install.exe` and choose OK.
5. Follow the directions on screen.
6. **CityTrees! Lite** will be installed in C:\CTLITE by default. You can change this destination if you wish.
7. Make sure both the Sample CTS Files and Text Data Files boxes have an X in them.

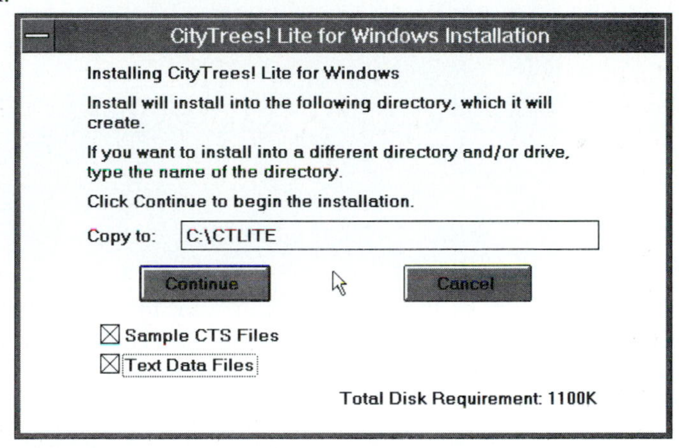

STARTING CITYTREES! LITE

The installation program will create a window containing an icon for the **CityTrees! Lite** Program. Start the program by double-clicking the left mouse button while pointing at the icon. The best way to learn how to use **CityTrees!** is to go through the tutorial described below.

A TUTORIAL

A file called `SAMPLE.CTS` has been provided with **CityTrees! Lite** so you can become acquainted with how the program works. We will now go through an entire simulation run in a step by step sequence.

- Click on the Continue Button on the opening screen.

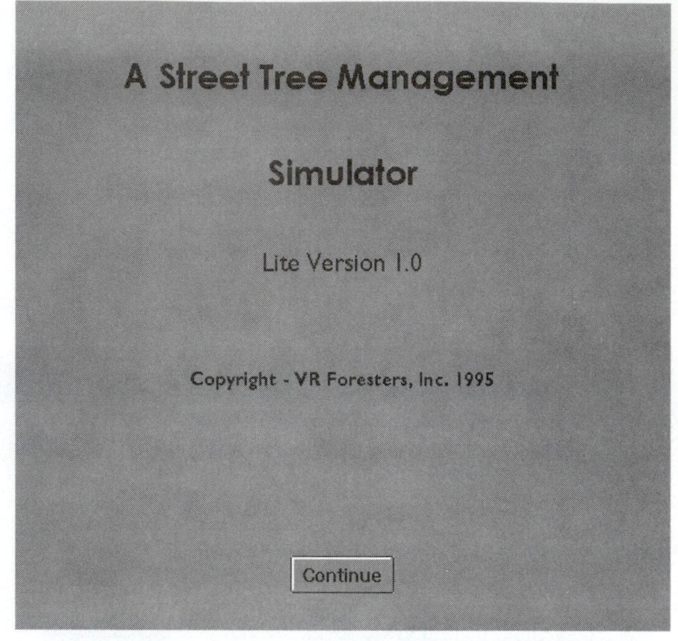

- Type in the letter of the drive that you want to save your files to. **CityTrees!** will look at this drive for its initialization files, any **CityTrees!** datafiles, inventory text files, and it will also use this disk to store Master (.mas) and Simulation (.sim) files.

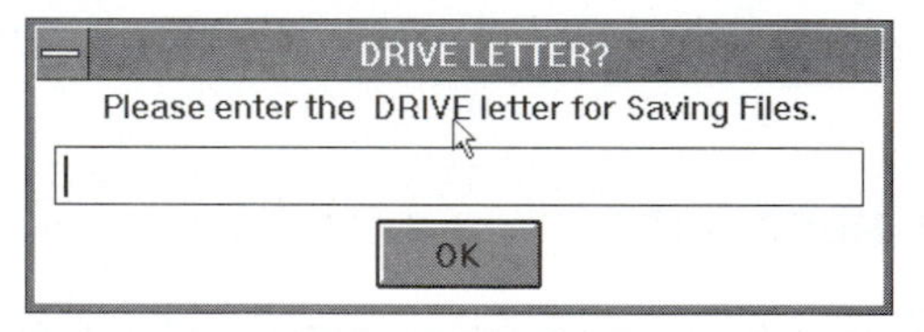

- Reading the data in: From the Menu Bar Select **File, Open**.

- Double Click on `sample.cts`.
- Click on the OK button after the file has been read.

- Now select **MAKE** from the menu bar and click on **GO! Create Master Files**.

- Press OK in response to the next message.

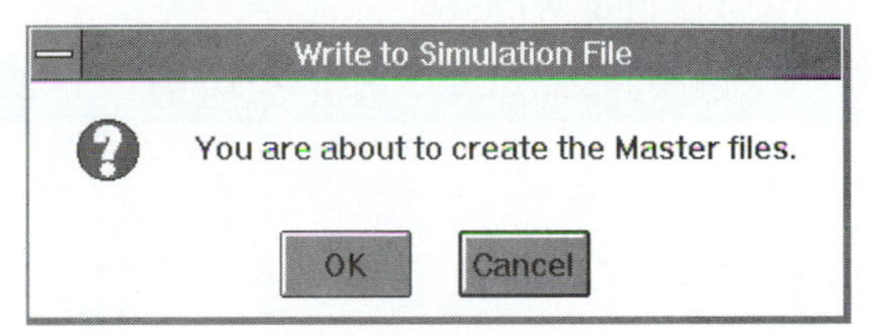

- Next choose **BUILD** and click on **GO!**.

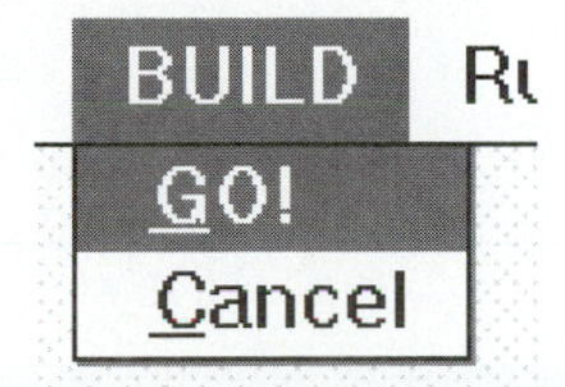

- Click OK in the next box.

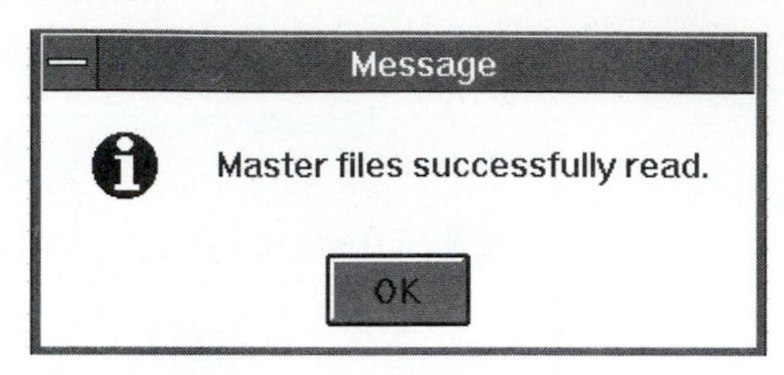

- Once the master files have been read in using **BUILD**, the simulation can begin. Select **Run Simulator** from the menu bar and then click on **GO!**.

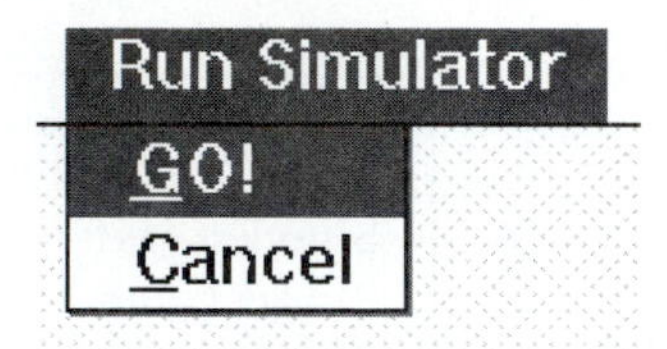

- Type in a name for this run. The name will serve to identify this run from other runs. Sample Run has been used as a name in this example.

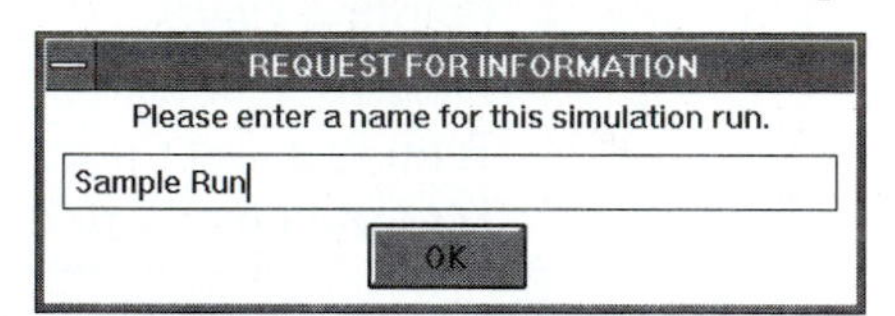

- The initial condition of the population of trees in a city is displayed in the **CityTrees!** Output Window.

CITYTREES! Lite - Output

16	206	2.3	66.5	1623	334401
18	232	2.6	64.6	1982	459802
20	229	2.6	69.3	2552	584402
22	220	2.5	65.1	2671	587654
24	185	2.1	64.8	3003	555519
26	150	1.7	68.6	3765	564720
28	90	1.0	66.0	3983	358484
30	52	0.6	69.2	4513	234700
32	19	0.2	69.5	5083	96576
34	15	0.2	70.6	6394	95913
36	10	0.1	68.0	5495	54947
38	2	0.0	70.0	6467	12934

- Choose **Run Simulator** and click on **GO!** to project what will happen to this population of trees a year from now. If the GAME MODE is turned on, various unexpected events might happen that would alter normal dynamics occurring in the population.

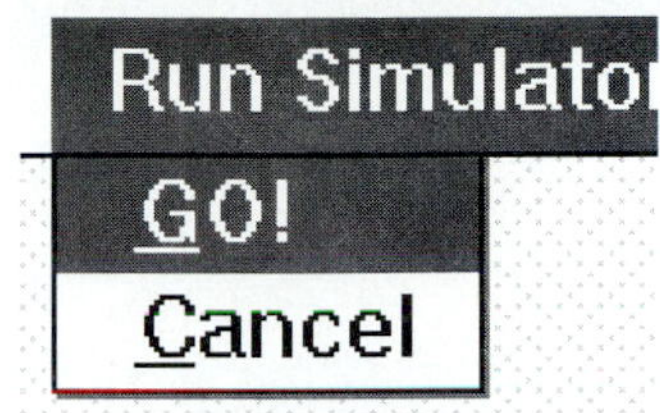

- You are asked if you want to prune trees. Respond **No** to this question for now. You may try this again later and respond yes.
- You are asked if you want to plant trees. Respond **No** for now.
- The results of your decisions are shown in the **CityTrees!** output window.
- Next you are asked if you want to continue. If you respond **Yes**, the simulation continues for another year. If you respond **No**, you are asked if you want to save the results so far, so that you can continue the simulation at a later time. Respond **No** and **No** for now.
- You end **CityTrees!** by selecting File from the Menu Bar and clicking on **Exit CityTrees!**.

THE DATA FILES THAT ARE USED WITH CITYTREES!

The following files are supplied on the install disk for **CityTrees! Lite:**

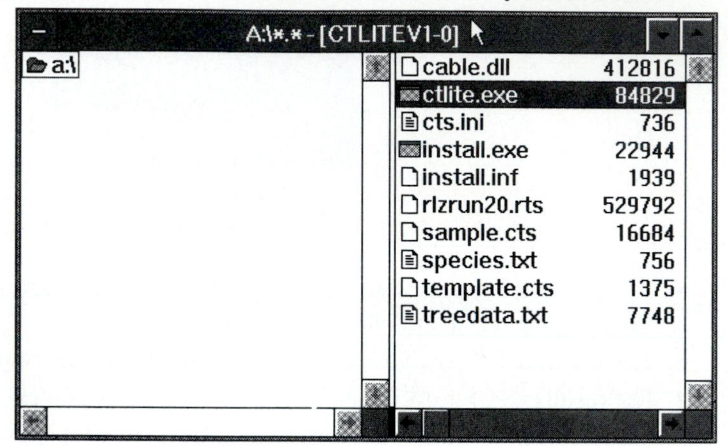

`Cable.dll` and `rlzrun20.rts` are required by WINDOWS to run **CityTrees!** `Install.ini` and `install.exe` are files used to install **CityTrees!** in WINDOWS. `Ctlite.exe` is the program file. `Cts.ini` contains all the default values used by **CityTrees!**.

The contents of `cts.ini`

```
[CITY TREES INITIALIZATION FILE]
[MAXIMUM LENGTH FOR SIMULATION PERIOD]
20
[LENGTH OF PRUNING CYCLE]
5
[YEAR TO START SIMULATION]
1995
[AVERAGE TREE VALUE]
19
[RATE OF CHANGE IN TREE CONDITION BECAUSE OF PRUNING]
0.50
[NUMBER OF YEARS TO GROW 1 DBH CLASS]
4
[NUMBER OF PLANTING LOCATIONS]
3000
[GAME MODE 1=YES 0=NO]
1
[TREE REMOVAL COST]
15
20
31
```

```
43
55
65
75
85
100
115
125
135
145
155
165
180
195
220
240
260
[TREE PRUNING COST]
14
14
14
14
21
21
21
30
30
30
35
35
35
40
40
40
45
45
45
50
[TREE MORTALITY TABLE]
1.5
1.2
1.4
0.5
0.7
0.9
1.1
1.3
1.5
```

```
1.6
1.8
2.0
2.2
2.4
2.6
2.8
3.0
3.2
3.4
3.6
[TREE PLANTING COST]
120
180
[PLANTING MORTALITY]
20
15
```

The `cts.ini` file can be edited by any text editor like NotePad in Windows. The format must be strictly adhered to. The items in brackets are the names of variables except for the first, which just serves to identify the contents of the file. The numbers that follow are the values of the variables. The nature of most of the variables are obvious from the name inside the brackets. Twenty values are entered for each of `TREE REMOVAL COST`, `TREE PRUNING COST`, and `TREE MORTALITY TABLE`. The twenty values correspond to the following diameter classes: 1,2,4,6,8,10,12,14, 16,18,20,22,24,26,28,30,32,34,36,38. Trees larger than 38" DBH are retained in that class. Cost is expressed in dollars and tree mortality is expressed in percent. `TREE PLANTING COST` and `PLANTING MORTALITY` consists of two entries that correspond to 1" DBH and 2" DBH respectively.

CityTrees! expects to find your `cts.ini` file in the main directory of the drive you specify in the opening dialog.

Files that have the three letter extension `cts` as part of the file name denote that the file is a **CityTrees!** data file. This file is created by **CityTrees!** by importing data from text files containing tree inventory data and can only be read by **CityTrees!** Two `cts` files are provided with **CityTrees!**: `sample.cts` and `template.cts`. `Sample.cts` contains data from a street tree inventory that you can use with the tutorial to become familiar with **CityTrees!** `Template.cts` is a file that contains data for only one temporary species. You can use this file together with the spreadsheets accessed from **MAKE** on the menu bar to enter tree inventory data. **Please note** that as you add species you must enter all corresponding data in the MAKE spreadsheets. Also be sure to use **SaveAs** from the **File Menu** to save under a different name to avoid overwriting `template.cts`.

CityTrees! also creates four work files that it writes to the disk you specify in the opening dialog. These files are `instru.mas`, `species.mas`, `instru.sim`, and `species.sim`.

THE MENU BAR CHOICES AND WHAT THEY DO

FILE

OPEN
Used to open `*.cts` files only.

CLOSE
Used to close open `*.cts` files.

SAVE
Used to save a `*.cts` file that retains the same name used to open the file.

SAVE AS
Used to save a `*.cts` file under a different name.

IMPORT
Used to import tree inventory data that has been prepared using a special text format. Two text files are used: a file containing a list of species, abbreviations, and relative value expressed as a percent; and a file containing specific inventory data. These files can be prepared using a text editor according to the specifications below.

SPECIES.TXT

Variable	Position in record
Species Id Number right justified	1-2
Condition Class right justified	3-6
Species Name Abbreviation rj	7-13
Full Species Name left justified	14-40

An example: SPECIES.TXT

```
 1  80 ASH-GRGREEN ASH
 2  60 ASH-WHWHITE ASH
 3  60 BASSWDBASSWOOD
 4  20 BOXEL BOXELDER
 5  80 PEAR CALLERY PEAR
 6  40 CATALPCATALPA
 7  80 CRABAPCRABAPPLE
 8  60 ELM-AMAMERICAN ELM
 9  20 ELM-SISIBERIAN ELM
10  20 POPLARPOPLAR SP
11 100 GINKGOGINKGO
12  80 HACKBEHACKBERRY
13  80 SYCAMOSYCAMORE
14  80 HONEY HONEYLOCUST
15  40 WILLOWWILLOW
```

```
16  40 AILANTTREE OF HEAVEN
17  60 HORSECHORSECHESTNUT
18  80 LIND-LLITTLELEAF LINDEN
19  60 BUCKEYBUCKEYE
20 100 MAP-NONORWAY MAPLE
21 100 MAPREDRED MAPLE
22  40 MAP-SLSILVER MAPLE
23 100 MAP-SUSUGAR MAPLE
24 100 MAP-CMMAPLE, NORWAY COLUMNAR
25 100 OAK-RDRED OAK
26 100 OAK-BUBUR OAK
27 100 OAK-ENENGLISH OAK
28 100 OAK-PNPIN OAK
29 100 OAK-WHWHITE OAK
30  40 H-MISCMISC HARDWOOD
```

TREEDATA.TXT

The first records in this file contain numbers of trees by species and DBH class. There is one row per species that matches the species in SPECIES.TXT. There are 21 columns of data. The first column occupies the first 2 positions in the row and contains the species id number right justified. Next there are 20 fields that are 6 positions wide. These fields correspond to the following DBH classes: 1,2,4,6,8, . . . 36,38.

Next there is a single record with a 0 recorded in the second position. The number of sites available for planting is recorded right justified in the next 6 positions. Zeros fill in the remaining 19 right justified fields that are 6 positions wide.

Next there is a record that begins with a –1 in the 2nd position. This is followed by fields that are 6 positions wide that contain the number of dead trees, right justified in the field, in each of the 20 diameter classes.

Finally, there follows condition class values by species and dbh class. Each row contains condition class values for a species. The first two positions contain the species id number right justified in the field. This is followed by 20 fields of 6 positions where condition class is recorded, right justified, with one decimal place.

Example: TREEDATA.TXT

```
 1     4    99   174   104    29    37    29    17    16     7    21    13     8     5     3     0     0     0     0     0
 2    14   648   321   182    36    23    12     7     8     7     5     2     7     2     3     0     0     0     0     0
 3     1    12    19    16     6     5    12    20    25    30    35    30    17    15     9     5     1     1     0     0
 4     0     0     4     0     0     0     0     0     1     0     1     0     4     3     0     1     0     0     1     1
 5     0     2     6     0     0     0     0     0     0     0     0     0     0     0     0     0     0     0     0     0
 6     0     0     0     0     0     1     2    13     9    20    22    31    27    21    10     6     3     0     2     0
 7     0     0     1     0     0     0     0     0     0     0     0     0     0     0     0     0     0     0     0     0
 8     1     3     4     6    11    22    58    81    75    73    57    56    46    37    17     8     7     5     2     1
 9     0     1     0     0     1     0     1     3     6     4     5     5     2     0     0     0     0     0     0     0
10    43     3     0     0     0     0     1     0     0     0     0     1     1     0     0     0     0     0     0     0
11     1     6     3     0     0     0     0     0     0     0     0     0     0     0     0     0     0     0     0     0
12     1    34    17     6     2     0     0     0     0     1     0     0     0     0     1     0     0     0     0     0
13     0     2     2     4     0     0     0     0     0     0     0     0     0     0     0     0     0     0     0     0
14     4   372   328   205    78    41    36    41     6     2     0     0     0     0     0     0     0     0     0     0
15     0     1     0     0     0     0     0     0     0     0     0     0     0     0     0     0     0     0     0     0
16     0     0     1     0     1     0     0     0     0     0     0     0     0     0     0     0     0     0     0     0
17     0     0     1     0     0     0     3     2     3     7     5     3     6     3     2     2     0     3     0     0
18     1   218   130    42     8     4     1     1     1     1     1     0     4     0     0     0     0     0     0     0
```

19	0	0	0	0	1	0	0	0	0	0	2	2	0	0	0	0	0	0	0	0
20	12	1019	858	441	223	153	90	28	26	34	30	16	9	10	4	1	0	0	0	0
21	0	1	0	0	0	1	0	0	0	1	0	1	0	0	0	0	0	0	0	0
22	0	5	5	4	4	0	5	6	26	36	40	56	52	51	40	27	8	6	5	0
23	0	129	113	64	8	2	4	4	4	8	5	3	0	0	0	1	0	0	0	0
24	3	363	141	42	27	18	0	1	0	1	0	1	2	2	1	0	0	0	0	0
25	0	9	1	0	0	0	0	0	0	0	0	0	0	0	0	0	0	0	0	0
26	0	26	5	0	0	0	0	0	0	0	0	0	0	1	0	0	0	0	0	0
27	0	6	3	0	0	0	0	0	0	0	0	0	0	0	0	0	0	0	0	0
28	0	2	0	0	0	0	0	0	0	0	0	0	0	0	0	0	0	0	0	0
29	0	6	3	2	0	0	0	0	0	0	0	0	0	0	0	1	0	0	0	0
30	7	6	7	1	0	0	0	0	0	0	0	0	0	0	0	0	0	0	0	0
0	2931	0	0	0	0	0	0	0	0	0	0	0	0	0	0	0	0	0	0	0
-1	28	221	27	3	2	1	0	0	1	2	2	1	1	0	1	1	0	0	0	0
1	35.0	68.0	71.9	76.5	70.3	75.6	73.7	69.4	65.0	60.0	68.5	67.6	65.0	64.0	60.0	0.0	0.0	0.0	0.0	0.0
2	52.8	72.6	73.6	77.9	77.2	77.3	56.6	62.8	67.5	71.4	68.0	70.0	68.5	70.0	66.6	0.0	0.0	0.0	0.0	0.0
3	20.0	71.6	81.0	76.2	80.0	68.0	70.0	65.0	73.6	72.6	73.1	66.6	68.2	74.6	73.3	76.0	80.0	80.0	0.0	0.0
4	0.0	0.0	65.0	0.0	0.0	0.0	0.0	0.0	60.0	0.0	60.0	0.0	75.0	80.0	0.0	80.0	0.0	0.0	80.0	60.0
5	0.0	80.0	80.0	0.0	0.0	0.0	0.0	0.0	0.0	0.0	0.0	0.0	0.0	0.0	0.0	0.0	0.0	0.0	0.0	0.0
6	0.0	0.0	0.0	0.0	0.0	40.0	50.0	60.0	66.6	55.0	66.3	61.2	60.7	62.8	66.0	56.6	80.0	0.0	60.0	0.0
7	0.0	0.0	60.0	0.0	0.0	0.0	0.0	0.0	0.0	0.0	0.0	0.0	0.0	0.0	0.0	0.0	0.0	0.0	0.0	0.0
8	60.0	60.0	60.0	66.6	61.8	63.6	65.1	65.4	67.2	63.8	71.5	67.8	61.7	71.3	61.1	60.0	62.8	72.0	70.0	80.0
9	0.0	100.0	0.0	0.0	80.0	0.0	40.0	53.3	56.6	60.0	60.0	56.0	60.0	0.0	0.0	0.0	0.0	0.0	0.0	0.0
10	46.9	53.3	0.0	0.0	0.0	0.0	80.0	0.0	0.0	0.0	0.0	100.0	80.0	0.0	0.0	0.0	0.0	0.0	0.0	0.0
11	80.0	76.6	80.0	0.0	0.0	0.0	0.0	0.0	0.0	0.0	0.0	0.0	0.0	0.0	0.0	0.0	0.0	0.0	0.0	0.0
12	80.0	68.2	70.5	83.3	80.0	0.0	0.0	0.0	0.0	80.0	0.0	0.0	0.0	0.0	80.0	0.0	0.0	0.0	0.0	0.0
13	0.0	70.0	70.0	80.0	0.0	0.0	0.0	0.0	0.0	0.0	0.0	0.0	0.0	0.0	0.0	0.0	0.0	0.0	0.0	0.0
14	55.0	71.5	75.1	76.3	78.2	76.5	77.7	78.5	83.3	70.0	0.0	0.0	0.0	0.0	0.0	0.0	0.0	0.0	0.0	0.0
15	0.0	40.0	0.0	0.0	0.0	0.0	0.0	0.0	0.0	0.0	0.0	0.0	0.0	0.0	0.0	0.0	0.0	0.0	0.0	0.0
16	0.0	0.0	80.0	0.0	80.0	0.0	0.0	0.0	0.0	0.0	0.0	0.0	0.0	0.0	0.0	0.0	0.0	0.0	0.0	0.0
17	0.0	0.0	80.0	0.0	0.0	0.0	66.6	70.0	73.3	77.1	84.0	73.3	80.0	73.3	90.0	80.0	0.0	73.3	0.0	0.0
18	80.0	70.4	77.0	80.9	82.5	95.0	60.0	80.0	80.0	60.0	60.0	0.0	75.0	0.0	0.0	0.0	0.0	0.0	0.0	0.0
19	0.0	0.0	0.0	0.0	80.0	0.0	0.0	0.0	0.0	0.0	70.0	80.0	0.0	0.0	0.0	0.0	0.0	0.0	0.0	0.0
20	68.3	71.2	74.6	76.1	77.4	80.1	79.7	72.8	63.8	66.4	66.0	70.0	68.8	74.0	60.0	60.0	0.0	0.0	0.0	0.0
21	0.0	60.0	0.0	0.0	0.0	60.0	0.0	0.0	0.0	60.0	0.0	80.0	0.0	0.0	0.0	0.0	0.0	0.0	0.0	0.0
22	0.0	76.0	80.0	85.0	75.0	0.0	48.0	63.3	56.9	62.2	67.0	61.0	64.2	65.4	66.5	72.5	70.0	66.6	68.0	0.0
23	0.0	66.5	72.9	78.1	82.5	100.0	90.0	75.0	75.0	57.5	72.0	60.0	0.0	0.0	0.0	60.0	0.0	0.0	0.0	0.0
24	60.0	68.5	77.4	81.9	85.1	88.8	0.0	40.0	0.0	60.0	0.0	60.0	60.0	70.0	40.0	0.0	0.0	0.0	0.0	0.0
25	0.0	66.6	20.0	0.0	0.0	0.0	0.0	0.0	0.0	0.0	0.0	0.0	0.0	0.0	0.0	0.0	0.0	0.0	0.0	0.0
26	0.0	72.3	64.0	0.0	0.0	0.0	0.0	0.0	0.0	0.0	0.0	0.0	0.0	80.0	0.0	0.0	0.0	0.0	0.0	0.0
27	0.0	50.0	33.3	0.0	0.0	0.0	0.0	0.0	0.0	0.0	0.0	0.0	0.0	0.0	0.0	0.0	0.0	0.0	0.0	0.0
28	0.0	70.0	0.0	0.0	0.0	0.0	0.0	0.0	0.0	0.0	0.0	0.0	0.0	0.0	0.0	0.0	0.0	0.0	0.0	0.0
29	0.0	60.0	66.6	80.0	0.0	0.0	0.0	0.0	0.0	0.0	0.0	0.0	0.0	0.0	0.0	80.0	0.0	0.0	0.0	0.0
30	77.1	80.0	68.5	80.0	0.0	0.0	0.0	0.0	0.0	0.0	0.0	0.0	0.0	0.0	0.0	0.0	0.0	0.0	0.0	0.0

PRINT

Print is used to print the contents of a window.

PRINTER SETUP

Printer setup is used to change or configure your printer.

EXIT CITY TREES!

Ends the **CityTrees!** program and returns to the Windows Operating System.

EDIT

The selections under edit function similarly to the usual Windows Edit function. Balance, however, has no application in this Version of **CityTrees!**.

MAKE

ENTER SIMULATION PARAMETERS

Here you have the opportunity to control how **CityTrees!** conducts the simulation procedure.

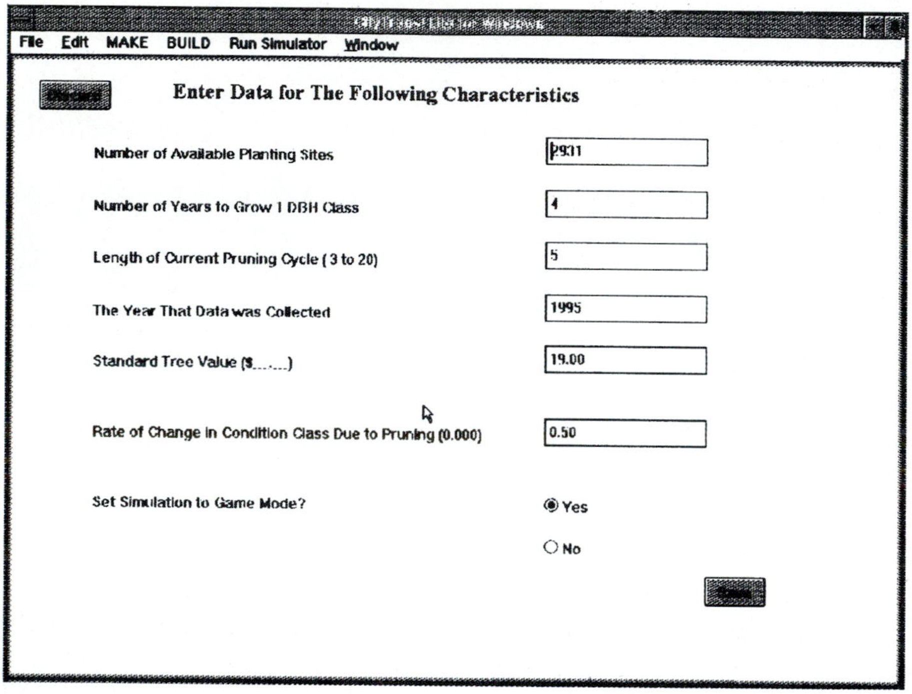

File Edit MAKE BUILD Run Simulator Window

Enter Data for The Following Characteristics

Number of Available Planting Sites 2931

Number of Years to Grow 1 DBH Class 4

Length of Current Pruning Cycle (3 to 20) 5

The Year That Data was Collected 1995

Standard Tree Value ($.....) 19.00

Rate of Change in Condition Class Due to Pruning (0.000) 0.50

Set Simulation to Game Mode? Yes No

If you make changes and want to save them, you must press the Save button. If you make changes but decide not to keep them, press the Discard button. The game mode default is yes. Under this mode, random events will occur that effect the **CityTrees!** population. Select **No** if you want to suppress these events.

The following nine choices bring up spreadsheet-like tables that allow you to make changes to your inventory data. These tables do not have the full capabilities of spreadsheets. They simply allow you to enter data. You may add species to the species list. If you do, you must include associated data in the stand table, tree condition, tree mortality, planting costs, and planting mortality. Be sure to save any changes by selecting **File Save** or **File SaveAs** if you wish to keep the changes permanently.

SPECIES LIST

STAND TABLE

TREE CONDITION

TREE MORTALITY

DEAD TREES

PLANTING COSTS

PLANTING MORTALITY

TREE PLANTING COSTS

TREE REMOVAL COSTS

GO! CREATE MASTER FILES

This creates the base files for the simulation run from the cts data files. These files contain the initial conditions at the beginning of the simulation.

BUILD

GO!

This step begins the simulation process by reading the master files and initializing the simulation run. The tree inventory at the beginning of the simulation is displayed. There can be only one set of master files (`species.mas` and `instru.mas`) on the working drive at any one time. Each time you create the master files from MAKE any previous `.mas` files are overwritten. Should you wish to retain master files for future use you must rename them. Then rename them again to `species.mas` and `instru.mas` when you want to use them again. **CityTrees!** will warn you if there are no Master Files to read.

CANCEL

Aborts the simulation.

RUN SIMULATOR

GO!

This choice uses data in the master file to simulate future inventory of the population of trees in your city. The inventory is update once a year. You have an opportunity to prune and plant trees. If you elect to end the simulation, you have an opportunity to save the current simulated inventory. If you elect to save, any previous simulation files will be overwritten. You must rename previous simulation files if you want to keep them. Simulation files have the names `species.sim` and `instru.sim`. When GO! is selected it looks for these files. If none are present BUILD or MAKE must be used.

CANCEL

Abort the procedure.

WINDOW

The following choices function similarly to the normal Windows actions except that SHOW SCHEDULER has no function in this version of **CityTrees! Lite.**

TILE

CASCADE

ARRANGE ICONS

SHOW PRINT LOG

SHOW SCHEDULER

THE STUDENT VERSION OF CITYTREES! LITE

The Student Version of **CityTrees! Lite** is the same as the regular version except the ability to create `*.cts` and `*.mas` files is disabled so the student version depends on having the regular version available to create the needed `species.mas` and `instru.mas` files. In addition to these `*.mas` files being on the student's work disk, the `cts.ini` file must also be present.

COMPONENTS OF THE MODEL

Inventory

An inventory file of approximately 9,000 trees is provided with the model. This file can be used to become familiar with **CityTrees!**, or it can be used if the intended use is as a training tool. City foresters can replace this file with an inventory summary table from their own community. The inventory file consists of tree data by species/diameter class, with each cell assigned a CTLA condition class used to compute tree value. The inventory file also includes the number of vacant tree spaces in the community. Each time the program is run the inventory file is altered by mortality, planting, pruning and growth, and is stored for the next run. See Appendix A for a sample inventory.

Mortality and Removals

Mortality is arrayed by species and diameter class. The table provided with the simulation is based on historic mortality rates in Wisconsin cities. A customized mortality matrix can be entered by the program user, and the matrix can be altered anytime between annual runs. All existing dead trees are removed at the start of each run. Trees killed by the mortality matrix are stored until the next run for removal. Removal costs are based on tree diameter, and these values can be manipulated by the user. See Appendix B for a sample mortality matrix.

Pruning

The program user can select a pruning cycle from two to twenty years, or can choose not to prune trees. Miller and Sylvester (1981) found that frequency of pruning affects tree condition and tree value, with longer pruning cycles lowering tree value and more frequent pruning increasing tree value. The pruning cycle selected by the user will affect tree condition and tree value over time by a value of –0.5 percent per year if the pruning cycle is lengthened, or +0.5 percent per year if the cycle is reduced (Miller and Sylvester 1981). The program user can change this value if desired. Pruning costs are based on tree diameter, and these values can be manipulated by the user. Annual pruning costs are calculated by calculating the cost to prune the entire population and dividing that value by the desired pruning cycle. Dead trees are not pruned.

Planting

A planting table with species is provided with the program. This table lists species available for planting and the user can choose to plant either 1-inch or 2-inch diameter trees. Different size planting stock has different planting mortality rates and different planting costs by species. The program user selects species and size of planting stock. A customized planting table can be entered for a community. See Appendix C for a sample planting table.

Growth

The growth rate in the tree data file is 0.5 inches in diameter per year, which is a value based on growth studies of street trees in Wisconsin. Program users can use this value or can select a value more descriptive of growth rates in their communities.

Random Events

A series of random events can be turned on if the program user wishes to use **CityTrees!** as a training tool. Values are randomly generated within the limits specified below.

1. Wind storm: Mortality rates in the mortality table are increased between 10 and 30% for one run. Larger trees are assigned higher rates.
2. Ice storm: Mortality rates in the mortality table are increased between 10 and 30% for one run. Within a given species all diameters are assigned the same rate.
3. New disease: A prominent species (exceeding 5% of the total population) is selected and its mortality rate increased to between 3 and 5.5% for all subsequent runs.
4. Drought: Planting mortality is increased to between 30 and 60% for one year.
5. Budget cut: The user is directed to cut his or her budget by 25 to 35% for one year.
6. Budget increase: The user is directed to increase his or her budget by 10 to 20% for one year.

Of the above events, only four will occur in a twelve-year period. These will consist of one major event (wind or ice storm), one minor event (drought or new disease), and two budget changes.

Cycles

CityTrees! runs on annual cycles, with each of the above components interacting with the inventory table each year. Costs for planting, pruning, and removals are summarized at the end of each run. These costs are subtracted from tree value to obtain the net value of the tree population. The output inventory table is stored for use in the next run. Each output table is stored for two years so the user can rerun a year if he or she desires to do so. The table is then erased. The initial table is not erased so the user can always start the simulation over again.

PROGRAM OUTPUT

After the simulation is run a table summarizing the altered inventory will appear on the screen (Appendix D). This table presents the inventory data summed by species and by diameter classes. An additional table summarizes changes in the number of trees and the value of the tree population, and the cost of planting, pruning, and removing trees. The net tree value (change in tree value − management costs) is also presented. This information can then be used to make additional management decisions and recommendations for the next run or cycle.

USES OF CITYTREES!

City Foresters

CityTrees! can be used to assist managers in determining the long term effects of management decisions. City foresters must enter a local species list that includes species names and species CTLA values, and summary tables of their street tree inventory by species, diameter class, and CTLA condition class. If data summaries are in fewer diameter classes than used by **CityTrees!,** these classes should be equally divided into the two-inch classes used by the simulator. If default values in the initialization file are not applicable to the community tree population, local data describing tree management costs, tree growth rates, date inventory data was gathered, standard tree value, effect of pruning on CTLA condition class, tree mortality, and planting mortality can be entered. The random event function must be turned off to simulate the effect over time of decisions regarding planting rates, species selection, and pruning cycle. **CityTrees!** summarizes the cost of management activities and the CTLA value of the tree population at the end of each run.

Comparisons of different management strategies can be made on the basis of the number of future trees, future species mix, and diameter distributions. Differences in planting stock sizes and species can be analyzed, and the population mean diameter growth rate over time determined. Management costs relative to future values of the tree population will help to select the most cost effective management strategies. **CityTrees!** can also be used to predict future pruning costs for various pruning cycles, and future planting and removal costs for different planting strategies.

Educational Uses

CityTrees! is also a teaching tool to provide students with an understanding of the long term effects of management decisions. Students can use the tree data files to prepare a one-year budget to meet long-term objectives such as maintaining a five-year pruning, and planting to fill all vacant spaces within a specified period of time. A budget using the data files provided with **CityTrees!** might resemble the following:

Salaries and fringes	$80,000
Operating expenses	10,000
Planting	80,000
Contract services	
Pruning	33,000
Tree and stump removal	11,000
Total	$214,000

The salaries and operating expenses in the above budget are provided by the instructor, while the student calculates planting, pruning, and removal costs. The instructor will then specify that the students can have only a percentage of what they requested (60 percent for example), and must adjust their budget accordingly without cutting salaries or removal expenses. This will leave only planting and pruning for budget adjustments. For each following run (year) students will be allowed to increase their budgets in real dollars by a specified percentage, but must give annual raises to employees as well as adjust pruning and planting schedules. Random events will challenge management skills as students cope with wind and ice storms, new tree diseases, and drought. Severe disruptions by storms may require the student to ask for emergency funding for clean up (charged under removals in the following run). Temporary adjustments in funding will force students to adapt to budget cuts and surpluses. Four random events will occur during each 12 simulations.

STUDENT DISK

A student disk is provided with **CityTrees!** for use by the instructor. When the student version is installed users will not have access to **MAKE** functions, and will not be able to read master data files other than those provided by the instructor. The instructor will use the standard version of **CityTrees! Lite** to provide either the existing master data files, edited master data files, or completely different master data files. The instructor can also edit default values in the initialization file that describe tree management costs, tree growth rates, year inventory data was gathered, standard tree value, effect of pruning on CTLA condition class, tree mortality, and planting mortality.

LITERATURE CITED

Council of Tree and Landscape Appraisers. 1994. *Guide for Plant Appraisal,* 8th Ed. Savoy, IL., International Society of Arboriculture.

MILLER, R. W., and M. S. MARANO. 1985. *URBAN FOREST Street Tree Management Simulation for Microcomputers.* Stevens Point, Wisc., Coll. of Natl. Res., U. Wisc.-Stevens Point.

MILLER, R. W., and W. A. SYLVESTER. 1981. An economic evaluation of the pruning cycle. *J. Arbor.* 7(4): 109-112.

APPENDIX B-1 Sample Inventory

Sample Run

1995 INITIAL STAND AND VALUE TABLE
COMBINED DBH

SPECIES NAME	NO OF TREES	% OF CITY	AVG DBH	AVER. COND	TREE AVG $	VALUE TOTAL
TREE OF HEAVEN	2	0.0	6.0	80.0	191	382
GREEN ASH	566	6.3	7.4	71.2	743	420733
WHITE ASH	1277	14.3	4.0	73.3	186	237218
BASSWOOD	259	2.9	16.5	72.0	2123	549741
BOXELDER	16	0.2	20.6	71.3	1196	19135
BUCKEYE	5	0.1	18.4	76.0	2482	12411
CATALPA	167	1.9	21.9	61.6	1875	313063
CRABAPPLE	1	0.0	4.0	60.0	115	115
AMERICAN ELM	570	6.4	18.2	66.0	2181	1243188
SIBERIAN ELM	28	0.3	17.5	59.3	567	15883
GINKGO	10	0.1	2.5	78.0	86	860
MISC HARDWOOD	21	0.2	2.5	75.2	37	777
HACKBERRY	62	0.7	3.8	71.2	280	17365
HONEYLOCUST	1113	12.4	4.9	74.6	319	355275
HORSECHESTNUT	40	0.4	21.2	77.0	3421	136824
LITTLELEAF LINDEN	412	4.6	3.6	74.1	199	81836
MAPLE, NORWAY COLUM	602	6.7	3.5	72.7	237	142780
NORWAY MAPLE	2954	33.0	5.3	73.9	499	1474223
SILVER MAPLE	376	4.2	22.5	65.0	2141	804876
SUGAR MAPLE	345	3.9	4.9	71.6	446	153749
RED MAPLE	4	0.0	13.0	65.0	2404	9614
BUR OAK	32	0.4	3.1	71.2	311	9960
ENGLISH OAK	9	0.1	2.7	44.4	46	418
PIN OAK	2	0.0	2.0	70.0	42	84
RED OAK	10	0.1	2.2	61.9	41	406
WHITE OAK	12	0.1	5.5	66.6	1025	12301
CALLERY PEAR	8	0.1	3.5	80.0	124	994
POPLAR SP	49	0.5	2.2	49.7	66	3244
SYCAMORE	8	0.1	4.5	75.0	214	1710
WILLOW	1	0.0	2.0	40.0	10	10
30	8961		7.2	72.2	672	6019173

APPENDIX B-1 Sample Inventory (*continued*)

COMBINED SPECIES

DBH CLASS	NO OF TREES	% OF CITY	AVER. COND	TREE AVG $	VALUE TOTAL
1	92	1.0	54.1	4	389
2	2973	33.2	70.7	36	107542
4	2147	24.0	74.4	155	331987
6	1119	12.5	77.0	356	398000
8	435	4.9	77.4	658	286284
10	307	3.4	78.1	1035	317843
12	254	2.8	72.8	1268	322145
14	224	2.5	68.5	1406	314929
16	206	2.3	66.5	1623	334401
18	232	2.6	64.6	1982	459802
20	229	2.6	69.3	2552	584402
22	220	2.5	65.1	2671	587654
24	185	2.1	64.8	3003	555519
26	150	1.7	68.6	3765	564720
28	90	1.0	66.0	3983	358484
30	52	0.6	69.2	4513	234700
32	19	0.2	69.5	5083	96576
34	15	0.2	70.6	6394	95913
36	10	0.1	68.0	5495	54947
38	2	0.0	70.0	6467	12934

APPENDIX B-2 Mortality Table

Species	d1	d2	d4	d6	d8	d10	d12	d14	d16	d18	d20	d22	d24	d26	d28	d30	d32	d34	d36	d38
AMERICAN ELM	1.5	1.2	1.4	0.5	0.7	0.9	1.1	1.3	1.5	1.6	1.8	2.0	2.2	2.4	2.6	2.8	3.0	3.2	3.4	3.6
BASSWOOD	1.5	1.2	1.4	0.5	0.7	0.9	1.1	1.3	1.5	1.6	1.8	2.0	2.2	2.4	2.6	2.8	3.0	3.2	3.4	3.6
BOXELDER	1.5	1.2	1.4	0.5	0.7	0.9	1.1	1.3	1.5	1.6	1.8	2.0	2.2	2.4	2.6	2.8	3.0	3.2	3.4	3.6
BUCKEYE	1.5	1.2	1.4	0.5	0.7	0.9	1.1	1.3	1.5	1.6	1.8	2.0	2.2	2.4	2.6	2.8	3.0	3.2	3.4	3.6
BUR OAK	1.5	1.2	1.4	0.5	0.7	0.9	1.1	1.3	1.5	1.6	1.8	2.0	2.2	2.4	2.6	2.8	3.0	3.2	3.4	3.6
CALLERY PEAR	1.5	1.2	1.4	0.5	0.7	0.9	1.1	1.3	1.5	1.6	1.8	2.0	2.2	2.4	2.6	2.8	3.0	3.2	3.4	3.6
CATALPA	1.5	1.2	1.4	0.5	0.7	0.9	1.1	1.3	1.5	1.6	1.8	2.0	2.2	2.4	2.6	2.8	3.0	3.2	3.4	3.6
CRABAPPLE	1.5	1.2	1.4	0.5	0.7	0.9	1.1	1.3	1.5	1.6	1.8	2.0	2.2	2.4	2.6	2.8	3.0	3.2	3.4	3.6
ENGLISH OAK	1.5	1.2	1.4	0.5	0.7	0.9	1.1	1.3	1.5	1.6	1.8	2.0	2.2	2.4	2.6	2.8	3.0	3.2	3.4	3.6
GINKGO	1.5	1.2	1.4	0.5	0.7	0.9	1.1	1.3	1.5	1.6	1.8	2.0	2.2	2.4	2.6	2.8	3.0	3.2	3.4	3.6
GREEN ASH	1.5	1.2	1.4	0.5	0.7	0.9	1.1	1.3	1.5	1.6	1.8	2.0	2.2	2.4	2.6	2.8	3.0	3.2	3.4	3.6
HACKBERRY	1.5	1.2	1.4	0.5	0.7	0.9	1.1	1.3	1.5	1.6	1.8	2.0	2.2	2.4	2.6	2.8	3.0	3.2	3.4	3.6
HONEYLOCUST	1.5	1.2	1.4	0.5	0.7	0.9	1.1	1.3	1.5	1.6	1.8	2.0	2.2	2.4	2.6	2.8	3.0	3.2	3.4	3.6
HORSECHESTNUT	1.5	1.2	1.4	0.5	0.7	0.9	1.1	1.3	1.5	1.6	1.8	2.0	2.2	2.4	2.6	2.8	3.0	3.2	3.4	3.6
LITTLELEAF LINDEN	1.5	1.2	1.4	0.5	0.7	0.9	1.1	1.3	1.5	1.6	1.8	2.0	2.2	2.4	2.6	2.8	3.0	3.2	3.4	3.6
MAPLE, NORWAY COLUMNAR	1.5	1.2	1.4	0.5	0.7	0.9	1.1	1.3	1.5	1.6	1.8	2.0	2.2	2.4	2.6	2.8	3.0	3.2	3.4	3.6
MISC HARDWOOD	1.5	1.2	1.4	0.5	0.7	0.9	1.1	1.3	1.5	1.6	1.8	2.0	2.2	2.4	2.6	2.8	3.0	3.2	3.4	3.6
NORWAY MAPLE	1.5	1.2	1.4	0.5	0.7	0.9	1.1	1.3	1.5	1.6	1.8	2.0	2.2	2.4	2.6	2.8	3.0	3.2	3.4	3.6
PIN OAK	1.5	1.2	1.4	0.5	0.7	0.9	1.1	1.3	1.5	1.6	1.8	2.0	2.2	2.4	2.6	2.8	3.0	3.2	3.4	3.6
POPLAR SP	1.5	1.2	1.4	0.5	0.7	0.9	1.1	1.3	1.5	1.6	1.8	2.0	2.2	2.4	2.6	2.8	3.0	3.2	3.4	3.6
RED MAPLE	1.5	1.2	1.4	0.5	0.7	0.9	1.1	1.3	1.5	1.6	1.8	2.0	2.2	2.4	2.6	2.8	3.0	3.2	3.4	3.6
RED OAK	1.5	1.2	1.4	0.5	0.7	0.9	1.1	1.3	1.5	1.6	1.8	2.0	2.2	2.4	2.6	2.8	3.0	3.2	3.4	3.6
SIBERIAN ELM	1.5	1.2	1.4	0.5	0.7	0.9	1.1	1.3	1.5	1.6	1.8	2.0	2.2	2.4	2.6	2.8	3.0	3.2	3.4	3.6
SILVER MAFLE	1.5	1.2	1.4	0.5	0.7	0.9	1.1	1.3	1.5	1.6	1.8	2.0	2.2	2.4	2.6	2.8	3.0	3.2	3.4	3.6
SUGAR MAPLE	1.5	1.2	1.4	0.5	0.7	0.9	1.1	1.3	1.5	1.6	1.8	2.0	2.2	2.4	2.6	2.8	3.0	3.2	3.4	3.6
SYCAMORE	1.5	1.2	1.4	0.5	0.7	0.9	1.1	1.3	1.5	1.6	1.8	2.0	2.2	2.4	2.6	2.8	3.0	3.2	3.4	3.6
TREE OF HEAVEN	1.5	1.2	1.4	0.5	0.7	0.9	1.1	1.3	1.5	1.6	1.8	2.0	2.2	2.4	2.6	2.8	3.0	3.2	3.4	3.6
WHITE ASH	1.5	1.2	1.4	0.5	0.7	0.9	1.1	1.3	1.5	1.6	1.8	2.0	2.2	2.4	2.6	2.8	3.0	3.2	3.4	3.6
WHITE OAK	1.5	1.2	1.4	0.5	0.7	0.9	1.1	1.3	1.5	1.6	1.8	2.0	2.2	2.4	2.6	2.8	3.0	3.2	3.4	3.6
WILLOW	1.5	1.2	1.4	0.5	0.7	0.9	1.1	1.3	1.5	1.6	1.8	2.0	2.2	2.4	2.6	2.8	3.0	3.2	3.4	3.6

APPENDIX B-3 Sample Planting Table

Planting	Available	Scheduled	Remaining	Selected
Sites	2931	200	200	0

			PlantCost		PlantMort		PlantList	
	Species	Rel Val%	1 inch	2 inch	1 inch	2 inch	1 inch	2 inch
1	AILANT	40	$120.00	$180.00	20.00	15.00	0	0
2	ASH-GR	80	$120.00	$180.00	20.00	15.00	0	0
3	ASH-WH	60	$120.00	$180.00	20.00	15.00	0	0
4	BASSWD	60	$120.00	$180.00	20.00	15.00	0	0
5	BOXEL	20	$120.00	$180.00	20.00	15.00	0	0
6	BUCKEY	60	$120.00	$180.00	20.00	15.00	0	0
7	CATALP	40	$120.00	$180.00	20.00	15.00	0	0
8	CRABAP	80	$120.00	$180.00	20.00	15.00	0	0
9	ELM-AM	60	$120.00	$180.00	20.00	15.00	0	0
10	ELM-SI	20	$120.00	$180.00	20.00	15.00	0	0
11	GINKGO	100	$120.00	$180.00	20.00	15.00	0	0
12	H-MISC	40	$120.00	$180.00	20.00	15.00	0	0
13	HACKBE	80	$120.00	$180.00	20.00	15.00	0	0
14	HONEY	80	$120.00	$180.00	20.00	15.00	0	0
15	HORSEC	60	$120.00	$180.00	20.00	15.00	0	0

APPENDIX B-4 Simulated Inventory

1996 PREDICTED STAND AND VALUE TABLE
COMBINED DBH

SPECIES NAME	NO OF TREES	% OF CITY	AVG DBH	AVER. COND	TREE AVG $	VALUE TOTAL $
TREE OF HEAVEN	1	0.0	6.0	80.0	172	172
GREEN ASH	485	5.9	9.4	71.2	1028	498405
WHITE ASH	1093	13.2	6.0	73.3	318	348038
BASSWOOD	216	2.6	18.5	72.0	2577	556703
BOXELDER	61	0.7	4.6	69.6	228	13929
BUCKEYE	76	0.9	3.2	69.7	221	16804
CATALPA	143	1.7	23.9	61.4	2206	315494
CRABAPPLE	1	0.0	6.0	60.0	258	258
AMERICAN ELM	500	6.0	20.2	66.0	2631	1315630
SIBERIAN ELM	23	0.3	19.9	57.7	710	16325
GINKGO	10	0.1	4.4	78.0	244	2435
MISC HARDWOOD	18	0.2	4.2	75.2	93	1682
HACKBERRY	59	0.7	5.8	71.4	461	27210
HONEYLOCUST	1040	12.6	6.9	74.5	532	553757
HORSECHESTNUT	36	0.4	23.1	77.3	4005	144182
LITTLELEAF LINDEN	377	4.6	5.6	74.1	365	137785
MAPLE, NORWAY COLUM	577	7.0	5.5	72.7	440	254126
NORWAY MAPLE	2808	34.0	7.3	73.9	777	2182527
SILVER MAPLE	306	3.7	24.5	65.1	2509	767611
SUGAR MAPLE	325	3.9	6.9	71.6	703	228404
RED MAPLE	3	0.0	12.0	60.0	1672	5016
BUR OAK	30	0.4	5.1	71.5	502	15056
ENGLISH OAK	9	0.1	4.7	44.4	139	1253
PIN OAK	2	0.0	4.0	70.0	167	334
RED OAK	9	0.1	4.2	61.4	153	1380
WHITE OAK	11	0.1	7.8	67.3	1414	15549
CALLERY PEAR	7	0.1	5.7	80.0	317	2217
POPLAR SP	33	0.4	3.2	49.3	70	2296
SYCAMORE	8	0.1	6.5	75.0	414	3315
WILLOW	1	0.0	4.0	40.0	38	38
30	8268		9.0	72.2	898	7427931

COMBINED SPECIES

DBH CLASS	NO OF TREES	% OF CITY	AVER. COND	TREE AVG $	VALUE TOTAL $
1	51	0.6	69.3	2	106
2	144	1.7	62.2	22	3127
4	2738	33.1	70.7	146	399349
6	1977	23.9	74.5	351	693019
8	1030	12.5	77.0	636	655501
10	403	4.9	77.4	1037	417832
12	284	3.4	78.3	1504	427273
14	229	2.8	72.9	1741	398642
16	202	2.4	68.8	1860	375801
18	179	2.2	66.6	2082	372688
20	202	2.4	64.6	2479	500784
22	198	2.4	69.4	3138	621264
24	186	2.2	65.0	3213	597698
26	157	1.9	64.7	3573	560968
28	129	1.6	68.6	4465	575942
30	77	0.9	66.0	4639	357178
32	44	0.5	69.1	5203	228936
34	16	0.2	69.2	5761	92168
36	13	0.2	70.7	7220	93866
38	9	0.1	67.6	6199	55789

SUMMARY	VALUE $	NUMBER
PRESENT FOREST	7427931	8268
PREVIOUS FOREST	6019173	8961
CHANGE IN VALUE	1408758	-693
PLANTING MORTALITY		35
OTHER MORTALITY		858
TOTAL MORTALITY		893
PLANTING EXPENSE	30000	200
REMOVAL EXPENSE	7186	291
PRUNING EXPENSE	0	0
NET GAIN OR LOSS	1371572	

Appendix C

Sample Municipal Tree Ordinance

TREE ORDINANCE
Public Works—Code of Ordinance
City of Rice Lake, Wisconsin

S 6-4-1	Statement of Policy and Applicability of Chapter Definitions
S 6-4-2	Definitions
S 6-4-3	Authorization of City Forester
S 6-4-4	Authority of City Forester to Enter Private Premises
S 6-4-5	Interference with City Forester Prohibited
S 6-4-6	Authorization of the Citizens Tree Management Advisory Board
S 6-4-7	Assessment of Costs and Abatement
S 6-4-8	Permit for Maintenance and Removal of Trees and Shrubs
S 6-4-9	Permit for Planting Trees and Shrubs
S 6-4-10	Trimming
S 6-4-11	Removal of Trees and Shrubs
S 6-4-12	Disclaiming Liability for Trees on Private or Public Property
S 6-4-13	Prohibitive Acts
S 6-4-14	Penalties on Non-Compliance

SEC 6-4-1 STATEMENT OF POLICY AND APPLICABILITY OF CHAPTER

(a) INTENT AND PURPOSE: It is the intention of this ordinance to sustain environmental health and enhance the economic well being of the neighborhoods of the city through aesthetics of beauty, tranquility, and integrity by promoting and encouraging the planting and maintenance of trees and shrubs on both public and private property. This ordinance establishes policy and guidelines for an orderly cost-effective system of achieving its intent with a minimum of danger or damage to persons, buildings, streets, curbs, sidewalks, overhead wires, and all underground utilities. It is also intended to guard against the spread of disease and the damage or unnecessary removal of trees and shrubs of either the public or private sector.

(b) The program goal is to maintain eligibility for annual recertification as a recognized "Tree City, USA".

(c) It is the policy of the city to encourage new tree planting on public and private property and to cultivate a flourishing urban forest.

SEC. 6-4-2 DEFINITIONS

Whenever the following words or terms are used in this chapter, they shall be construed to have the following meanings:

(a) PERSON. "Person" shall mean person, firm, association, or corporation.

(b) CITY. "City" is the City of Rice Lake, Wisconsin.

(c) PUBLIC AREAS. "Public Areas" includes all public parks and other lands owned, controlled, or leased by the city.

(d) PUBLIC TREES AND SHRUBS. "Public Trees and Shrubs" means all trees and shrubs located or to be located in or upon public areas.

(e) PUBLIC NUISANCE. "Public Nuisance" means any tree or shrub or part thereof which by reason of its condition interferes with the use of any public area: is infected with a plant disease; is infested with injurious insects or pests; is injurious to public improvements or endangers the life, health, safety or welfare of persons or property.

(f) BOULEVARD AREAS. "Boulevard Areas" means the land between the normal location of the street curbing and sidewalk. Where there is no sidewalk, the area four feet from the curbline shall be deemed to be a boulevard for the purpose of this chapter. Boulevard areas also refer to the center boulevard on center island streets.

(g) CLEAR-SIGHT TRIANGLE. "Clear-Sight Triangle" means a triangle formed by the curb lines of two intersecting rights-of-way and a third line connecting a full-view zone 30 feet from intersecting corners of street. On streets with no curb and gutter, the same triangle will be used and measurements made from where the curb should be.

(h) MAJOR ALTERATIONS. "Major Alterations" means trimming a tree beyond necessary to comply with this chapter.

(i) SHRUBS. "Shrubs" shall mean any woody vegetation or a woody plant having multiple stems and bearing foliage from the ground up.

(j) TREE. "Tree" shall mean any woody plant, normally having one stem or trunk bearing its foliage or crown well above ground level to heights of sixteen feet or more.

(k) TOPPING. "Topping" is the practice of cutting back large diameter branches of a mature tree to stubs. It is a particularly destructive pruning practice. It is stressful to mature trees and may result in reduced vigor, decline or even death of the tree. In addition, new branches that form below the cuts are only weakly attached to the tree, and are in danger of splitting out. Topped trees require constant maintenance to prevent this from happening, and it is often impossible to restore the structure of the tree crown after topping.

(l) TRIMMING. (pruning) Shall be accomplished according to current technical specifications of the National Arbor Association and current safety specifications of the American National Standards for Tree Care Operation, ANSI Z133.1. All specifications and procedures are available from the city forester.

SEC. 6-4-3 AUTHORIZATION OF CITY FORESTER

(a) The Parks, Recreation, and Cemeteries Board shall carry out the provisions of this section. They may designate one or more of their own employees to perform the duty of forester under Chapter 27, Wisconsin Statute, and authorize such forester to perform the duties and exercise the powers of the board imposed by this chapter.

(b) The city forester shall be properly trained and responsible to the Department of Parks, Recreation, and Cemeteries. In carrying out the duties of this section, the forester shall have the power to enforce rules, regulations, and specifications concerning the trimming, spraying, removal, planting, pruning, and protection of trees, shrubs, vines, hedges, and other plants upon the right-of-way of any street, alley, sidewalk, park, or other public place in the city.

SEC. 6-4-4 AUTHORITY OF CITY FORESTER TO ENTER PRIVATE PREMISES

(a) The city forester or any trained representative may enter upon private premises after notification of owner, at all reasonable times, for the purpose of examining any tree or shrub located upon or over such premises to carry out the provisions of this chapter.

SEC. 6-4-5 INTERFERENCE WITH CITY FORESTER PROHIBITED

(a) No person shall interfere with the city forester or any trained representative while he or she is engaged in carrying out work and activities authorized by this chapter.

SEC. 6-4-6 AUTHORIZATION OF THE CITIZENS TREE MANAGEMENT ADVISORY BOARD

(a) There is hereby created a Citizens Tree Management Advisory Board, which shall serve in an advisory capacity to the Department of Parks, Recreation, and Cemeteries, City of Rice Lake. The board shall advise the city on all matters pertaining to tree and landscape planning, maintenance, and development in the City of Rice Lake. The board shall consist of seven members appointed by the Board of Parks, Recreation, and Cemeteries to be composed of the following: one member from the Department of Parks, Recreation, and Cemeteries and six additional members representing the various civic groups and/or citizens of the Rice Lake area. Each board member will serve a three-year period, except that the initial board shall serve staggered terms of office as determined by the Board of Parks, Recreation, and Cemeteries. Board members shall serve without compensation, but all necessary expense shall be paid by appropriate council action. In the event a vacancy occurs, the successor shall be appointed by the Mayor or the Board of Parks, Recreation, and Cemeteries to fill the unexpired portions of the term.

(b) Removal of Citizen Management Tree Advisory Board member will be by the majority consent of the board members with consent of the Board of Parks, Recreation, and Cemeteries. Vote will be by roll call vote. Reasons for removal may include: conflict of interest, lack of meeting attendance or participation.

SEC. 6-4-7 ASSESSMENT OF COSTS

(a) The cost of treating or removing any tree or part thereof that is declared dangerous, nuisance, diseased, or dead that is located in or upon any park, boulevard, cemetery, or public ground shall be borne by the city.

(b) The cost of treating or removing any tree or part thereof that is declared dangerous, nuisance, diseased, or dead that is located in or upon private premises, when removed under the supervision of the forester, shall be assessed to the property owner.

(1) The forester or trained persons designated by the forester shall keep an accurate account of costs and forward the charges including description of work, land description, name and address of owner to the Director of Parks, Recreation, and Cemeteries on or before the designated monthly date.

(2) The Director of Parks, Recreation, and Cemeteries will forward a complete copy of the assessment to the City Clerk/Treasurer for processing and billing.

(3) The City Clerk/Treasurer shall mail notice of the assessment to the private property owner at the last known address, stating that unless paid within *30 days* such assessment will be entered on the tax roll as a tax against the property, and all proceedings in relation to the collection, return and sale of property for delinquent real estate taxes shall apply to such assessment.

(4) The private property owner reserves the right to a public hearing if requested with 7 days after re-receiving the assessment. See Spec. 6-4-15 Appeal From Determinations and Orders.

(5) The city hereby declares that in making assessments under this section, it is acting under its police power, and no damage shall be awarded to any owner for the destruction of any dangerous, nuisance, diseased or dead tree or wood or part thereof.

SEC. 6-4-8 PERMIT FOR MAINTENANCE AND REMOVAL OF TREES AND SHRUBS

(a) PERMIT REQUIRED. No person, except on order of the city forester, shall remove or do major alterations as determined by the city forester on a tree or shrub, in the public right-of-way, boulevard areas, other public areas, or cause such act to be done by others without written permit for such work from the city forester as herein provided. This includes all persons engaged in the business of cutting or removing trees or shrubs.

(b) PERMIT REQUIRED AND CONDITIONS. When an application for a permit is submitted, the city forester must decide that the proposed work described is necessary and in accord with the purpose of this chapter. The forester must take into account safety, health and welfare, location of utilities, public sidewalks, driveways, and street lights.

(c) PERMIT EXEMPTION. No permit shall be required to cultivate, fertilize or water trees and shrubs on the property identified as city boulevards, whenever such boulevard is between the normal curb and sidewalk area.

(d) PERMIT FORM; EXPIRATION, INSPECTION. Every permit shall be issued by the city forester on a standard form and include a description of the work to be done and shall specify the exact location. Any work under such permit must be performed in strict accordance with the provisions of this chapter. Permits issued under this section shall expire six (6) months after date of issuance. There is no charge for the permit.

(e) PERMITS TO PUBLIC UTILITIES

(1) A permit may be issued, under this section, to a public utility to remove, trim, prune, cut, disturb, alter, or do surgery on any public tree or shrub. The city forester shall limit the work to be done to the actual necessities of the utility and may assign an inspector to supervise the work done under the provisi)ns of the permit. The permitee shall adhere to the arboricultural specifications and standards of workmanship set forth in the permit. The expense of such inspection or supervision shall be charged to the utility at the usual city rate.

(2) A public utility may secure an annual working agreement with the city forester's office that gives the city forester the authorization to supervise and direct the work done associated with trees and shrubs.

SEC. 6-4-9 PERMIT FOR PLANTING TREES AND SHRUBS

(a) PLANTING. As of the date of adopting this ordinance all trees and shrubs planted on all public property, including boulevards, within the City of Rice Lake becomes the property of the City of Rice Lake.

(1) The planting of the following trees is prohibited within the City of Rice Lake.

a) Box elder
b) Cottonwood
c) Black locust

(2) There is no charge for the permit. The permit will be on a standard form. Permits issued under this section shall expire six (6) months after date of issuance.

(3) No person shall plant any tree, shrub or other vegetation in the public right-of-way boulevard area, center boulevard or any other public area without first obtaining a written permit for such work from the city forester. The request for such a permit shall specify the size, species and the variety of the tree, shrub, or other vegetation to be planted along with the planting procedure and shall be submitted to the city forester for approval before planting. Approved tree and shrub lists are available at the forester's office.

(4) There shall be a minimum distance of 18 feet and a recommended distance of twenty five (25) to fifty (50) feet between boulevard area trees. Mature size and landscape variables must be considered. Side boulevard trees shall be planted in equal distance between the sidewalk or proposed sidewalk and the back of the curb or proposed back of curb. On side boulevard areas that are less than three feet in width, only small specie plantings are permitted.

(b) No person shall plant on any public or private premises situated at the intersection of two or more streets or alleys in the city, any hedge, tree, shrub, or other vegetation device that may obstruct the view of the operator of any motor vehicle or pedestrian approaching such intersection. The new plantings shall not be allowed in the clear sight triangle as defined in SEC. 6-4-2(g). Shrubs shall not exceed 30 inches in height.

(1) Only small maturing trees or shrubs are authorized for planting under overhead utility wires. This applies to both public and private property.

(2) Only shallow root or small statured trees or bushes are authorized for planting over underground utilities. This applies to both public and private property.

(c) Public projects involving streets, boulevards, parks and public buildings shall have a landscape plan identifying the proposed planting, layout of trees and shrubs submitted to the Planning Commissions and city forester as a component of the preliminary application. This plan shall be approved by the City Council previous to start of project.

(1) Specific measure shall be made to include all existing trees as an integral part of the landscape development.

(d) Private industry including manufacturing, commercial, or retail shall have a landscape plan identifying the proposed planting and layout of trees and shrubs submitted to the Planning Commission and city forester as a component of the preliminary application. This plan shall be approved by the City Council previous to start of project.

(1) Specific measures shall be made to include all existing trees as an integral part of the landscape development.

SEC. 6-4-10 TRIMMING

(a) To insure high quality shade trees, shaped for maximum aesthetic appreciation and to prevent interference with traffic and utility wires, all newly planting trees in public areas, including boulevards, shall be trimmed and shaped for five (5) years after planting.

(b) Trimming (pruning) of all trees in public areas including boulevards shall be done according to current technical specifications of the National Arbor Association and current safety specifications of the American National Standard For Tree Care Operations, ANSI Z133.1. All specifications and procedures are available from the city forester. Trees shall not be topped.

(1) It shall be the duty and responsibility of all property owners adjacent to any public right-of-way or public area, excluding center boulevards, to maintain all trees and shrubs encroaching on such public areas trimmed (pruned), according to the provisions of this section.
(2) In residential areas and trimming (pruning) of trees on street boulevards adjacent to the resident is the duty and obligation of the owner. All directives, specifications, and procedures are available from the office of the city forester.
(3) The City of Rice Lake will maintain all trees, shrubs, and lawns on all center boulevards.

(c) TRIMMING. Trees and shrubs standing in or upon any boulevard, public area, or upon any private premises adjacent to any public right-of-way or public areas shall be kept trimmed so that the lowest branches projecting over the public street or alley provide a clearance of not less than fourteen (14) feet. The City Forester may waive the provision of this section for newly planted trees if determined that they do not interfere with public travel or endanger public safety. The City shall have the right to trim any tree or shrub in the City when it interferes with the safe use of streets or sidewalks or with the visibility of any traffic control.

(d) Clearance from the sidewalk to the lower branches shall not be less than eight (8) feet. All trees standing upon private property in the City, the branches of which extend over the line of the street, shall be trimmed (pruned) so that no branches shall grow or hang over the line of the sidewalk lower than eight (8) feet above the level of the sidewalk.

SEC. 6-4-11 REMOVAL OF TREES AND STUMPS

(a) DANGEROUS OBSTRUCTIVE AND INFECTED TREES. Any public tree or part thereof, whether alive or dead, that the City Forester shall find to be infected, hazardous, or a nuisance so as to endanger the public or other trees, plants, or shrubs grown within the City, or to be injurious to sewers, sidewalks, or other public improvements may be removed by the adjacent owner or the City after recommendation is given by a majority vote of the Citizens Tree Management Advisory Board. Any such tree may be trimmed or treated by the Forester without consulting the Board. Similarly affected trees on private property shall be removed, trimmed or treated by the owner of the property upon which the tree is located. See SEC. 6-4-4 for authority of City Forester to enter private premises. The City Forester shall give written notice to said owner to remedy the situation which shall be served personally or posed upon the infected tree. Such notice shall specifically state the period of time within which the action must be taken, which shall be within not less than twenty-four (24) hours nor more than fourteen (14) days as determined by the City Forester on the basis of the seriousness of the condition of the tree or danger to the public. If the owner shall fail to remove, treat or trim said tree within the time specified see SEC. 6-4-7 Assessment of Costs of Abatement.

(b) In cutting down trees located in public and boulevard areas, the tree must be removed with the root stump grubbed out, or ground out to the depth of at least seven (7) inches below grade measured in a straight line, normal grade of sidewalk to top of curb. All wood and debris must be removed from the street prior to the end of each working day, and all holes shall be filled to normal grade level with topsoil as soon as practical.

(c) All removed public trees shall be replaced with one or more trees, except for those that have been removed as a visibility hazard.

SEC. 6-4-12 DISCLAIMING LIABILITY FOR TREES ON PRIVATE OR PUBLIC PROPERTY

(a) To avoid accepting liability for any personal injury or property damage caused by trees on private or public property.

(1) Nothing in this ordinance shall be deemed to impose any liability for damages or duty of care and maintenance upon the City or upon any of its officers or employees. The person in possession of public property or the owner of any private property shall have a duty to keep the trees upon the property in a safe healthy condition. Any person who feels a tree located, possessed, controlled, or owned by them is a danger to the safety of themselves, others, or structural improvements on-site or off-site shall have obligations to secure around the tree or support the tree, as appropriate to safeguard both persons and improvements from harm.

SEC. 6-4-13 PROHIBITED ACTS

(a) DAMAGE TO PUBLIC TREES. No person shall, without the consent of the owner in the case of a private tree or shrub, or without written permits from the City Forester in the case of a boulevard tree, public tree, or shrub do or cause to be done by others any of the following acts:

(1) Secure, fasten or run any rope, wire sign, unprotected electrical installation, or other device or material to, around, or through a tree or shrub except to secure leaning or newly planted trees.

(2) Break, injure, mutilate, deface, kill, or destroy any tree or shrub or permit any fire to burn where it will injure any tree or shrub.

(3) Permit any toxic chemical, gas, smoke, oil, or other injurious substance to seep, drain, or be emptied upon or about any tree or shrub, or place cement or other solid substance around the base of the same.

(4) Remove any guard, stake, or other device or material intended for the protection of a public tree or shrub, or close or obstruct any open space about the base of a public tree or shrub designed to permit access of air, water, and fertilizer.

(5) Attach any sign, poster, notice, or other object on any tree, or fasten any guy wire, cable, rope, nails, screws, or other device to any tree; except that the City may tie temporary "no parking" signs to trees when necessary in conjunction with street improvement work, tree maintenance work, or parades.

(6) Cause or encourage any fire or burning near or around any tree.

(b) EXCAVATIONS. All trees on any boulevard of other publicly owned property near any excavation or construction of any building, structure, or street work, shall be sufficiently guarded and protected by those responsible for such work as to prevent any injury to said trees. No person shall excavate any ditches, tunnels, or trenches, or install pavement within a radius of ten feet from any public tree without a permit from the City Forester.

SEC. 6-4-14 PENALTIES FOR NON-COMPLIANCE

(a) The Director of Parks, Recreation, and Cemeteries is hereby charged with the responsibility for the enforcement of this ordinance and may serve notice to any person, partnership, firm, corporation, or other legal entity who violates any provision of this ordinance by the institution of legal proceedings as may be required, and the city attorney is hereby authorized to institute appropriate proceedings to that end.

(1) Offenses Against Property, Title 11, Chapter 3, Code of Ordinance, City of Rice Lake
Section 11-3-10 Damage to Public Property
Section 11-3-11 Disturbing Cemetery Property

(2) Chapter 943, Crimes Against Property, Wis Statute
89-90 943.01 Criminal Damage To Property
943.012 Criminal Damage to Religious and Other Property

(3) Section 11-3-12 Penalties, Title 11, Chapter 3, Code of Ordinance, City of Rice Lake
In addition to the general penalty of this code or any other penalty imposed for violation of any section of this chapter, any person who shall cause physical damage to or destroy any public property shall be liable for the cost of inventoried value or repairing such damaged or destroyed property. The parent or parents of any unemancipated minor child who violates Section 11-3-1 may also be held liable for the cost of replacing or repairing

such damaged or destroyed property in accordance with the Wisconsin Statutes.

SEC. 6-4-15 APPEAL FOR DETERMINATIONS AND ORDERS

(a) Any person who receives a determination or order under this chapter from the city forester and objects to all or any part thereof shall have the right to appeal such determination or order, subject to the provision of Chapter 88, Wis. Stat., to the City Council. The City Council shall hear such appeal within thirty (30) days of receipt of written notice of the appeal. After such hearing the City Council may reverse, affirm, or modify the order or determination appealed and grounds for its decision shall be stated in writing. The City Council shall by letter notify the party appealing the order or determination of its decision within ten (10) days after the hearing conclusion and file its written decision with the City Clerk/Treasurer.

SEC. 6-4-16 EVALUATION AND FUTURE UPDATE OF ORDINANCE

(a) A successful implemented management strategy must be monitored to insure that progress is being made and standards are being met. Evaluation provides the feedback necessary to determine whether the management strategy is working. Periodic evaluation also provides an opportunity to reassess the needs and goals of the community. The management strategy may need to be adjusted to reflect new or altered goals. By providing for regular evaluation as part of the management process, the need for change can be identified before a crisis develops. Evaluation methods include:

(1) Aerial and ground level photography

(2) Ground survey and record keeping

(3) Public opinion polling. Public attitudes serve an educational function by helping to keep urban forestry issues in the public eye.

SEC. 6-4-17 SEVERABILITY

(a) Severability. Should any part or provision of this ordinance be declared by a court of law to be invalid, the same shall not affect the validity of the ordinance as a whole nor any part thereof other than the part found invalid.

Appendix D

Tree Protection Ordinance City of Highland Park, Illinois

TREE PRESERVATION

SEC. 94.401 Intent and Purpose

While allowing for reasonable improvement of land within the City, it is the stated public policy of the City to add to the tree population within the City, where possible, and to maintain, to the greatest extent possible, existing trees within the City. The planting of additional trees and the preservation of existing trees in the City is intended to accomplish, where possible, the following objectives:

(A) To preserve trees as an important public resource enhancing the quality of life and the general welfare of the City and enhancing its unique character and physical, historical, and aesthetic environment;

(B) To preserve the essential character of those areas throughout the community that are heavily wooded and in a more natural state;

(C) To enhance and preserve the air quality of the City through the filtering effect of trees on air pollutants;

(D) To reduce noise within the City through the baffle and barrier effect of trees on the spread of noise;

(E) To reduce topsoil erosion through the soil retention effect of tree roots;

(F) To reduce energy consumption through the wind break and shade effects of trees;

(G) To preserve and enhance nesting areas for birds and other wildlife that in turn assist in the control of insects;

(H) To reduce storm water runoff and the costs associated therewith and replenish ground water supplies; and

(J) To protect and increase property values.

SEC. 94.402 Defintions

(A) The language of the text of Sections 94.401 through 94.407 shall be intrepreted in accordance with the following rules of construction:

- **(1)** The singular number includes the plural number, and the plural the singular;
- **(2)** The word "shall" is mandatory; the word "may" is permissive; and
- **(3)** The masculine gender includes the feminine and neuter.

(B) Whenever hereafter in Sections 94.401 through 94.407 the following words and phrases are used, they shall, for the purpose of the Sections aforesaid, have the meanings respectively ascribed to them in this Subsection, except when the context otherwise indicates.

- **(1)** "Building Activity Area" means that buildable area of a lot in which construction and building activities are to be limited and hence shall be the smallest possible area of a lot or parcel of land within which building activity may take place, including the entire area affected by building and grading activities related to the proposed construction, to be determined with maximum regard for existing mature trees.
- **(2)** "Caliper" means the diameter of a tree trunk six inches (6") above the existing grade or proposed planted grade. Caliper is usually used in reference to nursery stock for new plantings.
- **(3)** "City" means the City of Highland Park, Lake County, Illinois.
- **(4)** "City Forester" means the City Forester of the City.
- **(5)** "Coordinator of the Building Division" means the Coordinator of the Building Division of the City, or his designee.
- **(6)** "Diameter Breast Height" or "DBH" means the diameter of a tree measured at four and one-half feet (4 1/2') above the existing grade at the base of the tree.
- **(7)** "Lot" means a portion of platted territory measured, set apart, and subdivided as a distinct parcel having its principal frontage upon a street and shown upon a plat of subdivision or resubdivision approved by the Mayor and City Council of the City of Highland Park, and so recorded by the Recorder of Deeds of Lake County, Illinois.
- **(8)** "Remove" or "Removal" means the causing or accomplishing of the actual physical removal of a Tree, or the effective removal through damaging,

poisoning, or other direct or indirect action resulting in, or likely to result in, the death of a Tree.

(9) "Root Zone" means the area inscribed by an imaginary line on the ground beneath a tree having its center point at the center of the trunk of the tree and having a radius equal to one foot (1') for every inch of DBH.

(10) "Tree" means any self-supporting woody plant having a well-defined stem, a more or less well-defined crown and which has attained a height of at least eight feet (8') with a trunk diameter of not less than three inches (3"), or a cluster of main stems having an aggregate diameter of not less than three inches (3") measured at a point four and one-half feet (4 1/2') above the ground. Containerized Trees and nursery stock Trees kept for resale in licensed nurseries are exempt from the provisions of Sections 94.401 through 94.406.

(11) "Tree Preservation Area" means that area of a Lot or parcel of land within which all trees having a DBH of three inches (3") or larger shall be protected.

(12) "Tree Preservation Plan" means a written plan having text and/or graphic illustrations indicating the methods that are to be used to preserve existing trees during constructions.

(13) "Tree Removal Permit" means that permit required by this Chapter to be issued to remove any tree within the corporate limits of the city.

SEC. 94.403 General Regulations

(A) Unless otherwise specifically authorized in this Code, it shall be unlawful for any person without a written Tree Removal Permit from the City to remove, injure, destroy, or undertake any procedure the result of which is to cause the death or substantial destruction of any tree having a diameter of ten inches (10") DBH or larger, or having an aggregate diameter of fifteen inches (15") DBH or larger. Tree Removal Permits authorizing the removal of such trees may be issued by the City Forester for, but necessarily limited to, the following reasons:

(1) The tree is dead or dying;

(2) The tree is diseased;

(3) The tree is damaged or injured to the extent that it is likely to die or become diseased, or that it constitutes a hazard to persons or property;

(4) Removal of the tree is consistent with good forestry practice;

(5) Removal of the tree will enhance the health of remaining trees within the immediate vicinity; and/or

(B) In furtherance of the Goals contained in Subsection (A) above, the Coordinator of the Building Division shall issue a Tree Removal Permit where he finds that removal of the tree will avoid or alleviate an economic hardship or hardship of another nature on the lot or residence located on the lot.

(C) Upon receipt of a Tree Removal Permit, the permittee:

(1) Shall replace the tree so removed in each of the following instances:

(a) In the event the removal of a tree or trees will avoid or alleviate an economic hardship or hardship of another nature on the lot or the residence located on such lot.

(b) In the event a tree is damaged or injured by other than natural causes to the extent that it is likely to die or become diseased, or it constitutes a hazard to persons or property.

(2) Is encouraged to replace the tree, in the event the tree is diseased, dead or dying from natural causes, or in the event the tree is damaged or injured by natural causes where it is likely to die or become diseased.

(3) May not be required to replace the tree, in the event the City Forester determins that removal of the tree is consistent with good forestry practice, or in the event the removal of such tree will enhance the health of remaining trees within the immediate vicinity.

(4) Shall make all replacements with a new tree or trees of not less than two and one-half inches (2 1/2") caliper, the total aggregate caliper of which shall equal or exceed the diameter of the tree or trees so removed, provided however, that the aggregate caliper of replacement trees may be less than the diameter of the tree or trees so removed in the event the City Forester determines that full replacement would result in the unreasonable crowding of trees upon the lot. Such replacement shall be made within twelve (12) months of the date of the removal of any trees for which replacement is required.

SEC. 94.404 Tree Preservation During Construction

(A) In connection with projects involving the construction of new homes, additions, or detached accessory buildings requiring building permits, a Tree Preservation Plan shall be filed with the building permit application in order to assure that all buildings and other structures shall be located upon a lot or parcel of land in such a way as to minimize tree damage and/or removal, consistent with the various setback requirements of the Zoning Code of the City. The Tree Preservation Plan shall specify the following:

(1) Tree Preservation Area and Building Activity Area upon the Lot or parcel of land for which a building permit application has been filed. [Note: The Tree Preservation Area shall be protected physically from the Building Activity Area by a barrier to prevent impingement of construction vehicles, materials, spoils, and equipment into or upon the Tree Preservation Area]; and

(2) The general contractor, who shall be responsible for the construction, erection, and maintenance of temporary fencing or other physical barrier around Tree Preservation Areas so that all trees in the Tree Preservation Areas shall be preserved.

(B) A Tree Removal Permit shall be issued only in the event the City finds that all reasonable efforts have been undertaken in the architectural layout and design of the proposed development to preserve existing trees. No building permit shall be issued unless the Tree Preservation Plan has been filed with the building permit application and approved by the City Forester.

(C) During construction, all reasonable steps necessary to prevent the destruction or damaging of trees (other than those specified to be removed) shall be taken, including, but not limited to the following:

(1) No construction activity, movement and/or placement of equipment or material or spoils storage shall be permitted outside the Building Activity Area or within the Tree Preservation Area. No excess soil, additional fill, liquids, or construction debris shall be placed within the Root Zone of any tree that is required to be preserved;

(2) Crushed limestone, hydrocarbons, and other materials detrimental to trees shall not be dumped within the Root Zone of any tree, nor at any higher location where drainage toward the tree could conceivably affect the health of the tree;

(3) Appropriate protective fencing shall be temporarily installed for protection of remaining trees;

(4) All required protective fencing or other physical barrier must be in place and approved by the City prior to beginning construction. The fencing must remain in place during the entire construction period to prevent the impingement of construction vehicles, materials, spoils, and equipment into or upon the Tree Preservation Area. All fencing must be secured to metal posts driven into the ground spaced no further than ten feet (10') apart;

(5) No attachments, fences, or wires, other than those approved for bracing, guying or wrapping, shall be attached to trees during the construction period;

(6) Other measures such as construction pruning and root pruning of trees directly impacted by construction must also be indicated on the plan or on an accompanying sheet and approved by the City Forester; and

(7) Unless otherwise authorized by the Tree Removal Permit, no soil is to be removed from within the Root Zone of any tree that is to remain.

(D) It shall be unlawful for any person, firm or corporation to fail to abide by the terms of any Tree Preservation Plan pursuant to which a building permit or Tree Removal Permit has been issued.

(E) If, in the opinion of the City Forester, the necessary precautions as specified in the Tree Preservation Plan were not undertaken before construction commenced or are not maintained at any time during construction, a stop work order shall be issued by the Coordinator of the Building Division until such time as the permittee complies with these precautions.

SEC. 94.405 Application for Tree Removal Permits

(A) In the event a Tree Removal Permit is sought in connection with work for which no building permit is required, there shall be no charge for such permit. The application for such permit shall contain:

- **(1)** Name of applicant;
- **(2)** Commonly known address of Lot or property where said tree or trees sought to be removed is located;
- **(3)** A written statement indicating the reason for removal of the tree or trees;
- **(4)** A general description of other trees on the lot, including size and species; and
- **(5)** Name and address of contractor or other person who is proposed as having responsibility for Tree Removal.

(B) In the event a Tree Removal Permit is sought in connection with construction requiring a building permit, the application shall be accompanied by:

- **(1)** A Tree Removal Permit fee of $50;
- **(2)** A tree survey of the Lot, which shall be drawn on a scale of not less than 1"=30' and shall show trees having a DBH of ten inches (10") and larger or having an aggregate DBH of fifteen inches (15") or larger, including a listing of species. In the event construction activity is to take place in the Root Zone of such trees, protected trees on adjoining lots should be shown;
- **(3)** A Tree Preservation Plan in conformance with the requirements of Section 94.404; and
- **(4)** A report from a certified arborist if required by the City Forester.

SEC. 94.406 Emergencies

(A) In order to avoid danger or hazard to persons or property, during emergency conditions requiring the immediate cutting or removal of a tree or trees protected hereunder, a Tree Removal Permit will be issued by the City Forester without formal application.

(B) In the event of such an emergency, in the event neither the City Forester nor any member of his staff is available to issue such a Tree Removal Permit, it shall be lawful to proceed with the cutting of the tree or trees to the extent necessary to avoid immediate danger or hazard. In such event, the person causing the cutting shall report the action taken to the City Forester within forty-eight (48) hours thereof.

SEC. 94.407 Appeals

The property owner may appeal to the Zoning Board of Appeals of the City any decision made by the City Forester under the provisions of sections 94.401 through 94.406 within thirty (30) days of the decision being rendered.

Section Two: That Section 94.999 of "The Highland Park Code of 1968", as amended, be and the same is hereby amended by adding thereto a new subsection, Subsection (F), providing the penalty for violation of any of the provisions of Sections 94.401 through 94.407; so that hereafter the said Subsection (F) of Section 94.999 shall be and read as follows:

(F) "Whoever violates any of the provisions of Sections 94.401 through 94.407, or who shall interfere with the execution or enforcement thereof shall be fined not less than $50 nor more than $500. A separate and distinct violation shall be deemed to have occurred for each Tree unlawfully removed in violation of any of the sections aforesaid."

Section Three: That the City Clerk of the City of Highland Park be and is directed hereby to publish this ordinance in pamphlet from pursuant to the statutes of the State of Illinois, made and provided.

Section Four: That this ordinance shall be in full force and effect from and after its passage, approval, and publication in the manner provided by law.

AYES: Mayor Pierce and Councilmen O'Keef, Weiss, Abrahamson, and MacLeod
NAYS: None
ABSENT: Councilman Brenner
PASSED: March 25, 1991
APPROVED: March 25, 1991
PUBLISHED: March 26, 1991
ORDINANCE NO. 7-91

TREE REMOVAL PERMIT

This permit is required by Chapter 94.401–94.407 of "The Highland Park Code of 1968" for the removal of any tree within the corporate limits of Highland Park having a DBH larger than 10 inches.

Date of application ______________________________

A) Applicant information

Name: ______________________ Work address or location if different:

Telephone: ______________________ ______________________

Address: ______________________ ______________________

______________________ ______________________

B) The applicant seeks approval for removing the following trees:

SPECIES DBH GENERAL LOCATION

1. ____________ _____ ______________________

REASON FOR REMOVAL ______________________

2. ____________ _____ ______________________

REASON FOR REMOVAL ______________________

3. ____________ _____ ______________________

REASON FOR REMOVAL ______________________

C) General description of other trees of lot > 10" DBH (size and species):

____

____

____

D) Removal Contractor: ______________________________

______________________ ______________________

Signature Applicant *Date*

The following is to be filled out by the City Forester:

A) Number of diameter inches of tree replacement (minimum 2.5" caliper) required:

____

Number of diameter inches of tree replacement waived: ______________

B) Comments

____

____

____

______________________ ______________________

Larry G. King
City Forester Date

Permit # __________
Fee pd __________

TREE PRESERVATION ORDINANCE
PRECONSTRUCTION TREE PRESERVATION PERMIT

This completion of this permit is required by Chapter 94.404 of "The Highland Park Code of 1968" prior to the issuance of any building permit for construction work within the limits of Highland Park.

Date of application ______________________________

A) Applicant information

Name: ______________________ Work address or location if different:

Telephone: ______________________ ______________________

Address: ______________________ ______________________

______________________ ______________________

B) The Tree Survey of the Lot attached: Yes No

C) The Tree Preservation Plan attached: Yes No

D) Tree Removal required as part of construction Yes No

1) The Tree Removal Permit Fee of $50.00 attached: Yes No

2) The applicant seeks approval for removing the following trees as per the Tree Preservation Plan:

	SPECIES	DBH	GENERAL LOCATION
1.	____________	_____	______________________
2.	____________	_____	______________________
3.	____________	_____	______________________
4.	____________	_____	______________________

E) Certified Arborist: ______________________________

______________________ ______________________

Signature Applicant *Date*

The following is to be filled out by the City Inspector:

A) Permit granted Yes No

B) Tree replacement required Yes No

C) Number of trees required as replacement
(See Section 94.403 C4 of City Code) __________

D) Explanation of tree replacement requirements:

____

____

E) General comments:

____

____

____

____

__________________________	______________________
Signature *Inspector*	*Date*

Appendix E

*City of Lynnwood Ordinance No. 386**

AN ORDINANCE AMENDING TITLE 20 OF THE LYNNWOOD MUNICIPAL CODE BY ADDING TO VARIOUS CHAPTERS THEREOF CERTAIN REQUIREMENTS AS TO SITE SCREENING AND CRITERIA FOR GREENBELT AND PLANTINGS*

WHEREAS, the Planning Commission has considered certain planting and sight screening changes and has held public hearing relative thereto,

NOW THEREFORE, the City Council of the City of Lynnwood do resolve as follows:

Section 1. That Chapter 20.18 of Lynnwood Municipal Code be amended by adding thereto the following section;

20.18.070 Landscaping requirements for parking area in RM Districts:

(a) Purpose of the landscaping provisions:

(1) To break up the visual blight created by large expanses of barren asphalt that make up a typical parking lot.

*Olaf Unsoeld. 1980. *Analysis of Some Municipal Tree and Landscape Ordinances,* U.S.D.A. Forest Service, Southeastern Area.

(2) To encourage the preservation of mature evergreens and other large trees that are presently located on most of the potential apartment sites in this city.
(3) To provide an opportunity for the development of a pleasing visual environment in the apartment districts of this city from the viewpoint of the local resident and visitor passing through the districts. (A purpose of this ordinance as well as from the viewpoint of the apartment dweller; a purpose of the apartment developer.)
(4) To insure the preservation of land values in apartment districts by creating and insuring an environmental quality that is most compatible with the development of this land.
(5) To provide adequate control over the application of landscaping standards so that the above objectives (1, 2, 3, and 4) are accomplished in the most effective manner and to avoid the abuse of these intentions by placing the described landscaping in remote parts of the site or in recreational areas where they bear no relationship to these objectives.

(b) Parking areas fronting a street right-of-way shall provide a five (5) foot planting area along the entire street frontage except for driveways, provided that no bush or shrub shall be higher than thirty (30) inches and no tree shall have branches or foliage below six (6) feet.

(c) In addition, said planting may be placed on street right-of-way behind the sidewalk line if the property owner provides the City with a written release of liability for damages that may be incurred to said planting area from any public use of the right-of-way and an indemnity to the City against any injuries occurring within the portion of right-of-way so utilized.

(d) 10% of parking area to be landscaped (exclusive of landscaping on the street frontage), provided that:

(1) No landscaping area shall be less than 100 square feet in area or less than five (5) feet in width.
(2) No parking stall shall be located more than forty-five (45) feet from a landscaped area. Landscaping shall be proportionately distributed throughout the parking area in a manner that best fulfills the objectives of this ordinance. (See the above statement of purpose.)
(3) All landscaping must be located between parking stalls or between parking stalls and the property lines. Landscaping that occurs between parking stalls and apartments or between parking stalls and apartment recreation areas shall not be considered in the satisfaction of these landscaping requirements.

(e) Said planting area shall include liberal landscaping using such materials as trees, ornamental shrubs, gravel, river rock, driftwood, rockeries, lawn, or combination of such material.

Section 2. That the first paragraph of Chapter 20.16.060 which reads as follows:

> Procedure for requiring greenbelt. Whenever sight screening or a greenbelt are required by this title the building official may, prior to issuance of a building permit or occupancy permit, cause the matter to be presented to the Planning Commission for the purpose of determining standards for such sight screening or greenbelt.

should be repealed, revised and amended to read as follows:

> Procedure for requiring greenbelt. Whenever sight screening or a greenbelt are required by this title the building official may approve the landscaping plan if it complies totally with the requirements of the Lynnwood Municipal Code existing at the time of application of plans or in the discretion of the building official and prior to issuance of either a building permit or occupancy permit, cause the matter to be presented to the City Council for the purpose of determining standards for such sight screening or greenbelt.

Section 3. That Chapter 20.16 should be and the same is hereby amended by adding thereto a section as follows:

> Type I. Ornamental Landscaping of Low Plantings and High Plantings:
>
> Trees shall have a minimum height of 10 feet with branches eliminated to a height of 6 feet to prevent sight obstruction. Low plantings with a maximum height of 30 inches shall be provided so as to constitute a total ground cover within two years.
>
> Type II. Privacy Screening:
>
> Privacy screening is partially concealing, partially open type screen of wood, metal or concrete products that may not extend more than $\frac{1}{3}$the front or sideyard setback distance from the building, but most not become an objectionable sight, or a hazard to life, limb, or public health. (ord. 190 Art X pp10.5; March 12, 1964)
>
> Type III. Low Plantings, Trees, and Fencing:
>
> Conifer trees with a minimum height of 6 feet shall be spaced a minimum of 25 feet or center, backed by a 6-foot fence which forms and effective barrier to sight. The remainder of the planting strip shall be covered with low evergreen plantings so as to constitute a total ground cover within two years.
>
> Type IV. Mixed Trees, Shrubs, and Low Plantings:
>
> This planting strip shall be designed with a mix of evergreen plants which form an effective sight barrier and total ground coverage within two years. Minimum tree height shall be 6 feet.
>
> Type V. Wall of Trees:
>
> The purpose of this landscaping is to provide a sight, sound, and psychological barrier between land uses with a high degree of incompatibility. The living wall shall be composed of conifer trees with such maturity and spacing as to form an effective visual barrier within two years. Said trees shall be distributed so they cover

the full depth of the designated planting area. Minimum tree height shall be six (6) feet.

B. All landscaping plans shall bear the seal of a registered landscape architect or signature of a professional nurseryman and be drawn to a scale no less than 1"–20'.

All planting areas that fulfill City code requirements shall be covered by a performance bond with value equal to 20 cents per square foot of planting area, to be released after one full growing season if plants are in a healthy growing condition.

20.16.090. General Sitescreen Requirements for Side Yard and Rear Yards:

In order to reduce the incompatible characteristics of abutting properties with different land use classifications, the minimum landscaping standards of Section 20.16.050 shall apply to planting strips on the side and rear property lines (except for portions affected by building setbacks from streets) according to the following schedule:

RML adjacent to RS: Type IV Planting Strip, 5 feet in width.

RMHR adjacent to RML; Type IV Planting Strip, 5 feet in width.

RMM, RMH, and RMHR adjacent to RS: Type IV Planting Strip, 10 feet in width.

CG, BC, and BN adjacent to RS or RM: Type IV Planting Strip, 10 feet in width with a 4-foot sight obstruction fence in the center.

Manufacturing adjacent to RM or RS. Type V Planting Strip, 20 feet in width with a 6-foot fence along the manufacturing side of the planting strip.

Mobile home parks shall apply in the same manner as RMM.

Section 4. That Section 20.12.100 (e) of Lynnwood Municipal Code which reads as follows:

> (e) Number of dwelling units: The number of dwelling units permitted in any "R" zone shall be determined by dividing the net development area by the minimum lot area per dwelling unit required by the zone in which the area is located. Net development area shall be determined by subtracting the area set aside for churches, schools or commercial use from the total development area and deducting twenty per cent of the remainder for streets regardless of amount of land actually required for streets,

should be and the same is hereby repealed, and the same should be an is hereby amended to read as follows:

> (e) 1. Number of dwelling units: The number of dwelling units permitted in any "R" zone shall be determined by dividing the net development area by the minimum lot area per dwelling unit required in the zone in which the area is located. Reductions in lot size requirements when parking spaces are placed within the building structure shall not apply to planned unit developments. Net development

area shall be determined by subtracting the area set aside for churches, schools, or commercial use from the total development area.

2. Rezones to Planned Unit Development shall provide the opportunity to attain up to 50 percent increase in density beyond the densities designated by the Zoning Ordinance provided that accompanying explanatory text shall show that the objectives of the Comprehensive Plan and of the Planned Unit Development section of the Zoning Ordinance can be met; and that the public access, utilities, and facilities are adequate for the increased densities.

3. This subsection shall not be construed to require the City to permit a greater density than is otherwise permitted in the subject use zone but merely to authorize increased density in such zone in the sound discretion of the Lynnwood City Council.

Section 5. This ordinance shall be in full force and effect five (5) days after its passage, approval, and legal publication.

PASSED by the City Council of the City of Lynnwood, Washington, this 11th day of March, 1968, and signed in authentication of its passage this 11th day of March, 1967.

ATTEST: ______________________________
City Clerk

______________________________ Mayor

FORM APPROVED: ______________________

Appendix F

Sample Publication with Recommended Street Trees

A2014

Street Trees for Wisconsin

E. R. Hasselkus

Street trees bring nature into our everyday environment. They modify summer temperatures, control winds, deaden traffic noises and help purify polluted air. Trees soften the harsh lines of the man-made elements in our landscape. Colorful flowers and fruits, textural qualities of foliage, autumn foliage color and patterns of branches and bark provide an ever-changing seasonal interest. Loss of American elms to Dutch elm disease has forcefully demonstrated the aesthetic and economic value of street trees in numerous Wisconsin communities.

WHEN SELECTING STREET TREES . . .

Plant Several Species

This will help avoid monotony and insure your community against having most of its trees wiped out by an insect or disease invasion. Diversified street tree plantings may be accomplished with identical trees on one street and using other tree species on adjacent streets.

In areas with curved streets or with many existing trees, several kinds of trees might be combined if their forms, textures and colors are compatible.

Plant Species Suited to the Particular Site

Factors to watch for include:

- Winter hardiness
- Optimum height and spread at maturity. Fit the tree to the space available.
- Tolerance to pollutants. Trees vary in their tolerance to soot, gases, smoke, salt and other chemicals.
- Soil fertility, available moisture, soil alkalinity and a restricted root area may limit your choice of tree species.
- Tolerance of exposure to reflected heat, high winds or dense shade.

Plant Species Inexpensive to Maintain

These species include those with the following characteristics:

- Easy to transplant and establish.
- High resistance to serious diseases and insects.
- Long lived and strong wooded.
- Minimal litter, such as faded flowers, messy fruits or seed pods, which may be a nuisance.
- Roots that will not block sewers and drains or heave pavements.
- Tap root systems and filtered shade patterns allowing grass to grow.

Consider Proper Spacing

Keep space between trees at least one-half of their total spread. To allow motorists' visibility at intersections, plant trees no closer than 25 feet from the corner. Overhead utility lines may limit your choice to low-growing species.

If interplanting between existing trees, consider the shade cast by overhead branches as well as the competition from roots of adjacent trees.

(H. Armstrong Roberts)

Recommended Trees

BOTANICAL AND COMMON NAME	HEIGHT	SPREAD	GROWTH RATE	REMARKS
Acer platanoides Norway Maple	50′	50′	Medium	Greenish-yellow flowers before leaves. Avoid using in areas where turf is to be maintained; withstands salt and adverse city conditions.
cv. 'Cleveland' or 'Emerald Queen'	40–50′	30′	Fast	Upright oval-headed forms of the species with superior branching habit.
cv. 'Columnare' or 'Erectum' Columnar Norway Maples	40′	10–15′		Dense, columnar forms of the species.
cv. 'Globosum' Globe Norway Maple	20′	20′	Slow	Low-crowned globe from of the species. Not readily available from nurseries.
Acer rubrum Red Maple	50–60′	30–40′	Fast	Requires moist acid soil; bright red flowers before leaves; brilliant autumn foliage color; smooth gray bark. Intolerant of salt.
cv. 'Autumn Flame'	50–60′	30–40′		Develops an early display of scarlet autumn foliage.
cv. 'Armstrong' or 'Columnare' Columnar Red Maples	40′	10–20′		Columnar forms of the species.
cv. 'Bowhall' Bowhall Red Maple	40–50′	30′		Oval form of the species with orange autumn foliage color.
Aesculus hippocastanum 'Baumannii' Baumann Horsechestnut	30–60′	30′	Medium	Bears double white sterile flowers. Casts dense shade. Tolerant of salt. Not readily available from nurseries.
Celtis occidentalis Common Hackberry	30–50′	40′	Slow	Interesting pebbled bark; hard black fruits; "witch's broom" may be a problem; similar in habit to elm. Tolerant of both dry and wet soils. Sensitive to salt spray.
Crataegus phaenopyrum Washington Hawthorn	20–30′	15–20′	Medium	Bears thorns; white flowers, tiny orange fruits and red to orange autumn foliage display. Tolerant of adverse city conditions.
Fraxinus americana White Ash	50–80′	50′	Medium	Broad-headed tree with diamond-shaped fissures in bark and yellow to purple autumn foliage color.
cv. 'Autumn Purple' or 'Rosehill' Seedless White Ashes	50′	50′		Seedless forms of the species with orange-purple autumn foliage color.
Fraxinus pennsylvanica Green Ash	50–60′	30–40′	Fast	Dark green, glossy foliage turning yellow in autumn; variable in form. Tolerant of salt and both dry and wet soils.
cv. 'Marshall's Seedless' or 'Summit' Seedless Green Ashes				Seedless forms of the species; uniformly pyramidal in form.
**Ginkgo biloba* Ginkgo	60′	30–40′	Slow	Picturesque growth habit; fan-shaped leaves turn yellow in autumn; tolerant of city conditions.
*cv. 'Fastigiata' Sentry Ginkgo	60′	10–15′		Narrow columnar form of the species.

Recommended Trees

BOTANICAL AND COMMON NAME	HEIGHT	SPREAD	GROWTH RATE	REMARKS
Gleditsia triacanthos inermis cvs. 'Imperial' Honeylocust 'Shademaster' Honeylocust 'Skyline' Honeylocust	60'	40'	Fast	Podless and thornless varieties; fine textured foliage that casts a light shade. Tolerant of salt, adverse city conditions and both dry and wet soils.
Gymnocladus dioica Kentucky Coffeetree	60'	40'	Medium	A picturesque tree with unusual deep furrowed twigless branches.
Malus cvs. Flowering Crabapples:	20–30'	18–20'	Medium	Use only where fruits up to 3/4" in diameter can be tolerated.
'Centurion'				Rose-red flowers; glossy red, persistent fruits 5/8" in diameter; disease resistant
'Sentinel'				Pink flowers; red, persistent fruits 3/8" - 1/2 " in diameter; slightly susceptible to scab and fire blight.
Ostrya virginiana Hophornbeam or Ironwood	25'	20'	Slow	Similar in appearance to American elm, but much more refined; interesting fruits. Tolerant of dry soil and shade, but intolerant of salt.
Phellodendron amurense Amur Corktree	45'	30'	Fast	A sturdy tree with ash-like leaves with yellow autumn color and a thick corky bark. Tolerant of dry soils.
Pyrus calleryana 'Select' Select Callery Pear	30'	16'	Medium	Narrow pyramidal form; glossy green leaves turn red-orange to maroon in autumn; early, white flowers.
Quercus palustris Pin Oak	40–50'	30–40'	Medium	Symmetrical in form; fine autumn color; avoid using on alkaline soils. Tolerant of wet soils.
Quercus rubra Red Oak	60'	40–50'	Medium	Dark red autumn foliage color. Intolerant of heavy or poorly drained soils.
Tilia cordata Littleleaf Linden	40'	30'	Medium	Oval to pyramidal in form; tolerates adverse city conditions.
cv. 'Chancellor' or 'Greenspire' Pyramidal Littleleaf Lindens				Uniformly pyramidal in form.
Tilia x euchlora 'Redmond' Redmond Linden	50'	40–50'	Medium	Pyramidal form; coarse lustrous foliage.
*Tilia tomentosa Silver Linden	40'	30'		Dense pyramidal form; tolerant of hot, dry conditions.

*Adapted only to southern and eastern Wisconsin

AVOID THESE TREES:

Black Locust	Subject to borers.
Boxelder	Weak wooded, female trees attract the boxelder bug.
Catalpa	Litter of flowers, fruits and leaves.
Elms	Subject to Dutch elm disease.
Poplar	Roots block sewers, weak wooded, litter of fruits.
Silver Maple	Weak wooded, buttress roots heave pavements.
Sugar Maple	Intolerant of urban conditions
Sycamore	Subject to anthracnose disease.
Willows	Roots block sewers, weak wooded, litter of twigs.
Nut bearing trees	Litter of nuts.

Plant the Right Tree in the Right Place

Because of space limitations, some sites are not suitable for growing a tree. Other sites can support a tree only if special provisions are made.

If the width of the area between the curb and sidewalk is less than 3 feet, the terrace probably will not accommodate a tree. If the tree does survive, it may damage the sidewalk or curb.

The distance between the planting site and an adjacent building is also an important consideration. In situations where the distance between the curb and building is less than 12 feet, only plant trees with a columnar form.

A tree planted in an opening in concrete pavement requires special attention in order to survive in this confining site. Cast iron grates placed around the tree can benefit it in several ways. The grates allow for soil aeration and reduce the chance for soil compaction by providing a surface to walk on. The grates also facilitate watering and fertilization.

The list of trees recommended in this publication for planting in Wisconsin include trees of various sizes, forms, growth rates and environmental tolerance. Those marked with asterisks are adapted only to southern and eastern Wisconsin.

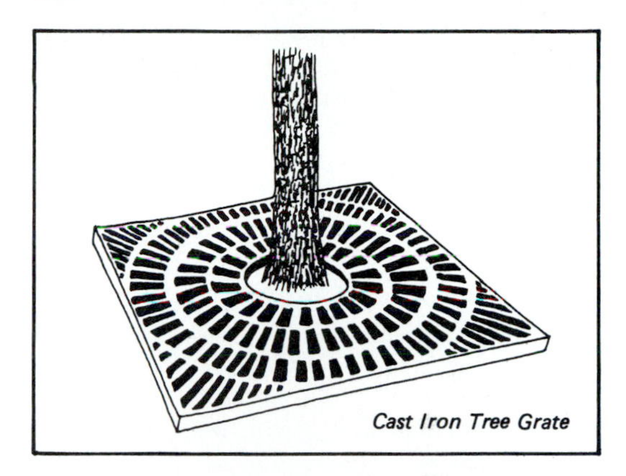
Cast Iron Tree Grate

For more information on insects, diseases, the dangers posed by construction and other tree care concerns, consult the following University of Wisconsin-Extension publications:

A7791025 *Urban Forestry: Bringing the Forest to Our Communities*
A3070 *How Trees Benefit Communities*
A3066 *Landscaping for Energy Conservation*
A3071 *Dutch Elm Disease—A Lesson in Urban Forestry Planning*
A3067 *Selecting, Planting and Caring for Your Shade Trees*
A1817 *Caring for Your Established Shade Trees*
A3072 *Preserving Trees During House Construction*
A3073 *Identifying Shade Tree Problems*
A2079 *Recognizing Common Shade Tree Insects*
A7791105 *Selecting an Arborist and Other Tree Professionals*
A7791106 *How to Develop a Community Forestry Program*
A2865 *A Guide to Selecting Landscape Plants for Wisconsin*
A2392 *Dutch Elm Disease in Wisconsin*
A2842 *Homeowner's Guide to Controlling Dutch Elm Disease with Systemic Fungicides*
A1771 *Deciduous Shrubs: Pruning and Care*
A1730 *Evergreens—Planting and Care*
G1609 *Landscape Plants that Attract Birds*
A2970 *Salt Injury to Landscape Plants*
A2308 *Fertilization of Trees and Shrubs*

Please contact your county Extension agent for films, slide sets, posters and other publications about trees.

This publication is part of the urban forestry series, developed to help in the reforestation of our towns and cities struck by Dutch elm disease.

Edward R. Hasselkus is professor of horticulture, College of Agricultural and Life Sciences, University of Wisconsin-Madison and the Divison of Economic and Environmental Development, University of Wisconsin-Extension.

COOPERATIVE EXTENSION PROGRAMS UWEX

UNIVERSITY OF WISCONSIN–EXTENSION

University of Wisconsin-Extension, Gale L. VandeBerg, director, in cooperation with the United States Department of Agriculture and Wisconsin counties, publishes this information to further the purpose of the May 8 and June 30, 1914 Acts of Congress; and provides equal opportunities in employment and programming including Title IX requirements. This publication is available to **Wisconsin residents** from county Extension agents. It's available to **out-of-state purchasers** from Agricultural Bulletin Building, 1535 Observatory Drive, Madison, Wisconsin 53706. **Editors,** before publicizing, should contact the Agricultural Bulletin Building to determine its availability. **Order by serial number and title;** payment should include prices plus postage.

APRIL 1980

10¢

A2014 STREET TREES FOR WISCONSIN This publication is revised.

Appendix G

*Storm Disaster Planning for Medium-Sized Communities Based on the Oak Park, Illinois Plan**

The Forestry Division (see Tables G–1 and G–2 of Oak Park, Illinois (population 54,900), has devised a major tree damage control plan which is presented and serves as a model for a medium-sized communities (Stankovich, 1991). Table G-2 summarizes the responsibilities of the Forestry Division staff.

TABLE G–1 OAK PARK, ILLINOIS, DEPARTMENT OF PUBLIC WORKS

Division	Number of Personnel
Automotive Fleet	14
Engineering/Administration	10
Forestry	14
Signs	3
Street Light	3
Streets	32
Water and Sewer	18
TOTAL	**94**

*From L. L. Burban and J. W. Andersen. 1994. *Storms Over the Urban Forest.* 2nd Ed. St. Paul, Minn.: USDA Forest Service.

TABLE G–2. OAK PARK, ILLINOIS, FORESTRY DIVISION RESPONSIBILITY SUMMARY

Village Forester

- Has overall direction for storm clean up efforts
- Reports to the Public Works Director on decisions relating to storm clean up efforts and advises on the need for outside assistance (contractors, other public works divisions)
- Is responsible for decisions relative to abandoning other divisional responsibilities in favor of storm damage clean up efforts
- Works with Communications Director for alerting media as to the progress and problems associated with the storm

Assistant Village Forester

- Coordinates the involvement of Forestry and other crews
- Inspects of damaged property (Village and private)
- Is responsible for maintaining other Divisional functions

Lead Forester

- Performs overall coordination of field activities related to storm damage
- Coordinates and assigns duties to personnel assisting the Forestry Division
- Is responsible for decisions related to staffing, equipment usage, and disposal of wood waste
- Reports to Assistant Village Forester on progress of storm clean up

Urban Forestry Technician I and II

- Performs actual clean up work
- Repairs minor equipment used in clean up efforts
- Responsible for care of equipment and tools

An outline of Oak Park's Major Tree Damage Control Plan is listed below. It consists of twelve points found under two major headings: "Early Warning" and "Immediate Reaction."

OAK PARK MAJOR TREE DAMAGE CONTROL PLAN

I. Early Warning
   A. Supervision of major tree damage control
   B. Weather warning
II. Immediate Reaction
   A. Tree damage clean up priority
   B. Public alley clearance
   C. Forestry Division and Street Division communications
   D. Equipment listing

E. Additional equipment and assistance
F. Brush removal from private property
G. Life threatening situations and property damage on private property
H. Brush disposal
I. Record keeping
J. Damage assessment

I. EARLY WARNING

A. Supervision of Major Tree Damage Control

The Village Forester, followed by the Assistant Forester (Public Works Foreman), will be in charge of storm clean up efforts. Duties and assignments will be discharged by these individuals to the Forestry Division Leadperson. Should the Forestry Division need the assistance of other Public Works Divisions (upon approval of the Public Works Director) the Forestry Division will contact the appropriate Division supervisors.

B. Weather Warning

In Oak Park, the Forestry Division is informed of potential severe weather by telephone and fax machine from a climatological consulting firm. Three to five hours lead time are usually provided before the onset of weather conditions that may cause tree damage. Note: The most critical period during the year for potential tree damage is when leaves are still on the trees (May through October); however, severe tree damage may occur at any time. Information from the weather consultant is relayed on a weather reporting form.

Based on weather condition information, the decision is made by Forestry Supervisory staff to alert the proper crews. In addition to the initial call from the weather consultant, a system for obtaining supplemental updates on weather conditions is at the disposal of the Forestry Division Supervisory staff. At this time, the equipment used to clean up tree damage should be readied for action.

A system for notifying Forestry Division staff should be established. In certain situations, the Forestry Division staff can be advised by the Police Department that tree damage has occurred in the Village. At other times, staff may be contacted using some form of a "branching" or "phone-tree" calling system, where each person is responsible for notifying another. In any case, this system must be established and kept up-to-date for a rapid and efficient storm response system.

In addition, the Forestry Division staff should be provided with a list of emergency telephone numbers that enumerates Village personnel and can assist with situations encountered during severe weather.

II. IMMEDIATE REACTION

Tree Damage Clean up Priority

CLASS I First, all life-threatening situations should be given priority. Supervisors should make an on-site visit to determine the severity of the damage in the event of multiple hazardous situations. Crews should remedy the situation to a point where it is no longer life threatening before proceeding to the next location. Final clean up should wait until all life threatening situations are resolved and all streets have been cleared.

CLASS II Second, all major property damage instances should be remedied to a point where the crisis is abated. Supervisors should personally inspect and determine the priority of the Forestry Division responses. Again, final clean up at those sites should wait until all streets and specialized areas are cleaned up.

CLASS III Third, preferential streets (considered to be all main thoroughfares) should be cleared of fallen trees and debris. State and county highway departments may be called to clear U.S., state and county routes. This should be followed by clearing residential streets and then parking lots, cul-de-sacs and other specialized areas, including parks. Because the specialized forestry skills required to abate life-threatening and property damage situations would be utilized immediately, the street clearance work (in case of widespread and severe damage) may not be undertaken by Forestry Division personnel until sometime well after the storm has passed. In this situation, the Village Forester should recommend to the Public Works Director that other public works crews be considered to assist in street clearance work. Immediate supervision of these supplementary crews would be under the direction of their respective divisions.

Public Alley Clearance

Many municipalities have the responsibility to provide clear passage through public alley thoroughfares. Often these alleys incorporate the majority of the utility line system within them. During severe weather, trees growing on private property can fall into the alleyway blocking portions of the alley roadway. Consequently, when trees or limbs fall from tree bordering alleys, they often become entangled with the utility lines. Most utilities prohibit anyone except their own crews from attempting to clear fallen trees from utility wires. Note: This policy varies with utility companies. Because many communities exist within an electrical service district that encompasses multiple municipalities, there is a great likelihood that it may take several days to respond to all tree and power line conflicts, particularly during violent weather producing widespread damage. In these instances, barricades should be set up to warn residents of the hazard.

Forestry Division and Street Division Communications

Constant communication during emergency situations is vital. Communication aids in improved response time, efficient crew scheduling, and in alerting emergency personnel of hazardous situations. All forestry vehicles should be equiped with two-way radios. Additionally, pagers are useful communication devices. However, during storm clean up work, the need often arises to contact people who do not use Public Works Department radios. These include, fire and police personnel, utility crews, contract crews employed for storm clean up, and others. Severe storms may destroy normal systems of communication such as radio towers and telephone lines making normal telephone service useless. These systems may not be repaired for days or several weeks. Identify several alternative communication systems for backup. Backup systems may include cellular phones, equipment from other agencies or nearby communities and ham radios.

Equipment Listing

A listing of Public Works Department equipment and vehicles available for tree clean up work should be developed and kept current. The list may include wood chippers, aerial bucket trucks, refuse packers, prentice loaders, supervisory vehicles, chain saws, barricade and lighting equipment, hand saws and pole pruners.

Additional Equipment and Assistance

When necessary, the municipal administrator may authorize the rental of additional equipment for storm clean up work. A list of potential vendors should be developed and kept current. Additionally, tree contractors may be authorized to work in the community to supplement municipal crews. As with the list of potential vendors, a list of potential contractors should be assembled. Depending on the path of a storm, there may be other municipalities in a geographic area that remain unaffected by the severe weather. Establish a system to contact these communities in the event that they could send staff and equipment to assist your municipality in its clean up efforts.

Brush Removal from Private Property

A system for handling tree debris from private property must be identified. Municipalities vary in both practice and policy regarding brush and related wood waste removal, as well as disposal from private property.

Authors' Note: If a major storm makes it difficult for private property homeowners to remove brush and debris, a decision should be made at the municipal level allowing for debris to be collected. Note that a community must determine if it has adequate equipment and staff available to accomplish this often enormous task. It is critical for the municipality to provide guidelines for residents specifying the types, amounts, and piling arrangement of the materials that will be accepted. Municipalities may wish to assist private homeowners who will contract with private companies for trimming and removal by providing information about companies that are fully licensed, professionally trained, and insured.

Life-Threatening Situations and Property Damage on Private Property

Normally, all tree work required on private property is the responsibility of the individual property owner. However, if in the opinion of police or fire department personnel the situation requires immediate attention, then forestry personnel should enter onto private property and take necessary action to solve the problem. Note: The municipality must be properly insured for this tree work to cover any potential lawsuit or liability due to personal injury or property damage. Clean up is the responsibility of the homeowner. Tree damage (during storms) is not limited to public property—it will also occur on private property.

Brush Disposal

The Forestry Division develops a budget for normal disposal costs associated with yearly tree maintenance tasks. Major tree debris disposal will require additional funding which may be authorized by the Village Manager.

Record Keeping

All Divisions that are involved with storm clean up should keep accurate and detailed records on equipment and staff hours. Their records will provide important information in the event of financial reimbursement from federal or state agencies, or in case of questions or confusion regarding use of staff, equipment or funds.

Damage Assessment

A critical tool to assist any emergency response is a current tree inventory of all publicly owned trees. Using the inventory, the Forestry Division can determine the actual damage to the urban forest. Accurate damage (in dollars) can be assessed and submitted for potential reimbursements. Specific costs can be developed for the repair of the urban forest (pruning, removal, cabling, and rodding) and for replanting efforts. Experience in other Illinois communities has shown that in the event of major storm damage, substantial monetary reimbursement can be given to communities that can produce accurate documentation of their loses. Federal funding may cover 75 percent of these costs, while State funds reimburse 12.5 percent. The community would then be responsible for the remaining 12.5 percent.

Authors' Note: An additional phase that evaluates *response* is high recommended after disaster activities have been completed. Time should be taken to assess response alternatives and activities that worked or failed, in order to minimize or enhance effectiveness during the next emergency.

LITERATURE CITED

Stankovich, M. R. 1991. Major tree damage control plan. Forestry Division, Oak Park, IL.

Appendix H

*Tree Walks in Madison, Wisconsin**

*From R. Bruce Allison. *Tree Walks in Madison and Dane County.* Madison, Wisconsin: Wisconsin Books.

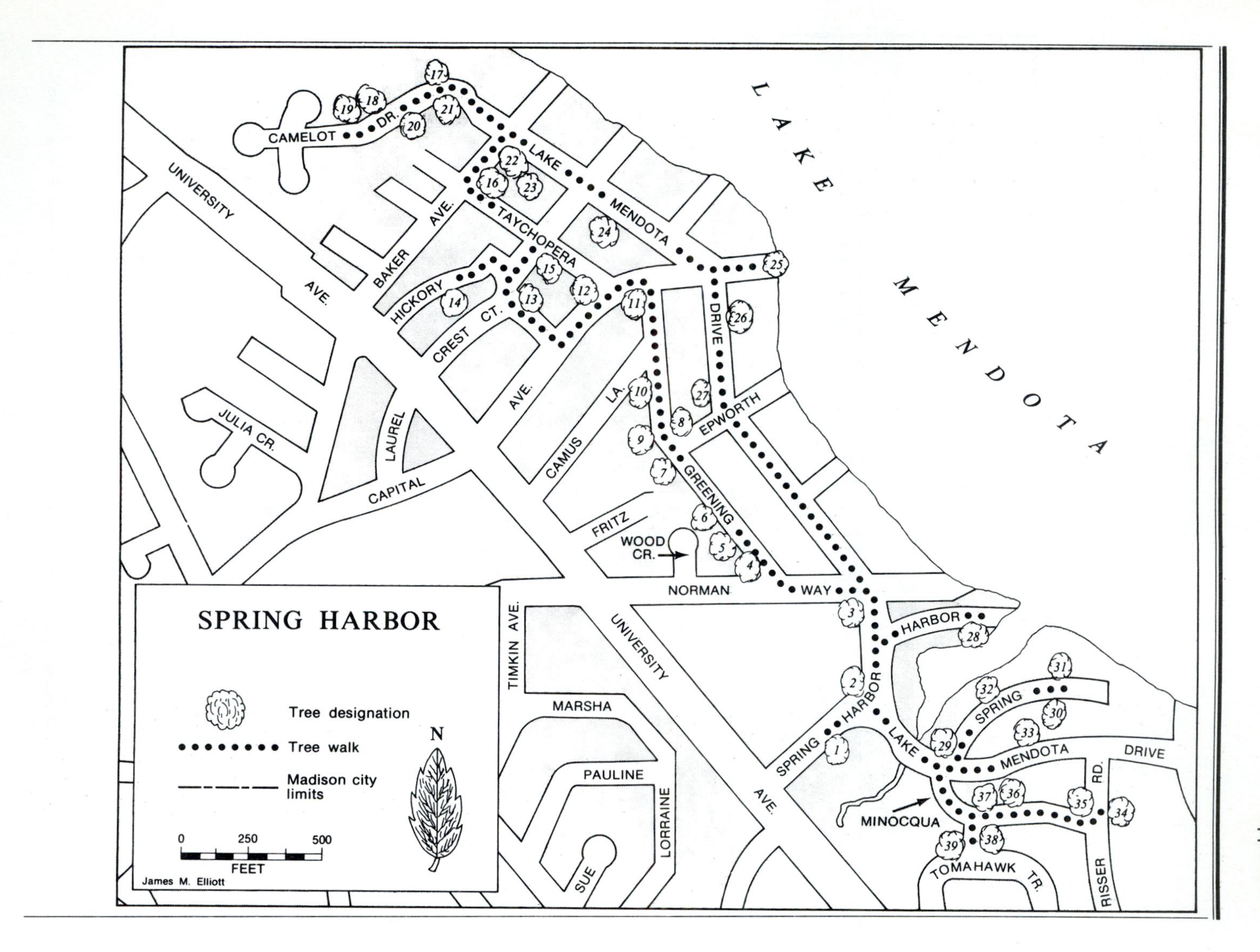

LAKE MENDOTA
SPRING HARBOR
Tree designation
Tree walk
Madison city limits
0
250
500
FEET
N
James M. Elliott
CAMELOT DR.
LAKE MENDOTA DRIVE
UNIVERSITY AVE.
BAKER AVE.
TAYCHOPERA
HICKORY CREST CT.
CAMUS LA.
AVE.
JULIA CR.
LAUREL
CAPITAL
EPWORTH
GREENING
FRITZ
WOOD CR.
NORMAN WAY
HARBOR
TIMKIN AVE.
MARSHA
PAULINE
LORRAINE
SUE
SPRING HARBOR
LAKE MENDOTA DRIVE
SPRING
MINOCQUA
TOMAHAWK TR.
RISSER RD.
1
2
3
4
5
6
7
8
9
10
11
12
13
14
15
16
17
18
19
20
21
22
23
24
25
26
27
28
29
30
31
32
33
34
35
36
37
38
39

SPRING HARBOR

1. Shagbark Hickory (*Carya ovata*) in Spring Harbor Park, numerous specimens, note shaggy bark on trunk and compound leaf.
2. Willow (*Salix alba*) line of several tall Willows on south side of Lake Mendota Drive near sidewalk.
3. Eastern Redcedar (*Juniperus virginiana*) 5321 Lake Mendota Drive, sideyard; note three Colorado spruce (*Picea pungens*) near street.
4. White Fir (*Abies concolor*) 5401 Greening Lane near intersection of Norman and Greening, sideyard.
5. Paper Birch (*Betula papyrifera*) 5409 Greening Lane, frontyard; note small Magnolia and Honeylocust (*Gleditsia triacanthos inermis*).
6. Arborvitae (*Thuja occidentalis*) a hedgerow in yard between 5431 and 5433 Greening Lane; note Japanese Yew (*Taxus cuspidata*) on opposite side of yard; across street tallest conifer is Norway Spruce (*Picea abies*).
7. Scots Pine (*Pinus sylvestris*) just behind mailbox at driveway of 5501 Greening Lane, odd-shaped specimen with young Canada Hemlock (*Tsuga canadensis*) in front and surrounded by Yews (*Taxus*).
8. Black Walnut (*Juglans nigra*) yard tree on Greening near Epworth intersection by fire hydrant.
9. White Pine (*Pinus strobus*) 5505 Greening, in frontyard with Red Pine (*Pinus resinosa*) near driveway. Note Russianolive (*Elaeagnus angustifolia*) across street near curb.
10. Black Walnut (*Juglans nigra*) at intersection of Greening and Camus beside street.
11. Japanese Larch (*Larix kaempferi*) intersection of Capital Ave and Greening, south side of Greening.
12. Siberian Elm (*Ulmus pumila*) 1832 Capital Ave. at curbside between mailbox and telephone pole.
13. Catalpa (*Catalpa speciosa*) 1801 Laurel Crest; note Norway Spruce (*Picea abies*) in intersection island.
14. Basswood (*Tilia americana*) across from 1721 Hickory, near street at end of split rail fence; in same yard of historic "Hickory House" is a Black Walnut from seed gathered at George Washington's Mt. Vernon, grown by Wisconsin conservationist Walter Scott. Also note two large Silver Maple (*Acer saccharinum*) in frontyard.
15. Saucer Magnolia (*Magnolia* x *soulangiana*) 5613 Taychopera Road, small tree in frontyard.
16. European Ash (*Fraxinus excelsior* 'Hessei') 5730 Taychopera at Baker St. intersection near driveway.
17. Cottonwood (*Populus deltoides*) 1822 Camelot, towering yard tree by garage.
18. Quaking Aspen (*Populus tremuloides*) 1802 Camelot, near front door; note Larch (*Larix decidua*) visible behind house.
19. Black Oak (*Quercus velutina*) 1750 Camelot Drive, in frontyard stonewell built to preserve tree when soil grade line was raised during construction.
20. Bur Oak (*Quercus macrocarpa*) 1813 Camelot, frontyard beside large Cottonwood.
21. Blue Colorado Spruce (*Picea pungens* 'Glauca') 1821 Camelot, two aging specimens in frontyard; note line of Bur Oak (*Quercus macrocarpa*) along sideyard border, survivors of construction and encroaching civilization.
22. Common Mulberry (*Morus alba*) 5725 Lake Mendota Dr. in frontyard.
23. White Ash (*Fraxinus americana*) 5709 Lake Mendota Dr. young yard tree on west side of drive; compare trunk and leaf characteristics to young Green Ash (*Fraxinus pennsylvanica*) on east side of frontyard.
24. Black Cherry (*Prunus serotina*) between 5631 and 5639 Lake Mendota Dr in yard; note small Buckthorn (*Rhamnus cathartica*) on east side of drive near street.
25. Willow (Salix alba) and Black Locust (*Robinia pseudoacacia*) cluster of trees at end of Capital Court on rocky lake edge; looking across the lake note heavy forestation of Shorewood Hills.
26. Basswood (*Tilia americana*) 5516 Lake Mendota Dr.
27. Apple (*Malus pumila*) 5511 Lake Mendota Dr., east sideyard.
28. Common Mulberry (*Morus alba*) 5209 Harbor Court, end of street, near lake.
29. American Sycamore (*Platanus occidentalis*) at junction of Spring Ct., Lake Mendota Dr. and Minocqua Crescent reaching out over street, identified by light colored, mottled bark.
30. Yellowbud Hickory (*Carya cordiformis*) north of house at 5121 Spring Court near asphalt driveway, a State Champion!
31. White Spruce (*Picea glauca*) 5132 Spring Court.
32. Bur Oak (*Quercus macrocarpa*) 5152 Spring Court, in frontyard; notice grove of Bur Oaks either side of street.
33. Bur Oak (*Quercus macrocarpa*) 5110 Lake Mendota Drive, with interesting shape.
34. Kentucky Coffeetree (*Gymnocladus dioica*) 1133 Risser, tallest tree in frontyard, identified by compound leaves; there are other unusual young plantings here including two White Pines (*Pinus strobus*) and Canada Hemlock (*Tsuga canadensis*) on the south side of driveway and on the north side White Fir (*Abies concolor*) Tamarack (*Larix larcina*) and American Sycamore (*Platanus occidentalis*).
35. Cutleaf Silver Maple (*Acer saccharinum* 'Wieri') 1130 Risser Road, in frontyard identified by its deeply lobed leaves.
36. Tuliptree (*Liriodendron tulipifera*) 5106 Minocqua Crescent, two specimens on west side of driveway among row of short Arborvitae (*Thuja occidentalis*).
37. White Oak (*Quercus alba*) 5130 Minocqua Crescent, frontyard near street, with large horizontal, spreading limbs.
38. Red Pine (*Pinus resinosa*) 5120 Minocqua Crescent, two specimens in yard; note across street two Austrian Pines (*Pinus nigra*) in yard, compare trunk and needle characteristics.

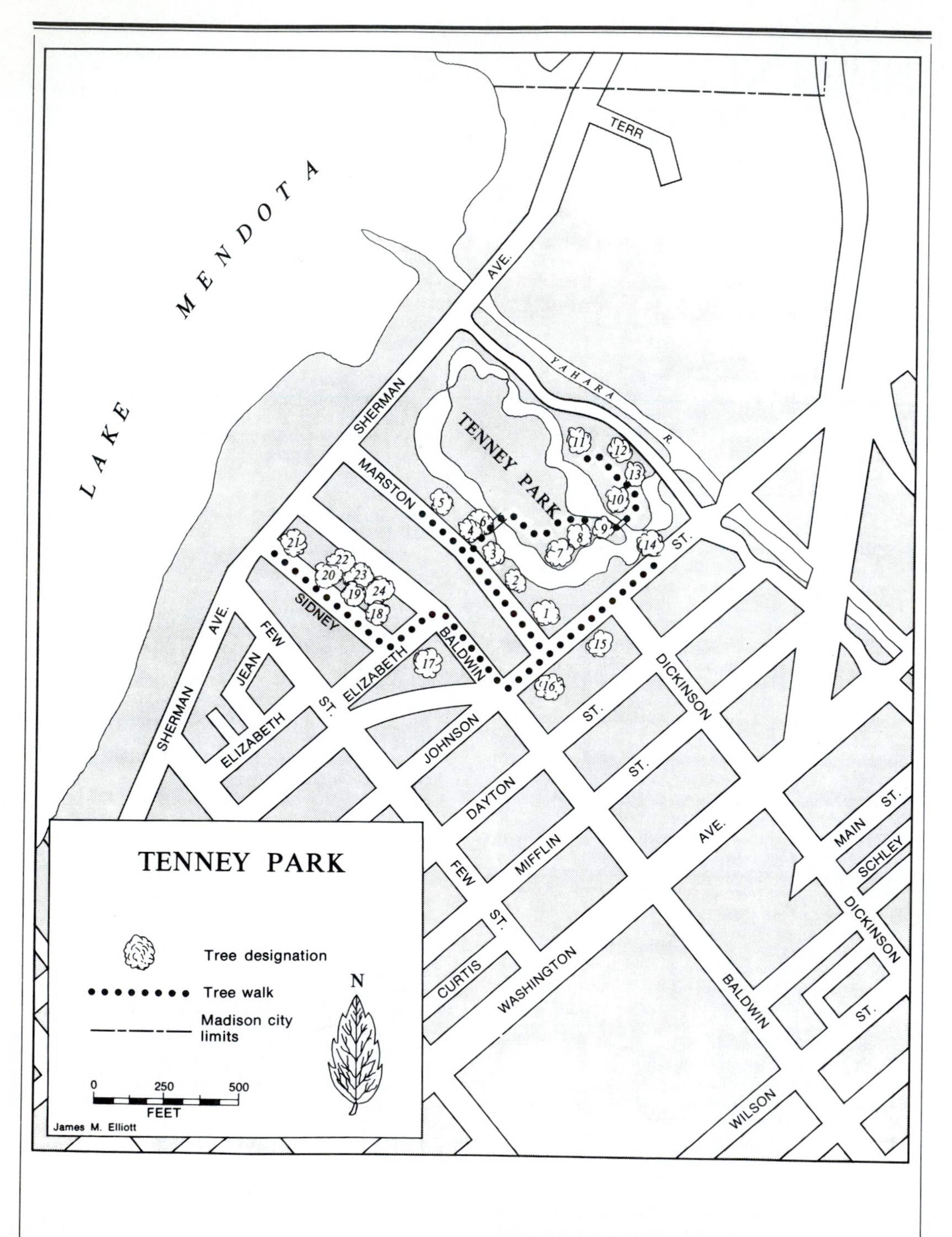
LAKE MENDOTA
TENNEY PARK
YAHARA R
TERR
SHERMAN AVE.
MARSTON
SIDNEY
FEW
JEAN
ELIZABETH ST.
BALDWIN
JOHNSON
DAYTON
MIFFLIN
FEW ST.
CURTIS
WASHINGTON
DICKINSON
AVE.
ST.
MAIN ST.
SCHLEY
DICKINSON
BALDWIN
WILSON
TENNEY PARK
Tree designation
Tree walk
Madison city limits
N
0 250 500
FEET
James M. Elliott

TENNEY PARK

1. Hackberry (*Celtis occidentalis*) across street from 326 Marston, 3 feet from road 5'8" CBH; note Red Oak (*Quercus rubra*) nearby.
2. Norway Maple (*Acer platanoides*) across from 408 Marston St.
3. Basswood (*Tilia americana*) across from 408 Marston, two specimens near Maple; largest 4'7" CBH.
4. Green Ash (*Fraxinus pennsylvanica*) group of four across from 426 and 422 Marston St.
5. American Sycamore (*Platanus occidentalis*) across from 450 Marston, 5'9" CBH, light colored, mottled bark and small, spiny spherical fruit; note Red Oak (*Quercus rubra*) nearby and Catalpa (*Catalpa speciosa*) across street, in front of 450 Marston.
6. European Larch (*Larix decidua*) along service road, left of bridge; 3'11" CBH, deciduous conifer with delicate needles; also note Black Locust (*Robinia pseudoacacia*) on right side of bridge.
7. Paper Birch (*Betula papyrifera*) on island, identified by white, paper-like bark.
8. Bur Oak (*Quercus macrocarpa*) twelve feet from water's edge in naturalized setting, 3'11" CBH.
9. European Alder (*Alnus glutinosa*) several along shore on island and elsewhere in park.
10. River Birch (*Betula nigra*) groups of young trees along water's edge near bridge and shelterhouse; peeling, cinnamon bark.
11. Swamp White Oak (*Quercus bicolor*) northwest of tennis courts, large, open grown specimen, 8'7" CBH.
12. Green Ash (*Fraxinus pennsylvanica*) two specimens north of tennis courts, largest 7'2" CBH.
13. Hackberry (*Celtis occidentalis*) north of tennis courts near road, 5'11" CBH.
14. Basswood (*Tilia americana*) in grove of trees and shrubs across from 1415 Johnson, 8'4" CBH.
15. Red Oak (*Quercus rubra*) frontyard 1343 E. Johnson near walkway to house, 12'3" CBH.
16. White Oak (*Quercus alba*) 1317 E. Johnson, frontyard, 8'3" CBH.
17. Silver Maple (*Acer saccharinum*) street tree, 1241 Elizabeth, 11' CBH; note small Sugar Maple (*Acer saccharum*) beside it.
18. Norway Maple (*Acer platanoides*) 409 Sidney, street tree.
19. Eastern Red Cedar (*Juniperus virginiana*) 429 Sidney St., two small evergreens on either side of porch in frontyard.
20. Blue Colorado Spruce (*Picea pungens* 'Glauca') 441 Sidney St., yard tree.
21. Yew (*Taxus*) and Canada Hemlock (*Tsuga canadensis*) at Sidney St. side entrance of corner house at Sherman Ave., near sidewalk.
22. White Spruce (*Picea glauca*) 450 N. Baldwin, yard tree.
23. Cottonwood Poplar (*Populus deltoides*) 430 N. Baldwin, in yard, 16'3" CBH.
24. White Oak (*Quercus alba*) 428 N. Baldwin, young street tree planting.

"Experimentation with nut trees is especially to be recommended for people in middle age and upward. One of the pains of advancing years is the declining circle of one's friends. One by one they leave the earth, and the desolating loneliness of old age is felt by the survivors. But the man who loves trees finds that this group of friends (trees) stays with him, getting better, bigger, and more lovable as his years and their years increase."

—J. Russell Smith
Tree Crops, 1953 (Devin-Adair, Old Greenwich, Conn.)

Appendix I

*Customer Newsletter, Published by a Commercial Arborist**

*Hendrickson: The Care of Trees, Inc. Editor and publisher: Larry Hall.

2371 South Foster Avenue Wheeling, Illinois 60090-0898

ARBOR TOPICS

Editor: Larry Hall

Spring/Summer '96

The purpose of this publication is to promote an increased awareness of trees and their value to our environment. To those of you that read and enjoy it and share our appreciation of trees, we dedicate this publication.

What's So Special About a Tree???

By Larry Hall, Sr. Vice President

Most people are aware that a tree is pretty to look at, provides shade and, as Joyce Kilmer suggested, wears a nest of robins in its hair. As we learn more about the chain of events that contributes to life on our planet, we recognize that the simple daily events often taken for granted are part of a web of immense complexity.

One of our political leaders once said "What's so special about a tree? You've seen one, you've seen them all". This is an indifference we can no longer afford. One acre of trees (approximately 78 trees) provides enough oxygen annually to sustain 18 people. Twenty mature trees are needed to offset the pollution caused by an auto driven 60 miles in one day.

The daily transpiration rate of an average size shade tree is reported to be 75-100 gallons of water. The evaporation of this much water requires about 225,000 kilocalories of energy, or the equivalent of 5 average size room air conditioning units. Obviously, a single tree would not effectively cool an immediate environment but many trees could contribute to a given micro-environment.

In many ways a city without trees is like a desert. The presence of trees and other vegetation helps cool our urban areas, acts as a buffer to heavy pollutants, acts as visual and sound barriers – all this and more.

Solar radiation beams through the atmosphere, striking vegetation, buildings, etc.

In This Issue:

Printed on Recycled Paper with Soy Inks

Summer radiation is very intense. Buildings and pavement absorb 90-95% of this incoming radiation. Most of this radiation is dispersed in the form of sensible heat which you can actually feel, and much is held in the concrete and buildings. The result is hotter summer days and nights.

The most apparent benefit from the urban tree is shade. Tree tops receive as much radiation as streets and buildings, but living green leaves absorb only 50-60% of the solar radiation and rid themselves of it efficiently. Some of the radiation is transmitted through the leaves and some is reflected. A slightly greater percent is removed in the evaporation process, which causes a cooling effect.

One wonders if it isn't sinful to justify the presence of trees on a piece of paper such as this. It is true that trees in some undefined way help to tranquilize our minds and make for a more pleasant existence. Simply sitting beneath a tree enjoying its shade and marveling at its massiveness, a person wonders who and how it was designed and structured - who decided its leaf form and placement.

Trees are special - they serve as symbols and have functions. We have been told we came from the trees; we came from the woods and forests; we came from the land.

Trees are relevant in contemporary culture and our attitude is reflected in the way we care for trees or any living plant.

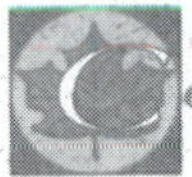

ontinuing Client Service

by Ron Rubin, Chief Operating, Mid-Atlantic Region and Certified Arborist

The Care of Trees is committed to providing superior service to our clients. To continue this outstanding care, our Arborist Representatives provide periodic inspections of client properties to check overall tree and shrub health. Trees and shrubs are living organisms which benefit from continued care and maintenance. Because they are living organisms growing outdoors, they are susceptible to the ever changing environment. To help catch many conditions before they become problems (being proactive) is our #1 job.

A few of the elements which can affect tree and shrub health:

- Extended periods of moisture
- Extended periods of drought
- Extreme temperature fluctuations
- Chemicals used near plants such as de-icing materials, swimming pool additives and weed killers
- High winds
- Heavy snow/ice loads on branches
- Insect and disease infestations
- Human activities such as soil compaction, shoveling of snow on, over or near plants, and air or water pollution

The Care of Trees' Certified Arborists are trained to take all of these and many additional factors into account during property inspections. We can often identify potential conditions which may affect the overall health and conditions of our clients' valuable plant material.

By contracting with the Care of Trees, our clients have entrusted us with the responsibility to help maintain and preserve their trees and shrubs. Many consider their landscape trees and shrubs irreplaceable. The Care of Trees does not take this responsibility lightly. Our recommendations help ensure that our clients receive professional care to protect their valuable plants.

The Care of Trees provides property inspections free of charge to our clients. May we help you care for your trees and shrubs? If you have not had one of our arborists out to your property in the last 6 months to a year, please fill out the post card or call our office.

oneylocust Hit Hard in 1995

by Rex Bastian, Ph.D., Director of Technical Services

During the spring of 1995, many honeylocust trees across the Midwest were severely injured by a complex of sap-sucking insects. The term "complex" is used when several different insects are present at the same time on a host and cause similar injury. In this case, both the honeylocust plant bug and the honeylocust leaf hopper were involved.

On many trees, both insect species were present. On others, only one of the two was present. The biology of both insects is similar. Each overwinters in the egg stage. The eggs are inserted into two or three year old twigs, usually around the nodes. During a typical year, the eggs hatch about one or two weeks following bud break. During 1995, however, egg hatch coincided with the flush of new growth. Large numbers of nymphs were present and their feeding activities resulted in quick and severe curling of the new shoots. Often, the damage was so severe that homeowners were convinced their honeylocusts had died over the winter because they were not leafing out.

Nymphs of both species continue to feed as they mature, favoring developing foliage. Development is completed by late June, when clouds of flying adults can be dislodged by a simple tap on a branch. Mating occurs and the females insert their eggs into young twigs. These eggs will hatch the following spring. By the end of June, the majority of the plant bugs and leafhoppers have disappeared. There is only one generation each year.

The Travelling Gypsy Moths

by Rex Bastian, Ph.D., Director of Technical Services

It is true. The gypsy moth does threaten our trees in Illinois. During the past years, state and federal agencies such as the Illinois Department of Agriculture, the USDA Forest Service and USDA Animal and Plant Health Inspection Service have done a tremendous job in keeping this insect at bay. Remember, the first gypsy moth was captured in Illinois back in 1973. Eradication efforts have restricted this insect from spreading since those early infestations.

Today, however, mass infestation looms near. Even though the number of moths trapped in Illinois during 1995 was only about half the number trapped in 1994 (2,138 vs. 4,672), large populations can be found in Wisconsin. A definite threat exists if travelers from the Chicago Metro area bring gypsy moths back with them from Wisconsin. Door, Sheboygan, Oconto and Marinette counties in Wisconsin are heavily infested with gypsy moth. In Door county alone, over 34,000 moths were trapped last year. While female moths cannot fly, and the caterpillars can only crawl short distances, the insect can "hitchhike" long distances on vehicles.

The female moths will lay their eggs on anything. Cars, trailers, recreational vehicles and camping gear are targeted while families are camping in infested areas. When the campers return home after a weekend in the woods, the egg masses come home with them. As the gypsy moth population climbs in Wisconsin, the chances of weekend travelers bringing egg masses back to Chicago will increase dramatically.

What can we do? First, carefully search your vehicle and camping equipment after vacationing in infested areas of Wisconsin or Michigan. This is most critical if you vacation during the time when the eggs are laid, usually during July and August. Remove and crush caterpillars, cocoons, or eggs that you find. Campgrounds in infested areas will have information on what to look for. If federal or state inspectors wish to look at your vehicle, please cooperate with them. It may result in a slight delay, but it could save literally thousands of Illinois trees from defoliation.

Long distance transport of the gypsy moth into Illinois is a real threat. It is up to all of us to help keep our beautiful trees safe.

Honeylocust trees continue to put on new foliage throughout the season as long as conditions are favorable for growth. By the middle or end of July, most trees have regained their normal appearance. What should we expect for 1996? We do not know how severe infestations will be. Damage during most previous years has been of little concern except in localized situations. Why the populations exploded last year is unclear, but may relate back to ideal weather conditions during mating, egg laying, and developmental periods.

The best strategy is to monitor honeylocust trees during and following bud break. Pay special attention to trees that were heavily infested last spring. Look for the rapidly moving nymphs of both insects. Leafhopper nymphs will have a humpbacked shape, whereas the plant bugs will be flat and oval. Nymphs of both species are light green and very small.

If nymphs are abundant, the trees can be treated by a variety of methods. Insecticidal soaps and oils may work against the youngest nymphs; however, repeat applications may be needed. If there are only a few nymphs present, consider not treating or delaying treatment applications for a few days. Honeylocusts are tough plants and can recover rather easily from this type of injury. Trees that look like they will surely die in the spring usually re-leaf and look little worse for the wear after a few weeks. Consider mulching, light fertilization, and good water management as an alternative to treating.

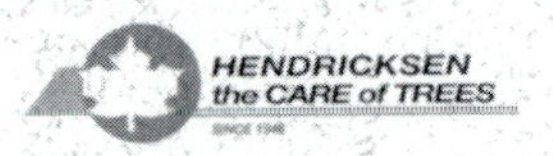

2371 South Foster Avenue
Wheeling, Illinois 60090 - 9898

ADDRESS CORRECTION REQUESTED

HENDRICKSEN, the CARE of TREES

RESIDENTIAL • COMMERCIAL • MUNICIPAL

Barrington	(847) 382-4120	Northwest Suburbs	(847) 394-4301
Crystal Lake	(815) 477-4414	Park Ridge	(847) 698-6599
Glen Ellyn	(708) 545-0606	Tree Preservation (Construction)	(847) 394-4226
Lake Bluff	(847) 918-8749		
Near North Shore	(847) 394-3903	West Chicago	(708) 584-0123

24 HOUR EMERGENCY PHONE: (708) 924-6244

National Arborist Association Awards Highest Honor

The National Arborist Association, a national trade association of commercial tree service firms, held its Annual Management Conference this past February in San Diego. At this conference, the NAA presented their Award of Merit to our President, John Hendricksen. This award, determined by nominations from peers, recognizes a person, company, or an institution for outstanding service in the field of commercial arboriculture.

While presenting the award, NAA president Arthur Batson had the following to say:

"This year the NAA Award of Merit is presented to a man who has given of himself unselfishly, for years, to his state arborist association, the National Arborist Association and the International Society of Arboriculture. He has always been willing to serve, to share information and, in many cases, lead to improve every aspect of this industry, from employee safety to financial management.

Our award winner is an NAA past president. He currently serves as a Trustee of the National Arborist Foundation, a member of the ANSI Z133 (the American National Standards Institute's current standard for tree care operations) Committee and Chairman of the International Society of Arboriculture's Certification Board."

Index

A

D

E

F

G

H

I

J

K

L

M

N

O

R

S

T

U

V

W

Y

Z